Springer Series in Wood Science

Editor: T.E. Timell

M.H. Zimmermann
Xylem Structure and the Ascent of Sap (1983)

J.F. Siau
Transport Processes in Wood (1984)

R.R. Archer
Growth Stresses and Strains in Trees (1986)

W.E. Hillis
Heartwood and Tree Exudates (1987)

S. Carlquist
Comparative Wood Anatomy (1988)

L.W. Roberts / P.B. Gahan / R. Aloni
Vascular Differentiation and Plant Growth Regulators (1988)

C. Skaar
Wood-Water Relations (1988)

J.M. Harris
Spiral Grain and Wave Phenomena in Wood Formation (1989)

B.J. Zobel / J.P. van Buijtenen
Wood Variation (1989)

P. Hakkila
Utilization of Residual Forest Biomass (1989)

J.W. Rowe
Natural Products of Woody Plants (1989)

K.-E.L. Eriksson / R.A. Blanchette / P. Ander
Microbial and Enzymatic Degradation of Wood and Wood Components (1990)

R.A. Blanchette / A.R. Biggs
Defense Mechanisms of Woody Plants Against Fungi (1992)

S.Y. Lin / C.W. Dence
Methods in Lignin Chemistry (1992)

R.A. Blanchette A.R. Biggs (Eds.)

Defense Mechanisms of Woody Plants Against Fungi

With 209 Figures

Springer-Verlag Berlin Heidelberg GmbH

Professor Dr. ROBERT A. BLANCHETTE
Department of Plant Pathology
495 Borlaug Hall
1991 Upper Buford Circle
University of Minnesota
St. Paul, MN 55108, USA

Professor Dr. ALAN R. BIGGS
West Virginia University
Fruit Research Laboratory
P.O. Box 609
Kearneyville, WV 25430, USA

Series Editor:

Professor T.E. TIMELL
State University of New York
College of Environmental
Science and Forestry
Syracuse, NY 13210, USA

Cover: Transverse section of *Pinus lambertiana* wood. Courtesy of Dr. Carl de Zeeuw, SUNY College of Environmental Science and Forestry, Syracuse, New York

ISBN 978-3-662-01644-2 ISBN 978-3-662-01642-8 (eBook)
DOI 10.1007/978-3-662-01642-8

Library of Congress Cataloging-in-Publication Data. Defense mechanisms of woody plants against fungi / R.A. Blanchette, A.R. Biggs (eds.). p. cm. – (Springer series in wood science) Includes bibliographical references and index. ISBN 3-540-54643-X (Springer-Verlag Berlin Heidelberg New York). – ISBN 0-387-54643-X (Springer-Verlag New York Berlin Heidelberg) 1. Woody plants – Disease and pest resistance. 2. Trees – Disease and pest resistance. 3. Phytopathogenic fungi – Host plants. 4. Fungal diseases of plants. /. Plant-fungi relationships. I. Blanchette, R.A. (Robert A.) II. Biggs, A.R. (Alan R.) III. Series. SB761.D35 1992 92-4617 582.1'5042326–dc20

Originally published by Springer-Verlag Berlin Heidelberg New York in 1992

Typesetting: Best-set Typesetters LTD, Hong Kong
31/3145-5 4 3 2 1 0 – Printed on acid-free paper

Preface

For the past decade, it has been apparent to both of us that a reference text covering all aspects of tree defense mechanisms to fungi was missing, needed and long overdue. Such a book would provide a clear, comprehensive overview of how living roots, stems and leaves respond to fungal pathogens. The need for such a book became increasingly clear to us from our conversations with each other, as well as from our interactions with students and colleagues who desired a sourcebook containing reviews of morphological, biochemical and physiological aspects of host-parasite interactions in trees. During a field trip sponsored by the Forest Pathology Committee of the American Phytopathological Society, on a bus from one site to another, we decided to take the responsibility to prepare a book of this type and began to plan its composition.

To adequately address the topic of this book as we had envisioned it, we believed that well-illustrated chapters were needed in order to reflect the important advances made by the many investigators who have examined the anatomical and physiological changes that occur when trees are attacked by fungi. We are grateful to Dr. Tore Timell, the Wood Science editor for Springer-Verlag, for supporting our efforts and for providing an avenue to publish such a profusely illustrated volume. The Springer Series in Wood Science is one of the finest quality book series published and readers of this book will find a treasury of photomicrographs that document the changes that occur in woody tissues in response to fungal pathogens. The numerous illustrations and quality of the publishing are not without their price, but the value they represent to those who use the book should surpass the increased publication costs. We hope this book will serve as a valuable reference to increase understanding of host-pathogen interactions and also to stimulate new interests that inspire continued study in this exceedingly important area of tree pathology.

Within this book, readers will find chapters that address anatomical and physiological responses of bark and xylem to mechanical injuries and invasion by fungi. Response mechanisms in angiosperm and gymnosperm roots, stems and leaves are presented in several chapters that integrate information on how trees, as whole organisms, defend and repair themselves. There are comprehensive chapters on specific tree pathogens that have received considerable research attention over the years, including *Ophiostoma ulmi, Cronartium quer-*

cuum f. sp. *fusiforme* and *Hypoxylon mammatum*. These diseases have served, and continue to serve, as model systems for studying vascular wilts, rust and canker diseases. In addition, important factors contributing to disease, such as plant-water relations and microenvironmental aspects of wounded and colonized tissues, are discussed within this volume. New approaches to studying tree defense reactions also are included and will be found in chapters covering molecular, tissue culture and immunocytochemical aspects. How trees respond to fungi associated with insects and the role of mycorrhizae in protecting tree roots against fungal pathogens are important contributions that expand the coverage of this book. The classic early work has been included in the first chapter with a historical perspective that reflects how much these early investigations still have to offer.

Researchers and students of tree pathology now have a sourcebook of information on defense mechanisms of woody plants against fungi. Of prime importance, inquiring students have a reservoir of information on how roots, stems and leaves respond to fungal pathogens. We hope these contributions will be of great value and assistance.

ROBERT A. BLANCHETTE and ALAN R. BIGGS

Contents

17 Water Relations in Trees and Responses to Infection
K. SUZUKI

18 Applications of Tissue Culture for Studying Tree Defense Mechanisms
M.E. OSTRY and D.D. SKILLING

19 Immunocytochemistry of Fungal Infection Processes in Trees

R.A. BLANCHETTE and A.R. ABAD

List of Contributors

ABAD, ANDRÉ R., Department of Plant Pathology, University of Minnesota, St. Paul, MN 55108, USA

ADASKAVEG, JIM E., Department of Plant Pathology, University of California, Davis, CA 95616, USA

BIGGS, ALAN R., University Experiment Farm, University of West Virginia, P.O. Box 609, Kearneysville, WV 25430, USA

BLANCHETTE, ROBERT A., Department of Plant Pathology, 495 Borlaug Hall, 1991 Upper Buford Circle, University of Minnesota, St. Paul, MN 55108, USA

BODDY, LYNN, School of Pure and Applied Biology, University of Wales, P.O. Box 915, Cardiff CFL 3 TL, Wales, United Kingdom

DUCHESNE, LUC C., Petawawa National Forestry Institute, Chalk River, Ontario, K0J 1JO, Canada

GEIGER, P. P., ORSTOM, laboratoire de Phytopathologie Tropicale, 2051 av. du val de Montferrand, 34000 Montpellier, France

GRIFFIN, DAVID H., College of Environmental Science and Forestry, State University of New York, Syracuse, NY 13210, USA

HUBBES, MARTIN, Faculty of Forestry, University of Toronto, Toronto, Ontario, Canada M5S 1A1

JENG, R. S., Faculty of Forestry, University of Toronto, Ontario M5S 3B3, Canada

JEWELL, DAPHNE C., School of Forestry, Louisiana Technical University, Ruston, LA 71272, USA

JEWELL, F. F. Sr., School of Forestry, Louisiana Tech University, P.O. Box 10138, Tech Station, Ruston, LA 71272, USA

KLEPZIG, KIER D., Department of Entomology, University of Wisconsin, Madison, WI 53706, USA

MANION, PAUL D., Department of Environmental and Forest Biology, SUNY College of Environmental Science and Forestry, 1 Forestry Drive, Syracuse, NY 13210, USA

MERRILL, WILLIAM, Department of Plant Pathology, 211 Buckhout Laboratory, Pennsylvania State University, University Park, PA 16802, USA

NANDRIS, DANIEL, ORSTOM, laboratoire de Phytopathologie, BP A5 Nouméa, Nouvelle-Calédonie, France

NICOLE, MICHEL, ORSTOM, laboratoire de Phytopathologie Tropicale, 2051 av. du val de Montferrand, 34000 Montpellier, France

OSTRY, MIKE E., USDA Forest Service, North Central Forest Experiment Station, Folwell Avenue, St. Paul, MN 55108, USA

OUELLETTE, G.B., Laurential Forest Research, Canadian Forestry Service, 1080 du Vallon, P.O. Box 3800, Sainte-Foy, Province de Québec, Canada G1V 4C7

RIOUX, DANNY, Laurentian Forest Research, Canadian Forestry service, 1080 du Vallon, P.O. Box 3800, Sainte Foy, Province de Québec, Canada, GIV 4C7

RAFFA, KENNETH F., Department of Forestry, University of Wisconsin, Madison, WI 53706, USA

SINCLAIR, WAYNE A., Department of Plant Pathology, Cornell University, Ithaca, NY 14853, USA

SKILLING, D.D., U.S.D.A. Forest Service, North Central Forest Experiment Station, St. Paul, MN 55108, USA

STROBEL, NORMAN E., Department of Plant Pathology, 334 Plant Science Building, Cornell University, Ithaca, NY 14853-5908

SUZUKI, KAZUO, Department of Forestry, Faculty of Agriculture, University of Tokyo, Yayoi, Bunkyo-ku, Tokyo 113, Japan

WOODWARD, STEPHEN, Department of Forestry, University of Aberdeen, St. Machar Drive, Aberdeen AB9 2UD, Scotland

YAMADA, TOSHIHIRO, Kansai Research Center, Forestry and Forest Products Research Institute, Momoyama, Fushimi, Kyoto 612, Japan

Chapter 1

Mechanisms of Resistance to Fungi in Woody Plants: A Historical Perspective

W. MERRILL

1.1 Preface

The intent of this chapter is to give an overview of some of the mechanisms postulated to be involved in the resistance of woody plants to fungal pathogens. Because most of the phenomena will be discussed in depth by subsequent authors, I will not provide an extensive list of pertinent literature, but will instead select papers that I believe were the primary publications on the subject, or papers that in some cases have been overlooked by more recent authors.

1.2 Introduction

Disease resistance is a dominant characteristic of life. If this were not so, most species would have died out during their evolution. Natural populations of wild plants are mixtures of diverse species and diverse genotypes. The physical mixture of different species slows down epidemic development, and the mixtures of diverse genotypes within the same species also slow epidemic development, reducing both X_o and r (van der Plank 1963). The pathogens attacking such host mixtures likewise consist of mixtures of diverse genotypes, each usually capable of affecting only a small portion of the host's population. Thus, populations of wild plants and their pests and pathogens fluctuate in an uneasy balance.

However, beginning with the Agricultural Revolution some 12000 years ago, Man has steadily interfered with this delicate balance. All devastating epidemics of plant diseases, from wheat stem rust in the Great Plains of North America to chestnut blight in southern Europe and the Appalachian Mts. of eastern North America, from the elm devastation wrought by *Ceratocystis ulmi* throughout Europe and the USA to rubber leaf blight in the Amazon to ergot in central Europe during the Middle Ages – these all have been the direct result of Man, in blissful ignorance, messing with Mother Nature.

Man has selected or bred susceptible strains of plants – cereal grains with their attendant smuts and rusts are glaring examples; has in ignorance or stupidity introduced virulent pathogens into crops that were highly susceptible, having never previously encountered these pathogens during their evolution – the North American elms, chestnut and soft pines, among others; has altered the growing conditions of the crop – the USA southern pinery and the attendant fusiform rust and annosus root rot problems, ad infinitum.

Resistance to pests and pathogens of necessity must be as old as pathogenicity. And Man's recognition of this phenomenon is far from new. Prior to 285 B.C. Theophrastus noted:

"Generally speaking, cereals are more liable to rust than pulses, and among these barley is more liable to it than wheat; while of barleys some kinds are more liable than others, and most of all . . . the kind called 'Achillean'".

This is the first written record of differences in varietal disease resistance! Yet little came of this for over two millenia, although farmers for centuries have noted and selected strains of cereal grains with greater disease resistance (Biffen and Engledow 1926). Mendel, if he had continued his studies in genetics, might have discovered this phenomenon. However, rejected by a leading botanist of the time (a notable example of the value of "peer" review!), his work lay buried for another generation. In 1900 Correns, Tschermak, and De Vries independently rediscovered and confirmed Mendel's work. It was only 5 years later that Biffen (1905) began to publish a series of papers showing that rust resistance in wheat was inherited according to Mendelian ratios. This was the beginning of systematic, scientifically based plant breeding and of scientific attention to the phenomena of resistance and susceptibility to disease. By 1919 Biffen's Little Joss wheat, the first scientifically bred disease-resistant hybrid, was in widespread production in England (Buller 1919).

Beginning with de Bary in 1853, attention of pathologists was focused almost exclusively on the pathogen. In the late 1800s, Sorauer and his followers adopted an opposite tack and placed all emphasis on the environment. It has been repeatedly demonstrated that environment greatly affects the interaction between suscept and pathogen. For example, the nutritional status of a plant may affect the susceptibility of its foliage to rusts and powdery mildews, as well as to diseases such as damping-off. Drought stress renders woody plants more susceptible to twig and stem cankers, and also to root rots. "Site stress" also may predispose plants to stem cankers. However, the literature on these topics is voluminous and should be considered in a separate chapter. Beginning with Biffen's work, real emphasis began to be placed on the suscept's response. A balance of these three viewpoints is brought together in the modern Triad Concept of Disease. However, as early as 1874 Robert Hartig had begun to describe and illustrate in great detail the pathological changes in affected plants, truly becoming not only the "Father of Forest Pathology", but also the "Father of Pathological Anatomy." Several phenomena now known to be associated with disease resistance were first illustrated or described by Hartig (1889), even though he was not fully aware of the underlying significance of his observations. Unfortunately, Biffen's first paper appeared 4 years after Hartig's death.

1.3 Terminology

Considerable confusion exists in the literature because of varying terminology. Many authors, at different times and often in the same paper, apply different meanings to the same term.

Neither "resistance" nor "susceptibility" is finite; they are reciprocals of a continuum that may vary over several orders of magnitude in terms of plant response as measured by the amount of damage caused by the pest or pathogen. Because "resistance" and "susceptibility" are relative terms, they can be measured only by comparing the response of two or more genetic strains of the suscept to the same strain of the pathogen.

"Pathogenicity" is the ability of an organism to cause disease in at least some members of the suscept population. It is an attribute of a species. An organism is either pathogenic or nonpathogenic, i.e., it can or cannot cause disease. "Virulence" is a measure of the amount of disease caused by a strain of a pathogen in a particular suscept; it is a *relative* term that can be measured only by the response of two or more strains of the pathogen attacking the same strain of the suscept. Virulence also is a continuum that has no easily defined lower limits – it may vary from almost infinitesimal cellular changes to mortality of the suscept.

"Escape" occurs when, by chance or by some external condition affecting the interaction between the suscept and pathogen, a normally susceptible individual does not become diseased. For example, in dense pine sapling stands, suppressed or overtopped pines often escape infection by *Cronartium ribicola* while nearby dominant trees become infected. This escape is due to the small area of spore-trapping surface of the suppressed trees compared with that of the larger trees. Hartig (1889) noted that late-developing spruces escaped rust infection. Buller (1919, p. 178) was the first to clearly differentiate "disease escape" from "disease resistance."

"Klendusity" has been used by some authors as synonymous with escape. Although it may be a special form of escape, klendusity is the resistance of the suscept to something other than the pathogen. Classic examples are brambles, highly susceptible to mosaic viruses, that rarely become infected in the field because they are resistant to the aphid vectors of those viruses (Schwartze and Huber 1936).

"Immunity" is a phenomenon unrelated to resistance. For example, pines are immune to the fungus causing corn smut. But this is of little interest and of no use to the breeder. For the pines to be considered "resistant," at least some members of the smut population would have to be able to infect and cause at least some cellular responses in some members of the pine population. It goes without saying that most plants and animals are immune to most pathogenic agents; for example, of over 6000 species of rust fungi, only about a score are able to infect pines and none are known to attack Man. Further, although resistance genes in the suscept plant create suscept-pathogen specificity, *the basis for this specificity lies in the susceptible plant, not in the resistant plant* (Vanderplank 1989).

Resistance to fungi may be passive or active. Passive resistance involves preformed morphological and/or chemical barriers to penetration or colonization. Active resistance involves induced morphological and/or chemical barriers to colonization that develop as the suscept plant responds to the presence of the pathogen.

1.4 Preformed Passive Barriers to Infection

The outer surfaces of the plant itself are resistant to penetration by many species of fauna and flora. The bark is suberized, may contain various toxic extractives, and in many species is periodically shed. The foliage is covered with cuticle. Few pathogenic organisms appear to be able to decompose these materials. These barriers may be breached by natural openings, such as lenticels and stomata. A great many pathogenic fungi, such as leaf rusts and needle blight fungi, have become adapted to utilize these natural openings through the suscept's defenses. Resistance to such pathogens usually involves active, induced internal barriers to subsequent colonization.

1.4.1 Bark and Root Cortex

Most pathogenic fungi, by themselves, cannot penetrate or consume intact bark. Thus, many pathogens of woody plants are wound parasites, as noted by Hartig (1889):

"In itself, however, the feebleness of old age [of a tree] is not a natural condition attributable to internal causes, but is a state induced by external influences. The older a tree is, so much the more numerous are the dangers through which it has had to pass, and so much the greater is the number of its injuries and wounds through which parasites and saprophytes can find entrance into its interior."

Bark of many tree species is slow to decompose; indeed, one often can find fallen branches or stems of *Betula papyrifera* where the cylinder of outer bark is still intact while the inner cortex and wood are totally decomposed. On the other hand, some fungi, such as *Fomes annosus* and *Armillaria mellea*, attack and penetrate through the intact "bark" of the roots, a phenomenon also first noted and studied experimentally by Hartig (1889).

The bark and root cortex defense system can be breached by any number of agents, including weather, fire, insects, higher animals including Man, and such accidental agents as falling trees and branches. It has long been known that in the USA wounds caused by wildfire are major infection courts for staining and decay fungi ("heartrot"). In addition, insects such as bark and ambrosia beetles breach the bark barriers carrying in or on their bodies numerous species of staining fungi. Some of these fungi may act as weak pathogens, killing stressed trees and thus mimicking the role of pneumonia as a killer of elderly humans who have been stressed by surgery. Siricid wood wasps actually inoculate a commensual decay fungus into the sapwood – and feed on the decaying wood. The success of such insect attack is often related to tree vigor; healthy conifers may be able to "pitch out" the insects by producing copious resin flow. The physiology of these processes is still not clear. Resistance to such wound parasites appears to be an active, induced response of the suscept.

1.4.2 Cuticle

The cuticular surface of the foliage is a barrier resistant to enzymatic attack by many pathogens. Hartig (1889) noted that young conifer seedlings became resistant to attack by *Phytophthora omnivora* once they had developed a cuticle; consequently, the fungus was pathogenic only on primary tissues. However, numerous foliage-attacking fungi have become adapted to mechanically breach the cuticular barrier via formation of appressoria and penetration pegs. A relatively low pressure exerted onto the end of a penetration peg with a cross-sectional area of about $0.01\,\mu m^2$ may result in pressures of thousands of kg/cm^2, thus allowing the fungus to easily breach the cuticle. Miyoshi (1895) showed that several foliar pathogens generated sufficient pressure with their penetration pegs to readily puncture thin metallic foils. Thus, cuticle thickness may come into play as a resistance mechanism. Melander and Craigie (1927) showed that the resistance of several species of *Berberis* to *Puccinia graminis* was related to cuticle thickness.

Foliar diseases are more common on understory plants or on lower, partially shaded branches. This phenomenon is usually attributed to a microclimate more favorable for disease development in the lower canopy. Yet, shade leaves have thinner cuticles than sun leaves; I know of no study in forest pathology regarding the effect of cuticular thickness on development of foliage diseases in the lower vs higher canopy.

1.4.3 Nature of the Cuticular Waxes

The chemical composition of the cuticle has attracted some attention, and attempts have been made to relate this factor to disease resistance. However, since at least some of the fungi investigated also are able to penetrate via the stomata, the argument seems spurious at best.

1.4.4 Vessel Diameter and Spacing

Ulmus sp. with widely spaced earlywood vessels of small diameter and length have been reported to be more resistant to *Ceratocystis ulmi*, presumably by restricting the rate of colonization by the pathogen (Pope 1943, Elgersma 1970). In a similar manner, individual *Pinus strobus* with small diameter terminals having closely spaced cortical resin ducts are resistant to attack by *Pissodes strobi*; female weevils are more likely to sever the resin ducts while cutting egg-laying niches, and the adults and/or larvae are "pitched out" of such trees (Stroh and Gerhold 1965). Crowdy (1949) and Biggs et al. (1983) noted that canker fungi circumvent the cortical defense barrier by penetrating through the phloem fibers, which are not occluded. It is possible that canker resistance might be greater in trees with fewer, smaller, and more widely spaced bundles of phloem fibers. Colonization through the phloem fibers may be the basis of Ashcroft's (1934) observation that penetration was faster in the phloem than in the cortex.

1.4.5 Heartwood Extractives

I am unable to find a "first" reference to the decay resistance of heartwood vs sapwood, or variation in decay resistance among tree species; yet this phenomenon was well known to ancient shipwrights and carpenters. In spite of this decay resistance, in nature even the most decay-resistant trees, such as redwood and black locust, are subject to extensive internal discoloration and decay. Hartig (1889) noted that *Trametes radiciperda* (= *Fomes annosus*) colonized only limited areas of the lower boles of *Pinus sylvestris* due to heavy resin deposition in those areas, whereas the fungus was able to colonize higher up the boles of *P. strobus* and *Picea* species because of the lower resin contents. Roots of American chestnuts survive the blight presumably because the extractive contents of their roots significantly reduce their colonization by the blight fungus (Graves 1926). On the other hand, Cook and Wilson (1916) found that *E. parasitica* could utilize tannins as carbon sources, producing copious mycelium on media containing up to 2% tannin, and producing pycnidiospores on media containing up to 2.4% tannin.

1.4.6 Xylem Decay Resistance

The preformed barrier to infection of xylem tissues by decay fungi that is overcome by a metabiotic succession of bacteria, yeasts, and staining fungi is an important passive resistance response of the tree, the nature of which is still unknown. However, some pioneer xylem-invading Fungi Imperfecti are able to utilize certain toxic wood extractives as sole carbon sources (Shriner 1972). Similarly, Bier and Rowat (1963) noted that bark saprophytes such as species of *Trichoderma* and *Aspergillus* altered *Populus* bark so that the pyrocatechol present in it no longer inhibited *Hypoxylon pruinatum*. Although Shigo (1965 onwards) did much to popularize the concept of this metabiotic succession, the initial, overlooked, and seldom-cited papers indicating the occurrence, nature, and importance of this succession are those of Basham (1958), Good (1959), and Good and Nelson (1962).

1.4.7 Branch Stub Occlusion

Hartig (1889, Fig. 144) described and diagrammed the resin deposition in the bases of naturally senescing conifer branch stubs which prevents their colonization by decay fungi. Oleoresin comprises up to 28% of the dry weight of such areas in *Pinus strobus* and consists of a great diversity of compounds (Shriner 1972). Nevertheless, a number of Fungi Imperfecti can colonize such tissues (Merrill 1970) and some of these can utilize various resin constituents as sole carbon sources (Shriner 1972). The pith may provide an avenue of penetration through this resin plug, as it contains very small amounts of pinosylvin in comparison with the rest of the dead branch base (Pawuk and Merrill unpublished results).

1.4.8 Host Form

A unique resistance mechanism is that postulated by Roth (1959), who suggested that pines with drooping branches and needles might be less prone to dwarf mistletoe infection than trees with more upright growth habits; mistletoe seeds would tend to be "shed" from the needles.

1.5 Active Mechanisms of Resistance

The most common active resistance mechanism in woody plants is the walling-out response. It occurs in roots, stems, and foliage, in cortex-phloem, and xylem. It is nonspecific, occurring in response to mechanical injury or to the activities of insects and fungal pathogens (Mullick 1977). Basically, it is the suscept's attempt to limit damage to the smallest possible volume of cells. Although much research has focused on the structural changes involved in this process, the underlying mechanism must involve chemical changes in the host tissues so that they resist enzymatic attack or become inhibitory or toxic to the pathogens.

1.5.1 Walling-Out in the Cortex-Phloem

The host's attempts to wall-out cortical pathogens were first described by Hartig (1889) in relation to larch canker and certain other diseases. Numerous others have since investigated details of this response, including, among others, Wiltshire (1922), Ashcroft (1934), Crowdy (1949), Lortie (1964), Mullick (1977), and Biggs et al. (1983). The mechanism appears to involve the formation of tissues resistant to enzymatic attack by the pathogen, but the presence of fluorescing compounds in the barrier zone suggests the possible involvement of host-produced toxins or phytoalexins. Bier (1959), in a series of papers, postulated that susceptibility to canker pathogens was determined by bark moisture. There is little doubt that stress can *predispose* trees to cankers and root rots, but this does not imply an effect of innate resistance.

1.5.2 Walling-Out in the Xylem

The xylem walling-out response, termed "compartmentalization," has been extensively studied and discussed by Shigo (Shigo and Marx 1977) and his coworkers. However, the essentials of this phenomenon had been described in the literature for decades prior to Shigo.

The restriction of discoloration and decay to those xylem tissues formed prior to wounding (Shigo 1965, Shigo and Marx 1977), Shigo's wall 4, was illustrated numerous times by Hartig (1889), for example his Fig. 159, although to be sure Hartig did not comprehend the full significance of his observations, thus falling victim to his own criticism of Ratzeburg (Hartig 1874, p. 74). On the other hand, he noted (1889) that silver fir formed a resinous barrier zone to contain the decay column, but that some decay fungi, such as *Polyporus robustus*, could grow through this barrier and form fruiting bodies on the tree surface. He obviously understood in part the nature of wall 4. Later, Wiltshire (1922) described the formation of what he termed "wound wood" in xylem rays to restrict the lateral movement of *Nectria*; this is Shigo's wall 3 (Shigo and Marx 1977).

1.5.3 Walling-Out in the Foliage

The first report I have found describing the walling-out response in foliage is that of Higgins (1914) who described the formation of the abcission layer in cherry leaves which results in "shot-hole" leaf spot. Heimburger (1962) reported that *Pinus griffithii* was resistant to *Cronartium ribicola* because a recessive gene or genes in the host caused infected needles to abscise and cast before the fungus reached the stem. Hoff and McDonald (1971) reported that in *P. monticola* one gene controlled such premature needle shedding. This response could be interpreted as a form of hypersensitive reaction. Because needle-shedding is often incomplete, the *rate* of response may be important, but I have found no studies in forest pathology relating to this.

Thus, the key phenomena of compartmentalization, as well as of metabiotic succession, have been long established in the literature, although perhaps not known to or appreciated by forest pathologists. The drawing together of all of the pieces into unified concepts and the popularization of the phenomena associated with both metabiotic succession and compartmentalization and their implications in a series of outstanding publications (e.g. Shigo and Marx 1977, Shigo 1979) will be long remembered. Even more important, however, was the proof that compartmentalization is under genetic control (Garrett et al. 1979). This provides the foundation upon which breeders can develop populations of trees with vastly reduced amounts of internal discoloration and decay, even if we do not yet understand the mechanisms of this walling-out process.

1.5.4 Xylem "Double Banding"

The "double band" phenomenon was first described by Schoeneweiss (1959). *Quercus alba* infected by *Ceratocystis fagacearum* developed a ring of vessels in the outer latewood after the earlywood vessels in that year's annual ring had been plugged by gums and tyloses. He postulated this to be a resistance mechanism, and later researchers focused on growth regulators involved in this syndrome. Lacasse (1966) later found that in *Q. prinus* wilting ceased 7 weeks after inoculation, whereas the double band formed 10 weeks after inoculation and hence was a result, not a cause of resistance.

1.5.5 Hypersensitive Responses

Hoff and McDonald (1971) found a gene in *Pinus monticola* that caused a necrotic reaction in the short shoots leading to death of *Cronartium ribicola* as it attempted to colonize through the short shoots into the stem. In infected current-year needles of *P. strobus*, the vacuoles fragmented and appeared to lose phenolics that diffused through the cytoplasm, leading to mortality of the fungus (Boyer 1964, Boyer and Isaac 1964). This was postulated to be the basis for previously reported "juvenile" needle resistance (e.g. Lachmund 1933). Gray and Amerson (1983) found that embryos of *Pinus taeda* exhibited a hypersensitive response to *Cronartium quercuum* f. sp. *fusiforme* similar to that active against rusts in agronomic crops. Jewell (1988) found a different type of response in resistant pine hybrids; the pathogen became established in the pith, but the host tissues from xylem to the cortex developed normally and were devoid of the fungus. In this case the nature of the resistance mechanism was not postulated. Thus there appear to be multiple types of resistant reactions toward the rusts – not surprising, as this has been found true with the cereal rusts where research has been far more intensive and extensive.

The phenomenon of juvenile-tissue susceptibility but mature-tissue resistance has been documented in agronomic crops for over half a century (Goulden et al. 1928), and there is no reason why the converse should not be true. It comes as no surprise, therefore, that some genetic lines of pines selected for resistance to fusiform rust in the seedling stage were later susceptible when outplanted (Schmidtling and Walkinshaw 1985), although it is not clear whether this was due to escape during the screening process or to changes in resistance level. Those authors reported that resistance (measured as percentage of plants infected) determined by the screening process was correlated with resistance after field planting, but the r^2 value was only 0.18.

1.5.6 Second Party Interactions

Mycorrhizae have been postulated to prevent root infection by pathogenic soil fungi via one or more of several mechanisms, including physical exclusion, formation of antibiotics, and consumption of chemo-attractant root secretions (Marx 1972). The work has been done in pure culture, and there are no confirming field studies, so this may be an experimental artifact. With modern gene-transfer techniques, it may be possible to actually tailor mycorrhizal fungi, incorporating genes from some of the very aggressive debris- and soil-inhabiting saprophytes that can restric development of root-rotting pathogens. On the other hand, biological control of plant pathogens has been a complete failure in practice, with a few rare and notable exceptions (potato scab and Texas root rot of cotton in some soils; crown gall of fruit trees), the first two of which still are not understood.

1.6 An Iconoclastic Viewpoint

There is a diverse array of resistant reactions in the suscept. The operative mechanism involved in each of these undoubtedly will be found to be biochemical and probably complex. Yet, the actual mechanism of the resistant response, although intellectually interesting, may be inconsequential.

The genetic nature of the response – whether it is monogenic, oligogenic, or polygenic, whether it is controlled by dominant or recessive genes, whether it is linked to deleterious genes – is the essential knowledge from a practical standpoint. Equally important, and to date nearly totally ignored for most tree pathogens, is knowledge of the genetic variability and plasticity in the pathogen's population. In the past, forest pathologists on the whole have tended to ignore all work done with agronomic or horticultural crops – *if it hasn't been done with a tree, then it hasn't been done!* This parochial outlook is only slowly changing. We ignore studies in other crops at our own peril and we could learn much, and probably avoid many mistakes, by studying this work. A brief anecdote will illustrate. A number of years ago the Forest Pathology Committee of the American Phytopathological Society held a discussion session relating to breeding for disease resistance. A keynote speaker stated there was no evidence that forest tree pathogens would undergo significant genetic changes similar to those that occur in pathogens of agronomic crops. The late Dr. William Loegering, who spent his life working on cereal rusts, took the speaker to task, pointing out the folly in this view and stating that based on his own experiences a new dominant resistance gene discovered in sugar pine (Kinloch et al. 1970) would not last 15 years before a new race of *Cronartium ribicola* would overcome it. The (forest pathologist) audience smiled politely at this field crops man who obviously did not understand forestry. Only 11 years later a race of *C. ribicola* was found overcoming this resistance gene (Kinloch and Comstock 1981). Another race of the fungus, more virulent than the normal wild type, has since been found (McDonald et al. 1984). The US Forest Service screening program for fusiform rust resistance in pine seedlings selects for vertical resistance, and this is fraught with grave dangers, as agronomists have discovered repeatedly. Resistance to one pathogen may be associated with or linked to susceptibility to another (for example, Wakely 1968). Again, the record of our agronomist and horticultural colleagues is replete with numerous examples, several of them disastrous, among others Victoria oats and corn hybrids with Texas male-sterility genes. We need to be far more cognizant of studies on other crops, and we need to pay far more attention to the genetic variability in the pathogens. As has oft been stated, "Those ignorant of history are condemned to repeat it."

References

Ashcroft JM 1934 European canker of black walnut and other trees. West Virg Agr Exp Sta Bull 261:52 pp

Basham JT 1958 Decay of trembling aspen. Can J Bot 36:491–505

Bier JE 1959 The relation of bark moisture to the development of canker diseases caused by native, facultative parasites 1. Cryptodiaporthe canker on willow. Can J Bot 37:229–238

Bier JE, Rowat MH 1963 Some inoculum and substrate factors in the cultural inhibition of Hypoxylon pruinatum (Klotzsche) Cke by pyrocatechol. Can J Bot 41:1585–1596

Biffen AH 1905 Mendel's laws of inheritance and wheat breeding. J Agr Sci 1:4–48

Biffen AH, Engledow FL 1926 Wheat breeding investigations at the Plant Breeding Institute Cambridge. HM Stationery Off, London, 114 pp

Biggs AR, Davis DD, Merrill W 1983 Histopathology of cankers in Populus caused by Cytospora chrysosperma. Can J Bot 61:563–574

Boyer MG 1964 Studies of white pine phenols in relation to blister rust. Can J Bot 42:979–987

Boyer MG, Issac PK 1964 Some observations on white pine blister rust as compared by light and electron microscopy. Can J Bot 42:1305–1309

Buller ARH 1919 Essays on wheat. Macmillan, New York, 339 pp

Cook MT, Wilson GW 1916 The influence of the tannin content of the host plant on Endothia parasitica and related species. New Jersey Agr Exp Sta Bull 291:3–45

Crowdy SH 1949 Observations on apple canker 3. The anatomy of the stem canker. Ann Appl Biol 36:483–495

Elgersma DM 1970 Length and diameter of xylem vessels as factors in resistance of elms to Ceratocystis ulmi. Neth J Plant Pathol 76:179–182

Garrett PW, Randall WK, Shigo AL, Shortle WC 1979 Inheritance of compartmentalization of wounds in sweetgum and eastern cottonwood. US For Serv Res Pap NE-443, 4 pp

Good HM 1959 Fungi associated with Fomes igniarius in decay of poplar. Proc 9th Int Bot Congr, Montreal, 2:136

Good HM, Nelson JI 1962 Fungi associated with Fomes igniarius var. populinus in living poplar trees and their probable significance in decay. Can J Bot 40:615–624

Goulden CH, Neatby KW, Welsh JN 1928 The inheritance of resistance to Puccinia graminis tritici in a cross between two varieties of Triticum vulgare. Phytopathology 18:631–658

Graves AH 1926 The cause of the persistent development of basal shoots from blighted chestnut trees. Phytopathology 16:615–621

Gray DJ, Amerson HV 1983 In vitro resistance of embryos of Pinus taeda to Cronartium quercuum f. sp. fusiforme: ultrastructure and histology. Phytopathology 73:1492–1499

Hartig R 1874 Wichtige Krankheiten der Waldbäume (Important diseases of forest trees, Transl W Merrill, DH Lambert, W Liese 1975) Phytopathology classic 12. Am Phytopathol Soc, St Paul, Minnesota, 136 pp

Hartig R 1889 Lehrbuch der Baumkrankheiten, 2nd edn (Textbook of tree diseases, Transl W Somerville 1894). Macmillan, London, 331 pp

Heimburger C 1962 Breeding for disease resistance in forest trees. For Chron 38:356–362

Higgins BB 1914 Contribution to the life history and physiology of Cylindrosporium on stone fruits. Am J Bot 1:145–173

Hoff RJ, McDonald GI 1971 Resistance to Cronartium ribicola in Pinus monticola: short shoot fungicidal reaction. Can J Bot 49:1235–1239

Jewell FF Sr 1988 Histopathology of fusiform rust-inoculated progeny from (shortleaf X slash) X shortleaf pine crosses. Phytopathology 78:396–402

Kinloch BB Jr, Comstock M 1981 Race of Cronartium ribicola virulent to major gene resistance in sugar pine. Plant Dis 65:604–605

Kinloch BB Jr, Parks GK, Fowler CW 1970 White pine blister rust: simply inherited resistance in sugar pine. Science 167:193–195

Lacasse NL 1966 The role of auxin in double band formation in chestnut oaks infected with Ceratocystis fagacearum. PhD Thesis, Pennsylvania State Univ, University Park, 83 p

Lachmund RG 1933 Resistance of the current season's shoots in Pinus monticola to infection by Cronartium ribicola. Phytopathology 23:917–922

Lortie M 1964 Pathogenesis in cankers caused by Nectria galligena. Phytopathology 54:261–263
Marx DH 1972 Ectomycorrhizae as biological deterrents to pathogenic root infections. Annu Rev Phytopathol 10:429–454
McDonald GI, Hansen EM, Osterhaus CA, Samman A 1984 Initial characterization of a new strain of Cronartium ribicola from the Cascade Mountains of Oregon. Plant Dis 68:800–804
Melander LW, Craigie JH 1927 Nature of resistance of Berberis spp. to Puccinia graminis. Phytopathology 17:95–114
Merrill W 1970 Microorganisms in the bases of dead branches of Pinus strobus (Abstr). Phytopathology 60:1303
Miyoshi M 1895 Die Durchbohrung von Membranen durch Pilzfaden (The penetration of membranes by fungus hyphae). Jahrb Wiss Bot (Pringsheim) 28:269–289
Mullick DB 1977 The non-specific nature of defense in bark and wood during wounding, insect and pathogen attack. Recent Adv Phytochem 11:395–441
Pope SA 1943 Some studies on the Dutch elm disease and the causal organism. PhD Thesis, Cornell Univ, Ithaca NY, 61 pp
Roth LF 1959 Natural emplacement of dwarf mistletoe seed on ponderosa pine. For Sci 5:265–369
Schmidtling RC, Walkinshaw CH 1985 Fusiform rust infection of loblolly pines that survived resistance screening and of their progeny. Plant Dis 69:491–493
Schoeneweiss DF 1959 Xylem formation as a factor in oak wilt resistance. Phytopathology 49:335–337
Schwartze CD, Huber GA 1963 Aphis resistance in breeding mosaic- escaping red raspberries. Science 86:158–159
Shigo AL 1965 Organism interactions in decay and discoloration in beech, birch and maple. Holz Organismen 1:309–324
Shigo AL 1979 Tree decay: an expanded concept. US Dep Agr, Agr Inf Bull 419, 73 pp
Shigo AL, Marx HG 1977 Compartmentalization of decay in trees. US Dep Agr, Agr Inf Bull 405, 73 pp
Shriner CR 1972 Resin constituents of Pinus strobus and their effects on the growth of selected wood-inhabiting fungi in vitro. MS Thesis, Pennsylvania State Univ, University Park, 66 pp
Stroh RC, Gerhold HD 1965 Eastern white pine characteristics related to weevil feeding. Silvae Genet 14:160–169
Theophrastus (date unknown) (Inquiry into plants and minor works on odours and weather signs, Transl A Hort 1916). Heinemann London, 2 vols
van der Plank JE 1963 Plant diseases: epidemics and control. Academic Press, New York, 149 p
Vanderplank JE 1989 A paradox as an aid to understanding host-pathogen specificity. Plant Pathol 38:144–145
Wakely PC 1968 Rust susceptibility in longleaf pine associated with brown-spot resistance and early commencement of height growth. For Sci 14:323–324
Wiltshire SP 1922 Studies on the apple canker fungus ii. Canker infection of apple trees through scab wounds. Ann Appl Bot 9:275–281

Chapter 2

Anatomical and Physiological Responses of Bark Tissues to Mechanical Injury

A.R. Biggs

2.1 Introduction

Studies of the defense systems of trees have focused mostly on xylem tissues because of their direct economic importance to the forest industry. Because bark tissues shield the xylem from the environment, containment of mechanical injuries and infectious microorganisms by bark tissues is of primary importance. The integrity of normal periderm and the ability of plants to form new periderms at wounds or injuries are essential characteristics for normal plant growth and development. However, in comparison with xylem tissues, responses of periderm and other bark tissues to injury and infection are inadequately defined.

Research on wound responses of trees is required in order to understand the processes that favor or impede the development of fungal infections in wood and bark. Many important and serious diseases of trees are caused by pathogens that initiate infections at wounds caused by insects, humans, fire, lightning, wind, hail, animals, and nutritional and physiological disorders. Therefore, it is possible that more precise information about wound responses could lead to innovative control measures based on a better understanding of the chronology of the wound response, how wound response may be influenced by external factors, or how the wound response could be modified for improved disease control. There have been many recent advances in the study of signal molecules exchanged between host and pathogen in wound tissues, and on the regulation of wound metabolism (Van Sambeek and Pickard 1976, Davies and Schuster 1981, Halverson and Stacey 1986). These and other studies have shown that plant responses to wounding and infection are often similar, suggesting the presence of common signal molecules. In trees, it is likely that the tissue regeneration process following wounding is also the defense process, given that structural responses often coincide with physiological processes that contribute to the biochemical foundation of resistant structures. In this context, the effort to distinguish between structural and physiological responses becomes misguided. Any distinctions made between physiological processes and structural wound responses in this chapter are done merely for convenience and for the purpose of discussion. It is the goal of this chapter to discuss recent findings on the anatomical and metabolic consequences of wounding in the bark of trees and to draw attention to investigations that are relevant to host-pathogen interactions.

2.2 Bark

The term "bark" is used most often in a nontechnical context and refers to all tissues external to the vascular cambium (Srivastava 1964, Esau 1965). Accordingly, the bark is an aggregation of organs and tissues that includes phloem and secondarily thickened tissues from the secondary plant body, as well as epidermis, cortex, and phloem derived from the primary plant body (Esau 1965). The term bark was used by earlier authors in a technical context in reference to all dead tissues exterior to a deep-seated periderm (de Bary 1884, Büsgen and Münch 1929). Contemporary authors refer to this aggregate of dead tissues, which consists of alternating layers of periderms and associated tissues, as "rhytidome", a term often considered synonymous with the term "outer bark" (Eames and MacDaniels 1947, Esau 1965). The living organs of the bark consist of the phloem and the living tissues of the innermost periderm, the phellogen and phelloderm. All living tissues have been collectively termed the "inner bark" (Eames and MacDaniels 1947). Borger (1973) provided an excellent discussion of the development and shedding of characteristic bark types. Trockenbrodt (1990) has provided an informative survey and discussion of terminology used in the bark anatomy literature.

Most pathogens are unable to penetrate directly the corky, suberized tissues of most outer bark and rhytidome. These outer layers represent constitutive defenses or preformed anatomical barriers to pathogen ingress. Given that the cutinized epidermis or suberized periderms of trees are the first tissues that potential pathogens encounter (Kolattukudy and Köller 1983), and given that the majority of trees remain alive for decades or centuries, these barriers are apparently very effective. This chapter will address anatomical aspects of wound response as it occurs in reaction to injuries of the living tissues of the inner bark and the role of wound response in resistance to fungal pathogens of trees.

2.2.1 Tissues of the Inner Bark

The living inner bark of a tree is composed of secondary phloem and periderm tissues (Fig. 2.1). In addition, cortical tissues fulfill an important role in bark of young stems. Periderm, the term first used by von Mohl (1845), is a protective tissue of secondary origin which replaces the epidermis in stems and roots that have continual secondary growth. Detailed descriptions of periderm formation are available (Srivastava 1964, Esau 1965, Fahn 1967). Roots, stems, and branches of gymnosperms, most dicotyledons, and a few monocotyledons develop periderm (Esau 1965). Herbaceous dicotyledons may form periderm, usually in the roots or oldest portions of the stem. In most coniferous and dicotyledonous trees, a periderm replaces the epidermis as the protective layer within the first year of growth. As trees age, sequent periderms may arise at successively greater depths, thus causing an accumulation of dead tissues on the surface of the stem or root and contributing to the formation of rhytidome on rough-barked species or simply outer bark on smooth-barked species.

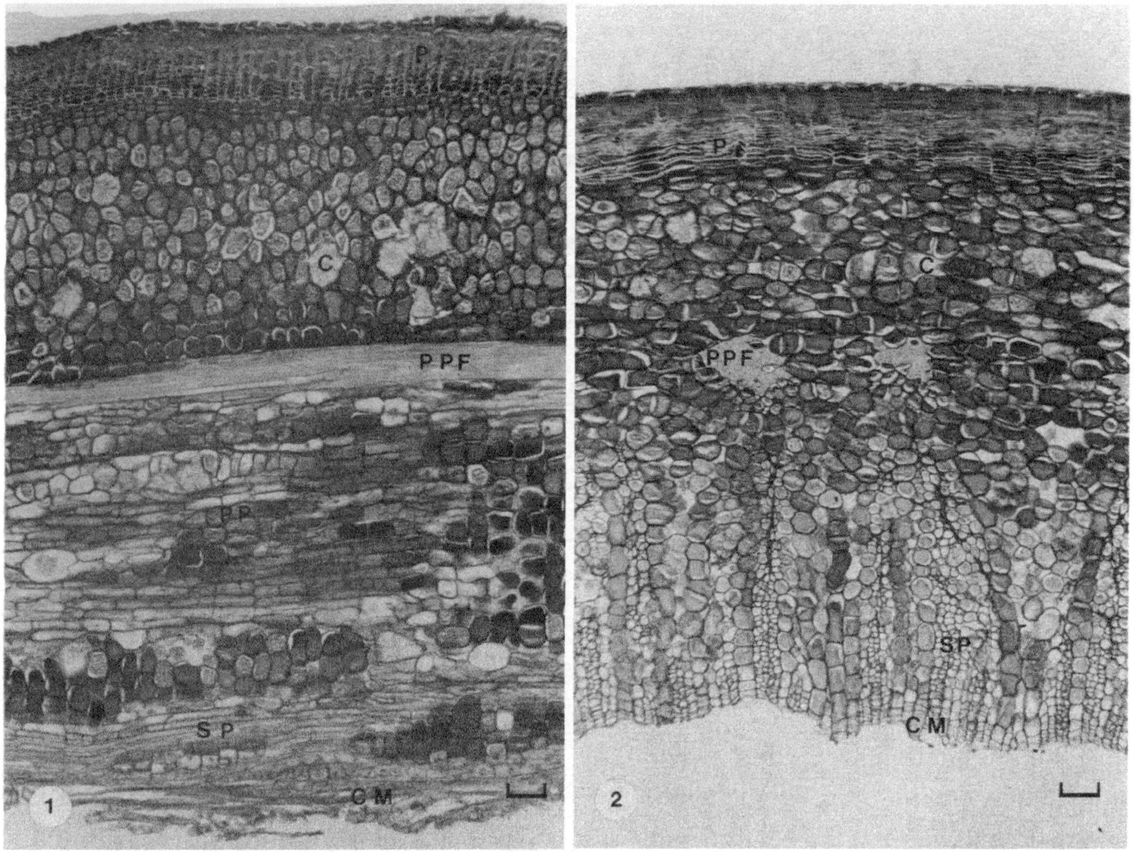

Fig. 2.1. Light micrographs (**1**, **2**) of longitudinal and transverse sections of 2-year-old nonwounded peach bark showing first periderm (*P*), cortex (*C*), primary phloem (*PP*), primary phloem fibers (*PPF*), secondary phloem (*SP*), and cambial region (*CM*). *Bar* = 20 μm

Briefly, the periderm consists of the following constituents as originally described by de Bary (1884): phellogen (cork cambium), the lateral meristem which produces the periderm; the phellem (cork), the suberized protective tissue formed outwardly by the phellogen; and the phelloderm, a living parenchyma formed inwardly by the phellogen. Phellogen cells generally appear oblong in transverse and radial sections and appear polygonal or irregularly shaped when examined in tangential section (Schneider 1955, Srivastava 1964, Esau 1965). Phellogen cells are characteristically thin-walled, have protoplasts, are vacuolated to varying degrees, and may contain ergastic substances, and chloroplasts. The phellogen may consist of only one layer, as in *Fraxinus pennsylvanica*, or may consist of a zone of meristematic cells, as in *Ailanthus altissima* (Borger and Kozlowski 1972a).

Phelloderm cells resemble cortical parenchyma cells in shape and content, although their radial arrangement distinguishes them from cortical cells. Walls of phelloderm cells may be thickened and intercellular spaces may be abundant. Phelloderm cells of *Populus tremuloides* are chlorophyllous and photosynthetically active (Strain and Johnson 1963).

There are two main types of phellem cells, suberized cork cells and lignified phelloid cells. Cork cells are radially shortened with relatively thick walls and phelloid cells are usually thin-walled and radially elongate (Grozditz et al. 1982). Both cells are dead at maturity and generally lack intercellular spaces (Esau 1965). The arrangement of phellem cells in bark varies according to species (Esau 1965).

Cortical tissues are found primarily in the bark of young stems. The first periderm in the stems of most species arises in the cortex, which is ultimately shed as new periderms arise. Phloem tissues are intimately involved in the development of bark structure (Borger and Kozlowski 1972a,b,c,d). The inner bark of smooth-barked species consists largely of phloem tissues. In addition, patterns of phloem element deposition in conjunction with particular patterns of periderm development are responsible for the structure of ring barks, scale barks, and furrowed barks (Kozlowski 1971). The living cells of the outer phloem give rise to deep-seated periderms.

2.2.2 Natural Periderms and Wound Periderms

According to Esau (1965), natural (including first and sequent periderms) and wound periderms are basically alike in method of origin and growth. The difference between them is mainly in timing of origin and restriction of wound periderm to the place of injury. Also, wound periderm is believed to differ from natural periderms in that the former is induced by a stimulus or injury, or by factors other than those responsible for the induction of natural periderms (Bloch 1941, 1952, 1953, Akai 1959, Srivastava 1964).

Mullick's studies of periderm formation in conifer bark revealed differences in pigment composition in different periderms and provided the first evidence that natural and wound periderms may be biochemically distinct (Mullick 1977). Based on these biochemical differences, he proposed a new nomenclature for woody plant periderms in order to distinguish the two types: (1) exophylactic periderms, which includes the first periderm and sequent periderms containing similar pigments, are thought to provide protection of living tissues from the environment, and (2) necrophylactic periderms, which include wound periderms, certain sequent periderms, and other periderms which are always found adjacent to dead tissue (nonsuberized impervious tissue or NIT), are thought to protect living tissues from the adverse effects of cell death. In angiosperms, wound periderms are distinct from the first periderms based on their dissimilar histochemical reactions to lignin reagents (Biggs 1984b, Rittinger et al. 1987) and the formation of a ligno-suberized boundary as a prerequisite to periderm differentiation (Mullick 1977, Soo 1977, Biggs 1984a,b).

2.3 Anatomy of Wound Response in Bark

Much is known about wound anatomy in trees and there is good documentation in the literature on the anatomical events that lead to boundary zone and wound periderm formation (Hartig 1894, Bramble 1934, Crowdy 1949, Butin 1955, Bloomberg and Farris 1962, Mullick 1977, Biggs et al. 1984). The concepts regarding the chronology of wound responses in plants were developed initially with sweet potato (Artschwager and Starrett 1931), and later studies confirmed and elaborated upon earlier observations (Morris and Mann 1955, Strider and McCombs 1958, McClure 1960).

2.3.1 Light Microscopy

Generally, the first indications of the response of a tree to mechanical wounding of the bark may be viewed with the light microscope as early as 24 h after wounding, although there may be species differences in the timing of events and other specific features, or there may be differences due to environmental effects on plant response. In injured peach bark tissues, for example, there may be extensive degradation of starch granules within the first 12 to 24 h. By 96 h, these granules will have disappeared entirely in the area of tissue that is undergoing dedifferentiation, although they may still be visible in the desiccated area near the wound surface and in the internal tissues some distance from the wound (Biggs 1984b). Cells in the incipient boundary zone undergo changes in the nucleus and the ability of the cytoplasm to take up morphological stains. In most cells of the incipient boundary zone, prophase nuclei are easily seen and the nucleoli are prominent 48 h after wounding (Fig. 2.2, 3).

In some species, including *Acer saccharum*, *Prunus persica*, and other *Prunus* spp., the deposition of a polysaccharide substance has been detected with periodic acid-Schiff's reagent in the walls of cells located in a zone about 300 μm from the wound surface (Biggs 1984b). Polysaccharide deposition occurred prior to the formation of a visible lignified zone. Although increased lignin can be detected with biochemical methods within the first 24 h after wounding (Doster

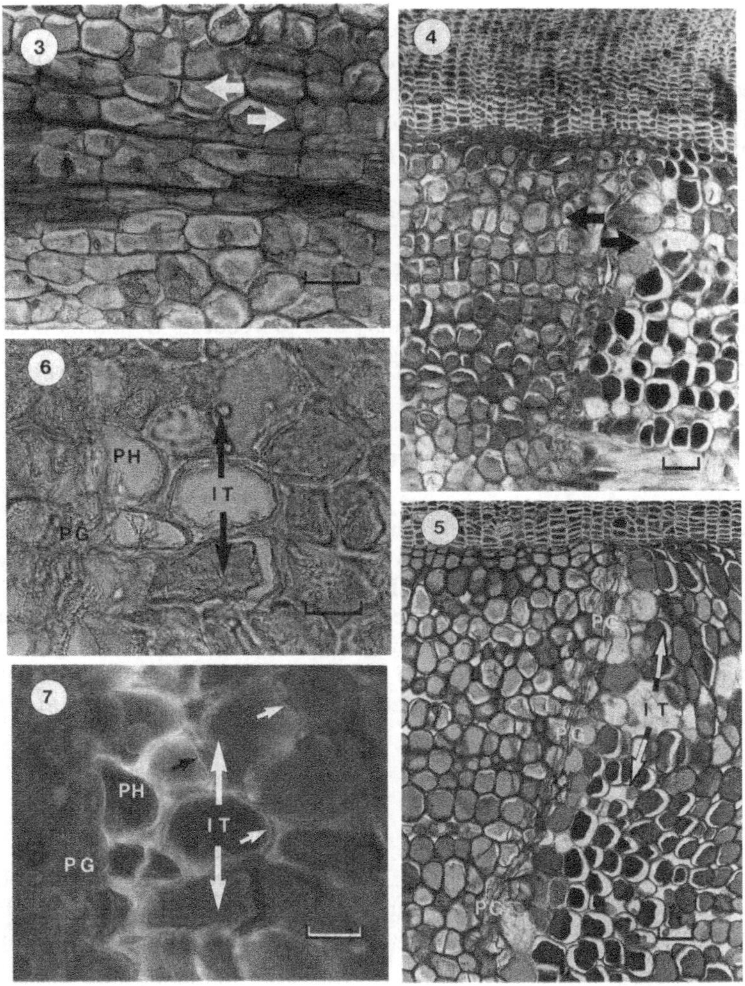

Fig. 2.2. Light micrographs (**3–7**) of longitudinal sections through tissue undergoing dedifferentiation during generation of new phellogen after mechanical wounding of bark of healthy 2-year-old peach bark. Outside bark in all figures is oriented toward the top of plate. *Bar* = 10 μm. **3** Section through the secondary phloem 48 h postwounding showing mitotic activity associated with tissue redifferentiation. Cells to the *left of the arrows* participate in phellogen generation; cells to the *right* are becoming hypertrophied and developing lignified walls and intracellular suberin linings. **4** Section through periderm and cortex 72 h postwounding showing two distinct tissue regions in the transition zone between healthy (*left*) and necrotic tissue (*right*). Tissue to the *right of the arrows* shows first positive reaction with phloroglucinol + HCl but is not yet impervious to fluid diffusion. **5** Section of differentiated phellogen (*PG*) and incipient impervious tissue (*IT* and *arrows*) 96 h postwounding. **6** and **7** Bark tissue stained with phosphine, under bright field and the same under ultraviolet excitation, respectively, showing impervious tissue (*IT*), first phellem (*PH*), and newly regenerated phellogen (*PG*) 7 days postwounding. Note in **7** the thin suberin lining of the cells in the impervious zone (*small arrows*)

and Bostock 1988a), the first signs of lignification detectable with histochemical reagents (phloroglucinol/HCl) are apparent within 72 h (Fig. 2.2, 4) and occur internal to the area of polysaccharide deposition if the latter occurs. Reports of lignification as a response to wounding in tree bark are numerous (Bramble 1934, Butin 1955, Mullick 1977, Soo 1977, Krähmer 1980, Biggs et al. 1983a, Biggs 1984a, 1985a,b). The first lignified cells can be detected in areas of the wound in closest proximity to the vascular cambium. Although most cell types exhibit lignification in response to wounding, the parenchymatous cells in phloem ray tissue are often the first to stain visibly with phloroglucinol/HCl.

Mullick, in a series of elegant studies (Mullick 1977), described a nonsuberized impervious tissue in conifer bark that was located in a position similar to the lignified zones reported in some of the earlier literature on tree wounds and infections. Soo (1977), a student of Mullick, reported a similar nonsuberized impervious tissue in angiosperm bark. However, results of studies reported from our laboratory on both angiosperms and gymnosperms have shown that the impervious layer that is formed prior to periderm regeneration is closely related to the formation of intracellular suberin linings in cells present at the time of wounding, and which have become lignified following wounding (Biggs 1984a, 1985a). Similar processes occur in wounded xylem parenchyma (Biggs 1987) and wounded tissues of various herbaceous and woody plant species and organs (Rittinger et al. 1987). Suberin deposition in lignified cells occurs within 24 to 48 h after visible lignification.

The ligno-suberized boundary zone is most often located approximately 0.8 to 1 mm inside the wound surface. Initial cells of this tissue can be detected within 4 to 7 days in wounds on actively growing trees in mid-summer (Fig. 2.2, 5; Biggs 1985b, 1986c) and usually occur in an area of the wound with closest proximity to the vascular cambium. A meristematic layer forms immediately internal to and abutting the primary ligno-suberized tissue and is usually detected 24–48 h after the formation of the latter tissue (Figs. 2.2, 6 and 7; 2.3, 8). Wound (necrophylactic) periderm may be well formed by 10 days postwounding (Fig 2.3, 9 and 10). Complete formation of the boundary zone and new periderm around the entire wound may take up to 28 days under ideal conditions (Fig. 2.4, 11 and 12); however, as boundary tissues continue to form in an outward direction, new phellogen cells form immediately internal to the established boundary tissue. As phellem is produced in an outward direction, the ligno-suberized boundary is crushed and diminishes in thickness (Biggs 1986c).

It is important to note that the presence of tissues in various stages of wound response can be observed by sampling wounded tissues at any one time, i.e., 7 to 24 days after wounding, depending upon species and inherent regenerative capacity. Ligno-suberized cells do not form synchronously into a distinct boundary zone (Biggs 1985b). The cells and subsequent tissues form first between the wound surface and the vascular cambium and last in the region of the original phellogen. One can often view wound tissues ranging from no visible reaction to those exhibiting complete periderm regeneration within the same histological section (Biggs 1985b).

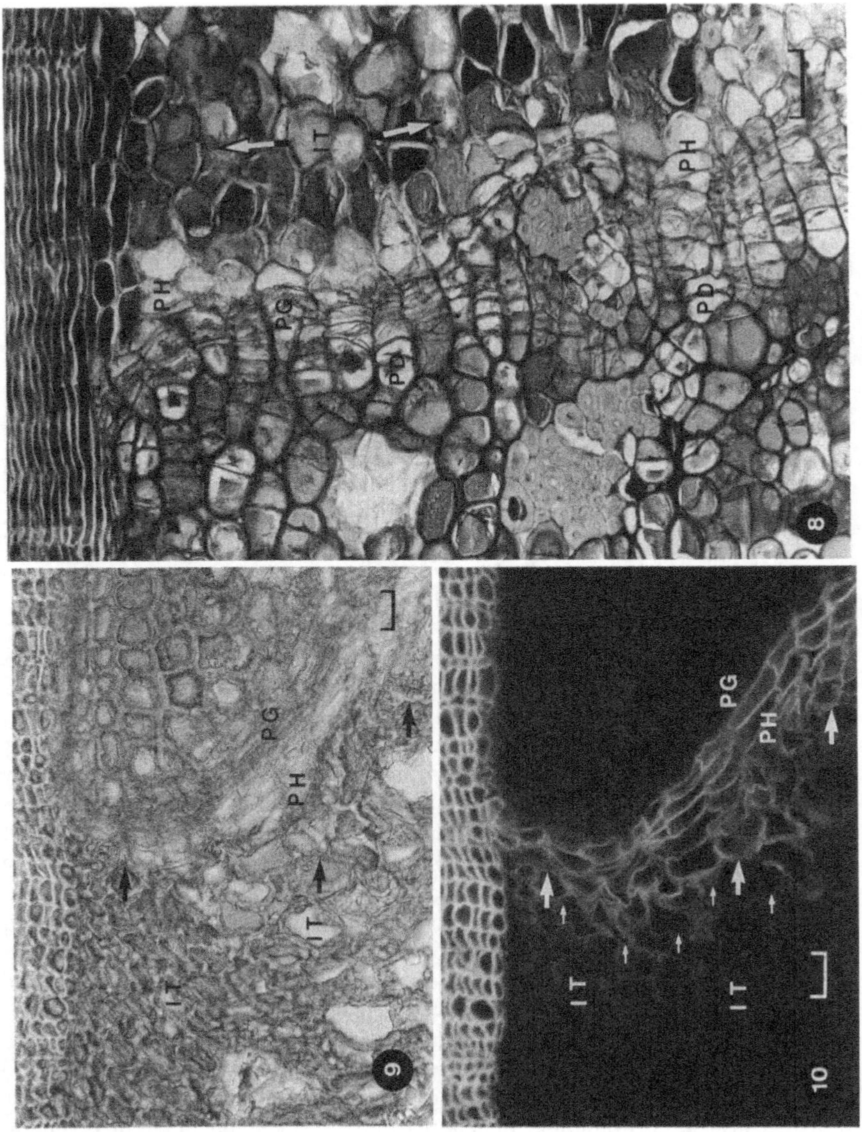

Fig. 2.3. Light micrographs (**8–10**) of transverse sections of wounded bark during the process of generation of new phellogen. *Bar* = 10 μm. **8** Stained transverse section of differentiated phellogen (*PG*) showing phellem (*PH*), phelloderm (*PD*), and impervious tissue (*IT*), 7 days postwounding. **9** and **10** Bark tissue stained with phosphine, under bright field and the same under ultraviolet excitation showing differentiated phellem (*PH*) and cells of the impervious zone (*IT*, its internal boundary delimited by *arrows*) exhibiting cell wall lignification and suberin linings (*small arrows* in **10**)

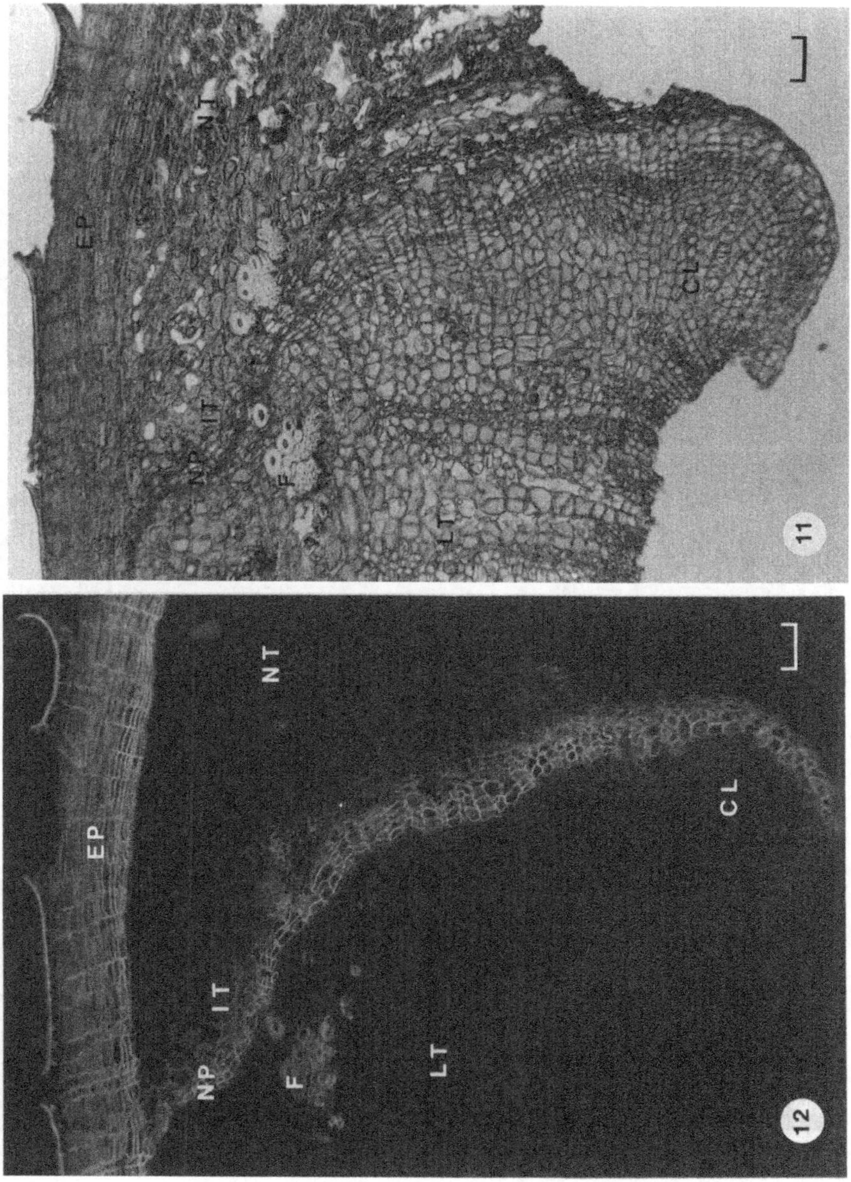

Fig. 2.4. Light and fluorescence micrographs (**11–12**), respectively, of transverse sections of 2-year-old peach bark at 28 days postwounding stained with phosphine GN showing necrophylactic periderm (*NP*) formed internal to a ligno-suberized boundary (*IT*). Note necrotic tissue (*NT*), living tissue (*LT*), exophylactic periderm (*EP*), primary phloem fibers (*F*), and wound callus tissue (*CL*). *Bar* = 20 μm

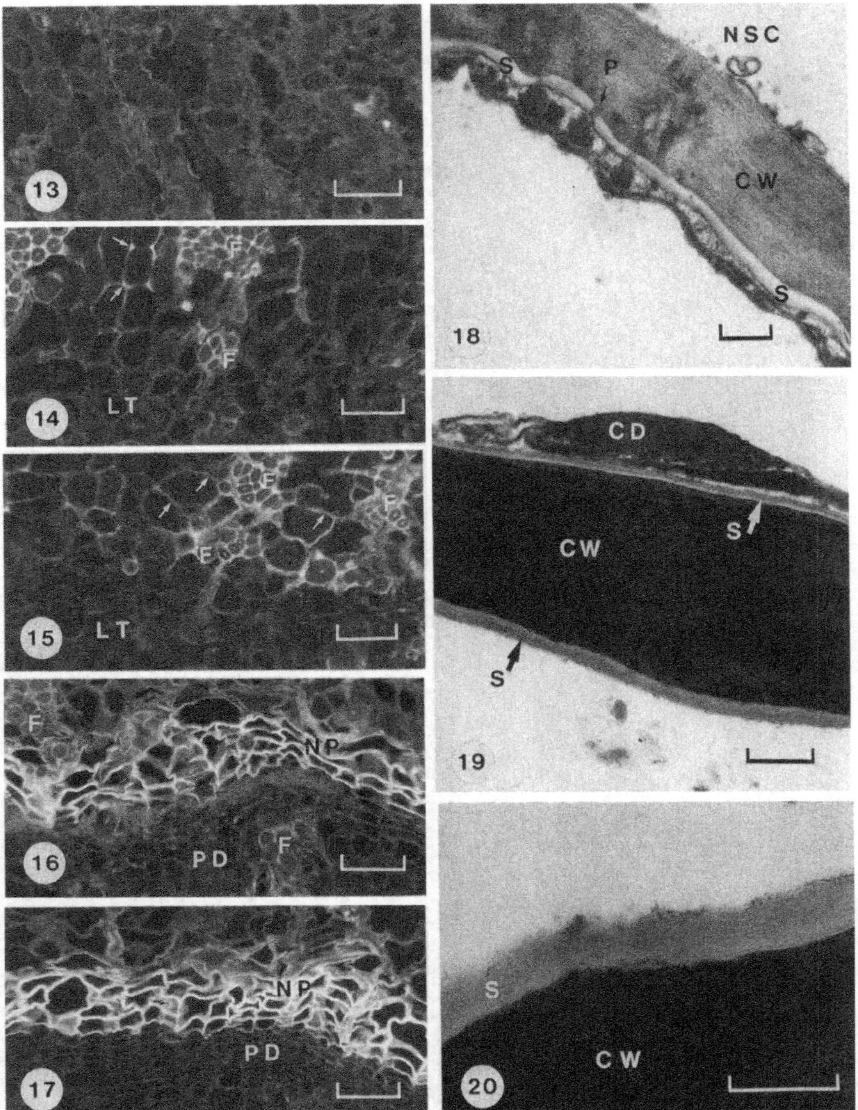
13
F
F
14
LT
F
F
F
F
15
LT
F
NP
16
PD
F
NP
PD
17
NSC
S
P
CW
S
18
CD
S
CW
S
19
S
CW
20

2.3.2 Electron Microscopy

At the ultrastructural level, dramatic changes are evident within the first 24 h in cells adjacent to the wound. Generally, these changes reflect the subcellular alterations visible with light and fluorescence microscopy described above. The most noticeable subcellular modifications are in the nucleus and include changes in nucleolar fine structure and chromatin organization, both of which suggest rRNA synthesis and active production of ribosomes. In the cytoplasm, there are increases in rough endoplasmic reticulum (ER), free ribosomes, and polysomes. Cells in the region of activity of the new phellogen show increases in the amount of cytoplasm, smooth ER, and dictyosomes (Barckhausen R 1978, Biggs and Stobbs 1986). All these changes reflect the intensified transcriptional, translational, and secretory activities of the responding cells. After 24 h, ultrastructural changes are limited to those cells either undergoing dedifferentiation to form the ligno-suberized boundary zone or those undergoing redifferentiation to form the new phellogen and its derivatives.

Ultrastructural evidence for cell wall suberization in wounded peach bark was observed at 8 days after wounding (Fig. 2.5, 13–20). Boundary zone cell walls were completely lined on the inside with an electron-lucid material corresponding to cell wall linings with the histochemical and autofluorescence characteristics of suberin. The suberin portion of the cell walls appeared, at first, electron-lucid, followed by the formation of many light and dark lamelations. The suberin lining in individual cells appeared to be uniform in thickness, although thickness of the lining varied from cell to cell (ca. 40–120 nm). Suberized cells in the boundary zone contained senescing cytoplasm with fragments of undifferentiated

---

Fig. 2.5. Transverse sections of peach bark examined with ultraviolet epifluorescence illumination (**13–17**, *bar* = 10 μm) and transmission electron microscopy (**18–20**, *bar* = 1 μm). In **13–16**, the wound surface is approximately 800 μ above the noted cellular changes. Note the presence of phloem fibers (*F*) in **14–16**. **13** Nonwounded control tissue showing ray parenchyma in the region of primary and secondary phloem. Note the lack of autofluorescence. **14** Tissue 6 days postwounding. First traces of lignin autofluorescence are in the cell corner-middle lamella region of boundary zone cells (*arrows*) immediately external to living tissues (*LT*). **15** Tissues 8 days postwounding. Note deposition of suberin linings (*arrows*) in boundary zone cells and the dedifferentiation of internal tissues in the process of forming new phellogen (*LT*). **16** Tissue 12 days postwounding showing a completely differentiated necrophylactic periderm with phelloderm (*PD*) and two to four layers of phellem cells (*NP*). **17** Tissues 14 days postwounding, showing a completely differentiated necrophylactic periderm with three to five phellem cell layers (*NP*) and newly differentiated phelloderm (*PD*). **18–20** Transmission electron micrographs of transverse sections of peach bark boundary zone tissues. **18** Portion of boundary zone cell 8 days postwounding with suberin lining (*S*) adjacent to a nonsuberized cell (*NSC*). Plasmodesmatal canals (*P*) in shared cell wall (*CW*) appear more electron dense than suberin. Suberin appears only slightly lamellate and is relatively electron lucid. **19** Shared cell wall of two adjacent cells in the boundary zone 12 days postwounding. The compound wall (*CW*) is very electron dense. Suberin linings (*S*) are of uniform thickness within individual cells and vary in thickness among cells. Note the fine light and dark lamellations of the suberized wall and the cytoplasmic debris (*CD*) appressed to the wall of the upper cell. **20** High magnification view of suberized cell wall in the boundary zone showing the fine light and dark lamelations characteristic of suberin

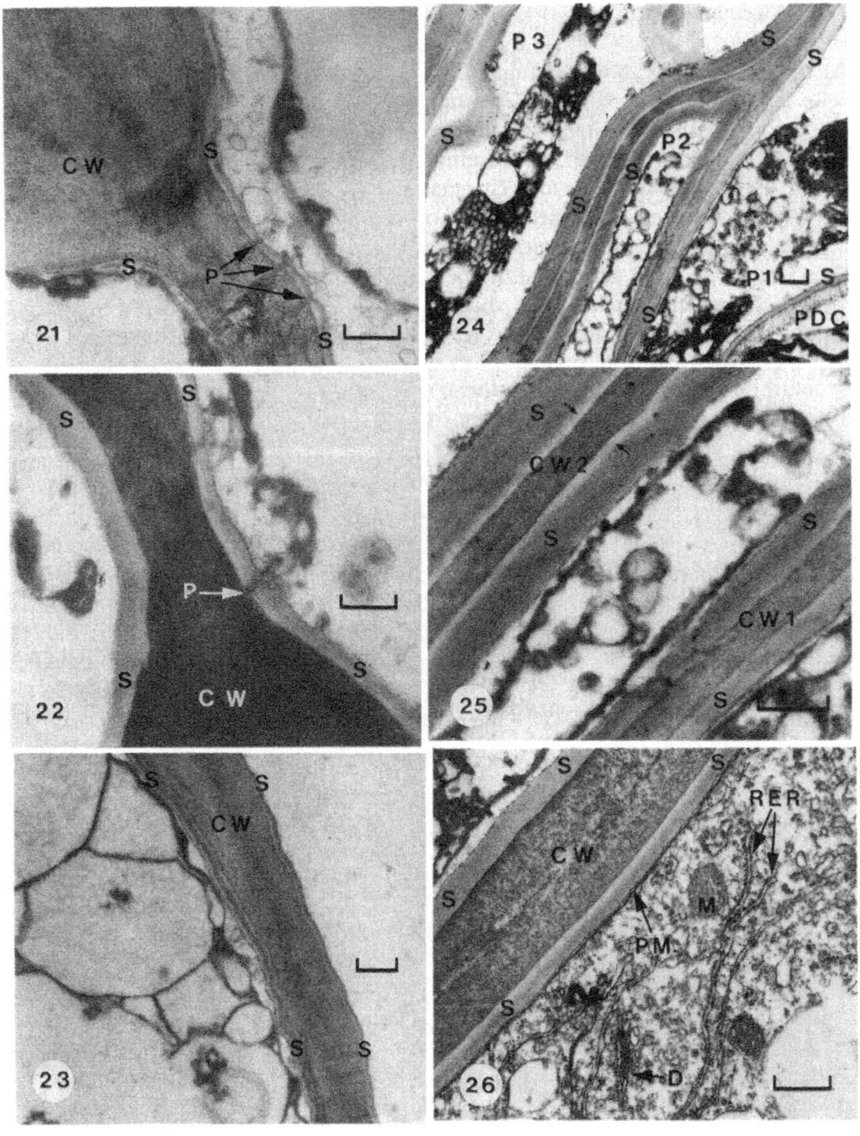
21
CW
S
P
S
S
22
S
S
P
CW
S
S
23
S
S
CW
S
S
24
P3
S
P2
S
S
P1
S
PDC
S
S
25
S
CW2
S
CW1
S
S
26
S
S
RER
CW
M
PM
S
S
D

dense material that formed a thin, discontinuous granular deposit inside the suberin layer. The granular, electron-dense materials likely resulted from the disintegration of the cytoplasmic ground substance and the various cell organelles. Between 8 and 12 days after wounding, the primary walls and the middle lamella in the boundary zone exhibited an increase in electron density. This is due probably to the deposition of phenolic polysaccharide material in the wall. In peach, these substances, usually referred to as gum, are produced nonspecifically in response to wounds or infections. Transmission electron microscopy of the boundary zone revealed that suberin linings were discontinuous over pit areas (Fig 2.6, 21 and 22; Biggs and Stobbs 1986). Therefore, it is likely that the impermeable nature of these primary ligno-suberized boundaries is due, to some extent, to both lignin and suberin. Vesicles of varying proportions were frequently associated with the periphery of the senescent cytoplasm (Fig. 2.6, 23).

Cells of the new necrophylactic phellem possess dense, granular cytoplasm with few distinct organelles (Fig 2.6, 24–26). In mature phellem, cell contents appear as a compact mass of electron-dense, amorphous material interspersed with electron-lucid deposits and dark bodies of various dimensions. Numerous vesicular elements were observed, the membrane elements appearing to be embedded randomly throughout the granular matrix. The plasmalemma was typically separated from the cell wall. Phellem cells possessed a compound middle lamella with an amorphous fine structure. This portion of the cell wall was red when stained with phloroglucinol/HCl. Suberin comprised the largest portion of the secondary wall and displayed fine light and dark lamellations. Thickness of phellem suberin layers (ca. 60–350 nm) increased with distance from the phellogen. Cell wall pits and plasmodesmata canals were not observed in the phellem. Phellem cells with intact organelles were detected infrequently and, when detected (Fig. 2.6, 26), were characterized by abundant mitochondria, rough endoplasmic reticulum, dictyosomes, and associated vesicles.

Fig. 2.6. Transmission electron micrographs (**21–26**) of transverse sections of the boundary zone and necrophylactic periderm of wounded peach bark. *Bar* = 1 μm. **21** Portion of a shared boundary zone cell wall (*CW*) 8 days postwounding showing a simple pit area, and associated plasmodesmata (*P*), and lamellate suberin linings (*S*). **22** High magnification view of plasmodesmatal opening through the suberized portion of the boundary zone cell wall. **23** Shared boundary zone cell wall (*CW*), at 12 days postwounding with lamellate suberin linings (*S*) of differing thickness and the multivacuolate appearance of a senescent boundary zone cell. **24** Portion of necrophylactic phellem showing at *bottom right* a portion of the lumen and shared wall of the external phellogen daughter cell (*PDC*), the most recently derived suberized phellem cell (*P1*) with thin suberized secondary walls (*S*), and two older phellem cells (*P2* and *P3*) showing thick suberized secondary walls (*S*) and senescing cell cytoplasm. **25** Closer view of shared necrophylactic phellem cells walls between P1 and P2 (*CW1*) and P2 and P3 (*CW2*) from **24**. Note the fine light and dark lamellations of the suberized secondary walls (*S*) and the nonlamellate electron-lucid region of the secondary wall adjacent to the primary wall (*arrows*). **26** Organelle-rich cytoplasm of a phellem cell of a necrophylactic periderm sampled 14 days postwounding. Note lignified primary wall (*CW*), suberized secondary walls (*S*), plasma membrane (*PM*), rough endoplasmic reticulum (*RER*), mitochondrion (*M*), dictyosome (*D*), and associated vesicles

2.4 Physiological and Biochemical Aspects of Wounding

In comparison with herbaceous plants, there is little information available on biochemical aspects of wounding in tree species, perhaps due to the inherent difficulty of conducting biochemical analyses of woody tissues. Much of our knowledge about trees in this area comes from histochemical investigations and is thus limited in scope. Generally, when a plant is injuried, a complex array of physiological and biochemical responses are elicited. Bostock and Stermer (1989) categorized these into immediate or rapid responses (e.g. depolarization of cell membranes, release of host or pathogen cell wall fragments) which occur within seconds or minutes after wounding and slow response (e.g. complex biosynthetic reactions, formation of boundary tissues) which occur over a period of hours, days, or even weeks. For convenience, wound responses can be separated among the cellular compartments where they occur, i.e., membranes, cytosol, and cell wall (Table 2.1). The separation of events according to spacial and temporal criteria is artificial given the intimate contact and free exchange of materials among compartments. What emerges following an injury is a series of predictable and coordinated events which concludes, in the bark of woody plants, with formation of a primary ligno-suberized boundary layer, cell division leading to the formation of wound periderm, the production of callus tissue, new vascular cambium, and eventual closure of the wound.

Stimulation of the shikimic acid pathway leading to enhanced production of phenolic derivatives is a ubiquitous response among plants when injured or infected (Vance et al. 1980). Phenolic substances derived from the shikimic acid pathway may be directly toxic to potential pathogens or may form the foundation for high molecular weight polymers, such as lignin, that are incorporated into walls of extant or newly differentiated cells. Fatty acid metabolism is involved in the synthesis of suberin and waxes (Kolattukudy 1984), and these substances are found also in extant and newly differentiated cells.

In trees, the rapid production of high levels of suberin may be a determining factor in resistance against fungal pathogens (Biggs and Miles 1988, Biggs 1989). Since suberization appears to be triggered by wounding, it is likely that some chemical signal generated soon after the wound is inflicted initiates the process leading to suberin deposition. Experiments with potato tubers have shown that suberization can be inhibited by washing the wound with water during the first 72 h (Soliday et al. 1978). Abscisic acid is removed by these washes and exogenous ABA applied to washed tissues partially restored suberin production. Abscisic acid presumably is an important intermediate in the signal transduction pathway leading to the induction of suberin (Kolattkudy 1984). Abscisic acid is probably not the direct inducer of suberin because suberin deposition can still be inhibited beyond the time that ABA can be washed out of the tissue. Kolattukudy maintains that ABA produced during the first 24–48 h after wounding triggers the formation of an as yet unidentified suberin-inducing factor at about 72 h after wounding. This is followed by the appearance of enzymes involved in suberin biosynthesis within 96 h and the presence of suberin soon thereafter.

There has been much recent interest in the stimulation of defense reactions in plants by polysaccharides of animal, plant, and fungal origin (Darvill and Albersheim 1984). Albersheim and his associates maintain that the synthesis of phytoalexins in some plants is elicited by cell wall oligosaccharides. When oligosaccharides from both the fungal and plant walls are used as a treatment, their effect on the production of phytoalexins is synergistic. Many other substances of host or microbial origin may be involved in the elicitation of defense reactions in plants. Traumatin, trans-2-dodecenedioic acid, or the oxidation product of its aldehyde, 12-oxo-trans-10-dodecenoic acid, have been found in healthy and wounded bean cells (Zimmerman and Coudron 1979). The eicosapolyenoic acids, arachidonic acid and eicosapentaenoic acid, are abundant in the lipids of *Phytophthora infestans* and related Oomycota (Bostock and Stermer 1989). When applied to potato tuber disks, these fatty acids cause marked changes in isoprenoid metabolism, especially when combined with fungal β-glucans, thus demonstrating that fungal substances can enhance and redirect wound metabolism.

Chitinase and β-1,3 glucanase activity increase in some plants after wounding or infection (Boller 1987). Indeed, Wargo (1975) extracted these enzymes from the phloem of *Acer saccharum* and several *Quercus* spp. and demonstrated that the enzymes could lyse the hyphal walls of the root rotting fungus *Armillaria mellea*. Hydrolase enzymes may act to inhibit fungal growth and degrade fungal cell walls, thus releasing elicitors during the early stages of colonization (Mauch et al. 1988). The release of vacuolar hydrolases in combination with secondary compounds and proteinase inhibitors (see Chap. 3) may act in the initial stages of a multicomponent system of defense.

If oligosaccharins, hormones, cations, or other substances are responsible for triggering wound responses, it may be possible to facilitate wound processes by treating tissues with putative messenger compounds. In experiments with peach bark, Biggs and Peterson (1990) applied 14 different chemical treatments to wounds, including acid extracts from peach leaf cell walls and fungal cell wall extracts (from *Leucostoma persoonii*) alone and in combination, ethylene, abscisic acid, chitosan, calcium ion, and cellobiose. Fungal cell wall extract and cellobiose stimulated bark lignin ten fold over the control. None of the other treatments affected lignin production significantly. Given the quantities of lignin observed in these treatments, it is possible that these polysaccharides play a role in triggering the processes that lead to the observed differences in the degree of lignin deposition in wounds versus infections (Biggs 1984b) (Chap. 3). None of the treatments stimulated suberin production and, in fact, many appeared inhibitory.

Cell division is associated with both periderm formation and callus production leading to wound closure. Yev-Ladun and Aloni (1990) hypothesized that auxin and ethylene are the major factors controlling first periderm formation in woody stems. They suggest that moderate auxin flow retards periderm formation, whereas high auxin levels promote ethylene production, which indirectly results in periderm formation. Their hypothesis is based on observations of the patterns of first periderm ontogeny in young woody stems. They observed that first periderm formation is inhibited below leaves and buds, probably due to polar

Table 2.1. Summary of membrane, cytosol, and cell wall responses to wounding (After Bostock and Stermer 1989)

Injury or wound stimulus → Membranes, Cytosol, Cell wall

	Membranes	Cytosol	Cell wall
	Depolarization of cell membranes	Cytoplasmic aggregation	
Immediate or rapid responses	Net K^+ Cl^- efflux	Generation of volatiles (aldehydes, ethane, ethylene, and other products of lipid peroxidation)	
	Net H^+ Ca^{2+} influx		
	Loss of compartmentation and release of lipid-degrading and oxidative enzymes	Wound respiration	
		Reactive oxygen species	
	Deacylation of membrane lipids	ABA synthesis	Stimulation of polysaccharidases (?), and basic peroxidases
	Generation of local and systemic wound signals (action potentials, traumatin, Ricca's factor (?)	Ethylene from ACC	
		Generation of other second messengers (phosphoinositides,	Release of oligogalacturonides (?)
	Stimulation of callose synthetase, cellulose synthetase, and ACC synthetase	Ca/calmodulin)	
		Enhanced transcriptional and translational activities	
	Repolarization of membrane potential	Activation of TCA cycle	

	Restoration of membrane structure	Elevated cyanide-insensitive respiration	
	Restoration of ion transport mechanisms	Accelerated catabolism of sugars by glycolytic and pentose pathways	
		Biosynthesis of phenylpropanoids, isoprenoids, and polyketides	Papillae and callose deposition
			Ca^{2+} and Si deposition
Slow responses		Induction of proteinase inhibitors, chitinase, glucanase, acidic peroxidases, phenoloxidases, and enzymes for suberin synthesis	Modification by phenolic esters, lignin, HRGPs
			Suberin deposition

Cell division, wound periderm formation and vascular occlusion auxin transport. In addition, they maintain that effects of light, humidity, oxygen, wounding, and pressure on periderm formation are related to the effects of these factors on the relative levels of auxin and ethylene in the tissues. Furthermore, they suggest the presence of a positive feedback control mechanism which promotes phellogen activity and rhytidome formation. The mechanism is based on the assumption that the first-formed periderm constitutes a barrier to the outward movement of ethylene from the inner tissues of the plant. Therefore, the bark inside the first-formed phellem accumulates relatively high ethylene concentrations which increases phellogen activity. This promotes the initiation of deeper phellogen and ultimately gives rise to rhytidome and the formation of visible bark scales. This interesting hypothesis awaits the actual determination of critical hormone levels in tissues undergoing these developmental changes. Chapter 3 contains additional discussions of the regulation of plant development, wound response, and resistance to pathogens in bark.

2.5 External Factors Influencing the Generation of New Bark Tissues

Many discussions on the role of wound healing in the resistance of plants to pathogens neglect the importance of external factors as determinants of the wound response rate and, indirectly, the host/pathogen interaction. Environmental factors may influence any or several of the factors listed in Table 2.1 and thereby alter quality and quantity of wound-related events. Additional factors other than those discussed below probably affect the wound response either directly or indirectly, including plant nutrient status, level of herbivory and history of defoliation (Wargo 1977), carbon/nitrogen ratio, acid rain, and air pollution.

Temperature has a strong influence on the rate of wound healing in both herbaceous and woody plant species (Bloch 1952, Krähmer 1980, Biggs and Northover 1985, Biggs 1986b, Morris et al. 1989). With apple, sweet cherry, and peach, a significant correlation exists between temperature and rate of boundary zone and periderm regeneration in wounded bark (Biggs 1986b). Trees wounded at various times during the growing season were examined for complete formation of the primary ligno-suberized zone and new periderm. Although the tissues could be detected with a 7- to 21-day time period after wounding, degree-days (base = 0 °C) accumulated during the postwounding period explained over 80% of the observed variation in wound response. Tree phenological stage did not appear to exert significant influence on the wound responses measured in this study.

In experiments to determine the formation of the primary ligno-suberized layer and phellogen following leaf abscission in peach, Biggs and Northover (1985) reported that plants maintained at 7.5, 12.5, and 17.5 °C showed first indications of the primary ligno-suberized layer at 18, 9 and 6 days, respectively. Subsequent generation of phellogen and the appearance of the first phellem cells were observed at 30, 18 and 12 days, respectively. Earlier research on the influence of temperature on wound-induced cell division has shown that, within, limits, the time required for the first cell division is linearly and inversely related to temperature (Lipetz 1970). Maximum and minimum temperature limits for wound responses have not been established for any tree species; however, in potato tubers, maximum suberization occurs at 20–25 °C (Artschwager 1927, Wigginton 1974, Thomas 1982, Morris et al. 1989). Krähmer (1980) found that periderm could not be detected microscopically in leaf scars of apple at temperatures below 8 °C. Accordingly, fruit scars remained susceptible to infection by *Nectria galligena* for more than 4 weeks at 6 °C.

When trees were wounded during the dormant stage, lignin content of the bark, determined with a thioglycolic acid assay, increased, although an increase in lignin detectable with the histochemical reagent phloroglucinol/HCl was not observed (Doster and Bostock 1988b). Wounds became more resistant to inoculation with *Phytophthora syringae*, however, thus suggesting resistance associated with lignin at very low tissue levels or mechanisms of resistance in addition to those associated with wound repair. Wound studies during the dormant season require long observation and sampling schedules, and it is possible that epiphytic microorganisms could colonize the wound site and contribute to altered rates of pathogen colonization, independent of the responses of the host.

Plants under water stress are generally considered more susceptible to invasion by weak pathogens (Schoeneweiss 1981), but few studies have been able to demonstrate the mechanism by which water stress increases the susceptibility of trees to fungal invasion (see Schoeneweiss 1981) for a review of this literature). Water deficit influences numerous physiological processes and pathways, and wound responses are no exception. Plant water status has been shown to influence the formation of the boundary zone, affect cell division in the wound periderm, and contribute to increased susceptibility of wounds to fungal pathogens.

Biggs and Cline (1986) examined the effects of irrigation treatments on the rate of boundary zone and wound periderm formation in wounds on peach limbs. No differences could be determined at 7 and 10 days after wounding between irrigated and nonirrigated trees for lignin autofluorescence, the intensity of suberin autofluorescence, or in the numbers of boundary zone cells; thus irrigation did not influence the formation of the primary ligno-suberized layer. Significant differences in suberin autofluorescence due to irrigation were measured 14 days after wounding and were related to increased numbers of suberized phellem cells in the wound periderm of irrigated trees relative to nonirrigated trees. Water stress, therefore, in the range of −0.65 to −0.80 MPa can inhibit periderm formation by diminishing the rate of cell division. Puritch and Mullick (1975) reported that formation of nonsuberized impervious tissue (NIT) (i.e., the primary ligno-suberized layer) in *Abies grandis* was inhibited by water stress less than −1.5 MPa. Where formation of NIT was retarded, subsequent generation of wound periderm was thought to be slowed also. Butin (1955) also determined that wound periderm formation was related to tissue water content. He found that susceptibility of poplar to *Cytospora chrysosperma* was increased where callusing of wounds was inhibited by increased water loss.

Relative humidity has an influence on the wound repair process in potatoes (Wigginton 1974, Morris et al. 1989) and, presumably on that of woody species also. In general, wound healing proceeds most rapidly at 20–25 °C with relative humidities between 70 and 100% and, at 10 °C, between 80 and 100%. As relative humidity approaches 100%, cell proliferation, rather than periderm formation, may occur. Relative humidities below 50% may inhibit wound healing, although it is possible that as humidity declines, the depth at which the plant forms new periderm increases as desiccation of wounded tissues occurs.

Other external factors also may influence the expression of wound responses in plants. Suberin deposition in primary ligno-suberized tissue in wounded *Pachysandra terminalis* was affected adversely by both exposure to solar radiation and deicing salts (Hudler et al. 1990). The influence of microorganisms as irritants (Kaufert 1936) or stimulating factors at the wound site (Blanchette and Sharon 1975) has received little attention. Studies in this area could reveal significant new information about fundamental aspects of wound responses in nature.

2.6 Regeneration of Vascular Cambium and Wound Closure

Given that many of the tissues involved in callus regeneration are part of the barrier zone in the process of compartmentalization, only a limited discussion will be presented here. Chapter 5 contains additional discussions on barrier zones. The production of callus tissue, the differentiation of new vascular cambium within callus, and the eventual closure of the wound, usually leading to the reestablishment of vascular cambium continuity, occurs following wounds inflicted to the depth of the xylem. The amount and rate of callus production following wounding varies according to tree species and selections within species (Gallagher and Syndor 1983, Martin and Sydnor 1987), the size of the wound, and location of the wound on the tree (Wensley 1966). The source of callus tissue also varies to some extent. In scoring experiments with poplar, silver maple, pear, and apple, callus formation was contributed mainly by living cells of vascular rays in the proximity of the cut, and, to a lesser extent, the longitudinal parenchyma of the phloem and xylem (Soe 1959). In linden, callus tissue forms from any active, newly formed cambial derivative rather than from any one particular cell type (Barker 1954). Generally, there is a diversity of opinion on the source of callus in the regeneration of new bark (Bloch 1941), although most researchers agree that the original vascular cambium does not contribute significantly to the formation of callus tissue. The initiation of phellogen regeneration in callus always precedes that of vascular cambium. The ventral region of growing callus, where the differentiating phellogen and vascular cambium are in close proximity, is nonlignified and nonsuberized and is especially susceptible to disruption by pathogenic fungi (Biggs 1986a, Biggs and Britton 1988).

Formation of callus over large diameter wounds may occur in a distinctive pattern (Fig. 2.7, 27; Shigo 1986). Usually, callus production is the greatest in

Fig. 2.7. Photographs (**27–29**) of normal wound responses during and after wound closure. **27** Callus ribs formed in a distinctive pattern on eastern hemlock with six bands of callus growing over a 6-year-old wound. Most of the time, the first callus rib is larger than the ribs produced subsequently. **28–29** Completely closed wound on a yellow birch showing external and internal appearance, respectively (Photographs courtesy of AL Shigo)

the first year, after which it declines within 3 to 5 years. Factors influencing the anatomical co-mingling of the new vascular cambia as the callus ribs converge on the center of the wound from the perimeter are not understood. It appears that complete closure is prevented if the vascular cambium turns inward. The periderm on the callus surface may prevent physically the co-mingling of vascular cambia in this situation, thus leading to a "bark inclusion", visible as a fine crack between the converged callus ribs. An incompletely closed wound may provide an excellent environment for fungal pathogens and it is possible that pathogenic microorganisms influence the direction of growth of new vascular cambia.

When wounds close via the co-mingling and junction of converging vascular cambia (Fig. 2.7, 28 and 29), active growth of microorganisms ceases is that area (Shigo 1986), probably due to altered gaseous regimes within the wound site. It is not clear whether organisms within closed sites actually succumb, remain latent, or are able to resume pathogenesis when conditions conducive to their growth re-occur (see Chap. 6).

2.7 Wounds as Infection Courts

In many host/pathogen interactions, the boundary setting process in the wound (the infection court) confers resistance to infection. Many researchers have demonstrated that wounds become increasingly less susceptible to infection with age (Butin 1955, Cline and Neely 1983, Russin and Shain 1984, Riffle and Peterson 1986, Bostock and Middleton 1987). This type of resistance to infection is thought to be related to nonspecfic plant responses leading up to and including formation of primary ligno-suberized tissues and secondary wound (or necrophylactic) periderm. The major structural components of these tissues are lignin and suberin (Kolattukudy 1984, Biggs 1985a). Definitive proof of the role and importance of ligno-suberized tissue and wound periderm in resistance to disease in trees has never been presented.

When an infection court is created for a wound pathogen, disease is most severe when the inoculum arrives at the infection court immediately. If the inoculum arrives later, disease frequency and severity decline with time until the wounded tissues express resistance comparable to that of noninjured bark (Biggs 1986c, 1989). The time it takes for complete resistance to become reestablished is dependent upon the factors discussed above as well as the pathogenic capabilities of the fungus. Histological studies with peach and the canker pathogen *Leucostoma persoonii* showed that wounds resistant to inoculation possessed a minimum of three phellem cells in the new periderm. At earlier stages, wounds were susceptible to the fungus, although severity of the symptoms declined beginning about 3 days after wounding (Biggs 1986c). The presence of periderm was more critical for inhibition of the pathogen than primary ligno-suberized tissues.

The relative number of cells in periderm tissue or the thickness of the suberized layers in wounded potato tubers generally reflect the relative resistance to pathogens (i.e., wounds with thicker suberized layers are more resistant to pathogens). Generally, this is true for wound periderms in woody plants when comparisons are being made within a single genotype (Biggs 1986c) or among different species over a time course study (Biggs 1986b). However, when comparing genotypes within a species, it is unlikely that numbers of wound phellem cells or the thickness of the suberized layer is of major importance in determining the resistance or susceptibility to wound pathogens. This statement is based upon the work of Biggs (1989), which demonstrated no significant correlation of these anatomical parameters with the field performance of various peach cultivars, canker length after inoculations, or the amount of accumulated suberin measured photometrically. For peach, resistance to *Leucostoma* spp. was correlated with an increased rate of suberin accumulation (Biggs and Miles 1985, 1988, Biggs 1989). Lignin accumulation, which is thought to play an important role in many host/pathogen interactions (Craft and Audia 1962, Vance et al. 1980), appears to serve a less important role than suberin in peach bark and perhaps the bark of other tree species. Periderm regeneration is probably only one of many possible types of resistance in peach bark to *Leucostoma* spp., and can be described as a type of rate-limiting or partial resistance.

Fungi inoculated into older wounds can survive in nature without causing immediate infection. Mycelium of *Hypoxylon mammatum* survived in wounds of aspen for periods of up to 2 years prior to development of symptoms (Ostry and Anderson 1983). Similarly, Russin and Shain (1984) reported that *Cryphonectria parasitica* remained viable for more than 36 weeks in inoculated wounds. Wounded peach bark inoculated with spores of *Botryosphaeria* spp. did not develop visible macroscopic symptoms until 8 weeks after inoculation (Biggs and Britton 1988). Formation of a ligno-suberized boundary zone, generation of new periderm, and callus tissue do not guarantee that wounds will not become infected, although the risk of infection is reduced greatly. Aspects of wound closure and microenvironment may help to explain the apparent latent colonization of canker-causing fungi.

2.8 Proposed Anatomical Model for Wound Responses in Woody Plants

The regeneration of periderm, callus tissue, and new vascular cambium is an energy intensive process that serves three purposes: (1) replace tissue and regenerate lateral meristems; (2) reestablish control over gas exchange and desiccation; and (3) prevent or restrict the ingress of pathogens. The model presented in Fig. 2.8 was originally developed by Mullick (1977) and was revised for this presentation to account for the presence of suberin in the primary ligno-suberized boundary zone which forms from extant cells prior to the formation of new phellem. The model describes, at the anatomical level, the nonspecific host responses associated with wounding, pathogen invasion, or insect injury.

Understanding the basis for wound responses in woody plants could lead to innovative control measures for tree diseases. Research in this area should

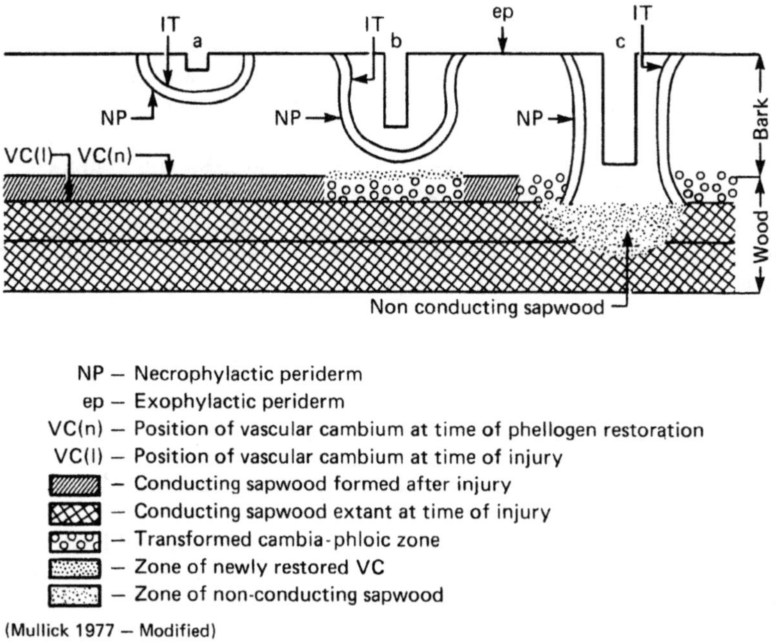

Fig. 2.8. Diagrammatic views of the anatomical model for nonspecific defense mechanisms which follow **a** injury limited to the vicinity of the living phellogen, **b** injury to the phellogen, cortex, phloem, and vascular cambium, and **c** injury to all the above tissues and the functional sapwood (After Mullick 1977)

include basic studies on the molecular regulation of wound metabolism in trees, pathological studies on the role of wound responses in resistance to bark and xylem pathogens, the role of nonpathogenic microorganisms on properties of wound tissues, and genetic experiments to determine the heritability of favorable wound response characters. Well-designed time-course studies are required in order to completely characterize wound related phenomena.

References

Akai S 1959 Histology of defense in plants. In: Horsfall J, Dimond A (eds) Plant pathology, vol II. Academic Press, New York, 391–434

Artschwager E 1927 Wound-periderm formation in the potato as affected by temperature and humidity. J Agr Res 35:995–1000

Artschwager E, Starrett RG 1931 Suberization and wound periderm formation in sweet potatoes and gladiolus as affected by temperature and relative humidity. J Agr Res 43:353–364

Barckhausen R 1978 Ultrastructural changes in wounded plant storage tissue cells. In: Kahl G (ed) Biochemistry of wounded plant tissues. Walter de Gruyter, Berlin New York, 1–42

Barker WG 1954 A contribution to the concept of wound repair in woody stems. Can J Bot 32:486–490

Biggs AR 1984a Intracellular suberin: occurrence and detection in tree bark. IAWA Bull 5:243–248

Biggs AR 1984b Boundary zone formation in peach bark in response to wounds and Cytospora leucostoma infection. Can J Bot 62:2814–2821

Biggs AR 1985a Detection of impervious tissue in tree bark with selective histochemistry and fluorescence microscopy. Stain Technol 60:299–304

Biggs AR 1985b Suberized boundary zones and the chronology of wound response in tree bark. Phytopathology 75:1191–1195

Biggs AR 1986a Comparative anatomy and host response of two peach cultivars inoculated with Leucostoma cincta and L. persoonii. Phytopathology 76:905–912

Biggs AR 1986b Wound age and infection of peach bark by Cytospora leucostoma. Can J Bot 64:2319–2321

Biggs AR 1986c Phellogen regeneration in injured peach tree bark. Ann Bot 57:463–470

Biggs AR 1987 Occurrence and location of suberin in wound reaction zones in xylem of seventeen tree species. Phytopathology 77:718–725

Biggs AR 1989 Temporal changes in the infection court following wounding of peach bark are associated with cultivar variation in infection by Leucostoma persoonii. Phytopathology 79:627–630

Biggs AR, Britton KO 1988 Presymptom histopathology of peach trees inoculated with Botryosphaeria obtusa and Botryosphaeria dothidea. Phytopathology 78:1109–1118

Biggs AR, Cline RA 1986 Influence of irrigation on wound response in peach bark. Can J Plant Pathol 8:405–410

Biggs AR, Davis DD, Merrill W 1983a Histology of cankers on Populus caused by Cytospora chrysosperma. Can J Bot 61:563–574

Biggs AR, Davis DD, Merrill W 1983b Stage of growth and restriction of wound-associated infections in Populus. Can J Plant Pathol 5:269–272

Biggs AR, Merrill W, Davis DD 1984 Discussion: response of bark tissues to injury and infection. Can J For Res 14:351–356

Biggs AR, Miles NW 1985 Suberin deposition as a measure of wound response in peach bark. HortScience 20:902–905

Biggs AR, Miles NW 1988 Association of suberin formation in noninoculated wounds with susceptibility to Leucostoma cincta and L. persoonii in various peach cultivars. Phytopathology 78:1070–1074

Biggs AR, Northover J 1985 Formation of the primary protective layer and phellogen following leaf abscission in peach. Can J Bot 63:1547–1550

Biggs AR, Peterson CA 1990 Effects of chemicals applied to peach bark wounds on accumulation of lignin and suberin and susceptibility to Leucostoma persoonii. Phytopathology 80:861–865

Biggs AR, Stobbs LW 1986 Fine structure of the suberized cells walls in the boundary zone and necrophylactic periderm in wounded peach bark. Can J Bot 64:1606–1610

Blanchette RA, Sharon EM 1975 Agrobacterium tumefaciens, a promoter of wound healing in Betula alleghaniensis. Can J For Res 5:722–729

Bloch R 1941 Wound healing in higher plants. Bot Rev 7:110–146

Bloch R 1952 Wound healing in higher plants. II. Bot Rev 28:655–679

Bloch R 1953 Defense reactions in plants to the presence of toxins. Phytopathology 43:351–354

Bloomberg WJ, Farris SH 1962 Cytospora canker of poplars: bark wounding in relation to canker development. Can J Bot 41:303–310

Boller T 1987 Hydrolytic enzymes in plant disease resistance. In: Plant-microbe interactions. In: Kosuge T, Nestor EW (eds) MacMillan, New York, 385–413

Borger GA 1973 Development and shedding of bark. In: Kozlowski TT (ed) Shedding of plant parts. Academic Press, New York London, 205–236

Borger GA, Kozlowski TT 1972a Early periderm ontogeny in Fraxinus pennsylvanica, Ailanthus altissima, Robinia pseudoacacia, and Pinus resinosa seedlings. Can J For Res 2:135–143

Borger GA, Kozlowski TT 1972b Effect of growth regulators and herbicides on normal and wound periderm ontogeny in Fraxinus pennsylvanica seedlings. Weed Res 12:190–194

Borger GA, Kozlowski TT 1972c Effects of cotyledons, leaves and stem apex on early periderm development in Fraxinus pennsylvanica seedlings. New Phytol 71:691–702

Borger GA, Kozlowski TT 1972d Wound periderm ontogeny in Fraxinus pennsylvanica seedlings. New Phytol 71:709–712

Bostock RM, Middleton GE 1987 Relationship of wound periderm formation in resistance to Ceratocystis fimbriata in almond bark. Phytopathology 77:1174–1180

Bostock RM, Stermer BA 1989 Perpsectives on wound healing in resistance to pathogens. Annu Rev Phytopathol 27:343–371

Bramble WC 1934 Reaction of chestnut bark to invasion by Endothia parasitica. Am J Bot 23:89–95

Büsgen M, Münch E 1929 The structure and life of forest trees. John Wiley, New York, 436 pp

Butin H 1955 Über den Einfluß des Wassergehaltes der Pappel auf ihre Resistenz gegenüber Cytospora chrysosperma (Pers.)Fr. Phytopathol Z 24:245–264

Cline MN, Neely D 1983 Wound-healing processes in geranium cuttings in relationship to basal stem not caused by Phythium ultimum. Plant Dis 67:636–638

Craft CC, Audia V 1962 Phenolic substances associated with wound barrier formation in vegetables. Bot Gaz 123:211–219

Crowdy SH 1949 Observations on apple canker. II. The anatomy of the stem canker. Ann Appl Biol 36:483–495

Darvill AG, Albersheim P 1984 Phytoalexins and their elicitors – a defense against microbial infection in plants. Annu Rev Plant Physiol 35:243–275

Davies E, Schuster A 1981 Intercellular communication in plants: evidence for a rapidly generated, bidirectionally transmitted wound signal. Proc Natl Acad Sci USA 78:2422–2426

de Bary A 1884 Comparative anatomy of the vegetative organs of the phanerogams and ferns. Oxford Univ Press (Clarendon), London New York

Doster MA, Bostock RM 1988a Quantification of lignin formation in almond bark in response to wounding and infection by Phytophthora species. Phytopathology 78:473–477

Doster MA, Bostock RM 1988b Effects of low temperature on resistance of almond trees to Phytophthora pruning wound cankers in relation to lignin and suberin formation in wounded bark tissue. Phytopathology 78:478–483

Eames AJ, MacDaniels LH 1947 Introduction to plant anatomy, 2nd edn. McGraw-Hill, New York, 427 pp

Esau K 1965 Plant anatomy, 2nd edn. John Wiley, New York, 767 pp

Fahn A 1967 Plant anatomy. Pergamon, New York Oxford, 454 pp

Gallagher PW, Sydnor TD 1983 Variation in wound response among cultivars of red maple. J Am Soc Hortict Sci 108:744–746

Grozditz GA, Godkin SE, Keith CT 1982 The periderm of three North American conifers, pt 1: anatomy. Wood Sci Technol 16:305–316

Halverson LJ, Stacey G 1986 Signal exchange in plant-microbe interactions. Microbiol Rev 50:193–225

Hartig R 1894 Diseases of trees. In: Ward HM (ed) Diseases of trees. McMillan, London New York, 225–269

Hudler GW, Neal BG, Banik MT 1990 Effects of growing conditions on wound repair and disease resistance in Pachysandra terminalis. Phytopathology 80:272–277

Kaufert F 1936 Factors influencing the formation of periderm in aspen. Am J Bot 24:24–30

Kolattukudy PE 1984 Biochemistry and function of cutin and suberin. Can J Bot 62:2918–2933

Kolattukudy PE, Köller W 1983 Fungal penetration of the first line defensive barriers of plants. In: Callow JA (ed) Biochemical plant pathology. John Wiley, New York, 79–100

Kozlowski TT 1971 The growth and development of trees. Academic Press, New York London, 560 pp

Krähmer H 1980 Wound reactions of apple trees and their influence on infections with Nectria galligena. J Plant Dis Prot 87:97–112

Lipetz J 1970 Wound-healing in higher plants. Int Rev Cytol 27:1–28

Martin JM, Sydnor TD 1987 Differences in wound closure rates in 12 tree species. HortScience 22:442–444

Mauch F, Mauch-Mani B, Boller T 1988 Antifungal hydrolases in pea tissue. II. Inhibition of fungal growth by combinations of chitinase and β-1,3 glucanase. Plant Physiol 88:936–942

McClure TT 1960 Chlorogenic acid accumulation and wound healing in sweet potato roots. Am J Bot 47:227–280

Morris LL, Mann LK 1955 Wound healing, keeping quality and compositional changes during curing and storage of sweet potatoes. Hilgardia 24:143–183

Morris SC, Forbes-Smith MR, Scriven FM 1989 Determination of optimum conditions for suberization, wound periderm formation, cellular desiccation and pathogen resistance in wounded Solanum tuberosum tubers. Physiol Mol Plant Pathol 35:177–190

Mullick DB 1977 The non-specific nature of defense in bark and wood during wounding, insect and pathogen attack. Recent Adv Phytochem 11:395–441

Ostry ME, Anderson NA 1983 Infection of trembling aspen by Hypoxylon mammatum through cicada oviposition wounds. Phytopathology 73:1092–1096

Puritch GS, Mullick DB 1975 Effect of water stress on the rate of nonsuberized impervious tissue formation following wounding in Abies grandis. J Exp Bot 26:903–910

Riffle JW, Peterson GW 1986 Thyronectria canker of honeylocust: influence of temperature and wound age on disease development. Phytopathology 76:313–316

Rittinger PA, Biggs AR, Peirson DR 1987 Histochemistry of lignin and suberin deposition in boundary layers formed after wounding in various plant species and organs. Can J Bot 65:1886–1982

Russin JS, Shain L 1984 Initiation and development of cankers caused by virulent and cytoplasmic hypovirulent isolates of the chestnut blight fungus. Can J Bot 62:2660–2664

Schneider H 1955 Ontogeny of lemon wood bark. Am J Bot 42:893–905

Schoeneweiss DF 1981 The role of environmental stress in diseases of woody plants. Plant Dis 65–308–314

Shigo AL 1986 A new tree biology. Shigo and Trees, Durham, NH, 595 pp

Soe K 1959 Anatomical studies of bark regeneration following scoring. J Arnold Arbor 40:260–267

Soliday CL, Dean BB, Kolattukudy PE 1978 Suberization: inhibition by washing and stimulation by abscisic acid in potato disks and tissue culture. Plant Physiol 61:170–174

Soo BVL 1977 General occurrence of exophylactic and necrophylactic periderms and nonsuberized impervious tissues in woody plants. PhD Thesis, Univ British Columbia, Can

Srivastava LM 1964 Anatomy, chemistry, and physiology of bark. Int Rev For Res 1:204–277

Strain BR, Johnson PL 1963 Corticular photosynthesis and growth in Populus tremuloides. Ecology 44:581–584

Strider DL, McCombs CL 1958 Rate of wound phellem formation in the sweet potato. Proc Am Soc Hortic Sci 72:435–442

Thomas P 1982 Wound-induced suberization and periderm development in potato tubers as affected by temperature and gamma irradiation. Potato Res 25:155–164

Trockenbrodt M 1990 Survey and discussion of the terminology used in bark anatomy. IAWA Bull 11:141–166

Vance CP, Kirk TK, Sherwood RT 1980 Lignification as a mechanism of disease resistance. Annu Rev Phytopathol 18:259–288

Van Sambeek JW, Pickard BG 1976 Mediation of rapid electrical, metabolic, transpirational, and photosyntheic changes by factors released from wounds. Can J Bot 54:2642–2650

von Mohl H 1845 Untersuchungen über die Entwicklung des Korkes und der Borke auf der Rinde der baumartigen Dicotylen. Verm Schr Bot Inh:212–228

Wargo PM 1975 Lysis of the cell wall of Armillaria mellea by enzymes from forest trees. Physiol Plant Pathol 5:99–105

Wargo PM 1977 Wound closure in sugar maple: adverse effects of defoliation. Can J For Res 7:410–414

Wensley RN 1966 Rate of healing and its relation to canker of peach. Can J Plant Sci 46:257–264

Wigginton MJ 1974 Effects of temperature, oxygen tension and relative humidity on the wound-healing process in the potato tuber. Potato Res 17:200–214

Yev-Ladun S, Aloni R 1990 Polar patterns of periderm ontogeny, their relationship to leaves and buds, and the control of cork formation. IAWA Bull 11:289–300

Zimmerman DC, Coudron CA 1979 Identification of traumatin, a wound hormone, as 12-oxo-trans-10-dodecenoic acid. Plant Physiol 63:536–541

Chapter 3

Responses of Angiosperm Bark Tissues to Fungi Causing Cankers and Canker Rots

A.R. BIGGS

3.1 Introduction

Cankers on trees are the visible manifestation of necrotic periderm, cortex, phloem, and vascular cambium tissues (Fig. 3.1, 1–4). In most shade tree or forest pathology textbooks, cankers are classified according to types or classes to facilitate instruction and discussion (Boyce 1948, Tattar 1978, Manion 1981). Generally, cankers can be either annual or perennial in occurrence and, if perennial, can be either diffuse or target-shaped. In reality, canker symptomatology is extremely diverse, often with gradations among the distinct types. The biological basis for the varying expression of disease symptoms is due to the interaction of several factors, including pathogen virulence, host resistance, influence of environment, and time.

Cankers may be incited by fungi from the Ascomycetes, Deuteromycetes, Basidiomycetes, and, less commonly, Oomycetes. Generally, stem cankers are caused by Ascomycetes, Deuteromycetes, and Oomycetes, whereas canker rots are caused by Basidiomycetes. Most fungi that cause stem cankers are restricted to bark and xylem tissues that succumb due to the effects of toxins or secreted enzymes. Such organisms are termed necrotrophic or facultative parasites and include many of the more extensively studied canker pathogens (e.g. *Nectria galligena*, *Cryphonectria parasitica*, *Leucostoma persoonii*). Many stem canker pathogens invade and colonize xylem tissues in the vicinity of the canker without causing decay. Basidiomycetous fungi that cause cankers and also extensively invade the xylem, simultaneously causing wood decay, are termed canker rot pathogens (e.g. *Cerrena unicolor*). This distinction between stem cankers and canker rots is, in many cases, artificial because many ascomycetous fungi that cause cankers can also extensively degrade the xylem (e.g. Strumella, Xylaria, Hypoxylon, and many others).

In evaluating abnormal plant tissues, not only a comparison of normal and abnormal forms is of great importance, but a comparison of the abnormal forms with one another is also essential. With this premise in mind, this chapter provides points of comparison and contrast with several of the various stem cankers and canker rots caused by fungi. Normal tissues are not discussed extensively in this chapter since this information is summarized in Chapters 2 and 5. The terminology of bark tissues discussed by Trockenbrodt (1990) is utilized in this chapter. Chapter 2 offers a brief discussion on bark anatomy.

Fig. 3.1. 1 Perennial canker on sugar maple caused by *Eutypella parasitica*. **2** Conspicuous infection caused by the canker rot fungus *Inonotus obliquus* on sweet birch. Note the typical "bowling pin" shape of the infected tree and the black mass of sterile fungus tissue in the center of the canker. **3** Canker caused by *Urnula craterium* (anamorph = *Strumella corynoidea*) at the base of an oak tree (*arrow*). Strumella cankers are found almost exclusively near dead branch stubs. **4** Formation of xylem inclusions in bark (*single arrow*) of oak infected by *S. corynoidea*. Note the visible margins of the canker in the bark and xylem (*double arrows*) (Photographs courtesy of A.L. Shigo)

3.2 Infection Courts for Canker Pathogens

Most fungi causing stem cankers and canker rots colonize the host plant via open wounds, dead branches, branch stubs, twigs, leaf scars or, less commonly, through leaves. The importance of dead branches and branch stubs is not based so much on quantitative data, although the limited data available do confirm these observations (Biggs 1989b); rather it is based on the observation of the dead branch or stub in the center of the canker (Fig. 3.1, 3). Much is known about the anatomy of bark wounds (Chap. 2), but considerably less is known about the anatomy of senescing and dead, but still attached, branches (Shigo 1985). Direct penetration of intact bark or epidermis has not been demonstrated for any of the facultative parasites that cause most stem cankers. However, some pathogens, e.g. *Macrophoma tumefaciens* on poplar (Kaufert 1937, Zalasky 1964), *Apiosporina morbosa* on plum (Koch 1935), which initiate galls and other hypertrophic tissues on stems, may directly penetrate the epidermis of current years' shoots or via lenticels. A possible exception that has been reported recently, although not confirmed histologically, is the direct penetration of peach bark lenticels by *Botryosphaeria dothidea* when trees had been exposed to water stress conditions following inoculation (Pusey 1989). One potentially significant observation that deserves further investigation is the often superficial location of the phellogen in branch/trunk junctions (Shigo 1985). This raises the possibility of direct penetration of bark tissues by any or all of the canker-causing fungi. Accumulation of snow and ice also may wound easily this apparently sensitive area which could help explain the high frequency of canker initiation by *Nectria galligena* in branch axils of New England hardwoods observed by Grant and Spaulding (1939).

3.3 Colonization and Establishment of Canker Pathogens

Most anatomical studies of the colonization and establishment of canker fungi have used artificial wounds and massive amounts of hyphal inoculum, on agar plugs or infested wood dowels or grain, to initiate infection. Under these conditions, the frequency of infection is nearly 100%, and colonization of bark tissues proceeds at a rapid rate. Typically, a portion of bark is pulled back or removed with a scalpel or cork borer and the inoculum is put in place and covered with the original bark, and the site is sealed with tape or a plastic wrapping. Very few detailed histopathological studies on the initial events of host colonization have been conducted with spore inoculations because of the generally low frequency of successful infection. The following examples provide some information on the early stages of infection.

3.3.1 *Valsa* and *Leucostoma* Canker of Peach and Various Hardwoods

Butin (1955) and Biggs et al. (1983) reported that the advancing margin of *Cytospora chrysosperma* in poplar bark (Fig. 3.2, 5 and 6) is associated with an

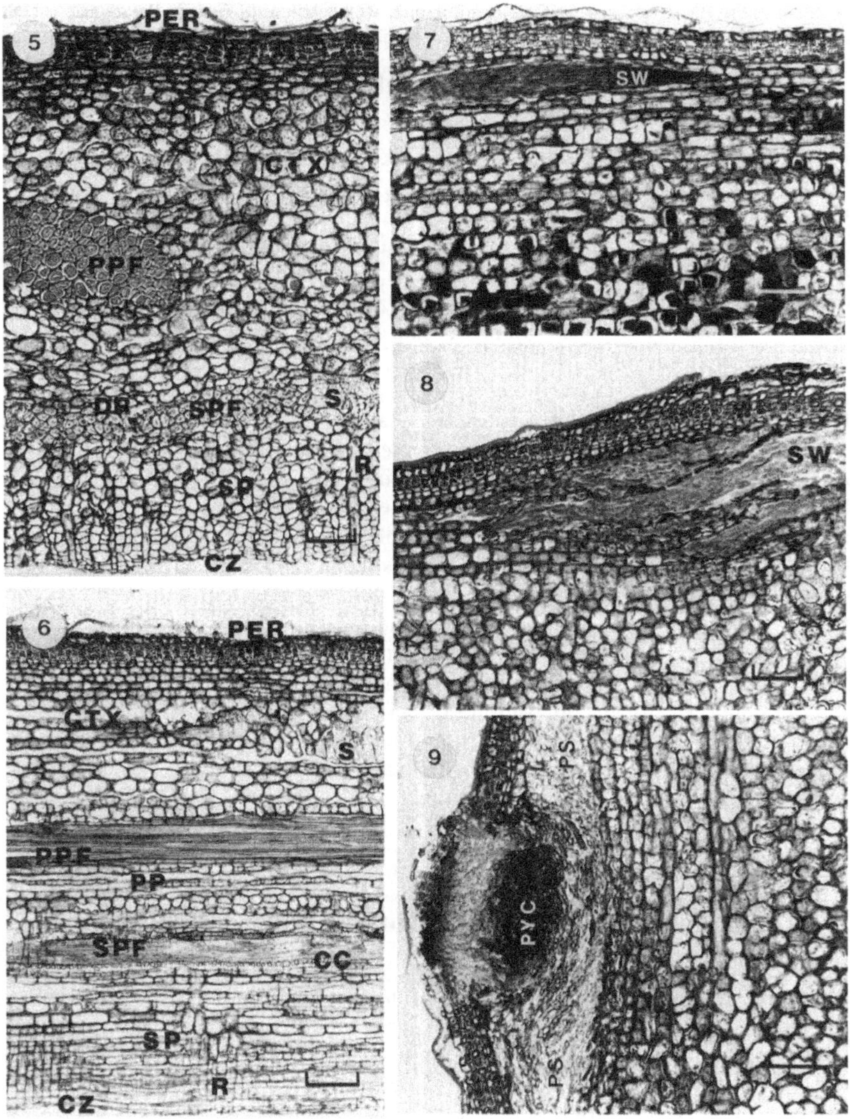

Fig. 3.2. 5, 6 Light micrographs of transverse and longitudinal section, respectively, through cambial zone and bark of healthy 1-year-old hybrid poplar cutting showing cambial zone (*CZ*), secondary phloem (*SP*), secondary phloem fibers (*SPF*), primary phloem (*PP*), primary phloem fibers (*PPF*), sclerids (*S*), ray cells (*R*), dilated ray (*DR*), crystal-containing cells (*CC*), cortex (*CTX*), and periderm (*PER*). **7–9** Longitudinal sections through the outer bark of poplar cutting inoculated with *Cytospora chrysosperma*. **7** Section through a portion of bark 1 week after inoculation, showing intercellular stromatic wedge (*SW*) composed of large-diameter hyphae extending through the inner periderm region. **8** Section through a portion of bark 2 weeks after inoculation, showing intercellular stromatic wedge (*SW*) composed of multiple advancing tips extending through the inner periderm and cortical regions. **9** Section through a portion of bark 3 weeks after inoculation, showing smaller-diameter intracellular hyphae and loosely woven pycnidial stroma (*PS*) and immature pycnidium (*PYC*) breaking through the outer bark. *Bar* = 40 μm in all micrographs

intercellular aggregation of hyphae that appears in the outer cortex and inner periderm or, more rarely, in the phloem region (Fig. 3.2, 7 and 8). Behind the margin and in the obviously necrotic area, hyphae become less tightly woven, smaller in diameter, and more diffuse (Fig. 3.2, 9). Colonization in this area is by both intercellular and intracellular hyphae that are present in the cortex, phloem, xylem, and ray cells. In cankers sampled 6 to 12 months after inoculation, mycelial aggregates are observed adjacent to phloem fiber bundles (Biggs et al. 1983). The authors suggested that the fungus could penetrate into living tissues through the fibers and/or along the interface of fiber bundles and necrophylactic periderm.

Single hyphae of *Valsa ceratosperma*, the causal fungus of apple canker in Japan, invade bark slowly during the dormant season. As temperatures increase in the summer, the fungus forms mycelial aggregates which penetrate and destroy wound periderm tissues (Tamura and Saito 1982). In peach bark inoculated with *Leucostoma persoonii*, fungal colonization was initially diffuse, although mycelial aggregations formed within 3 weeks (Wisniewski et al. 1984). These authors also observed the penetration of wound periderm tissues by mycelial aggregates. Other researchers have reported the presense of mycelial aggregations of *L. persoonii* (Tekauz and Patrick 1974, Biggs 1984b, 1986a) and *L. cincta* in peach bark (Tekauz and Patrick 1974, Biggs 1986a).

3.3.2 *Nectria* Canker of Apple and Various Hardwoods

Nectria galligena is able to enter the host through wounds inflicted by pruning, frost injury, breakage caused by ice and snow (Lortie 1964), and small dead branch stubs (Grant and Childs 1940, Zalasky 1968). In apple, *N. galligena* enters the host through pruning wounds, frost injury, infections caused by *Venturia inaequalis* and *Neofabrea malicorticis*, and leaf scars (Wiltshire 1921, Crowdy 1949). It is not known if forest trees are infected through leaf scars. For the fungus to become well established, wounds must be to the depth of the vascular cambium. The host is able to confine the pathogen to the cortex when wounds are shallow. The fungus is able to colonize all of the tissues that it penetrates. In the cortex and phloem, hyphae are intercellular in the early stages of infection but intracellular hyphae were observed during the later stages. The fungus grows vigorously in the phloem region and enters the xylem via the rays. Although the pathogen is unable to penetrate suberized cell walls (Krähmer 1980), it grows freely in the vessels and tracheids and spreads in the xylem through the pits. Intracellular phenolic compounds, gum in the xylem, and wound lignin at the levels observed in this host/pathogen interaction do not deter fungal colonization.

3.3.3 *Hypoxylon* Canker of Aspen

Hypoxylon mammatum enters aspen trees through wounds caused by insects, axe cuts, snow and ice breakage, and dead branches. The periderm is not invaded, but the hyphae are found in the phloem, cortex, and cambium tissues (Bier 1940). Hyphae penetrate the cell walls directly, forming bore holes, and spread within the tissue is intracellular. Cambial cells, sieve tubes, and phloem parenchyma cells are destroyed by the fungus, but phloem fibers and other sclerified cells are not colonized (Rogers and Berbee 1964). These authors described two hyphal types associated with bark colonization. Primary hyphae averaging 1.2 μm in diameter were observed immediately beyond the advancing margin of the canker, while the secondary hyphae, averaging 2–5 μm in diameter, were present in greatest abundance in previously killed tissue. *Hypoxylon* cankers are one of the stem cankers caused by ascomycetous fungi that cause extensive decay of the xylem, often resulting in breakage of the trunk at the canker.

3.3.4 *Cryphonectria* Canker of American Chestnut

The fungus *Cryphonectria parasitica* can become established through any type of wound in the bark that is deeper than the outer green cortex (Boyce 1948). The initial penetration of chestnut bark tissue by the fungus also is accomplished by the mass action of hyphae (Keefer 1914, Bramble 1934). The fungus mycelium accumulates in the wound, expanding the initial lesion for about 9 days (Hebard et al. 1984). Lesion enlargement ceases and then resumes at about 26 days after inoculation, persisting until the tree is girdled or until environmental factors limit growth. The fungus makes its initial penetration into surrounding tissue by means of flattened masses of mycelium termed "mycelial fans" (Bramble 1934). Mycelial fans cleave and partially destroy cells during penetration of the bark. Secondary invasion of individual cells occurs behind the advancing margin, similar to that described for *Cytospora chrysosperma*. The advancing margin of hyphae may invade just beneath the periderm, as also described for *C. chrysosperma*, and also into the internal cortical tissues. Individual hyphae are not found in advance of mycelial fans in living bark, but host cells in advance of the fungus appear dead before being colonized by the fungus (Hebard et al. 1984).

3.3.5 *Eutypella* Canker of Maple

The fungus *Eutypella parasitica* enters its host mostly through dead branch stubs and, less often, through dead twigs and other wounds deep enough to penetrate the bark (French 1969). The colonization of bark tissues occurs initially in the functional phloem. The first indication of infection is a zone of necrotic tissue in advance of the mycelium, suggesting the action of a toxin or other secreted substance. The necrotic cells are then invaded by hyphae and then mycelial fans form in the necrotic zone that was initially infected. As the fungus ramifies in the phloem, hyphae also spread radially into the phelloderm, cambium, and xylem tissues. The fungus can degrade completely sclerified cells in bark. It enters xylem and phloem cells through the pits, later causing wall erosion and forming bore holes. In thick bark, multiple mycelial fans often form and may coalesce to form larger mycelial aggregates. The bark of *Eutypella* cankers remains firmly attached to the face of the canker, unlike that of *Nectria* cankers.

3.3.6 *Botryosphaeria* Canker of Peach

Inoculation of peach bark wounds, inflicted with a cork borer, with spores of *Botryosphaeria obtusa* and *B. dothidea* provided an interesting contrast to the mycelial inoculation studies described above. Initial colonization by the fungus occurred on the wound surface, in necrotic bark tissue on the wound margin, and in necrotic xylem tissue to a depth of about 200 μm (Biggs and Britton 1988). In inoculated trees, few macroscopic symptoms of infection were present during the 8-week course of the study, and normal wound healing responses appeared macroscopically and microscopically to be proceeding as with the noninoculated wounds. Histologically, no differences in lignin and suberin deposition, periderm formation, and callus formation were observed between noninoculated wounds and wounds inoculated with either fungus after 1 and 2 weeks incubation. However, by 28 days after inoculation for both fungi, new callus tissue was being colonized by fungal hyphae located on the surface of the xylem beneath the nonsuberized ventral callus surface. The nonsuberized ventral region of callus was primarily parenchymatous with a ligno-suberized outer layer, and was the focal point for fungal pathogenesis. By 8 weeks, the breakdown of tissue in this region was directly associated with fungal hyphae in close ventral proximity. Direct fungal penetration of the callus, although only occasionally observed, was not required to induce gum pockets, cambial alterations, and tissue breakdown. Intracellular hyphae of both *Botryosphaeria* spp. were observed colonizing cortical and callus parenchyma, and xylem ray parenchyma, vessels and tracheids. Intercellular hyphae were observed in phloem fibers, necrotic cortical parenchyma external to wound periderm, and necrotic ligno-suberized parenchyma on the callus surface.

3.3.7 *Cerrena unicolor* Canker Rot of Maple and Birch

Little is known about how canker rot pathogens penetrate and become established in their hosts. Following mycelial inoculations, *Cerrena unicolor*, the cause of canker rot and decay in sugar maple and paper birch, colonized both phloem and xylem tissues (Enebak and Blanchette 1989). In phloem observed 6 months after inoculation, the fungus had formed hyphal wedges which grew intercellularly, resulting in splitting and subsequent death of the tissue. Similar patterns of colonization occur for other canker rot fungi (Fig. 3.3, 10) (Shigo 1969).

3.4 Delimitation of Canker Pathogens in Bark Tissues

The delimitation of fungal pathogens in tree bark ultimately occurs via the formation of ligno–suberized boundary tissue derived from cells extant at inocula-

Fig. 3.3. 10 Penetration of the bark of yellow birch by the canker rot pathogen *Inonotus obliquus*. The pathogen produces a wedge of fungus material into the bark. The *larger white arrows* show the boundary of the wedge that penetrated the bark 3 years before the tree was dissected. The *smaller white arrows* show an earlier boundary. The *double black arrows* in the bark show the tip of the wedge. The *white pointer* indicates the position where the wedge meets the living cambium. The *black arrows* in the xylem indicate the barrier zone formed in response to infection. **11** Transverse section of a *Strumella* canker on red oak showing the series of annual callus rings typical of most perennial target-shaped cankers. The *smaller, black arrows* drawn on the wood show the xylem boundary formed during the seasons before this section was cut. The *white arrows* show the pathogen during the current season, circumventing the bark boundaries and invading previously healthy bark. **12** Section of aspen with *Hypoxylon* canker showing a xylem boundary in the bark (*white arrow*). **13** Section of aspen with *Hypoxylon* canker taken from the vertical tip of the canker and showing the old infected bark beneath the back of the *white arrow* (Photographs courtesy of A.L. Shigo)

tion and necrophylactic periderm derived from newly differentiated phellogen (Mullick 1977, Biggs 1984a,b, Biggs et al. 1984) (Chap. 2). For annual cankers, these barriers appear to be effective in preventing continued colonization by the fungus, and diseased tissues are usually sloughed. For perennial cankers, bark boundary tissues, which also may include extensive xylem formation (Figs. 3.3, 11–13; 3.4, 14), are either directly penetrated or circumvented annually, resulting in a series of concentric callus ridges. The direct involvement of the ligno-suberized layer and/or wound periderm, or by-products generated by the tree during the differentiation of these tissues, in the resistance of woody plants to fungi has never been demonstrated conclusively. Several lines of evidence indicate that periderms function to prevent desiccation of internal layers, prevent inward movement of pathogen toxins or enzymes, and serve as physical barriers to colonization. The process of generating new periderm also may result in lignification of pathogen hyphae (Vance et al. 1980, Biggs 1986a) or in the production of fungistatic compounds which act as phytoalexins. Additional discussion of resistance in tree bark to canker pathogens is presented in Section 3.6.

Several recent papers have described the formation of necrophylactic periderm and associated tissues in response to pathogens in considerable detail (Biggs 1984b, 1986a, Biggs and Britton 1988). The formation of ligno-suberized boundary tissue and necrophylactic periderm in inoculated bark is similar to that in noninoculated wounds (Chap. 2), although there are many notable differences.

Pathogens affect the position or location of periderm and related tissues in wounded bark relative to their location in the absence of pathogens. In wounds, new tissues are regenerated usually within 1–2 mm of the wound surface (Biggs 1984a). When wounds are colonized by aggressive pathogens, the formation of new tissues may be several centimeters from the infection court.

Pathogens affect the morphology and differentiation of new tissues and may retard or inhibit the formation of periderm and/or the deposition of suberin in wound periderm and xylem barrier zones ((Goring 1975, Biggs 1986a, Biggs and Britton 1988). Necrophylactic periderm that formed in peach bark cortex inoculated with *Leucostoma persoonii* and *L. cincta* occasionally exhibited phellem cells with altered morphology (Fig. 3.4, 14). Altered cells appeared thin-walled and rounded compared with the relatively thick, rectangular phellem produced at control wounds. Bramble (1934) also reported that wound periderms in American chestnut inoculated with *Chryphonectria* (*Endothia*) *parasitica* varied in their structure from that of normal periderm. In wound periderms, as many as ten layers of thin-walled phellem formed prior to the formation of thick-walled phellem, in comparison with normal periderm, which formed a minimum of three layers of thin-walled cells. Wound periderm in chestnut also possessed additional phelloderm layers relative to normal periderm (Bramble 1934, Hebard et al. 1984).

Wounds inoculated with *Leucostoma* spp. and *Botryosphaeria* spp. both showed delayed or inhibited suberin formation, beyond that which occurred naturally, in the ventral callus surface (Biggs 1986a, Biggs and Britton 1988). This region was often a site for continued colonization of bark and xylem tissues

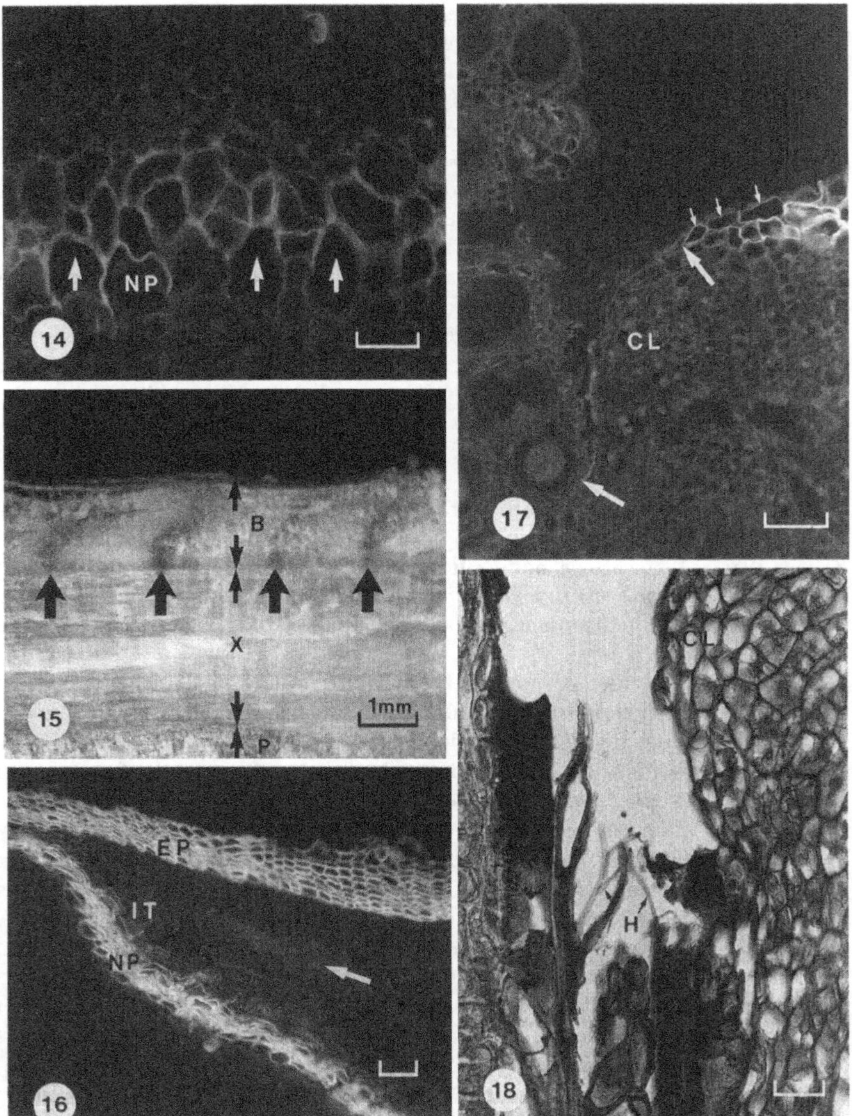
14
NP
15
B
X
P
1mm
16
EP
IT
NP
17
CL
18
CL
H

by these pathogens. When xylem tissues were examined for suberin formation following wounding, large areas of suberized ray parenchyma were observed in the wall 3 region based on the CODIT model of compartmentalization (Shigo 1979) (Chap. 5). Such regions were absent in inoculated xylem (Biggs 1986a).

Once infection is established, few hosts are able to restrict the initial invasion of a pathogen with the formation a single ligno-suberized boundary zone and necrophylactic periderm. Instead, the tree often forms a series of incompletely formed periderms within a single season in response to the advancing fungus (Fig. 3.4, 15). The fungus will continue to spread in the host tissue until conditions become more favorable for the host to form a complete periderm and less favorable for the pathogen to penetrate further. This is why many canker lesions on juvenile stems have visible concentric areas of discoloration during the initial establishment phase. For perennial target-shaped cankers, the tree also produces annual rings of callus tissue which represent the successful formation of necrophylactic periderm and callus tissue for that year but which gives no clue as to the number of attempts during that particular year.

In bark of peach cultivars inoculated with *Leucostoma persoonii* and *L. cincta*, necrophylactic periderm formed at 21–28 days after inoculation and was associated with an effectively delimited lesion in both cultivars inoculated with *L. cincta*, the less pathogenic fungus (Fig. 3.4, 16). For cultivars inoculated with *L. persoonii*, the wound periderm was not as thick, had fewer cells, and was associated with only a slight inhibition of lesion expansion, followed by a second rapid increase in canker length as the periderm was breached between 28 to 35 days after inoculation (Biggs 1986a). In chestnut, Hebard et al. (1984) observed that periderms were produced by the host and then breached by the fungus at 18 or 28 days after inoculation, depending on the virulence of the isolate. In resistant chestnuts, some virulent isolates continued to expand lesions up to 30 days after inoculation (Russin and Shain 1984).

←

Fig. 3.4. 14 Transverse section of a portion of the ligno-suberized boundary zone and new periderm (*NP*) adjacent to peach bark tissue colonized by *Leucostoma persoonii* (*top*) showing an altered phellem cell shape (*arrows*). **15** Longitudinal section through pith (*P*), xylem (*X*), and bark (*B*) (tissues are between oppositely oriented *arrows*), sampled 7 days after inoculation with *Leucostoma persoonii*, showing four polysaccharide-rich reaction zones (*large arrows*). Colonization from *left to right* had proceeded in the bark through the first zone and was almost to the second. **16** Fluorescence illumination of the longitudinal margin of a 56-day-old quiescent canker on peach caused by *Leucostoma cincta*. The section has been treated with phloroglucinol + HCl and shows a complete intact suberizecl impervious tissue (*IT*) and necrophylactic periderm (*NP*) contiguous with the original exophylactic periderm (*EP*). Note aggregation of fungus mycelium (*arrow*) exhibiting weak autofluorescence. **17** Transverse section of a 14-day-old mechanical wound on peach bark, treated with phloroglucinol + HCl, and showing the nonsuberized ventral region (*between large arrows*) of new callus (*CL*). The residual autofluorescence is due to suberin in the outer cell layers of the callus (*small arrows*). **18** Longitudinal section of a 14-day-old mechanical wound on peach bark inoculated with *Botryosphaeria dothidea*, showing hyphae (*H*) colonizing the ventral region of the callus (*CL*) and the xylem surface. For all figures, *bar* = 10 μm except where indicated

3.5 Breaching of Host Boundaries by Canker Pathogens

Very few studies on hardwood cankers have presented evidence to show conclusively how the fungus invades the phellogen layer to reinfect the phloem after the canker has stopped expanding for the season. For many cankers, the main mode of reinfection is by hyphal penetration of the wound periderm. Penetration may be either direct via mass action of hyphae or indirect via hyphal penetration of fibers or sclereids which occasionally bridge wound periderms and connect diseased and healthy tissues. Penetration of periderms by mass action has been described for *Cytospora* canker of peach (Wisniewski et al. 1984), *Nectria* canker of apple (Crowdy 1949), and canker rot of maple and birch caused by *Cerrena unicolor* (Enebak and Blanchette 1989). *Cryphonectria parasitica*, which forms diffuse cankers, grows under wound periderm before it forms (Bramble 1934) and thus can rapidly girdle chestnut stems. Hebard et al. (1984) showed that mycelial fans of the fungus also could penetrate developing wound periderms directly.

Other avenues of reinfection occur in natural and pathogen-induced discontinuities along the wound periderm and new callus tissue produced by the host in response to invasion. For example, Biggs (1986a) observed *Leucostoma persoonii* to circumvent wound periderm by colonizing the nonsuberized ventral callus region (Fig. 3.4, 17 and 18) or by growing between the original periderm and wound periderm external to the cortex, or between the wound periderm and phloem fibers. Similar routes of reinfection were described for *Valsa* canker of apple (Tamura and Saito 1982).

The lumens of phloem fibers also may provide a means of entry for fungi into healthy tissues delimited by wound periderm. Poplar bark colonized by *Cytospora chrysosperma* (Biggs et al. 1983) and apple colonized by *Nectria galligena* (Crowdy 1949) may be reinvaded in this manner. Similarly, *Eutypella parasitica* may reinvade maple bark through sclereids or other discontinuities in the wound periderm (French 1969).

Ashcroft (1934) proposed that fungi such as *N. galligena* which colonize the xylem may reinvade the bark by growing through the medullary rays and into the wound wood (CODIT wall 4) and ray parenchyma into healthy bark that was not delimited by wound periderm. This has never been demonstrated, although French (1969) mentions it as a possibility for *Eutypella parasitica*.

3.6 Expression of Resistance to Canker Pathogens in Bark Tissues

Discussions of resistance include two basic concepts: constitutive, or passive defense, and induced, or active, counterattack defense (Cowling and Horsfall 1980). Constitutive defenses are critical in the protection of trees against fungal pathogens. These defenses encompass the plethora of structures and chemicals that trees have evolved over the millennia to exclude pathogens and avoid or tolerate herbivory (Berryman 1988), and include leaf trichomatous structures, resin-covered or resin-filled structures in conifers, cell walls impregnated with lignin or phenolic compounds, cells containing phenolic compounds or other preformed toxic substances (Tattar and Rich 1973, Parker 1977), and heavily cutinized or suberized tissues. Fungi that cause canker diseases most often encounter the latter in the form of cutinized epidermis and suberized exophylactic (first) or necrophylactic (sequent) periderm. Given that the majority of trees are free of canker infections, it appears that constitutive defenses are very effective.

Preformed or constitutive defenses against fungi that cause cankers are breached by wounds, thus initiating the processes of induced defense. This type of defense consists of the production of chemicals and/or reestablishment of barriers (wound periderms) to pathogen ingress in response to infection. Phytoalexins have been detected in the bark of woody angiosperms, although these substances have not been as extensively characterized as those found in herbaceous plants. Their production in the barks of *Populus tremuloides* (Flores and Hubbes 1980) and *Morus alba* (Shirata 1978) may have a defensive function prior to the formation of new periderms. The new periderms are distinct from the preformed constitutive barriers based on (1) their dissimilar histochemical reactions to lignin reagents (Biggs 1984b, Rittinger et al. 1987); (2) the formation of a ligno-suberized boundary as a prerequisite to periderm differentiation (Mullick 1977, Soo 1977, Biggs 1984 a,b); and (3) biochemical differences in phellem cell contents (Mullick 1977, Soo 1977).

The induction of defensive reactions in plants is multifaceted. Two types of induced defensive reactions, localized and systemic, have been described for plants and several hypotheses have been proposed as possible mechanisms for their induction (Sequeira 1983, Darvill and Albersheim 1984, Ryan 1988). The localized response occurs in the vicinity of the site of wounding or infection and is characterized by the necrosis of plant cells and the accumulation of phytoalexins. Phytoalexins, substances produced de novo which act nonspecifically to restrict colonization by microorganisms, are accumulated and released by plant cells in response to various elicitors released from pathogen or plant cell walls (Hahn et al. 1981, Nothnagel et al. 1983, Darvill and Albersheim 1984, Ryan 1988, Bostock and Stermer 1989). The localized response is expressed in variable intensity depending upon the extent of host cell necrosis triggered by the wound itself and the colonization of the wound by microorganisms.

Systemic defense reactions are observed in the plant at sites other than the point of injury or infection and may be present well after its induction. Studies have shown that the systemic reaction in bean, which is not induced by wounding alone, makes a plant more resistant to subsequent inoculations following the original inducing inoculation (Kuc 1983). It is not certain what role, if any, systemic resistance (sensu Kuc) has in defense against canker pathogens. Other investigations of systemic resistance (sensu Ryan and others) have shown the induction of various defensive chemicals by wounding, pathogen colonization, and fungal elicitors (see below). Indeed, the discussion of intraplant systemic communication takes on much broader implications with the recent discovery of interplant communication and the possible role of airborne volatile substances, such as methyl jasmonate (Farmer and Ryan 1990), acting to induce defensive genes in nonwounded plants located adjacent to wounded plants.

The occurrence of wounds and the reaction of the healthy plant to wounds are part of the natural life process for trees. The plant's response to wounding leads to the restoration of its natural permeability barriers after they are broken. Plant responses to wounding occur as part of normal developmental processes and responses to infection are essentially similar or identical to wound responses. Given the greater likelihood of ingress furnished by wounding, the delimitation of pathogens by the same or similar processes that restore the plant's natural permeability barriers and meristems is strategic, from an evolutionary point of view, in its conservation of response repertoires. What are the differences then between the tree's response to wounding and infection, and what factors regulate these differences?

Induction of the wound response is characterized by chemical biosynthesis regulated by a relatively small number of cells in which relatively small quantities of defensive chemicals are produced by living cells in close proximity to the point of injury. Cell death is minimal because toxic substances within the cells are membrane-bound, limited to intercellular spaces, impregnated into cell walls, or are synthesized in quantities that are nontoxic. Some induction of systemic responses may occur following wounding in woody plants (Parsons et al. 1989, Farmer and Ryan 1990). Simple wound response may be triggered by host cell wall oligosaccharides, oligosaccharides from nonpathogenic fungi, lipids, peptides, or toxins secreted or released during the host/pathogen interaction, or wound responses may be related to hormones (e.g. ethylene, ABA) transported via the vascular tissue and regulated by barriers to outward diffusion (Li and Cui 1985, Yev-Ledun and Aloni 1990) (Chap. 2).

Wound periderm formation by itself is an effective resistance mechanism, which is highly dependent on suberin biosynthesis (see Chap. 2). This assertion is based on the strong statistical correlation between the amount of suberin in the infection court at the time of inoculation and resistance in peach cultivars to *Leucostoma persoonii* (Biggs and Miles 1988, Biggs 1989a). Relationships between resistance and lignin accumulation, rate of periderm formation, thickness of periderm, and number of cells in periderm were not significant in these experiments. It is likely that suberin accumulation prior to the arrival of inoculum at the infection court is only one component of resistance. It functions epidemiologically as a type of rate-limiting resistance and should be considered as a form of partial resistance. Suberin accumulation in wounds is also a type of active resistance characterized by chemical biosynthesis regulated by a relatively small number of cells.

When canker-causing fungi are placed into a wound, induction of defense becomes characterized by chemical biosynthesis regulated by a relatively large number of cells. Large quantities of defensive chemicals are produced as a result of the autonomous process of cell death triggered by pathogen metabolites or by the induction of the host's own self-destruct mechanisms (e.g. deregulation of host metabolism by pathogen products that are not directly toxic). The reactions of living cells away from the site of pathogen ingress are characterized by the rapid conversion of starches and sugars into molecules with multiple functions, i.e., lignin, suberin, their precursors, and breakdown products act as physical and chemical barriers leading to reduced water loss, restricted toxin diffusion, restricted nutrient availability, and increased antibiotic potential. In addition, processes are triggered which lead to cell dedifferentiation, redifferentiation, and the eventual formation of heavily suberized periderm tissues. Effective defense against fungi most likely occurs from combinations of responses acting in concert to reestablish the continuity of tissues and restrict pathogen ingress.

Only a few studies have been conducted to explore the elicitation of defense reactions in woody species (Miller et al. 1986, Lieutier and Berryman 1988, Parsons et al. 1989, Biggs and Peterson 1990, Farmer and Ryan 1990). In studies on the defense reactions of seven conifer species, Lieutier and Berryman (1988) made inoculations with *Ceratocystis clavigera* and two chemical elicitors, chitosan and the solanaceous proteinase inhibitor-inducing factor (PIIF). Chitosan is a glucosamine polymer from fungal cell walls that has been shown to trigger lignin accumulation (Pearce and Ride 1982) and PIIF is a pectic oligomer from plant cell walls that has been shown to trigger a systemic response in herbaceous plants (Green and Ryan 1972, Ryan 1988, Farmer et al. 1989). Inoculation with fungus or chitosan induced similar reactions including resin soaking in the phloem and sapwood. The plant reaction to PIIF and the buffer control were similar. Although PIIF induces hypersensitive reactions in some herbaceous plants (Nothnagel et al. 1983), it did not do so in conifers.

In studies of the elicitation of the defense reaction in peach bark, Biggs and Peterson (1990) exposed bark wounds to several potential trigger chemicals, including cell wall extract from *Leucostoma persoonii* and *L. cincta*, cell wall extract from peach leaf cells, leaf and fungal cell wall extracts combined, chitosan, calcium ion, chitosan + calcium, and the disaccharide cellobiose. Only the fungal cell wall extract and cellobiose induced a significantly greater defensive reaction than the control, based on thioglycolic acid analyses of lignin produced after wounding and treatment with elicitor. Although treatment with wall extracts from both fungi stimulated lignin production and resulted in smaller canker size after inoculation, only treatment with the extract from *L. persoonii* resulted in decreased percent infection rates. These results suggest that a mechanism(s) in addition to lignification is involved in the resistance of peach to these two fungi. It is likely that suberin may be more important than lignin (Biggs and Miles 1988, Biggs 1989a), but a correlation with suberization could not be established in the present study due its relatively short time course. Although the involvement of other factors, such as the production of phytoalexins, has not been investigated for the peach system, their presence is highly likely, given that a level of increased resistance to inoculations with various fungi has been observed in bark of *Prunus* spp. prior to the formation of ligno-suberized tissue and new periderm (Biggs 1986b, Bostock and Middleton 1987, Doster and Bostock 1988b).

One of the more promising chemical treatments in terms of reducing both infection frequency and canker size was cellobiose. It stimulated lignin production in both the field and laboratory experiments. This suggests that small saccharides may be responsible for triggering the processes that lead to the observed differences in degree of lignin deposition in wounds versus infections (see Biggs 1984b). Another chemical treatment included in these experiments was calcium. Although this ion did not stimulate the production of lignin or suberin, it was the most effective chemical for reducing both infection frequency and size of cankers. The mode of action of calcium could be related to the stimulation of the synthesis of phytoalexins and/or phenols as suggested by Kohle et al. (1985), although our analyses did not confirm this hypothesis. A second possibility is that calcium may reduce fungal polygalacturonase activity by forming cation cross-bridges between pectic acids in the plant, thus making the plant cell walls more resistant to degradation. The latter possibility in supported by the observation that breakdown of the pectins present in peach bark cell walls and the sequestration of calcium by fungal oxalic acid have been shown to occur during pathogenesis by *Leucostoma persoonii* (Biggs 1984b, Traquair 1987).

3.7 Molecular Regulation of Plant Response: Normal Development, Wounding, and Pathogenesis

There is a large body of literature on the regulation of wound responses, and responses to herbivores and pathogens, by wound-responsive genes in plant systems. The production of phytoalexins, hydroxyproline-rich glycoproteins, pathogenesis-related (PR) proteins, chitinases, glucanases, and proteinase inhibitors may all be subject to the expression of defensive genes in plants. A greater understanding of the molecular regulation of the defensive chemicals listed above could lead to augmentation of natural defense mechanisms via direct manipulation of the tree genome to produce trees with increased resistance to pathogens.

In hybrid poplar, Parsons et al. (1989) have demonstrated the existence of two wound-inducible mRNAs present in unwounded upper leaves following mechanical wounding of the lower leaves. Several classes of transcripts accumulated in response to repeated wounding, although the two cDNAs that were characterized and sequenced showed a high degree of similarity to gene families which encode chitinases from bean, tobacco, and barley. Wargo (1975) has demonstrated the presence of chitinases in phloem tissues of *Acer saccharum* and several *Quercus* spp. Pathogenesis-related proteins identified as chitinases are known to be induced by pathogen colonization, fungal elicitors, ethylene, and mechanical wounding in *Phaseolus vulgaris* (Broglie et al. 1986). Systemic occurrence after wounding is also characteristic of the proteinase inhibitors in solanaceous plants (Ryan 1988). In this system, systemic gene activation can be elicited by host cell wall oligogalacturonides, although it is not likely that these substances move systemically within the plant because of their relatively large size (Dixon and Lamb 1990). Their function may be as local elicitors, either potentiating or transducing the systemic signal. Furthermore, their presence may be germane to nonsystemic wound reactions and/or responses which lead to the delimitation of lesions. Plant hormones are well known to mediate effects in plants over long distances and it is well known that ethylene (Broglie et al. 1986) and ABA (Sánchez-Serrano et al. 1990) promote gene activation of some defense relate processes. Recently, a heat stable non-carbohydrate substance has been isolated which acts to phosphorylate specific plasma membrane polypeptides (Farmer et al. 1989). These workers suggest that protein kinases may play an important role in the signaling mechanism for proteinase inhibitor synthesis in response to wounding.

From the above, it appears that a broad array of defensive chemicals are induced by a diversity of signal events or molecules, and it is possible that the initiation of transcripts stimulated by wounding or infection is initiated at different genomic sites regulated by different promoters. For example, the activation of the genes for cell wall hydroxyproline-rich glycoproteins (HRGP) show markedly different patterns of transcript accumulation under conditions of wounding or infection (Corbin et al. 1987). The presence of several distinct intercellular stress systems would provide plants with the machinery to activate similar arrays of defensive responses under different sets of circumstances related to the expression of different forms of resistance, including periderm regeneration and prevention of infection after mechanical injury, induced systemic resistance, and resistance involved in the delimitation of fungal pathogens.

Interestingly, systemic wound-inducible and developmental (i.e., preformed in seeds and tubers) expression of potato proteinase inhibitor II was shown to be mediated by a single member of the proteinase inhibitor II gene family. This finding contrasts well with multiple genomic hydroxyproline-rich glycoprotein induction described above and suggests a possible linkage of wound responses with normal plant developmental processes. Indeed, one common feature appears to be that the phloem tissue serves as the transmitting organ used in both developmental and wound-induced gene activation (Keil et al. 1989).

The above examples describe single components that comprise a portion of the massive changes in gene expression following wounding or pathogen invasion. These changes underlie the induction of an array of defensive responses, including phytoalexin accumulation, synthesis of cell wall HRGP, and lignin and suberin biosynthesis and deposition. In trees, it is likely that many defensive responses act in concert or in tandem to prevent colonization of wounds and stimulate generation of new tissues. Given the varied nature of fungal canker and canker rot diseases, it is surprising to find a high degree of similarity among pathosystems in the resistance and boundary setting processes. Future research on the resistance in tree bark to fungal pathogens should include well-designed time course studies on the host-pathogen interaction. Noninoculated wounds should be included as controls. Studies with biochemical inhibitors will help identify the pathways that are critical to resistance in trees. Genetic studies on the inheritance of nonspecific types of resistance are needed to determine the potential for success in conventional breeding programs. Studies in molecular biology are needed to determine the potential for in vitro approaches to augmenting natural mechanisms of defense in trees.

References

Ashcroft JM 1934 European canker of black walnut and other trees. West Virginia Agr Exp Stn Bull 261

Berryman AA 1988 Towards a unified theory of plant defense. In: Mattson WJ, Levieux J, Bernard-Dagan C (eds) Mechanisms of woody plant defense against insects. Springer, Berlin Heidelberg New York, 39–55

Bier JE 1940 Studies in forest pathology. III. Hypoxylon canker of poplar. Can Dep Agr Tech Bull 27, 40 pp

Biggs AR 1984a Intracellular suberin: occurrence and detection in tree bark. IAWA Bull 5:243–248

Biggs AR 1984b Boundary zone formation in peach bark in response to wounds and Cytospora leucostoma infection. Can J Bot 62:2814–2821

Biggs AR 1986a Comparative anatomy and host response of two peach cultivars inoculated with Leucostoma cincta and L. persoonii. Phytopathology 76:905–912

Biggs AR 1986b Wound age and infection of peach bark by Cytospora leucostoma. Can J Bot 64:2319–2321

Biggs AR 1989a Temporal changes in the infection court following wounding of peach bark are associated with cultivar variation in infection by Leucostoma persoonii. Phytopathology 79:627–630

Biggs AR 1989b Pruning technique affects Leucostoma infection and wound closure in peach trees. Plant Dis 73:771–773

Biggs AR, Britton KO 1988 Presymptom histopathology of peach trees inoculated with Botryosphaeria obtusa and Botryosphaeria dothidea. Phytopathology 78:1109–1118

Biggs AR, Davis DD, Merrill W 1983 Histology of cankers on Populus caused by Cytospora chrysosperma. Can J Bot 61:563–574

Biggs AR, Merrill W, Davis DD 1984 Discussion: response of bark tissues to injury and infection. Can J For Res 14:351–356

Biggs AR, Miles NW 1988 Association of suberin formation in noninoculated wounds with susceptibility to Leucostoma cincta and L. persoonii in various peach cultivars. Phytopathology 78:1070–1074

Biggs AR, Peterson CA 1990 Effects of chemicals applied to peach bark wounds on accumulation of lignin and suberin and susceptibility to Leucostoma persoonii. Phytopathology 80:861–865

Bostock RM, Middleton GE 1987 Relationship of wound periderm formation in resistance to Ceratocystis fimbriata in almond bark. Phytopathology 77:1174–1180

Bostock RM, Stermer BA 1989 Perspectives on wound healing in resistance to pathogens. Annu Rev Phytopathol 27:343–371

Boyce JS 1948 Forest pathology, 2nd edn. McGraw-Hill, New York, 550 pp

Bramble WC 1934 Reaction of chestnut bark to invasion by Endothia parasitica. Am J Bot 23:89–95

Broglie KE, Gaynor JJ, Broglie RM 1986 Ethylene-regulated gene expression: molecular cloning of the genes encoding an endochitinase from Phaseolus vulgaris. Proc Natl Acad Sci USA 83:6820–6824

Butin H 1955 Über den Einfluß des Wassergehaltes der Pappel auf ihre Resistenz gegenüber Cytospora chrysosperma (Pers.)Fr. Phytopathol Z 24:245–264

Corbin DR, Sauer N, Lamb CJ 1987 Differential regulation of a hydroxyproline-rich glycoprotein gene family in wounded and infected plants. Mol Cell Biol 7:4337–4344

Cowling EB, Horsfall JG 1980 How plants defend themselves. In: Horsfall JG, Cowling EB (eds) Plant disease: an advanced treatise. Academic Press, New York, 1–16

Crowdy SH 1949 Observations on apple canker. III. The anatomy of the stem canker. Ann Appl Biol 36:483–495

Darvill AG, Albersheim P 1984 Phytoalexins and their elicitors – a defense against microbial infection in plants. Annu Rev Plant Physiol 35:243–275

Dixon RA, Lamb CJ 1990 Molecular communication in interactions between plants and microbial pathogens. Annu Rev Plant Physiol Mol Biol 41:339–367

Doster MA, Bostock RM 1988a Effects of low temperature on resistance of almond trees to Phytophthora pruning wound cankers in relation to lignin and suberin formation in wounded bark tissue. Phytopathology 78:478–483

Doster MA, Bostock RM 1988b Quantification of lignin formation in almond bark in response to wounding and infection by Phytophthora species. Phytopathology 78:473–477

Enebak SA, Blanchette RA 1989 Canker formation and decay in sugar maple and paper birch infected by Cerrena unicolor. Can J For Res 19:225–231

Farmer EE, Ryan CA 1990 Interplant communication: airborne methyl jasmonate induces synthesis of proteinase inhibitors in plant leaves. Proc Natl Acad Sci USA 87:7713–7716

Farmer EE, Pearce G, Ryan CA 1989 In vitro phosphorylation of plant plasma membrane proteins in response to the proteinase inhibitor inducing factor. Proc Natl Acad Sci USA 86:1539–1542

Flores G, Hubbes M 1980 The nature and role of phytoalexin produced by aspen (Populus tremuloides Michx.). Eur J For Res 10:95–103

French WJ 1969 Eutypella canker on Acer in New York. State Univ Coll For Tech Publ 94, Syracuse NY, 56 pp

Goring M-C 1975 Zur Ätiologie der Valsa-Krankheit an Süßkirsche. Histologische Untersuchungen über Eintrittspforten und Ausbreitung von Leucostoma persoonii (Nits.) Hohnel. Mitt Biol Bundesanst Land Forstwirtsch, Berlin-Dalhem 162

Grant RJ, Childs TW 1940 Nectria canker of northeastern hardwoods in relation to stand improvement. J For 38:797–802

Grant TJ, Spaulding P 1939 Avenues of entry for canker forming Nectrias of New England hardwoods. Phytopathology 29:351–358

Green TR, Ryan CA 1972 Wound-induced proteinase inhibitor in tomato leaves. Plant Physiol 51:19–21

Hahn MG, Darvill AG, Albersheim P 1981 Host-pathogen interactions. XIX. The endogenous elicitor, a fragment of plant cell wall polysaccharide that elicits phytoalexin accumulation in soybeans. Plant Physiol 68:1161–1169

Hebard FV, Griffin GJ, Elkins JR 1984 Developmental histopathology of cankers incited by hypovirulent and virulent isolates of Endothia parasitica on susceptible and resistant chestnut trees. Phytopathology 74:140–149

Kaufert F 1937 Factors influencing the formation of periderm in aspen. Am J Bot 24:24–30

Keefer WE 1914 Pathological histology of the Endothia canker of chestnut. Phytopathology 4:191–200

Keil M, Sanchez-Serrano JJ, Willmitzer L 1989 Both wound-inducible and tuber specific expression are mediated by the promoter of a single member of the potato proteinase inhibitor II gene family. EMBOJ 8:1323–1330

Koch LW 1935 Investigations on the black knot of plums and cherries. IV. Studies in pathogenicity and pathological history. Sci Agr 15:729–744

Kohle H, Jeblick W, Poten F, Blaschek W, Kauss H 1985 Chitosan-elicited callose synthesis in soybean cells as a Ca^{2+}-dependent process. Plant Physiol 77:544–551

Krähmer H 1980 Wound reactions of apple trees and their influence on infections with Nectria galligena. J Plant Dis Prot 87:97–112

Kuc J 1983 Induced systemic resistance in plants to diseases caused by fungi and bacteria. In: Bailey JA, Deverall BJ (eds) The dynamics of plant defense. Academic Press, New York London, 255–271

Li CL, Cui KM 1985 The effects of several exogenous hormones on bark regeneration in Eucommia ulmoides. Acta Bot Sin 27:1–6

Lieutier F, Berryman AA 1988 Elicitation of defensive reactions in conifers. In: Mattson WJ, Levieux J, Bernard-Dagan C (eds) Mechanisms of woody plant defenses against insects. Search for pattern. Springer, Berlin Heidelburg New York, 313–319

Lortie M 1964 Pathogenesis in cankers caused by Nectria galligena. Phytopathology 54:261–263

Manion P 1981 Tree disease concepts. Prentice-Hall, Englewood Cliffs, 399 pp

Miller RW, Berryman AA, Ryan CA 1986 Biotic elicitors of defense reactions in lodgepole pine. Phytochemistry 255:611–612

Mullick DB 1977 The non-specific nature of defense in bark and wood during wounding, insect and pathogen attack. Recent Adv Phytochem 11:395–441

Nothnagel EA, McNeil M, Albersheim P, Dell A 1983 Host-pathogen interactions. XXII. A galacturonic acid oligosaccharide from plant cell walls elicits phytoalexins. Plant Physiol 71:919–926

Parker J 1977 Phenolics in black oak bark and leaves. J Chem Ecol 3:489–496

Parsons TJ, Bradshaw HD Jr, Gordon MP 1989 Systemic accumulation of specific mRNAs in response to wounding in poplar trees. Proc Natl Acad Sci USA 86:7895–7899

Pearce RB, Ride JB 1982 Chitin and related compounds as elicitors of the lignification response in wounded wheat leaves. Physiol Plant Pathol 20:119–123.

Pusey LW 1989 Influence of water stress on susceptibility of nonwounded peach bark to Botryosphaeria dothidea. Plant Dis 73:1000–1003

Rittinger PA, Biggs AR, Peirson DR 1987 Histochemistry of lignin and suberin deposition in boundary layers formed after wounding in various plant species and organs. Can J Bot 65:1886–1982

Rogers JD, Berbee JG 1964 Developmental morphology of Hypoxylon pruinatum in bark of quaking aspen. Phytopathology 54:154–162

Russin JS, Shain L 1984 Initiation and development of cankers caused by virulent and cytoplasmic hypovirulent isolates of the chestnut blight fungus. Can J Bot 62:2660–2664

Ryan CA 1974 Assay and biochemical properties of the proteinase inhibitor-inducing factor, a wound hormone. Plant Physiol 54:328–332

Ryan CA 1988 Oligosaccharides as recognition signals for the expression of defensive genes in plants. Biochemistry 27:8879–8883

Sánchez-Serrano JJ, Peña-Cortés H, Willmitzer L, Prat S 1990 Identification of potato nuclear proteins binding to the distal promoter region of the proteinase inhibitor II gene. Proc Natl Acad Sci USA 87:7205–7209

Sequeira L 1983 Mechanisms of induced resistance in plants. Annu Rev Microbiol 37:51–79

Shigo AL 1969 How Poria obliqua and Polyporus glomeratus incite cankers. Phytopathology 59:1164–1165

Shigo AL 1979 Tree decay – an expanded concept. US Dep Agr, For Serv, Agr Inf Bull 419, 73 pp

Shigo AL 1985 How tree branches are attached to trunks. Can J Bot 63:1391–1401

Shirata A 1978 Production of phytoalexin in cortex tissue of mulberry shoot. Ann Phytopathol Soc Jpn 44:485–492

Soo BVL 1977 General occurrence of exophylactic and necrophylactic periderms and nonsuberized impervious tissues in woody plants. PhD Thesis, Univ British Columbia, Can

Tamura O, Saito I 1982 Histopathological changes of apple bark infected by Valsa ceratosperma (Tode ex Fr.) Maire during dormant and growing periods. Ann Phytopathol Soc Jpn 48:490–498

Tattar TA 1978 Diseases of shade trees. Academic Press, New York London, 361 pp

Tattar TA, Rich AE 1973 Extractable phenols in clear, discolored, and decayed woody tissues and bark of sugar maple and red maple. Phytopathology 63:167–169

Tekauz A, Patrick ZA 1974 The role of twig infections on the incidence of perennial canker of peach. Phytopathology 64:683–688

Traquair JA 1987 Oxalic acid and calcium oxalate produced by Leucostoma cincta and L. persoonii in culture and in peach bark tissues. Can J Bot 65:1952–1956

Trockenbrodt M 1990 Survey and discussion of the terminology used in bark anatomy. IAWA Bull 11:141–166

Vance CP, Kirk TK, Sherwood RT 1980 Lignification as a mechanism of disease resistance. Annu Rev Phytopathol 18:259–288

Wargo PM 1975 Lysis of the cell wall of Armillaria mellea by enzymes from forest trees. Physiol Plant Pathol 5:99–105

Wiltshire SP 1921 Studies on the apple canker fungus. I. Leaf scar infection. Ann Appl Biol 8:182–192

Wisniewski M, Bogle AL, Wilson CL 1984 Histopathology of canker development on peach tress after inoculation with Cytospora leucostoma. Can J Bot 62:2804–2813

Yev-Ladun S, Aloni R 1990 Polar patterns of periderm ontogeny and their relationship to leaves and buds. IAWA Bull 11:289–300

Zalasky H 1964 The histopathology of Macrophoma tumefaciens infections in black poplar. Can J Bot 42:385–391

Zalasky H 1968 Penetration and initial establishment of Nectria galligena in aspen and peachleaf willow. Can J Bot 46:57–60

Chapter 4

Responses of Gymnosperm Bark Tissues to Fungal Infections

S. Woodward

4.1 Introduction

Although the bark of trees provides the initial barrier to agents with the potential to damage the economically important internal woody tissues, it has received comparatively little attention in terms of responses to wounding and infection when compared with the wood itself. Until relatively recently, the majority of studies on bark concentrated on taxonomic comparisons (Eremin 1976) and the relationship between physical properties and commercial uses in consolidated board or reinforced plastic (e.g. Krahmer and Wellons 1973, Wellons and Krahmer 1973, Miller et al. 1974). Examination of bark extractives was based on their potential industrial and pharmaceutical uses (Hergert 1960, Rogers 1967) and little attention was paid to the role of these compounds in the tree itself, despite the fact that, from investigations on other plants, many were known to possess antimicrobial activity. During the past 15 years, however, the importance of the bark as a barrier to pests and diseases has become more widely recognized, and significant advances have been made in our understanding of the nature of responses to infection in these tissues (Mullick 1977, Biggs et al. 1984, Biggs 1985, Pearce 1987, 1989).

4.2 Physical Barriers

4.2.1 Structure of the Periderm and Bark Cortex

The bark of trees forms an important durable barrier, protecting the underlying conducting tissues from the surrounding environment. As in angiosperms, conifer bark is composed of a periderm or series of sequent periderms, the rhytidome, comprising the outer tangential bands of thick and thin-walled phellem, the meristematic phellogen, and the inner phelloderm (Srivastava 1964, Esau 1977) (Chap. 2). These tissues overlie the secondary bark cortex and, due to the presence in the phellem of lignin and suberin, are highly resistant to damage by both biotic and abiotic agents. In addition to these structural components, the bark contains many water and lipid-soluble secondary metabolites, some of which are active against bacteria, fungi, and phytophagous insects.

In healthy trees, any breach of the bark tissues results in the triggering of an active response, the ultimate aim of which is to restore rapidly the integrity of the plant surface and thereby restrict the entry of any potentially destructive

Table 4.1. Examples of important fungal pathogens capable of colonizing conifer bark tissues

Pathogen	Type of growth[a]	Mode of transmission[b]	Penetration of intact bark[c]
Root diseases			
Armillaria mellea	N	{ Rhizomorphs	+
Armillaria ostoyae	N	{ (+root-root)	+
Heterobasidion annosum	N	Basidiospores/root-root	+
Inonotus tomentosus	N	{ Basidiospores	−
Inonotus circinatus	N	{	−
Phellinus weirii	N	Basidiospores/root-root	+
Rhizina undulata	N	Ascospores/soil-borne	+
Stem diseases			
Cronartium spp.	N	Basidiospores	−
Endocronartium spp.	N	Aecidiospores	−
Melampsora caryophyllacearum	B	Basidiospores	−
Peridermium spp.	B	Aecidiospores	−
Atropellis piniphilla	N	Ascospores	+
Crummenulopsis soraria	N	Ascospores	?
Fusarium moniliforme var. *subglutinans*	N	Conidia	−
Lachnellula willkommii	N	Ascospores	−
Seiridium cardinale	N	Conidia	−

[a] Type of growth within host tissues. N, Necrotrophic; B, biotrophic.
[b] Mode of transmission: manner in which pathogen normally encounters a new host.
[c] Penetration of intact bark. +, Pathogen proven to enter bark through intact periderm; −, pathogen not proven to enter through intact periderm.

pests or pathogens. Success or failure of the response depends on many internal and external factors, such as the pathogenicity of any microorganisms involved, the host genotype, and the presence or absence of other external stresses affecting the physiological status of the tree.

Durability of the bark tissues notwithstanding, several important pathogens of conifers are capable of sustained necrotrophic or biotrophic growth within the bark cortex, and a number have been demonstrated to produce enzymes capable of breaking down the polymeric compounds forming the outermost rhytidome layers of the periderm (Table 4.1). Also of importance is the ability of certain fungi to metabolize antifungal compounds present in the bark to relatively non-toxic breakdown products. These pathogenic fungi, therefore, are able to overcome the normal defense responses within the bark, either through tolerance or through disruption of the stress metabolism of the host plant.

4.2.2 Role of Suberized and Lignified Tissues in Preventing Entry of Potential Pathogens

Suberin is a complex and variable polymer, composed of both lipid and phenolic components (Kolattukudy 1981). The presence of lipids, which include long chain fatty acids and fatty alcohols, *w*-hydroxy fatty acids and dicarboxylic acids, renders the polymer hydrophobic, a property which contributes to the relatively inert nature of suberized plant tissues. Equally, lignin, in conifers a polymer predominantly composed of coniferyl alcohol residues (Vance et al. 1980), is highly resistant to microbial degradation (Kirk 1971). Precisely how lignin inhibits attack by potential pathogens is uncertain, but several possible mechanisms have been postulated (Friend 1976, Ride 1978) including inhibition of penetration by the physical strength of the polymer, restriction of diffusion of pathogen-produced cell wall degrading enzymes, the physical shielding of host cell wall polysaccharides from attack by fungal enzymes, and the effects of the phenolic precursors of lignin on the pathogens themselves.

Despite the lack of information directly relating to gymnosperms, it is clear that the suberized and lignified tissues comprising the periderms and rhytidomes of these plants are highly resistant to attack by microorganisms.

4.3 Chemical Barriers

4.3.1 Antifungal Phenolic Compounds in the Bark

The antifungal activity of heartwood extractives from both gymnosperms and angiosperms has been the subject of extensive study (Rudman 1962, 1963, 1965, Hart and Shrimpton 1979) (Chaps. 7 and 8). The possibility of similar antimicrobial activity being present in the bark of trees, however, has received little attention, despite the presence in the tannin component of these tissues of high levels and complex mixtures of phenolics (Hergert 1960), many of which are known, from studies on heartwood extractives and responses of herbaceous angiosperms to infection, to possess antimicrobial properties. For example, compounds known to be inhibitory to the growth of *Heterobasidion annosum* in vitro (Alcubilla et al. 1971), such as taxifolin (3,3′,4′,5,7-pentahydroxyflavanone), D-catechin, quercetin (3,3′,4′,5,7-pentahydroxyflavone), and various stilbene derivatives, are present in the barks of many economically important genera of conifers, including *Cedrus*, *Larix*, *Picea*, *Pinus*, *Pseudotsuga*, *Thuja*, and *Tsuga* (Hergert 1960). Of these compounds, the group which has received the most attention in relation to their possible role in defense, both in the wood and the bark, is the stilbenes.

Stilbenes are compounds based around a 14-carbon phenolic structure (Fig. 4.1). Many derivatives of this basic structure are known from a diverse range of plants (Hart 1981), and in healthy bark the stilbenes are often present as glucosides, presumably because in these forms the compounds are much less biologically active (Woodward and Pearce 1988a). In Sitka spruce (*Picea sitchensis*) bark, the major antifungal compounds present were the stilbenes astringin (5,3′,4′-trihydroxystilbene-3-β-D-glucoside) and rhaponticin (5,3′-dihydroxy-4′-methoxystilbene-3-β-D-glucoside) (Fig. 4.1a,b) (Woodward and Pearce 1988a), where they typically comprise more than 3% of the fresh weight of the bark (Drago-Toscano 1989). These stilbenes have been reported as major bark components in several *Picea* species (Manners and Swan 1971). In *P.*

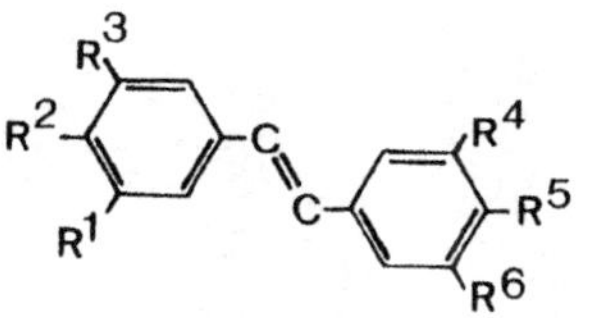

	R1	R2	R3	R4	R5	R6
a – Astringin	glucose	H	OH	OH	OH	H
Astringenin	OH	H	OH	OH	OH	H
b – Rhaponticin	glucose	H	OH	OH	OCH_3	H
Isorhapontigenin	OH	H	OH	OH	OCH_3	H
c – Resveratrol glucoside	glucose	OH	H	OH	H	OH
d – Pinosylvin	OH	H	OH	H	H	H

Fig. 4.1.a–d Structures of stilbenes found in the bark of *Picea* and *Pinus* spp.

sitchensis, staining with diazotized o-tolidine of sections of bark cut on a freezing microtome to determine the distribution of phenolic compounds (Shain 1967), principally stilbenes, revealed the localization of these compounds in the lumina of distinct layers of cells within the parenchyma (Drago-Toscano 1989). The proportion of positively staining cells increased in the outer cortical cells, with the outermost three to five layers directly below the periderm responding uniformly to the stain (Fig. 4.2).

Along with resveratrol glucoside (4,3′,5′-trihydroxystilbene-3-β-D-glucoside) (Fig. 4.1c), pinosylvin (3,5-dihydroxystilbene) (Fig. 4.1d) has also been reported from the inner bark of the Siberian pine (*Pinus sibirica*) (Gromova et al. 1975).

In assays against fungal growth in vitro, astringin and rhaponticin inhibited *Phaeolus schweinitzii* and *Sparassis crispa*, decay-causing species of relatively low pathogenicity, whereas the more aggressive species, *Armillaria ostoyae*, *Heterobasidion annosum*, *Rhizina undulata*, and *Stereum sanguinolentum* either tolerated the presence of the stilbenes or were markedly stimulated (Woodward and Pearce 1988a). Culture filtrates obtained from these tests showed that at the end of the incubation period, the weak pathogens had metabolized the stilbene glucosides to the aglycones astringenin (5,3,3′,4′-tetrahydroxystilbene) and isorhapontigenin (5,3,3′-trihydroxy-4′-methoxystilbene) (Fig. 4.1a,b). The stilbene aglycone isorhapontigenin was approximately six times more fungitoxic than the respective glycone, rhaponticin to *P. schweinitzii*. In contrast, filtrates of the more pathogenic species contained relatively low levels of both the stilbene glycones and aglycones, suggesting an ability to degrade these compounds to less toxic products.

Taxifolin has been shown to inhibit totally the growth of *Heterobasidion annosum* when incorporated into malt agar at levels of 0.45% (Kennedy 1956). Actual levels of this compound in healthy bark of conifers have not been reported however, and the possible roles of taxifolin and related compounds of the quercetin group in resistance of these tissues to fungal attack remain to be determined.

Levels of inhibitory phenolic compounds, including astringin, rhaponticin, and taxifolin, present in the inner bark of *Picea abies*, were found to decrease in

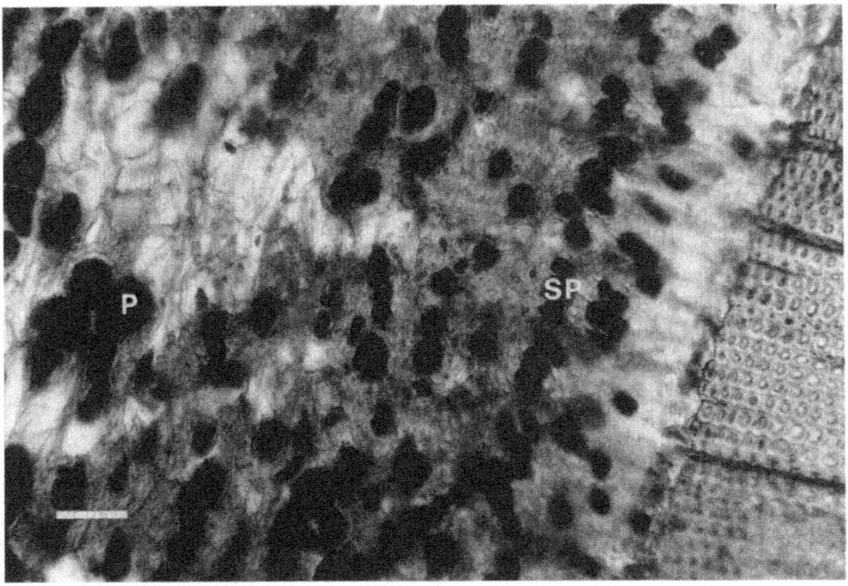

Fig. 4.2. Localization of phenolic compounds in transverse section of bark of *Picea sitchensis* using diazotized o-tolidine stain. Staining of the cell lumen is localized in distinct layers of the parenchyma (*P*) derived from the secondary phloem (*SP*), and in the secondary phloem itself. *Bar* = 60 μm (Photograph courtesy of Dr. C. Drago-Toscano)

trees stressed through drought, deficiency of manganese or excessive nitrogen (Alcubilla et al. 1971). The trees growing on the stressed sites were more prone to attack by *H. annosum*, though in the absence of data relating to other factors known to be associated with increased susceptibility to *Heterobasidion*, a direct relationship between these particular bark compounds and resistance to pathogens cannot be proven.

In a detailed study of the resistance of Sitka spruce bark to penetration by *Phaeolus schweinitzii* and *Heterobasidion annosum*, no correlations were found between the constitutive levels of astringin and rhaponticin and the extent of tissue necrosis and hyphal penetration (Drago-Toscano 1989), although limited evidence was obtained to suggest a genetic effect on colonization of the bark.

The presence in the bark of compounds inhibitory to fungal growth has received little attention in other conifers. Although chloroform and acetone extracts of the inner bark and phloem tissues of *Pinus sylvestris* were active against *Heterobasidion annosum*, the compounds responsible for this inhibition were not determined (Ladejtschikova and Pasternak 1982).

4.3.2 Resin Terpenes

There have been many reports in the literature concerning the effects of both the monoterpene and diterpene (resin acid) components of conifer resins on the growth of tree pathogens (Cobb et al. 1968, Hintikka 1970, Shain 1971, De Groot 1972, Chou and Zabkiewicz 1976, Flodin and Fries 1978, Ennos and Swales 1988). It is not the intention in this chapter to provide a review of the antimicrobial activities of these compounds (see Kuc and Shain 1977, Schuck 1982). However, circumstantial evidence for the importance of resins in resistance of pines to *Heterobasidion annosum* root rot has been presented (Gibbs 1968), and components of the monoterpene fraction, particularly α-pinene and β-phellandrene, have been suggested as markers for resistance in Sitka spruce against the same pathogen (Forrest 1982), and in pines against *Cronartium quercuum* f.sp. *fusiforme* (Rockwood 1973, 1974) and *Gremmeniella abietina* (Aitken 1986). Conflicting reports as to the effectiveness of these compounds in inhibiting the growth of fungi pathogenic to conifers are found in the literature but may simply reflect the use of various assay methods and different fungal strains. The overall consensus, however, is that saturated atmospheres of the monoterpenes, which are likely to be found in the vicinities of fresh wounds on conifers, will reduce or prevent the growth of potentially pathogenic fungi (Hintikka 1970, Schuck 1982).

4.4 Responses of Bark to Wounding and Pathogen Attack

4.4.1 Physical Alterations, "Walling-Off" of Wounds and Restoration of the Periderm (Necrophylactic Periderm Formation)

4.4.1.1 Responses to Nectrotrophic Pathogens

Penetration of the intact periderm of conifers by pathogenic fungi has been demonstrated in very few instances; the majority of species capable of growth within the bark or xylem require the presence of wounds to gain entry. Species such as *Armillaria ostoyae*, *Heterobasidion annosum*, and *Phellinus* (*Inonotus*) *weirii* may penetrate the periderm directly (Thomas 1934, Wallis and Reynolds 1962, Peek et al. 1972), each requiring a substantial food base from which to mount the sustained attack necessary for the breakdown of the suberized tissues. It has been suggested that the canker-forming ascomycete *Atropellis piniphila* penetrates undamaged aerial bark of *Pinus contorta* at branch nodes (Hopkins 1963), although the mechanism of this action was not described.

The mode of infection by *Armillaria* has been described in detail (Thomas 1934, Woeste 1956). The base for attack in this species is usually an old colonized stump from which the pathogen spreads through the surrounding soil by the growth of rhizomorphs; fresh infections of wounds via basidiospores are a relatively rare event (Rishbeth 1970). Once in contact with the host root bark, the rhizomorph produces several small branches, which penetrate the bark scales and form mycelial wedges. In successfull attacks, the mycelium then goes on to reach the cambium. *Armillaria* may produce enzymes capable of breaking down suberin (suberinase) (Zimmerman and Seemüller 1984), which aid in the physical process of penetration. In resistant reactions in the root bark of *Picea*

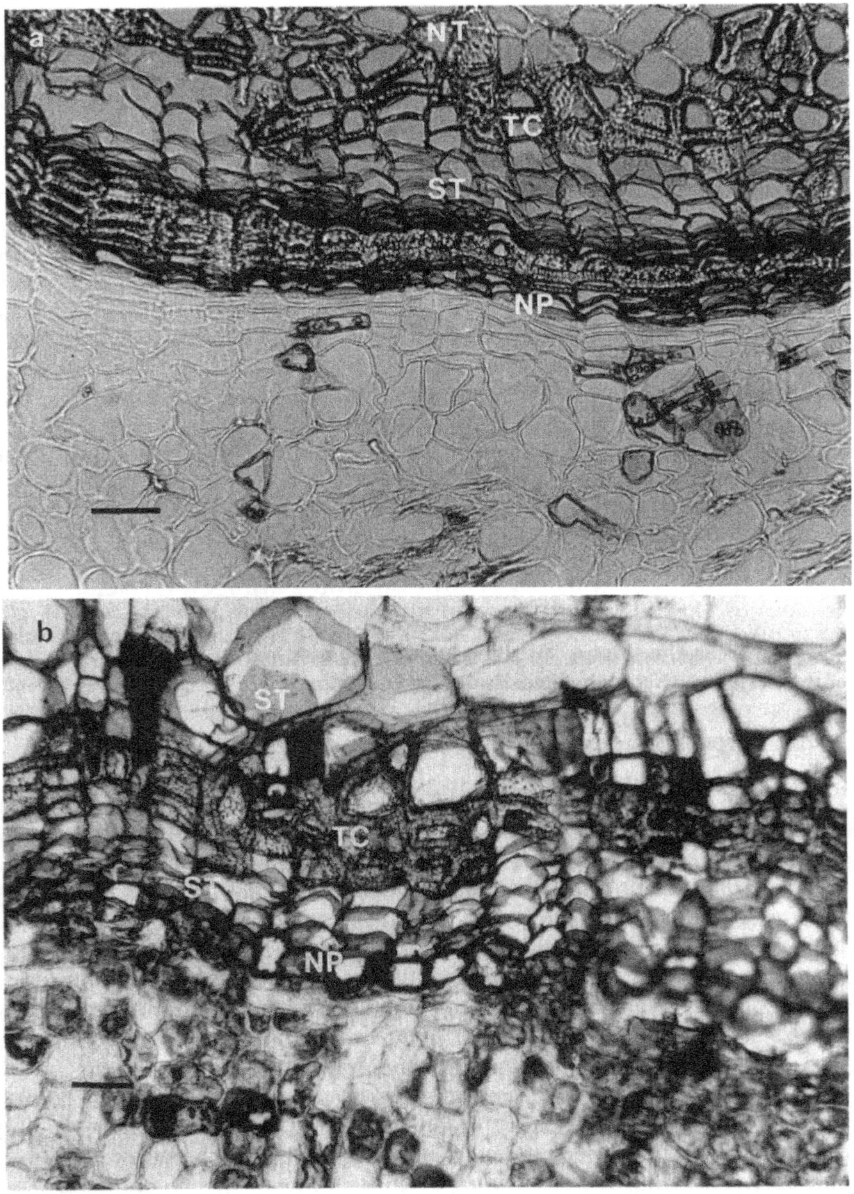

Fig. 4.3.a,b Structure of the necrophylactic periderm formed in the response of **a** *Picea sitchensis* to attack by *Armillaria obscura*; **b** as a non-specific response, 72 days after wounding. Stained with Sudan IV after extraction with chlorine dioxide to remove phenolic compounds. *NT* Necrotic tissue; *TC* thickened cells; *ST* suberized tissues; *NP* necrophylactic periderm. *Bar* = 60 μm. (Photograph **a** courtesy of Dr. R.B. Pearce, Oxford Forestry Institute)

sitchensis, small (ca. 10 mm diameter) necrotic regions developed in the superficial bark tissues around the sites of attempted penetration by *Armillaria ostoyae* (Pearce 1989). Mechanical attachment of these sites to the host was weak, resulting in the ready detachment of the necrotic regions from the bark. Histological examinations showed that the detachment occurred in the plane of the necrophylactic periderm, where cell wall alterations were similar to those found in artificial wounds made into the root bark of *Picea sitchensis* (Fig. 4.3a, b) (Woodward and Pearce 1988b), emphasizing the generalized, non-specific nature of the response, (Mullick 1977, Biggs et al. 1984) (Chap. 2).

Within a stand, transmission of other root-invading pathogens with low competitive saprophytic abilities, such as *Heterobasidion annosum* and *Phellinus weirii*, is by root to root contact between infected and healthy individuals. Hyphae of these species grow on the bark surface of infected roots and are transferred to neighboring trees at points of root contact (Rishbeth 1951a,b, Wallis 1976, Bloomberg and Reynolds 1982). Both of these pathogens are capable of penetrating the bark tissues in order to reach the internal woody tissues, where in older trees, decay is caused.

Pathogens which invade the bark of the aerial parts of the tree display different modes of infection. Generally, these species require wounds to enter the bark cortex and are not capable of penetrating directly through the suberized and lignified tissues of the rhytidome, a possible exception to this rule being *Atropellis piniphila* mentioned above. Other canker-causing pathogens, such as *Fusarium moniliforme* var. *subglutinans*, *Seiridium cardinale*, and *Lachnellula willkommii*, all require bark wounds in order to gain entry (Buczacki 1973, Dwinell and Barrows-Broaddus 1981, Ponchet and Andréoli 1989).

The histology of the response of various pine species to the pitch canker pathogen *Fusarium moniliforme* var. *subglutinans* has been described in detail by Barrows-Broaddus and Dwinell (1983), and provides further evidence for the non-specific nature of bark responses to injury and infection. In susceptible genotypes, particularly of *Pinus elliotii* var. *elliotii* and *P. virginiana*, hyphae of *F. moniliforme* var. *subglutinans* rapidly grew intercellularly in the cortex, the infection becoming diffuse in the absence of necrophylactic periderm formation. Resistance was characterized by the formation of several layers of regenerative parenchyma and by lignification, with necrophylactic periderm eventually delimiting the lesions. The infected region was then sloughed off as a result of continued growth of the host. Similar responses have been described for *Seiridium* (*Coryneum*) canker of cypress caused by *Seiridium cardinale* (Ponchet and Andréoli 1989), where a resistant host response was correlated with the ability to form a necrophylactic periderm in advance of the pathogen.

Repeated breaching of necrophylactic periderms occurs in diseases which typically give rise to target cankers, such as in larch infected by *Lachnellula wilkommii* (Buczacki 1973, Yde-Anderson 1979). This pathogen enters the bark through cracks made during freezing, and kills the cambium. Each spring, a fresh necrophylactic periderm is formed on the healing callus growth in the live bark adjacent to the canker. Further freezing weakens the live bark cells surrounding the canker, and, during periods of higher winter temperatures, *L. willkommii* penetrates the thin necrophylactic periderm and kills the underlying weakened cells. Annual repetition of these events results in the formation of the typical target-type cankers characteristic of this disease.

4.4.1.2 Responses to Biotrophic Pathogens

The stem rust pathogens (*Cronartium* spp., *Peridermium* spp., *Melampsora caryophyllacearum*) usually enter their hosts through needles, either via wounds or through stomata, and reach the secondary bark tissues after growing down the succulent young shoots. In certain circumstances, entry may be achieved through wounds in the bark itself. Reactions to the presence of these fungi within the bark again reflect the non-specific nature of the response (Kinloch 1982).

The stem rust pathogens grow intercellularly within the bark, causing swelling and the accompanied formation of galls (Krebill 1968). Hyphae are found up to 55 mm in advance of visible symptoms within these tissues (Ehrlich and Opie 1940), depending on the genotype of the host and the rust-causing pathogen. Haustoria develop from the intercellular hyphae into cells of the bark cortex, the phloem, and into ray parenchyma cells of the wood (Krebill 1968, Woo and Martin 1981). Resistance in the bark to the stem and gall rusts is characterized by hypersensitive responses based on the rapid necrosis of the host cells, and the development of necrophylactic periderms (Bingham et al. 1960, Hoff and McDonald 1972, Miller et al. 1976, Kinloch 1982) through which the pathogens rarely penetrate.

4.4.2 Role of Antimicrobial Chemicals in the Wound Healing Response

Resin canals are a normal feature of the bark of many conifers, particularly in *Pinus* and *Picea*. In addition, wounding may result in the formation of traumatic resin ducts, even in *Abies* or *Tsuga* where ducts are not present in healthy tissues, giving the potential for further resin production in the injured area. Any wounding of these tissues, therefore, results in resin flow, a process which ultimately leads to the formation of a impervious layer of dried oleoresin. This dried resin creates a hydrophobic environment in the vicinity of the wound that is highly resistant to penetration by potential pests and pathogens (Gibbs 1968, Prior 1976).

The monoterpene content of resins formed in response to wounding may differ from that found in intact tissues. Higher levels of myrcene and Δ-3-carene were found in wounds in *Abies grandis* (Russel and Berryman 1976), whereas in *Picea abies* quantities of β-pinene decreased markedly in the immediate vicinity of the wounds (Schuck 1982). The significance of these changes in relation to resistance to fungal pathogens has not been determined.

Following challenge of Sitka spruce root bark with root pathogens or decay-causing fungi, the levels of astringin and rhaponticin decreased markedly (Woodward and Pearce 1988a). At the same time, the respective aglycones of these compounds, astringenin and isorhapontigenin, were found to accumulate in the challenged tissues. Release of astringenin and isorhapontigenin from the glucosides was mediated by the action of fungal β-glucosidases, an increase in which was concomitant with the reduction in levels of stilbene glucosides. The higher levels of antifungal activity shown by the aglycones (see Sect. 4.3.1) suggest that their release creates a more inhibitory environment in the challenged tissues.

Knowledge of the physiological responses of conifer bark tissues to attack by other pathogenic fungi is limited. For example, the blister rusts (*Cronartium* spp.) are some of the most damaging diseases of pines in the world, especially in the southeastern United States, yet little is known of the chemical nature of the defense response in the bark of resistant or susceptible host plants. Preliminary work in this laboratory has suggested that enzymes capable of attacking the fungal cell wall are produced in the bark cortex of *Pinus sylvestris* during the development of *Peridermium pini* cankers (Nsolomo and Woodward, unpublished), though the significance of these enzymes in this interaction has yet to be determined.

4.5 Conclusions

It is apparent that a combination of both physical and chemical events contribute to the wound closure and resistance responses in the bark tissues of conifers. Susceptibility occurs where an aggressive pathogen is able to circumvent these normal responses. A key event in the non-specific response to wounding is the early formation of an impervious zone in the vicinity of the damage (Mullick 1975, 1977, Biggs et al. 1984, Rittinger et al. 1987, Woodward and Pearce 1988b) (Chap. 2). This zone had been considered as non-suberized (Mullick 1975, 1977), but improved tissue preparation techniques have demonstrated the suberization of cells present at the time of wounding (Rittinger et al. 1987, Woodward and Pearce 1988b). Deposition of lignins and cell wall thickening occur concomitantly in adjacent cell layers, resulting in the formation of thick- and thin-walled necrophylactic phellem external to the newly differentiated phellogen. The important point to emphasize is that the early suberization and lignification occur in cells *extant at the time of wounding*. These cells form the important impervious barrier required to protect the underlying cells which are redifferentiating to form the necrophylactic periderm. It is possible that the signals which initiate this process are released from the walls of dying cells in the wounded tissues (Bloch 1953) in a similar manner to that described for the constitutive elicitors of phytoalexins in herbaceous angiosperms (Bailey 1982).

A model for the non-specific healing of Sitka spruce bark tissues following wounding and attack by potential pathogens has been proposed recently (Woodward and Pearce 1988b). In this model, the stilbene glucosides, coupled with volatile terpenes, provide the first line of defense, following breaching of the bark surface. The action of β-glucosidases produced by fungi in the vicinity of the wound releases stilbene aglycones, which inhibit the growth of most potential pathogens. The stilbenes and resins provide a temporary barrier to invasion, pending the deposition of lignin and suberin that form the suberized impervious zone, a process first detectable between 10 and 13 days after wounding. The necrophylactic periderm then differentiates in the tissues immediately internal to the suberized impervious tissue. This process results in the reestablishing of an intact periderm surface, from which the necrotic region may be separated as a bark scale during normal growth.

Responses of conifer bark to fungal attack and infection have received relatively little attention in the past, especially when considering the economic importance of some of the groups of diseases associated with the periderm. Further in-depth and critical studies of factors contributing to resistance within the bark will assist in devising improved methods for the control of pathogens capable of breaching these tissues and in the future may provide important indicators for consideration in forest tree selection and breeding programs.

Acknowledgements. The author wishes to thank Dr. Ray Pearce of the Oxford Forestry Institute for his critical review of this chapter, and for the use of Fig. 4.3a. Dr. Claudia Drago-Toscano supplied the photograph for Fig. 4.2.

References

Aitken EAB 1986 Brunchorstia pinea on conifers. PhD Thesis, Dep For, Univ Aberdeen

Alcubilla M, Diaz-Palacio MP, Kreutzer K, Laatsch W, Rehfuess KE, Wenzel G 1971 Beziehungen zwischen dem Ernährungszustand der Fichte (Picea abies Karst.), ihrem Kernfäulebefall und der Pilzehemmwirkung ihres Basts. Eur J For Pathol 1:100–114

Bailey JA 1982 Mechanisms of phytoalexin accumulation. In: Bailey JA, Mansfield JM (eds) Phytoalexins. Blackie, Glasgow London, 289–318

Barrows-Broaddus J, Dwinell LD 1983 Histopathology of Fusarium moniliforme var. subglutinans in four species of southern pines. Phytopathology 73:882–889

Biggs AR 1985 Suberized boundary zones and the chronology of wound response in tree bark. Phytopathology 75:1191–1195

Biggs AR, Merrill W, Davis DD 1984 Discussion: response of bark tissues to injury and infection. Can J For Res 14:351–356

Bingham RT, Squillace AE, Wright JW 1960 Breeding blister rust resistant western white pine. Silvae Genet 9:33–41

Bloch R 1953 Defense reactions of plants to the presence of toxins. Phytopathology 43:351–354

Bloomberg WJ, Reynolds G 1982 Factors affecting transfer and spread of Phellinus weirii mycelium in roots of second growth Douglas fir. Can J For Res 12:424–427

Buczacki ST 1973 A microecological approach to larch canker biology. Trans Br Mycol Soc 61:315–329

Chou CKS, Zabkiewicz JA 1976 Toxicity of monoterpenes from Pinus radiata cortical oleoresin to Diplodea pinea spores. Eur J For Pathol 6:354–359

Cobb FW, Krstic M, Zavarin E, Barber HW 1968 Inhibitory effects of volatile oleoresin components on Fomes annosus and four Ceratocystis species. Phytopathology 58:1327–1335

De Groot RC 1972 Growth of wood-inhabiting fungi in saturated atmospheres of monoterpenoids. Mycologia 64:863–870

Drago-Toscano CD 1989 Variation in defence mechanism activities in Sitka spruce (Picea sitchensis (Bong.) Carr.) and its relation to disease resistance and susceptibility. PhD Thesis, Green Coll, Univ Oxford

Dwinell LD, Barrows-Broaddus J 1981 Pitch canker in seed orchards. In: Proc 16th Southern Forest Tree Improvement Conference. VA Polytech Inst State Univ, Blacksburg, VA, 234–240

Ehrlich J, Opie RS 1940 Mycelial extent beyond blister rust cankers on Pinus monticola. Phytopathology 30:611–620

Ennos RA, Swales KW 1988 Genetic variation in tolerance of host monoterpenes in a population of the ascomycete canker pathogen Crumenulopsis sororia. Plant Pathol 37:407–416

Eremin VM 1976 Anatomiya kory vidov Picea Sovetskogo Souza. Bot Z 61:700–709

Esau K 1977 Anatomy of seed plants, 2nd edn. John Wiley, New York, 550 pp

Flodin K, Fries N 1978 Studies on volatile compounds from Pinus sylvestris and their effect on wood-decomposing fungi. II. Effects of some volatile compounds on fungal growth. Eur J For Pathol 8:300–310

Forrest GI 1982 Preliminary work on the relation between resistance to Fomes annosus and the monoterpene composition of Sitka spruce resin. In: Heybroek HM, Stephan BR, Weissenberg K (eds) Resistance to diseases and pests in forest trees. PUDOC, Wageningen, 194–197

Friend J 1976 Lignification in infected tissue. In: Friend J, Threlfall DR (eds) Biochemical aspects of plant-parasite relationships. Academic Press, New York London, 291–303

Gibbs JN 1968 Resin and the resistance of conifers to Fomes annosus. Ann Bot 32:649–665

Gromova AS, Tyukavkina NA, Lutskii VI, Kalabin GA, Kushnarev DF 1975 (Hydroxystilbenes of the inner bark of Pinus sibirica.) Khim Prirod Soed 6:677–682

Hart JH 1981 Role of stilbenes in decay and disease resistance. Annu Rev Phytopathol 19:437–458

Hart JH, Shrimpton DM 1979 Role of stilbenes in resistance of wood to decay. Phytopathology 69:1138–1143

Hergert HL 1960 Chemical composition of tannins and polyphenols from conifer wood and bark. For Prod J 10:610–617

Hintikka W 1970 Selective effect of terpenes on wood-decomposing hymenomycetes. Karstenia 11:28–32

Hoff RJ, McDonald GI 1972 Resistance in Pinus armandii to Cronartium ribicola. Can J For Res 2:303–307

Hopkins JC 1963 Atropellis canker of lodgepole pine: etiology, symptoms, and canker development rates. Can J Bot 41:1535–1545

Kennedy RW 1956 Fungicidal toxicity of certain extraneous components of Douglas-fir heartwood. For Prod J 6:80–84

Kinloch BB 1982 Mechanisms and inheritance of rust resistance in conifers. In: Heybroek HM, Stephan BR, Weissenberg K (eds) Resistance to diseases and pests in forest trees. PUDOC, Wageningen, 119–129

Kirk TK 1971 Effects of micro-organisms on lignin. Annu Rev Phytopathol 9:185–210

Kolattukudy PE 1981 Structure, biosynthesis, and biodegradation of cutin and suberin. Annu Rev Plant Physiol 32:539–567

Krahmer RL, Wellons JC 1973 Some anatomical and chemical characteristics of Douglas fir cork. Wood Sci 6:97–105

Krebill RG 1968 Histology of canker rusts in pines. Phytopathology 58:155–164

Kuc J, Shain L 1977 Antifungal compounds associated with disease resistance in plants. In: Seigel MR, Sisler HD (eds) Antifungal compounds, vol 2. Dekker, New York, 497–535

Ladejtschikova EI, Pasternak GM 1982 Biochemical aspects of the resistance of Pinus sylvestris to Fomes annosus. In: Heybroek HM, Stephan BR, Weissenberg K (eds) Resistance to diseases and pests in forest trees. PUDOC, Wageningen, 198–205

Manners GD, Swan EP 1971 Stilbenes in the barks of five Canadian Picea species. Phytochemistry 10:607–610

Miller DJ, Wellons JC, Krahmer RL, Short PH 1974 Reinforcing plastics with Douglas fir bark fiber. For Prod J 24:18–22

Miller T, Cowling EB, Powers HR, Blalock TE 1976 Types of resistance and compatibility in slash pine seedlings infected by Cronartium fusiforme. Phytopathology 66:1229–1235

Mullick DB 1975 A new tissue essential to necrophylactic periderm formation in the bark of four conifers. Can J Bot 53:2443–2457
Mullick DB 1977 The non-specific nature of defense in bark and wood during wounding, insect and pathogen attack. Recent Adv Phytochem 11:395–442
Pearce RB 1987 Antimicrobial defence in perennial plants. In: Pegg GF, Ayres, PG (eds) Fungal infection of plants: establishment, progress and outcome of infection. Cambridge Univ Press, 219–238
Pearce RB 1989 Cell wall alterations and antimicrobial defense in perennial plants. In: Lewis NG, Paice MG (eds) Plant cell wall polymers: biogenesis and biodegradation. Am Chem Soc Symp Ser 399, Washington, DC, 346–360
Peek R-D, Liese W, Parameswaran N 1972 Infektion und Abbau der Wurzelrinde von Fichte durch Fomes annosus. Eur J For Pathol 2:104–115
Ponchet J, Andréoli C 1989 Histopathologie du chancre cortical du cypres à Seiridium cardinale. Eur J For Pathol 19:212–221
Prior C 1976 Resistance by Corsican pine to attack by Heterobasidion annosum. Ann Bot 40:261–279
Ride J 1978 The role of cell wall alterations in resistance to fungi. Ann Appl Biol 89:302–306
Rishbeth J 1951a Observations on the biology of Fomes annosus with particular reference to East Anglian pine plantations. II. Spore production, stump infection, and saprophytic activity in stumps. Ann Bot 15:1–21
Rishbeth J 1951b Observations on the biology of Fomes annosus with particular reference to East Anglian pine plantations. III. Natural and experimental infection of pines, and some factors affecting severity of the disease. Ann Bot 15:221–246
Rishbeth J 1970 The role of basidiospores in stump infection by Armillaria mellea. In: Tousson TA, Bega RV, Nelson PE (eds) Root diseases and soilborne pathogens. Univ Cal Press, Berkeley, 141–146
Rittinger PA, Biggs AR, Peirson DR 1987 Histochemistry of lignin and suberin deposition in boundary layers formed after wounding in various plant species and organs. Can J Bot 65:1886–1892
Rockwood DL 1973 Monoterpene-fusiform rust relationships in loblolly pine. Phytopathology 63:551–553
Rockwood DL 1974 Cortical monoterpene and fusiforme rust resistance relationships in slash pine. Phytopathology 64:976–979
Rogers IH 1967 A review of the wood and bark extractives of spruces. For Prod Lab, Vancouver, Inf Rep VP-X16
Rudman P 1962 The causes of natural durability in timber. IX. The antifungal activity of heartwood extractives in a wood substrate. Holzforschung 16:74–77
Rudman P 1963 The causes of natural durability in timber. XI. Some tests of the fungitoxicity of wood extractives and related compounds. Holzforschung 17:54–57
Rudman P 1965 The causes of natural durability in timber. XVIII. Further notes on the fungitoxicity of wood extractives. Holzforschung 19:57–58
Russel CE, Berryman AA 1976 Host resistance to the fir engraver beetle. I. Monoterpene composition of Abies grandis pitch blisters and fungus-infected wounds. Can J Bot 54:14–18
Schuck HJ 1982 Monoterpenes and resistance of conifers to fungi. In: Heybroek HM, Stephan BR, Weissenberg K (eds) Resistance to diseases and pests in forest trees. PUDOC, Wageningen, 169–175
Shain L 1967 Resistance of sapwood in stems of loblolly pine to infection by Fomes annosus. Phytopathology 57:1034–1045
Shain L 1971 The response of sapwood of Norway Spruce to infection by Fomes annosus. Phytopathology 61:301–307
Srivastava LM 1964 Anatomy, chemistry, and physiology of bark. Int Rev For Res 1:203–277
Thomas HE 1934 Studies on Armillaria mellea (Vahl) Quél., infection, parasitism and host resistance. J Agr Res 48:187–218
Vance CP, Kirk TK, Sherwood RT 1980 Lignification as a mechanism of disease resistance. Annu Rev Phytopathol 18:259–288
Wallis GW 1976 Growth characteristics of Phellinus (Poria) weirii in soil and on root and other surfaces. Can J For Res 6:229–232

Wallis GW, Reynolds G 1962 Inoculation of Douglas fir roots with Poria weirii. Can J Bot 40:637–645

Wellons JD, Krahmer RL 1973 Self bonding in bark composites. Wood Sci 6:112–122

Woeste U 1956 Anatomische Unterschungen über die Infektionswege einiger Wurzelpilze. Phytopathol Z 26:225–272

Woo JY, Martin NE 1981 Hypha-like haustoria of the haploid stage of Cronartium ribicola in Pinus monticola observed with the scanning electron microscope. Eur J For Pathol 11:193–199

Woodward S, Pearce RB 1988a The role of stilbenes in the resistance of Sitka spruce (Picea sitchensis (Bong.) Carr.) to entry of decay fungi. Physiol Mol Plant Pathol 33:127–149

Woodward S, Pearce RB 1988b Wound-associated responses in Sitka spruce root bark challenged with Phaeolus schweinitzii. Physiol Mol Plant Pathol 33:151–162

Yde-Anderson A 1979 Lachnellula willkommii – canker formation and the role of microflora. A literature review. Eur J For Pathol 9:347–355

Zimmerman W, Seemüller E 1984 Degradation of raspberry suberin by Fusarium solani f.sp. pisi and Armillaria mellea. Phytopathol Z 110:192–199

Chapter 5

Anatomical Responses of Xylem to Injury and Invasion by Fungi

R.A. Blanchette

5.1 Introduction

Anatomical responses of xylary cells to injury and subsequent invasion by microbes are profound changes that represent a dynamic sequence of events facilitating tree defense. These morphological transformations are only a part of the multifaceted alterations that occur within the xylem of living trees. Although this chapter deals specifically with anatomical changes, it is important to consider all aspects of tree defense mechanisms so that a comprehensive understanding of the processes involved is obtained. Biochemical, microenvironmental, and physiological changes, presented in other chapters of this book, must also be integrated for a complete analysis of tree response mechanisms. Anatomical responses in the xylem clearly function for protection. It is debatable whether this defense system is produced as an active mechanism for resistance (Shigo and Marx 1977, Shigo 1984), a nonspecific response designed to maintain sapwood function (Boddy and Rayner 1983, Rayner 1986), or a combination of the two depending on the type of wound and invading organism. Previous investigators have suggested that there may be a dual function for the mechanisms of protection; to prevent the entrance of air and loss of moisture as well as a barrier to microorganisms (Sucoff et al. 1967, Sharon 1974, Shain 1979, von Aufsess 1984). Recently, Liese and Dujesiefken (1990) have suggested a modified interpretation of the compartmentalization process that modifies Shigo's model (Shigo and Marx 1977, Shigo 1984) to include wound responses that also protect against desiccation and disfunction due to the inflow of air (Boddy and Rayner 1983, Rayner 1986, Rayner and Boddy 1988). Whatever the mechanism in operation, the end result is an effective system that limits decay in living trees.

5.2 Anatomy of Sound Wood

Several texts on wood technology have been published that provide extensive information concerning the anatomy of sound wood (Panshin and de Zeeuw 1970, Core et al. 1979, Carlquist 1988). The relationship of woody cell structure to microbial colonization and degradation has also been reviewed recently (Eriksson et al. 1990). The micromorphological aspects of sound wood, therefore, are only briefly presented in this chapter to provide a general overview so that comparisons can be made with altered cell structures.

Gymnosperm wood is relatively homogeneous in structure and consists primarily of tracheids and ray parenchyma cells (also axial and epithelial parenchyma in some genera) (Esau 1977) (Figs. 5.1a–c). Tracheids are closed cells that overlap adjacent cells. Bordered pits are found throughout the tracheid but are usually numerous on the radial plane and at the ends of the cells that overlap other tracheids. Pits may be simple, half-bordered, or bordered (Panshin and de Zeeuw 1970). A cross-section of a bordered pit (Fig. 5.1c) shows the presence of a pit membrane within the pit chamber. When functional, the pit membrane is positioned in the center of the chamber. The circular inner region of the membrane is the torus surrounded by a network of radiating strands called the margo. The margo is porous, allowing fluids to move through it as well as substances that are less than $0.2\,\mu m$ in diameter (Frey-Wyssling 1968). Resin ducts, axial and radial, are found in the woods of some genera, such as *Picea*, *Pinus*, and *Larix*.

Dicotyledonous wood consists largely of vessels, fibers, and parenchyma cells (Fig. 5.1d–f). Large variation in number and distribution of cell types exists among genera of dicotyledons. Diffuse and ring porous species exhibit major anatomical differences. In diffuse porous wood, such as *Betula*, the vessels are relatively uniform in size and are distributed throughout a growth ring, whereas ring porous wood, such as in *Quercus* and *Ulmus*, has extremely large vessels in the earlywood. Ring porous wood conducts water in the outermost growth ring at a speed ten times as fast as that in diffuse porous woods (Esau 1977). Parenchyma cells function in the radial transport of assimilates and store carbohydrate reserves. Ray cells are connected to adjacent xylary elements by pits. Axial and ray parenchyma located next to vessels are called contact cells. These cells deliver carbohydrates to the vessels for transport during active growth periods (Sauter et al. 1973).

In angiosperms and gymnosperms, secondary changes occur in wood, resulting in heartwood formation. The transformation of sapwood to heartwood is accompanied by the necrosis of parenchyma cells, depletion of nutrient reserves, and formation of cell deposits that impart a darkening of the xylem (Shigo and Hillis 1973, Hillis 1987). The amount and rate of heartwood formation differ among different species and the dark color is not always easily discernible. Extractives accumulate in the heartwood that consist of polyphenols and other secondary compounds (Hillis 1987). Heartwood differs from discolored wood in that heartwood is age-altered wood, whereas discolored wood is injury-altered wood (Shigo 1986a).

5.3 Changes in Preexisting Xylem After Wounding

The first macroscopically visible change that occurs in the extant xylem is the formation of discolored wood. Injury-induced discoloration is a relatively rapid response with most anatomical changes occurring within a few weeks. The first color changes are often lighter in color than sound sapwood, but these cells become rapidly darker than the sapwood (Shigo 1986b). Discolored wood con-

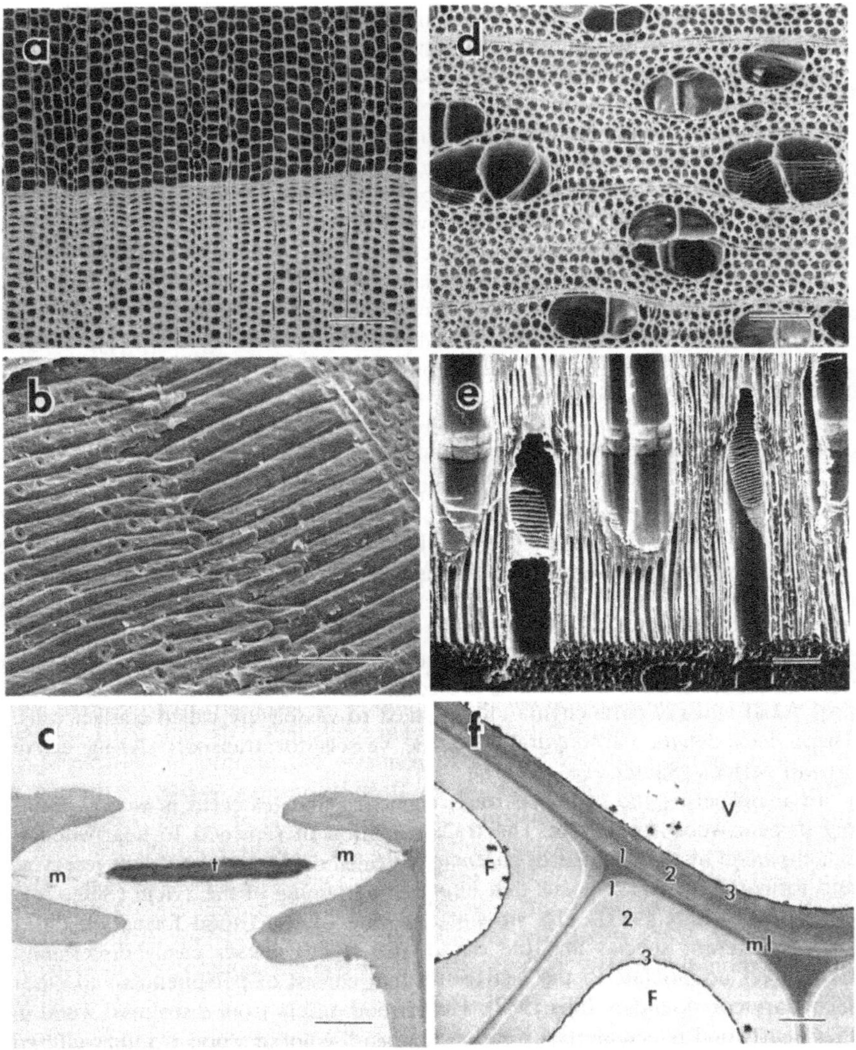

Fig. 5.1a–f. Sound wood typical of gymnosperms (*Pinus*) (**a** to **c**) and a diffuse porous angiosperm (*Betula*). **a** Latewood and earlywood tracheids. **b** Ends of tracheids showing overlapping cells. Tracheids are connected by bordered pits. **c** Section through a bordered pit showing pit aperture, margo (*m*) and torus (*t*). **d** Vessels, fibers, and ray parenchyma cells can be seen. **e** Vessels with scalariform perforation plates connecting one vessel element to another surrounded by fibers and parenchyma cells. **f** Secondary cell wall layers (*1*, *2*, and *3*) and middle lamellae (*ml*) in fibers (*F*) and vessel (*V*). **a,d** Transverse section, SEM; **b,e** radial section SEM; **c,f** transverse section, TEM. *Bar* = 100 μm in **a,b,c,d**; 2 μm in **c,f**

sists of xylary cells with occlusions that fill cell lumina. Concomitant with cell occlusion, starch granules disappear from parenchyma cells (Wardell and Hart 1970, Sharon 1974, Blanchette and Sharon 1975). High respiratory activity has been found associated with parenchyma cells in the immediate vicinity of the wound. Wardell and Hart (1970) observed formazan deposits from reduced triphenyl tetrazolium chloride (TTC), indicating actively metabolizing cells, in the injured sapwood of *Quercus bicolor* 4–6 days after wounding. No cells were capable of reducing TTC 12 days after wounding. Sharon (1974), using nicotinamide adenine dinucleotide, showed intense diformazan deposits following wounding in parenchyma cells in contact with vessels. Deposition occurred in contact cells proximal to the wound in all annual rings of sapwood traversed by the wound. Thirteen days after wounding, diaphorase activity in the cells decreased and all cellular activity ceased after the 18th day. This loss of activity coincided with the depletion of starch from parenchyma cells and the accumulation of vessel plugs (Sharon 1974). As diaphorase activity decreased with time after wounding, vessel occlusions increased in number. In reaction zones of conifers, all ray and epithethial parenchyma failed to reduce nitro-blue tetrazolium, indicating they were no longer alive (Shain 1967). Starch was depleted and phenols, fatty acids, and resin acids increased.

The production of pigmented substances also corresponds to the degeneration of cell nuclei and cytoplasm (Sucoff et al. 1967, Wardell and Hart 1970). Two days after wounding of *Populus tremuloides*, some nuclei in cells near the wound were reduced in size and at 8 days many ray parenchyma cells had degenerated nuclei and were dead (Sucoff et al. 1967). The length of time cells actively metabolize after wounding apparently varies with the species of tree, severity of the wound, and type of injury (Hart 1965, Wardell and Hart 1970, Basham 1978).

The amount and intensity of discolored wood that develops after wounding is dependent upon the size and depth of the wound within the xylem. Drill wounds that bore into the xylem have appreciably more discoloration than surface wounds (Shigo and Sharon 1968, Basham 1978, Shortle and Cowling 1978, Shortle 1979a,b). In all cases, interactions of xylary tissues with air appear necessary for discoloration to occur. In studies where the xylem was injured without breaking the bark, no xylem discoloration occurred (Sucoff et al. 1967). The presence of microorganisms in the xylem also is not necessary for discoloration processes (Sharon 1974). However, the presence of bacteria and fungi may intensify the formation of cell occlusions, resulting in greater amounts of discoloration (von Aufsess 1975, Blanchette and Sharon 1975, Shortle et al. 1978, Blanchette 1982c).

Discoloration can be delayed or inhibited when various chemicals are placed in contact with exposed xylem. Paraformaldehyde has been used in tapholes of *Acer saccharum* to increase and prolong the yield of sap. When paraformaldehyde is used, ray parenchyma cells are killed quickly and the normal discoloration processes, including the formation of vessel occlusions, are altered (Walters and Shigo 1978a,b). Five-year-old wounds that received paraformaldehyde treatment had more decay than nontreated wounds. Trees injected with various compounds (e.g. elms injected to control Dutch elm disease) have also shown altered patterns of discoloration that ultimately lead to greater internal defects (Shigo and Campana 1977, Andrews et al. 1982). In *Pinus* species, paraquat has been used for many years to increase the oleoresin flow from wounds made in the xylem. Paraquat applications result in an immediate conversion of starch and synthesis of oleoresin (Hillis 1987). All affected parenchyma exude copious amounts of oleoresin into contiguous tracheids. This increased oleoresin flow occurs during a period of gradual parenchyma cell lysis (Brown et al. 1976).

The time of year that wounds occur and the species of tree can influence the degree of discoloration. Wounds made in late autumn or winter had reduced amounts of discoloration compared with those made during the growing season (Wardell and Hart 1970, Shigo 1976, Basham 1978). Significantly less discoloration was found in stems of *Populus deltoides* clones wounded at the end of a growing season (August) than in those wounded in May (Shain and Miller 1988). However, discoloration associated with wounds in May were significantly less than those associated with wounds made in November. This study is similar to an investigation of wounded *Acer rubrum*, where greater volumes of discolored wood were associated with wounds made in autumn than in wounds made in spring (Leben 1985). In this study, greater phloem dieback occurred around the autumn wounds, thereby increasing the wound size. The larger area of exposed xylem apparently resulted in larger volumes of discolored wood.

Moisture loss from the tree is also an important consideration that directly influences bark dieback and withdrawal of water from the xylem. Loss of moisture and replacement with air in xylem can affect discoloration processes and subsequent invasion by fungi (Chap. 6), (Boddy and Rayner 1983, Rayner 1986). The extent to which air will enter the xylem can vary at different times of the year depending upon whether water columns are under positive or negative hydrostatic pressure when the xylem is wounded (Zimmermann 1983, Rayner 1986). These differences appear to be important considerations when the season of wounding is being examined.

When wounded xylem from different tree species was examined and compared, the extent of discoloration was found to differ among the tree species examined (Shigo and Sharon 1968, Shigo 1986b). Bauch et al. (1980) showed that the anatomical structure of a tree species and the functional behavior of the xylem cells after wounding influenced the amount of discoloration. The longitudinal extent of discoloration was greater in *Birch* than in *Acer*. The vessels of *Betula* have relatively small pit apertures connecting adjacent contact cells and therefore remain free of occlusions. Accessory compounds, however, accumulated in ray parenchyma and were found in fibers that have larger pit apertures. In vessels, only the scalariform perforation plates had encrustations. In *Acer*, however, vessel occlusions were frequent and substances apparently diffused readily from rays through large pits into the lumina of vessels. The intense initial reaction of *Acer* was suggested to restrict the area of discoloration, whereas in *Betula* a weaker response was observed, allowing greater volumes of discolored wood to occur (Bauch et al. 1980).

In another histological study, greater vessel occlusions were observed in birch wood after inoculation with *Inonotus obliquus* compared to noninoculated wounds, indicating that subsequent colonization of wounds of aggressive decay fungi can influence the intensity of vessel plugging (Blanchette 1982c). The amount of accessory compounds produced, their location within cells of the xylem, and rate of production after injury all appear to influence defense barriers that function to restrict loss of moisture from the tree, entrance of air into the xylem, and subsequent colonization by microorganisms. Small areas of intense discolorations in the wood that occlude all xylary cells immediately adjacent to the injury appear to be more effective barriers than diffuse zones of occlusions that occupy large volumes of wood around an injury (Fig. 5.2).

When wood becomes discolored, many changes take place in addition to physical alterations. Discolored wood has higher pH and higher moisture, ash, and mineral contents (Hart 1965, Shigo and Sharon 1970, Tattar et al. 1971, Safford et al. 1974, Shortle 1979a,b). Phenols also increase and the altered wood becomes resistant to colonization by most Basidiomceyetes that cause decay (Hart and Johnson 1970, Shigo and Sharon 1970, Tattar et al. 1971, Shigo 1974, Shortle and Cowling 1978, Shortle 1979a,b). Phenol-tolerating organisms, however, are often found associated with discolored xylem (Shigo 1967, Tattar et al. 1971, Shortle 1979b, Blanchette et al. 1981). Other factors, such as the high mineral content of discolored wood can also influence the growth of microorganisms (Shigo 1970a, 1974, Shigo and Hillis 1973).

Discolored wood is often made up of gums and gels as well as tyloses (Fig. 5.3). Tyloses consist of hypertrophied parenchyma cell walls that balloon out into the lumen of an adjacent vessel. They are not only common in vessels of certain species of angiosperms after wounding, but can also be found in large earlywood vessels of nonwounded xylem (Carlquist 1988). Tylosis formation appears to result from loss of water pressure in vessels, and is characteristically found in vessels that fill with air. They also form as a response to vascular infection and chemical stimuli (Beckman 1964). Tyloses can subsequently develop secondary thickenings, and when large or abundant, they may completely occlude vessels (Dimond 1970). Although tyloses may impede movement of propagules within vessels, slow the colonization process, and localize the infection (Beckman 1964), many decay fungi can grow readily through these structures (Fig. 5.3e,f).

5.4 Compartmentalization

All of the processes that occur in xylem extant at the time of wounding function to form barriers that resist adverse effects and protect the living tree (Figs. 5.2 and 5.4). These barriers were recognized by Hartig (1878) and Hepting (1935), and studied by Dr. Alex Shigo and others in numerous investigations (Chap. 1) (Shigo 1984) to determine how trees form barriers to isolate injured tissues and resist pathogens. An expanded concept of discoloration and decay in living trees has been proposed as the compartmentalization of decay in trees concept (Shigo 1970b, 1976, 1984, Shigo and Hillis 1973, Shigo and Marx 1977). This conceptual model, called CODIT, has two parts. One refers to events that take place within

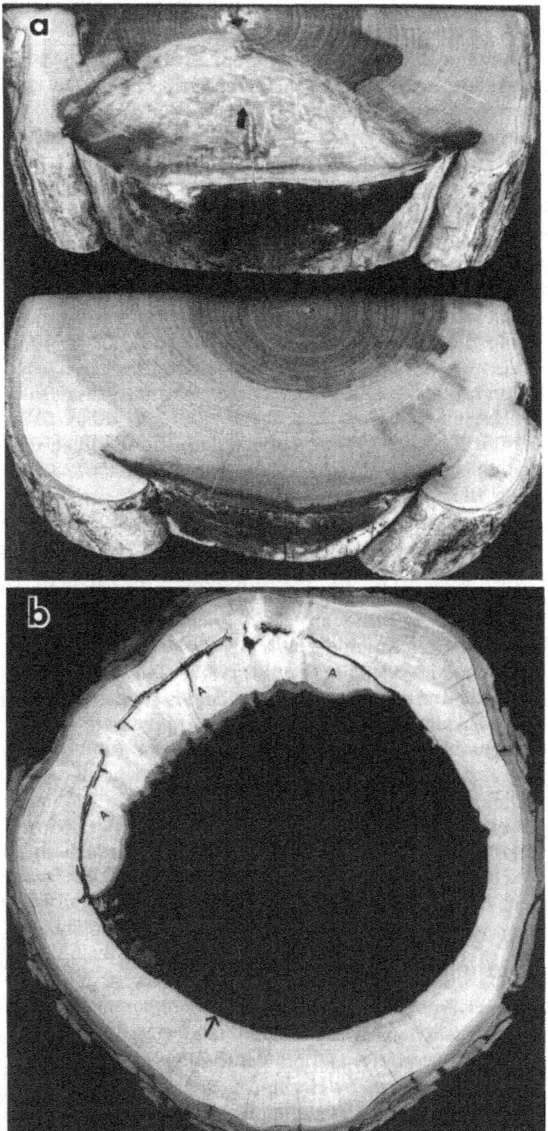

Fig. 5.2a,b. Compartmentalization of decay in trees. **a** Cross-sections from two trees (*Acer rubrum*) with large wounds. The top disk shows weak compartment formation with extensive decay present in wood behind the wound (*arrow*). Bottom disk shows strong compartment formation and no decay. **b** A hollow within a tree due to degradation of all xylem present at the time of wounding. A barrier zone (*arrow*) separates the new wood formed after wounding. The new xylem forms a protective barrier to prevent entrance of decay fungi. Areas around the wound have new callus tissue that has enrolled into the hollow (*A*) (Shigo 1986b)

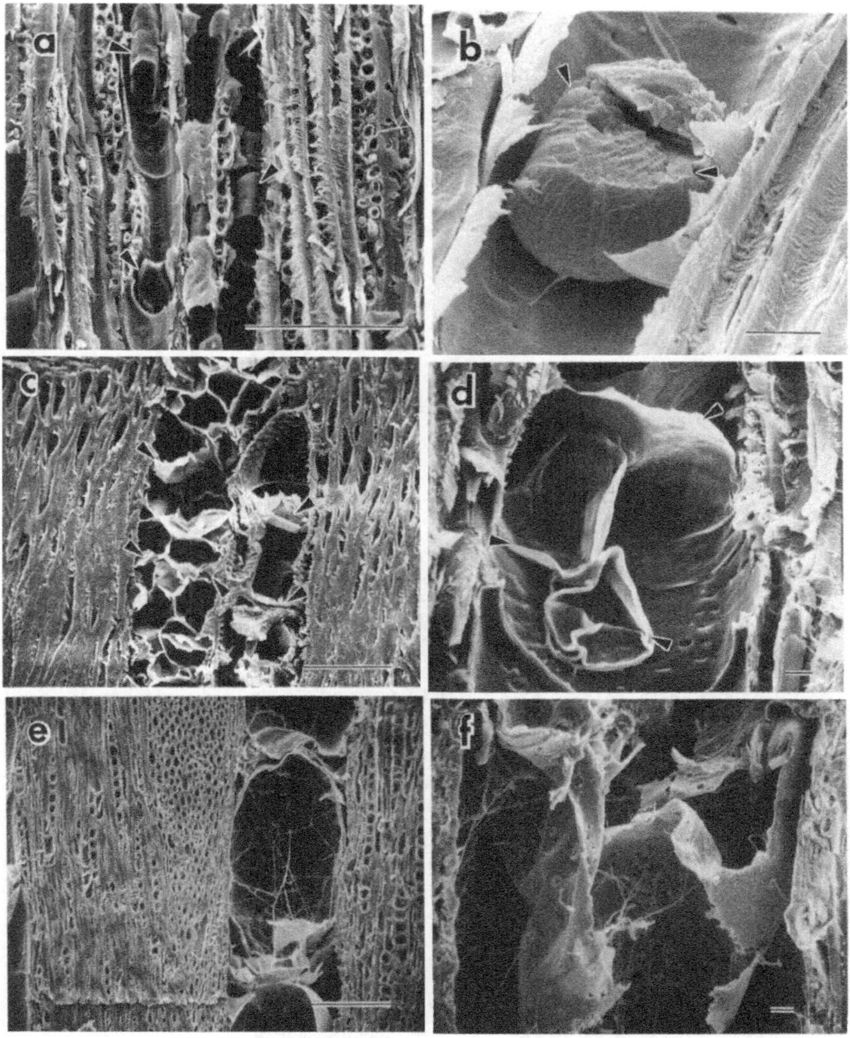

Fig. 5.3a–f. Vessel occlusions and tyloses in discolored wood. **a,b** Vessel plugs (*arrowheads*) in discolored wood of *Malus*. Vessel occlusions fill lumina of vessels obstructing movement of fungal hyphae and spores. **c,d** Tyloses (*arrowheads*) in vessels of discolored *Ulmus* wood. Multiple tyloses fill lumina of vessels. Collapsed tyloses (**d**), due to preparation techniques, show distribution in vessel lumen. **e,f** Tyloses in *Quercus* showing fungal hyphae of the decay fungus, *Inonotus dryophilus*, colonizing vessels and penetrating the tyloses. Tangential sections, SEM. *Bar* = 100 μm in **a,c,e**; 10 μm in **b,d,f**. (**a,b** from Blanchette 1979, **e,f** from Otjen and Blanchette 1982)

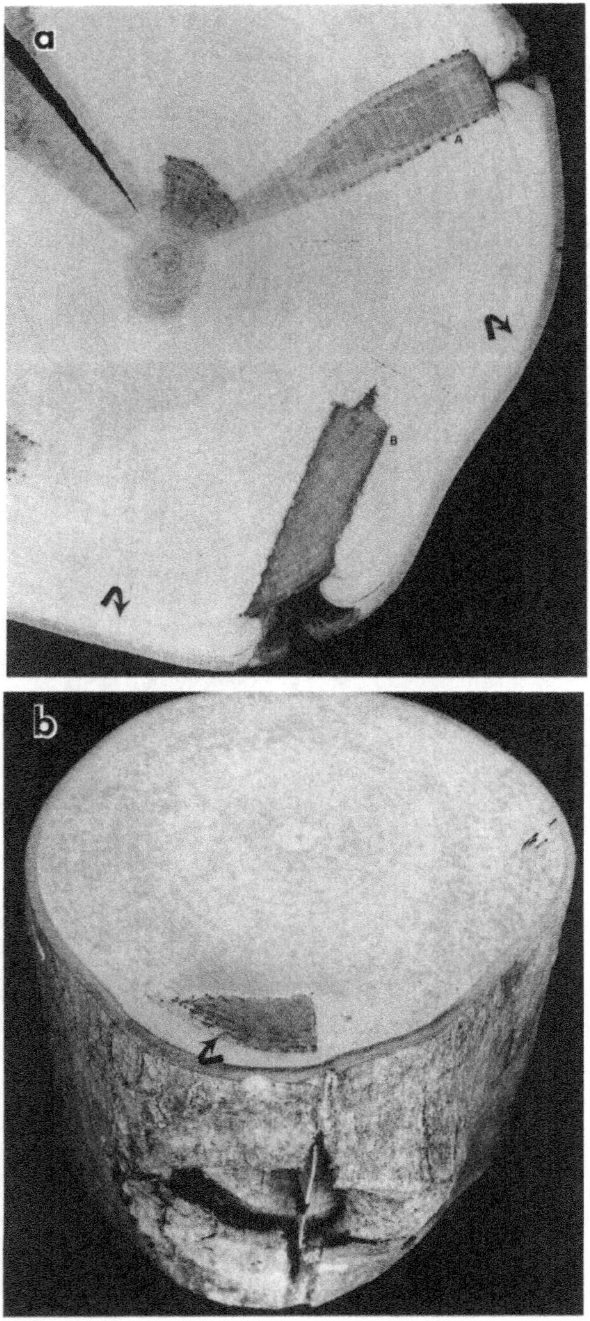
a
A
B
b

the xylem at the time of wounding and the other to the cells formed after wounding.

The reaction zone, in cells present within the wood at the time of injury or infection, results from processes, such as those previously discussed in this chapter, associated with discoloration of xylem (Figs. 5.2 and 5.4). Three barrier walls, or boundaries, serve to compartmentalize the affected zone in all planes. Vertical walls form from occlusions in longitudinal xylary elements, tangential movement is hindered by the annual rings within the xylem (i.e., latewood cells serving as a barrier to movement toward the pith), and side walls appear to follow ray parenchyma cells to the pith, preventing radial movement within the tree. These barrier walls are not always effective and alternating "breakout" and new compartment formation occurs (Shigo 1984). Boundaries that form in injured trees always follow the anatomical features of the tree. As Shigo (1979, 1984) states "a unique feature of trees is their highly ordered, woody, compartmented structure." Investigations of dissected trees with various injuries provided visual evidence of compartmentalization (Shigo 1965), and a framework for the conceptual processes that have been described and illustrated to help understand the system (Shigo 1979, Shigo and Marx 1977).

Part 2 of the CODIT model, or the fourth wall, is the boundary established by cells formed after the injury. This region effectively separates the new tissues from those existing at the time of injury. Few pathogens or parasitic microorganisms can penetrate this barrier (Shigo 1984).

There is significant variation among tree species and within a species in the production of barriers that effectively compartmentalize injuries and infections. Studies involving clones of *Populus* and *Liquidambar* have shown that compartmentalization is under genetic control (Shigo et al. 1977, Garrett et al. 1979, Shain and Miller 1988) (Fig. 5.2). A recent investigation suggests that it is possible to screen trees for superior compartmentalizing genotypes (Shain and Miller 1988). In this study, the production of ethylene produced by wounded trees during different times of the year was proposed as an indicator of the better compartmentalizing clones. Assays for ethylene production could be done rapidly and large numbers of genotypes could be screened in comparison to the arduous tasks of wounding experiments and dissections that would take years to complete.

Fig. 5.4a,b. Compartmentalization of wounded xylem. **a** Drill wounds in *Acer saccharum* showing two compartmentalized wounds. Wound *A* has increased amounts of discolored wood where the discoloration processes induced by the new wound have joined an older internal column. Wound *B* shows the compartment formation in the shape of the drill bit. New xylem formed after injury (*arrows*) forms a protective barrier. **b** A cross-section of *Acer rubrum* above a wound. A metal plate (*arrow*) was forced into the wound separating it into two equal parts. The left side was treated, whereas the right side served as a control. The treated half of the wound had cambial dieback and increased discoloration. The right side of the wound shows good wound closure and no discoloration above the wound. Area of new xylem formed after wounding is indicated by a *curved arrow*. Treatments that affect the wound responses of the tree may cause increased cambial death, prevent wound closure, and increase internal discoloration and decay processes (Leben et al. 1982, Shigo 1986b)

5.5 Formation of the Barrier Zone

When a tree is wounded, the cells generated after wounding close the wound (Figs. 5.2 and 5.4). Callus tissue is formed around the margins of the wound by the vascular cambium. Callus forms rapidly during active growth periods for the first few years after injury, but then decreases (Shigo 1986b). Small wounds may close within one to several years, whereas large wounds take decades to close completely. There is no indication that wound closure is directly correlated with the amount of discoloration and decay in the wood beneath the wound (Shigo 1986a, Shain and Miller 1988). However, complete wound closure should inhibit microbial processes dependent upon gaseous exchange with the open wound.

Investigations of anatomical characteristics of xylem formed after wounding have shown the barrier zone to consist of abnormal cells. One of the first studies of this zone (Hepting and Blaisdell 1936) demonstrated that in fire-wounded trees a zone of new xylem formed, separating the decayed wood beneath the wounds from the cells that formed after wounding. The zone served to protect the new cells from decay. Histological investigations have shown that in angiosperms, the first 10 to 50 cells formed after wounding were irregularly shaped and thick-walled (Sharon 1973, Blanchette and Sharon 1975, Moore 1978, Pearce and Rutherford 1981, Rademacher et al. 1984).

Greater numbers of ray and axial parenchyma cells and fewer fibers and vessels were present in the barrier zone compared with extant xylem. The drastic change in cell composition was clearly shown in a study by Rademacher et al. (1984). In *Acer*, xylem normally consisted of 15% vessels, 15% parenchyma, and 70% fibers, whereas a zone of cells formed after wounding consisted of 95% parenchyma 5% vessels, and no fibers. The length and width of cells formed after wounding was also altered. The disoriented axial cells had distorted shapes and were significantly reduced in length (Scurfield 1966, Moore 1978, Rademacher et al. 1984). However, ray parenchyma cells increased in width (Rademacher et al. 1984).

In some tree species, wounding initiated the production of traumatic gum canals, such as those produced in barrier zones of *Liquidambar* (Moore 1978) and *Eucalyptus* (Wilkes 1985, 1986, Tippett 1986). In *Eucalyptus*, the gum canals or veins are called kino veins. These structures can form in the phloem or may be included in the xylem (Tippett 1986). Kino veins develop by periclinal cell division in the cambial zone following injury. Broad bands of parenchyma cells are produced and differentiation of lacunae occurs. The veins are filled with a reddish-brown exudate that consists of various polyphenols. An interesting aspect of sapwood with kino veins is that it may not darken upon heartwood formation, leaving areas of included sapwood within the heartwood (Wilkes 1986). The barrier zone in *Eucalyptus* inhibits subsequent heartwood developmental processes by apparently disrupting the centripetal translocation of carbohydrates that are precursors to heartwood phenols (Wilkes 1986). A general characteristic of barrier zones in all trees is the nonconducting nature of the zones, which not only inhibits translocation inward from the outer sapwood and cambium, but also interferes with vertical transport (Mulhern et al. 1979).

Histochemical investigations of barrier zones have shown the presence of pigmented materials within parenchyma cells (Sharon 1973, McGinnes et al. 1977, Moore 1978), and these zones have high phenol contents (Shortle 1979a). Extracts made from barrier zones have been found to inhibit growth of decay fungi (Moore 1978). In addition, axial parenchyma cells formed after wounding have been found to contain suberin (Pearce and Rutherford 1981, Pearce and Holloway 1984). Histochemical tests and chemical analyses have confirmed the presence of suberin in xylem cells of *Quercus* formed after wounding. The suberized zone was shown to be resistant to attack by *Stereum gausapatum*, a fungus that usually can invade and decay oak sapwood (Pearce and Rutherford 1981). Suberin production has been frequently associated with antimicrobial defense reactions of plants (Aist 1976, Akai and Fukutomi 1980, Biggs 1984, 1987, Rittinger et al. 1987), and its presence in the barrier zone of xylem of *Quercus* implicates suberization as a general component of barrier zones in trees. Suberin within these cells would impart additional resistance to microbial colonization. Additional investigations of barrier zones in different species of trees are needed to show that suberized, traumatic parenchyma cells are common. It has already been demonstrated that suberized impervious tissues occur in the phloem and extant xylem of many tree species after wounding (Biggs 1985, 1987).

The combined chemical and morphological barriers produced in this protective zone after wounding are most prevalent immediately adjacent to the wounded area. As the distance increases from the injured zone, cells are of normal morphology and chemical composition. The return to normal xylary tissue occurs more quickly circumferentially than longitudinally (Moore 1978, Shortle 1979b). It appears that effective barriers are dependent on the quantity of cells making up the barrier zone, spatial distribution around the wood, and the quality of components produced within the zone. The factors that regulate the size and effectiveness of the barrier are not well understood.

Barrier zones also are produced in conifers (Fig. 5.5a,b). The cells formed after wounding contained distorted tracheids, abundant axial parenchyma cells that accumulate phenols and tangential bands of traumatic resin ducts (Fahn and Zamski 1970, Tippett and Shigo 1981, Blanchette 1982b, Tippett et al. 1982, Shigo 1986b). Barrier zones form in response to mechanical wounds as well as in response to xylem injury caused by pathogens (Tippett and Shigo 1981, Blanchette 1982a,b, Tippett et al. 1982) (Fig. 5.5a,b). Parenchyma cells in barrier zones are larger than those in extant xylem and appear compacted among misaligned tracheids. The tracheids associated with wounds formed after pathogen attack often have characteristics of cells usually found in compression wood. Tracheids exhibit rounded outlines containing thick secondary wall layers (S_2) with helical cavities and no S_3 layer, intercellular spaces, and a high lignin content (Timell 1986). Many fungal pathogens, such as *Sphaeropsis sapinea*, and various *Cronartium* species, cause profound changes to occur in the phloem, cambial zone, and xylem of the host (Blanchette 1982a, Timell 1986) (Chap. 12). In cells that are formed in the xylem after pathogen attack, tracheids are replaced by parenchyma cells and dark deposits form in the rays (Marks and Minko 1969, 1970). Timell (1986) in his treatise on compression wood, noted

a

b

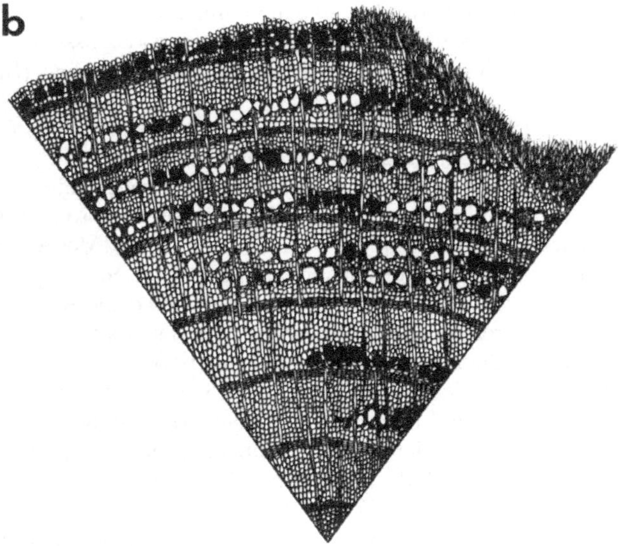

Fig. 5.5. a Cross-section of white fir with perennial canker caused by *Phellinus pini* var. *cancriformans*. **b** Illustration drawn from a composite of transverse sections obtained from wood within the circled area in **a**. Tangential series of resin ducts and parenchyma cells are present in xylem formed after injury. Repeated attacks by the fungus resulted in numerous barrier zones formed in response to the injuries. Each time the cambium was killed, a series of resin ducts were formed in the new xylem adjacent to the canker (Blanchette 1982b)

that one of the most remarkable features of xylem adjacent to infected tissue was the presence of tracheids with characteristics of compression wood. The formation of compression wood has also been associated directly with injury by the balsam woolly aphid (*Adelges piceae*), and indirectly to injury by many other forest insects (Timell 1986).

The formation of traumatic resin ducts after mechanical wounding or injury from pathogenic fungi occurs in all conifers after injury. In genera such as *Abies*, *Tsuga*, and *Cedrus* resin ducts are not formed in normal xylem, but these trees produce tangential series of resin ducts in xylem formed after injury (Bannan 1936, Blanchette 1982b, Tippett et al. 1982) (Fig. 5.5a,b). Investigations of traumatic resin ducts have shown them to have the same anatomy as those produced intermittently in normal xylem of *Pinus*, *Picea*, and *Larix* (Fahn and Zamski 1970). Traumatic resin ducts are usually localized to the area around the wound, and the number of ducts and their distribution are influenced by the severity of the wound (Bannen 1936, Fahn and Zamski 1970). Experiments with auxins on xylem development showed IAA, NAA, and 2,4-D to induce and enhance vertical resin duct formation (Fahn and Zamski 1970).

The formation of traumatic resin ducts can occur repeatedly if injury reoccurs in subsequent years. In a study of canker formation by *Phellinus pini* var. *cancriformans*, one or more series of traumatic resin ducts formed each growing season after the fungus invaded the cambial region (Blanchette 1982b). The barriers formed after wounding are usually effective in stopping decay fungi from moving into the new xylem, but in the case of canker rot fungi that reinvade from the phloem side of the wood (Blanchette 1982b,c, Shigo 1986b, Enebak and Blanchette 1989) (Chap. 3), the barriers are not positioned in an appropriate place to afford protection, or cannot withstand the mechanical pressure generated by hyphal aggregations.

5.6 Xylem Decay Processes

Compartmentalization processes in trees produce barriers that resist colonization by decay fungi. The morphological and chemical barriers that are produced, however, do not make the tree immune to microbial colonization. As the wound remains open, microorganisms become established and microenvironmental conditions change, resulting in compartment failure (Fig. 5.2). New barriers are then established at greater depths and distances from the wound once again to compartmentalize the defect. Wood that becomes discolored immediately after wounding is not conducive for most wood-destroying fungi to proliferate. The high phenol content, increased pH, high concentrations of microelements, and physical obstructions in xylary cells are not suitable for colonization and rapid growth of fungi in the Basidiomycotina (Shigo and Hillis 1973, Shortle 1979a,b, Mercer 1982, Shigo 1984). Although some Basidiomycotina may be present in wounded tissues and trigger increased host defense reactions (Shortle and Cowling 1978, Shortle et al. 1978, Mercer 1982, Rizzo 1986), other fungi, considered pioneer species (Shigo 1967), such as *Phialophora melinii*, are not affected by the tree's nonspecific defense mechanisms or defense responses initiated by Basidiomycotina (Shigo 1977). In addition to *Phialophora*, many different species of bacteria and nondecay-causing fungi have been isolated from discolored wood (Shigo 1967, Shigo and Hillis 1973, Shortle and Cowling 1978, Roll-Hansen and Roll-Hansen 1980, Blanchette et al. 1981, Mercer 1982). As

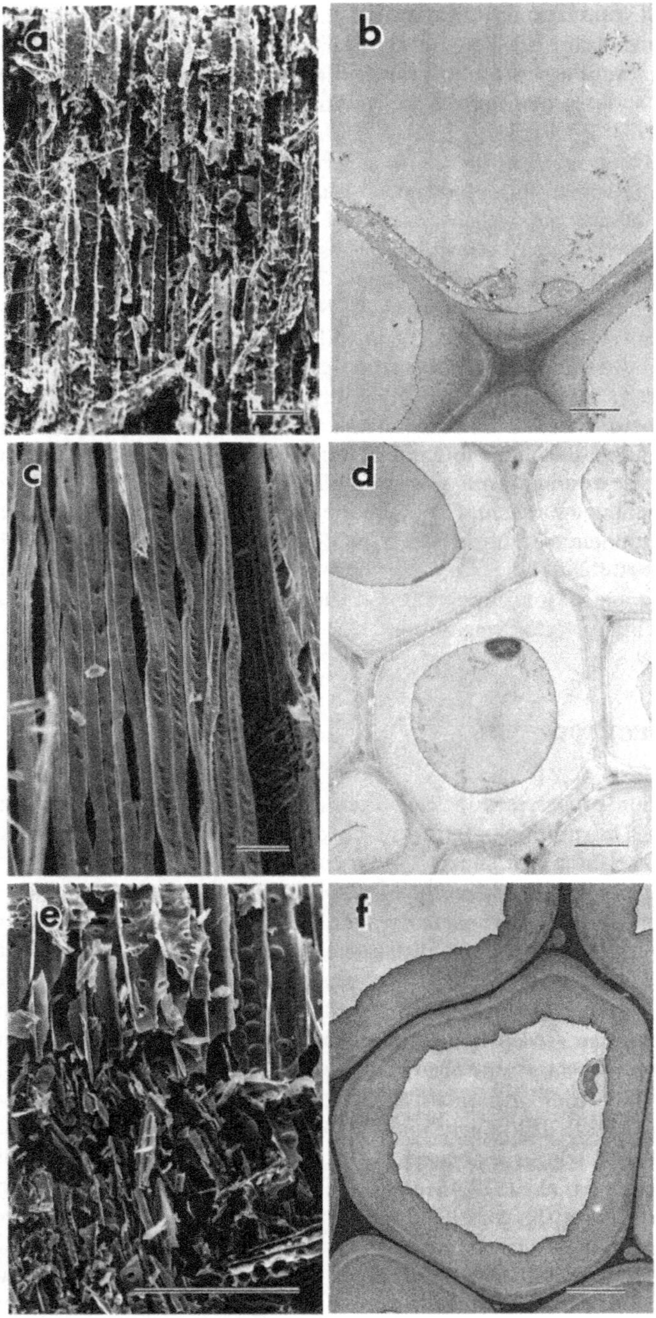
a
b
c
d
e
f

these organisms grow in discolored xylem, they may alter the substrate and detoxify the reaction zone as well as remove physical obstructions (Shigo 1967, 1977, Blanchette 1979, Merrill and Shigo 1979, Blanchette et al. 1981). Once conditions are suitable for growth of Basidiomycotina, these decay-causing fungi will inevitably become established. The sequence of events and changes in microbial populations appear to follow a succession (Shigo 1967). Various theories have been proposed concerning the order that microorganisms follow during the succession in discolored wood (i.e., bacteria and nonhymenomycete followed by decay organisms or decay organisms followed by bacteria and nonhymenomycetes) (Shigo 1967, Shortle and Cowling 1978, Mercer 1982). Another view of the successional sequence is that the first fungi to enter wounded xylem are those that have prolific spore production and are able to grow on easily assimilable substrates, followed by more specific communities containing combative organisms which replace the pioneer species (Rayner and Boddy 1988). The first fungi to appear are those that get there first and they may not have special abilities with regard to detoxifying tree protective chemicals (Boddy and Rayner 1983). It is also possible for some species of Basidiomycotina to enter wounds as pioneer colonizing species. This is frequently observed for canker rot fungi such as *Inonotus obliquus* and *Cerrena unicolor* (Blanchette 1982c, Enebak and Blanchette 1989), and appears to be true for other pathogens such as *Armillaria*, *Heterobasidion*, and *Phellinus* (Shain 1967, Blanchette 1982b, Rizzo 1986, Rayner and Boddy 1988). Wood-destroying Ascomycotina, such as *Hypoxylon* and *Xylaria*, may also aggressively colonize xylary tissues.

When wood-destroying fungi enter the xylem, and conditions are appropriate for growth, cell wall decomposition begins as soon as most easily assimilated nutrients are depleted (Adaskaveg and Ogawa 1990). The morphological characteristics of cell wall degradation are different among various species of fungi (Fig. 5.6a–f). Two major types of degradation found in standing trees are white and brown rots. White rot fungi have the capacity to degrade all cell wall components, resulting in complete erosion of the cell wall or preferential degradation of one component over another. Fungi that preferentially degrade lignin often cause a white-pocket rot form of decay, consisting of localized areas of white zones that consist primarily of cellulose. A great deal of variation exists among the decay patterns produced by white rot fungi, depending on the amount of lignin, hemicellulose, or cellulose that is ultimately degraded (Otjen

←

Fig. 5.6a–f. Wood degradation by white and brown rot fungi. **a,b** Eroded cell walls caused by a white rot that simultaneously degraded all cell wall layers. Cell walls were progressively degraded from the lumen toward the middle lamellae. Erosion troughs coalesced to form large holes within the cell walls. **c,d** Selective degradation of lignin from cell walls by the white rot fungus, *Phellinus pini*. The middle lamellae have been degraded, resulting in a separation of cells. The remaining cell walls consisted of delignified secondary wall layers. **e,f** Wood decayed by brown rot fungi consisted of cells that were extremely fragile. Slight pressure applied to the wood resulted in collapse and fragmentation of the cells. Cell walls appeared swollen and consisted primarily of lignin. **a,c,e** Tangential section, SEM; **b,d,f** transverse section, TEM. *Bar* = 100 μm in **a,c,e**; 3 μm in **b,d,f** (Blanchette 1980, Blanchette et al. 1990) and Blanchette 1986, Blanchette 1987, Eriksson et al. 1990). Brown rot fungi degrade the polysaccharide portions of the cell but do not degrade appreciable amounts of lignin. Modification of lignin, however, does occur. Characteristics of both white and brown rot decay of wood are shown in Fig. 5.6a–f.

Wood also may be degraded by soft rot fungi or by bacteria. These types of organisms are not commonly associated with cell wall decay in wood of standing trees but their mode of action and degradation patterns have been extensively studied in relation to deterioration of wood products. (Blanchette 1987, Blanchette et al. 1990, Eriksson et al. 1990).

References

Adaskaveg JE, Ogawa JM 1990 Wood decay pathology of fruit and nut trees in California. Plant Dis 74:341–352

Aist JR 1976 Papillae and related wound plugs of plant cells. Annu Rev Phytopathol 14:145–163

Akai S, Fukutomi M 1980 Preformed internal physical defenses. In: Horsfall JG, Cowling EB (eds) Plant diseases: an advanced treatise, vol 5. Academic Press, New York London, 139–159

Andrews MW, Blanchette RA, French DW 1982 Effects of benzimidazole compounds for Dutch elm disease control on wood surrounding elm injection sites. Plant Dis 66:495–498

Bannan MW 1936 Vertical resin ducts in the secondary wood of the Abietinae. New Phytol 35:11–46

Basham JT 1978 Early sugar maple stem discoloration and microorganism invasion in simulated wounds of felling and fire scars. Phytopathology 68:1693–1699

Bauch J, Shigo AL, Starck M 1980 Wound effects in the xylem of Acer and Betula species. Holzforschung 34:153–160

Beckman CH 1964 Host responses to vascular infection. Annu Rev Phytopathol 2:231–252

Biggs AR 1984 Barrier zone formation in peach bark in response to wounds and Cytospora leucostoma infection. Can J Bot 62:2814–2821

Biggs AR 1985 Detection of impervious tissue in tree bark with selective histochemistry and fluorescence microscopy. Stain Technol 60:299–304

Biggs AR 1987 Occurrence and location of suberin in wound reaction zones in xylem of 17 tree species. Phytopathology 77:718–725

Blanchette RA 1979 A study of progressive stages of discoloration and decay in Malus using scanning electron microscopy. Can J For Res 9:464–469

Blanchette RA 1980 Wood decomposition by Phellinus (Fomes) pini: a scanning electron microscopy study. Can J Bot 58:1496–1503

Blanchette RA 1982a Phellinus (Fomes) pini decay associated with sweetfern rust in sapwood of jack pine. Can J For Res 12:304–310

Blanchette RA 1982b Decay and canker formation by Phellinus pini in white and balsam fir. Can J For Res 112:538–544

Blanchette RA 1982c Progressive stages of discoloration and decay associated with the canker-rot fungus, Inonotus obliquus, in birch. Phytopathology 72:1272–1277

Blanchette RA 1987 New directions in forest products pathology. Can J Plant Pathol 9:361–369

Blanchette RA, Nilsson T, Daniel G, Abad A 1990 Biological degradation of wood. In: Rowell, RM, Barbour RJ (eds) Archaeological wood: properties, chemistry and preservation. Advances in chemistry series 225. American Chemical Society, Washington, DC, 141–174

Blanchette RA, Sharon EM 1975 Agrobacterium tumefaciens, a promoter of wound healing in Betula alleghaniensis. Can J For Res 5:722–730

Blanchette RA, Sutherland JB, Crawford DL 1981 Actinomycetes in discolored wood of living silver maple. Can J Bot 59:1–7

Boddy L, Rayner ADM 1983 Origins of decay in living deciduous trees: the role of moisture content and a reappraisal of the expanded concept of tree decay. New Phytol 94:623–641

Brown CL, Clason TR, Michael JL 1976 Paraquat induced changes in reserve carbohydrates, fatty acid and oleoresin content of young slash pines. Lightwood Res Coor Counc Proc Annu Meet Jacksonville FL, 8–19

Carlquist S 1988 Comparative wood anatomy: Systematic, ecological, and evolutionary aspects of dicotyledon wood. Springer, Berlin Heidelberg New York, 436 pp

Core HA, Côté WA, Day AC 1979 Wood structure and identification. Univ Press, Syracuse, 182 pp

Dimond AE 1970 Biophysics and biochemistry of the vascular wilt syndrome. Annu Rev Phytopathol 8:301–322

Enebak SA, Blanchette RA 1989 Canker formation and decay in sugar maple and paper birch infected by Cerrena unicolor. Can J For Res 19:225–231

Eriksson K-E, Blanchette RA, Ander P 1990 Microbial and enzymatic degradation of wood and wood components. Springer, Berlin Heidelberg New York, 407 pp

Esau K 1977 Anatomy of seed plants. John Wiley & Sons, New York, 550 pp

Fahn A, Zamski E 1970 Influence of pressure, wind, wounding, and growth substances on the rate of resin duct formation in Pinus halepensis wood. Isr J Bot 19:429–446

Frey-Wyssling A 1968 The ultrastructure of wood. Wood Sci Technol 2:73–83

Garrett PW, Randall WK, Shigo AL, Shortle WC 1979 Inheritance of compartmentalization of wounds in sweetgum (Liquidambar styraciflua L.) and eastern cottonwood (Populus deltoides Bartr.). US For Serv Res Pap NE-443, 4 pp

Hart JH 1965 Formation of discolored sapwood in three species of hardwoods. Mich State Univ Agr Exp Stn Q Bull 48:101–116

Hart JH Johnson KC 1970 Production of decay-resistant sapwood in response to injury. Wood Sci Technol 4:267–272

Hartig R 1878 Die Zersetzungserscheinungen des Holzes der Nadelholzbäume und der Eiche. Springer, Berlin Heidelberg New York, 151 pp

Hepting GH 1935 Decay following fire in young Mississippi Delta hardwoods. US Dep Agr Tech Bull 494, 32 pp

Hepting GH, Blaisdell DJ 1936 A protective zone in red gum fire scars. Phytopathology 26:62–67

Hillis WE 1987 Heartwood and tree exudates. Springer, Berlin Heidelberg New York, 268 pp

Leben C 1985 Wound occlusion and discoloration columns in red maple. New Phytol 99:485–490

Leben C, Shigo AL, Hall TJ 1982 A method for evaluating tree wound treatment materials. Can J For Res 12:115–117

Liese W, Dujesiefken D 1990 Principles of compartmentalization in trees. IAWA Bull 11:129–130

Marks CC, Minko G 1969 The pathogenicity of Diplodia pinea to Pinus radiata D Don. Aust J Bot 17:1–12

Marks CC, Minko G 1970 The resistance of Pinus radiata to infection by Macrophoma pinea. Aust J Bot 18:55–65

McGinnes EA Jr, Phelps JE, Szopa PS, Shigo AL 1977 Wood anatomy after tree injury–a pictorial study. Res Bull Missouri Agr Exp Stn, Columbia, 1025

Mercer PC 1982 Basidiomycete decay of standing trees. In: Frankland JC, Hedges JN, Swift MJ (eds) Decomposer basidiomycetes: their biology and ecology. Univ Press, Cambridge, 143–160

Merrill W, Shigo AL 1979 An expanded concept of tree decay. Phytopathology 69:1158–1160

Moore KE 1978 Barrier-zone formation in wounded stems of sweetgum. Can J For Res 8:389–397

Mulhern J, Shortle WC, Shigo AL 1979 Barrier zones in red maple: an optical and scanning microscope examination. For Sci 25:311–316

Otjen L, Blanchette RA 1982 Patterns of decay caused by Inonotus dryophilus (Aphyllophorales: Hymenochaetaceae), a white-pocket rot fungus of oaks. Can J Bot 60:2770–2779

Otjen L. Blanchette RA 1986 A discussion of microstructural changes in wood during decomposition by white rot Basidiomycetes. Can J Bot 64:905–911

Panshin AJ, de Zeeuw C 1970 Textbook of wood technology, vol 1, McGraw-Hill, New York, 705 pp
Pearce RB, Holloway PJ 1984 Suberin in the sapwood of oak (Quercus robur L.): its composition from a compartmentalization barrier and its occurrence in tyloses in undecayed wood. Physiol Plant Pathol 24:71–81
Pearce RB, Rutherford J 1981 A wound-associated suberized barrier to the spread of decay in the sapwood of oak (Quercus robur L.) Physiol Plant Pathol 19:359–369
Rademacher P, Bauch J, Shigo AL 1984 Characteristics of xylem formed after wounding in Acer, Betula, and Fagus. IAWA Bull 5:141–151
Rayner ADM 1986 Water and the origins of decay in trees. In: Ayres PG, Boddy L (eds) Water, fungi and plants. Univ Press, Cambridge, 321–341
Rayner ADM, Boddy L 1988 Fungal decomposition of wood. Its biology and ecology. John Wiley & Sons, New York, 587 pp
Rittinger PA, Biggs AR, Peirson DR 1987 Histochemistry of lignin and suberin deposition in boundary layers formed after wounding in various plant species and organs. Can J Bot 65:1886–1892
Rizzo DM 1986 Wind damage and root disease of red spruce and balsam fir in the subalpine zone of the white mountains of New Hampshire. MS Thesis, Univ New Hampshire, Durham, 164 pp
Roll-Hansen F, Roll-Hansen H 1980 Microorganisms which invade Picea abies in seasonal stem wounds. I. General aspects. Hymenomycetes. Eur J For Pathol 10:321–339
Safford LO, Shigo AL, Ashley M 1974 Gradients of cation concentrations in discolored and decayed wood of red maple. Can J For Res 4:435–440
Sauter JJ, Iten W, Zimmermann MH 1973 Studies on the release of sugar into the vessels of sugar maple. Can J Bot 51:1–8
Scurfield G 1966 The cankers of Acacia mucronata Willd. ex H. Windl. Aust J Bot 14:293–301
Shain L 1967 Resistance of sapwood in stems of loblolly pine to infection by Fomes annosus. Phytopathology 57:1034–1045
Shain L 1979 Dynamic responses of differentiated sapwood to injury and infection. Phytopathology 69:1143–1147
Shain L, Miller JB 1988 Ethylene production by excised sapwood of clonal eastern cottonwood and the compartmentalization and closure of seasonal wounds. Phytopathology 78:1261–1265
Sharon EM 1973 Some histological features of Acer saccharum wood formed after wounding. Can J For Res 3:83–89
Sharon EM 1974 An altered pattern of enzyme activity in tissues associated with wounds in Acer saccharum. Physiol Plant Pathol 4:307–312
Shigo AL 1965 The pattern of decay and discoloration in northern hardwoods. Phytopathology 55:648–652
Shigo AL 1967 Successions of organisms in discoloration and decay of wood. Int Rev For Res 2:237–299
Shigo AL 1970a Growth of Polyporus glomeratus, Poria obliqua, Fomes igniarius, and Pholiota squarosa-adiposa in media amended with manganese, calcium, zinc and iron. Mycologia 62:604–607
Shigo AL 1970b An expanded concept of decay in living trees. Laval Univ Bull 13:7–9
Shigo AL 1974 Effects of manganese, calcium, zinc, and iron on growth and pigmentation of Trichocladium canadense, Phialophora melinii, Hypoxylon rubiginosum, Daldinia concentrica and Cytospora decipiens. Mycologia 66:339–341
Shigo AL 1976 Compartmentalization of discolored and decayed wood in trees. Mat Org Suppl 3:221–226
Shigo AL 1977 Phialophora melinii: inoculations in wounded red maple. Phytopathology 67:1333–1337
Shigo AL 1979 Tree decay: an expanded concept. US For Serv Inf Bull 419, 73 pp
Shigo AL 1984 Compartmentalization: a conceptual framework for understanding how trees grow and defend themselves. Annu Rev Phytopathol 22:189–214
Shigo AL 1986a A new tree biology dictionary. Terms, topics, and treatments for trees and their problems and proper care. Shigo and Trees Assoc, Durham, New Hampshire, 132 pp

Shigo AL 1986b A new tree biology, facts photos, and philosophies on trees and their problems and proper care. Shigo and Trees Assoc, Durham, New Hampshire, 595 pp

Shigo AL, Campana R 1977 Discolored and decayed wood associated with injection wounds in American elm. J Arboricult 3:230–235

Shigo AL, Hillis WE 1973 Heartwood, discolored wood, and microorganisms in living trees. Annu Rev Phytopathol 11:197–222

Shigo AL, Marx HG 1977 Compartmentalization of decay in trees. US For Serv Agr Inf Bull 405

Shigo AL, Sharon EM 1968 Discoloration and decay in hardwoods following inoculations with Hymenomycetes. Phytopathology 58:1493–1498

Shigo AL, Sharon EM 1970 Mapping columns of discolored and decayed tissue in sugar maple, Acer saccharum. Phytopathology 60:232–237

Shigo AL, Shortle WC, Garrett PW 1977 Genetic control suggested in compartmentalization of discolored wood associated with tree wounds. For Sci 23:179–182

Shortle WC 1979a Compartmentalization of decay in red maple and hybrid poplar trees. Phytopathology 69:410–413

Shortle WC 1979b Mechanisms of compartmentalization of decay in living trees. Phytopathology 69:1147–1150

Shortle WC, Cowling EB 1978 Development of discoloration, decay and microorganisms following wounding of sweetgum and yellow-poplar trees. Phytopathology 68:609–616

Shortle WC, Menge JA, Cowling EB 1978 Interaction of bacteria, decay fungi, and live sapwood in discoloration and decay of trees. Eur J For Pathol 8:293–300

Sucoff E, Ratsch H, Hook DD 1967 Early development of wound initiated discoloration in Populus tremuloides Michx. Can J Bot 45:649–656

Tattar TA, Shortle WC, Rich AE 1971 Sequence of microorganisms and changes in constituents associated with discoloration and decay of sugar maple infected with Fomes connatus. Phytopathology 61:556–558

Timell TE 1986 Compression wood in gymnosperms, vol 3. Springer, Berlin Heidelberg New York, 2150 pp

Tippett JT 1986 Formation and fate of kino veins in Eucalyptus L'Herit. IAWA Bull 7:137–143

Tippett JT, Bogle AL, Shigo AL 1982 Response of balsam fir and hemlock roots to injuries. Eur J For Pathol 12:357–364

Tippett JT, Shigo AL 1981 Barriers to decay in conifer roots. Eur J For Pathol 11:51–59

von Aufsess H 1975 Über die Bildung einer Schutzsperre an der Astbasis von Laub- und Nadelbäumen und ihre Wirksamkeit gegen das Eindringen von Pilzen in das Kernholz lebender Bäume. Forstwiss Cbl 94:140–152

von Aufsess H 1984 Some examples of wood discolorations related to mechanisms for potential protection of living trees against fungal attack. IAWA Bull 5:133–138

Walters RS, Shigo AL 1978a Tapholes in sugar maples: what happens in the tree. US For Serv Tech Rep NE-47, 12 pp

Walters RS, Shigo AL 1978b Discoloration and decay associated with paraformaldehyde-treated tapholes in sugar maple. Can J For Res 8:54–60

Wardell JF, Hart JH 1970 Early responses of sapwood of Quercus bicolor to mechanical injury. Can J Bot 48:683–686

Wilkes J 1985 Host attributes affecting patterns of decay in a regrowth eucalypt forest. I. Patterns of natural decay. Holzforschung 39:17–22

Wilkes J 1986 Host attributes affecting patterns of decay in a regrowth eucalypt forest. VI. Included sapwood. Holzforschung 40:79–83

Zimmermann MH 1983 Xylem structure and the ascent of sap. Springer, Berlin Heidelberg New York, 143 pp

Chapter 6

Microenvironmental Aspects of Xylem Defenses to Wood Decay Fungi

L. Boddy

6.1 Introduction

In nonwoody plants there is considerable evidence that living host cells can respond to fungal invasion and are frequently able to prevent or limit subsequent colonization. Similarly, recent concepts of decay in living trees (e.g. Shigo and Marx 1977, Shigo 1979, Shortle 1979) have implied an active host defense against infection in sapwood and even heartwood (Shigo and Shortle 1979). It is not disputed that living cells in wood can and do respond to tissue damage, and perhaps to fungal invasion, but such responses may be non-specific and may serve to mainatain sapwood function. Moreover, few or no living cells are present in heartwood, and even in sapwood living cells usually account for less than 10% of volume (Hillis 1977), thus it is likely that xylem defense does not rely specifically and directly upon host response mechanisms but on indirect mechanisms.

It has been argued recently that development of decay in living angiosperms can best be explained on the basis that functional sapwood is unsuitable for mycelial establishment owing to its high moisture content and lack of easily assimilable nutrients, and that mechanisms which maintain sapwood function will necessarily prevent establishment of mycelia of decay and stain fungi (Boddy and Rayner 1983b, Rayner 1986, Rayner and Boddy 1988). These and other microenvironmental aspects of xylem defense will be considered in this chapter.

6.2 Xylem Microenvironment

The living tree imposes considerable pressures militating against invasion of its interior, due primarily to limitation of accessibility because of the presence of physical barriers, chemical and ionic barriers, general lack of readily available nutrients, water saturation, and adverse gaseous regimes. These are discussed briefly below.

6.2.1 Resource Composition

6.2.1.1 Xylem Anatomy: Routes of Access

The first priority of a decay fungus is to gain access to the nutrients held within wood. In order to achieve this, the fungus has to pass through external barriers and then grow within the wood. Once within the wood, spread is achieved within the natural system of passages provided by the various cellular elements in the wood cylinder. These passages are provided by the axial and radial systems of interconnecting cells, which distribute water and nutrients within the xylem of the living tree. A brief discussion is presented below; for further consideration, see Rayner and Boddy (1988).

The distribution, size, and degree of interconnection between these cells are crucial in determining patterns of colonization, since, although wood-decay fungi have the capacity for directly penetrating wood cell walls, the natural tendency is to extend most rapidly along avenues of least resistance. Thus, colonization patterns might be expected to be different, but predictable, in different trees. For example, axial spread is likely to be particularly rapid in angiosperms, where numerous long vessels (up to 15–18 m in *Quercus* and *Fraxinus* (Greenidge 1952)) provide effective channels. In ring-porous wood, the wider vessels in the earlywood facilitate axial and, to a lesser extent, tangential colonization in this region. Diffuse porous wood, with an even distribution of vessels, may limit penetration at early stages. By contrast, in conifers tracheids form the main axial system and, since they are usually only 3–4 mm long, they provide a less effective pathway for spread. This may, however, be offset partly by their lack of separation with interstial tissues, ordered arrangement, and regular alteration of size in earlywood and latewood, in a manner analogous to ring porosity.

The rays provide the major system for radial access, although limited radial access is afforded by pits. When dead, ray cells initially provide a source of readily assimilable nutrients and a direct radial route for colonization. Tangential spread is usually slower than in axial or radial directions since, in most woods, there is no ordered system of elements in this direction. Pits, and to a limited extent ring porosity, can facilitate tangential spread to a small degree, and the presence of a dead cambial layer, allowing subcortical spread, can provide a route for rapid tangential spread.

This system of channels in the wood cylinder is sealed from the outside by an outer layer of bark and periderm which contains suberized tissue. This barrier can be transgressed as a result of mechanical injury or by active pathogenic mechanisms, but typically such damage elicits responses that effectively protect undamaged tissue from further disruption. Such responses include the production of tyloses (outgrowths of parenchyma through pits in the lateral walls of vessels) in angiosperms, which may become lignified and suberized, covered in gums and deposits, and occlude the entire lumen, thus restricting axial access of liquids, gases and, incidentally, hyphae (see Chap. 5). Tyloses are commonly formed in heartwood of many angiosperm trees at the boundary of damaged and undamaged tissues, and indeed also when vascular wilt pathogens invade (Chap. 13). In conifers, gums and resins are the major devices for sealing damaged from undamaged conducting tissues (see Chap. 8). Suberized tissues are formed in positions where they effectively take over the role of damaged bark, e.g. abscission zones, callus tissue, necrophylactic periderm (Mullick 1977), and barrier zones that formed after wounding (Pearce and Rutherford 1981, Pearce and Holloway 1984).

6.2.1.2 Nutrients

Readily accessible, assimilable substrates, including soluble sugars, lipids, peptides, and starch, are the only available carbon sources for nondecay microorganisms and are probably crucial for the establishment of decay fungi. However, except in trees with seasonally sugary sap (e.g. *Acer*), they occur in relatively small amounts (<10% by dry weight) and are almost exclusively found in parenchyma cells, which in functional sapwood will be living and hence unavailable. The amounts of these substances may vary seasonally and with location in the wood. For example, starch sometimes occurs maximally in the middle sapwood and declines to zero at the boundary with heartwood (Hillis 1977). Apart from these compounds, the major available carbon sources are the relatively refractory major structural wood cell wall polymers, i.e., cellulose, hemicelluloses, and lignin. Mineral nutrients are also essential for fungal growth, and attention has focused almost exclusively on the distribution and abundance of nitrogen. Nitrogen is generally regarded as being limiting to fungal growth in wood, although, due to its insoluble form, presence in discrete packages, and the sequestering and reutilizing abilities of wood decay fungi, this may not be as severe as sometimes thought (Rayner and Boddy 1988). It varies spatially within the wood cylinder, and is greatest where parenchyma cells are most abundant, namely, in the cambium and immature xylem, where it declines rapidly, and least in inner heartwood. It is also greater in root wood than stem wood of gymnosperms, and in nonwounded than discolored tissues (Platt et al. 1965, Merrill and Cowling 1966, Allen et al. 1974, Basham and Cowling 1975). Nitrogen varies temporally, is redistributed toward the surface upon drying (King et al. 1974), and increases relative to carbon as decomposition proceeds (Swift 1977).

Within the standing tree, spread will depend on accessibility to these substrates and patterns of colonization will be at least partly determined by their location. Access to these nutrients will depend on microclimatic variables, in particular water and gaseous regimes (Sect. 6.2.2), and on the barriers and the interconnected system of channels already described.

6.2.1.3 Extractives

In addition to the carbon compounds and mineral nutrients outlined above, wood contains a wide range of extractives that vary in type, quantity, and distribution according to tree species. These extractives, which can be removed with water or neutral organic solvents, include waxes, fats, fatty acids and alcohols, steroids, higher carbon compounds, and resin mixtures, e.g. terpenes, lignans, stibenes, and flavonoids (Hillis 1962, 1987, Buchanan 1963, Fengel and Wegener 1984, Rowe 1989). Some of them have fungitoxic or fungistatic properties, and, therefore, influence fungal growth and confer decay resistance against some fungal species (Sect. 3.3).

These extractives characteristically are found in heartwood, where they may cause a distinct color change compared with sapwood, and changes in their content can be very abrupt at the heartwood periphery. They are formed at the heartwood transition zone in situ from carbohydrates and lipid substrates via, for example, the acetate and shikimic acid pathways (Hillis 1962, 1971, 1972, 1977). There are sometimes radial gradients in extractive concentration within heartwood and/or localized differences; extractives are found largely in parenchyma cells but can occur also in tracheid, fiber, and vessel lumina, and sometimes in specialized cells (Hillis 1971, 1972). In some trees, e.g. *Eucalyptus*, extractives are broken down progressively in heartwood such that concentrations are least in inner heartwood (Rudman 1964, 1965, 1966).

Extractives also form and accumulate in localized regions in sapwood, particularly at the boundary between damaged and functional tissues (e.g. Jorgensen 1961, Shain 1967, Hart et al. 1975, Shortle 1979; see Chaps. 7 and 8). The formation of gums and resins as sealants has already been mentioned (Sect. 6.2.1.1).

6.2.1.4 Wood Acidity

The pH of wood typically lies between 4 and 6 (Gray 1958), although there can be wide variation between tissues, particularly following discoloration and microbial colonization. For example, uninfected heartwood of oak (*Quercus*) and Douglas fir (*Pseudotsuga menzeisii*) has a lower pH than sapwood (Hartley et al. 1961), and pH tends to be higher in discolored tissues (Good et al. 1955, 1968, Hart 1968, Tattar et al. 1971).

The optimum pH for growth of many wood decay fungi lies between 4 and 6, i.e., the same range as the pH of wood of many species, although how gross measurements of pH actually relate to pH at the hyphal level is uncertain. The pH of wood may nonetheless have selective influences on colonizing fungi. For example, in oak the pH is particularly low (sometimes <pH 3; Packman 1960), and in elm (*Ulmus* spp.) it is relatively high (>pH 7; Gray 1958).

6.2.2 Microclimate: Water and Gaseous Regime

Although temperature is a major determinant of rate of fungal growth, temperatures are unlikely ever to be lethal. Thus, since temperature only affects germination and colonization rate, its role in xylem defense will only be related to the rate of events occurring. Temperature can, however, influence cavitation and rate of gaseous diffusion (Sect. 6.2.2.2).

6.2.2.1 Problems of Measurement and Expression of Water Content

Traditionally, water content has been expressed gravimetrically as a percentage of dry weight, and occasionally as a percentage of fresh weight. The latter term is of little use since the denominator of the expression is not constant for different water contents (see Boddy 1986, Rayner and Boddy 1988). Unfor-

Table 6.1. Average moisture content of sapwood and heartwood of various living trees (Skaar 1972)

Species	Moisture content (% dry weight)	
	Sapwood	Heartwood
Conifers		
Lodgepole pine (*Pinus contorta*)	120	41
Ponderosa pine (*Pinus ponderosa*)	148	40
Western red cedar (*Thuja plicata*)	249	58
Douglas fir (*Pseudotsuga menziesii*)	112–154	30–37
Sitka spruce (*Picea sitchensis*)	142	41
Western hemlock (*Tsuga heterophylla*)	170	85
Grand fir (*Abies grandis*)	136	91
Larch (*Larix*)	119	54
Mean for 27 conifer species	149	55
Angiosperms		
Aspen (*Populus tremuloides and P. grandidentata*)	113	95
Beech (*Fagus grandifolia*)	72	55
Paper birch (*Betula papyrifera*)	72	89
Sugar maple (*Acer saccharum*)	72	65
White oak (*Quercus* spp.)	78	64
American elm (*Ulmus americana*)	92	95
Mean for 34 angiosperm species	83	81

tunately, neither can the former term be used for comparative purposes, except with reference to exactly the same sample, since density may vary. Expression as a proportion of wood volume allows direct comparison between different samples, but woods of different densities have different water-holding capacities, and hence the relationship between water content (as a percentage of volume) and air-filled void space will differ. Thus, for each sample, it is necessary to know how the moisture content term is related to actual availability (water potential), when low water contents obtain, or to rates of gaseous diffusion (see Sect. 6.2.2.2) at high water contents. Unfortunately, this is unclear for the majority of available data, and most estimates are described in terms of percentage moisture content alone. Further, in angiosperms, it is difficult to obtain reliable measurements of functional sapwood moisture content, since severence of vessels under hydrostatic tension leads to immediate withdrawal of water from the damaged elements.

6.2.2.2 Gaseous Regime and Distribution of Water in Xylem

Despite the comments made in the previous section, available data do provide a general indication of the distribution of water in xylem, and demonstrate considerable variation both between and within trees (Table 6.1). Typically, in intact, unstressed boles and branches, there are large differences between the youngest sapwood, which is functional in water conduction and may be virtually saturated, and older tissues which are not (Table 6.1). These differences may

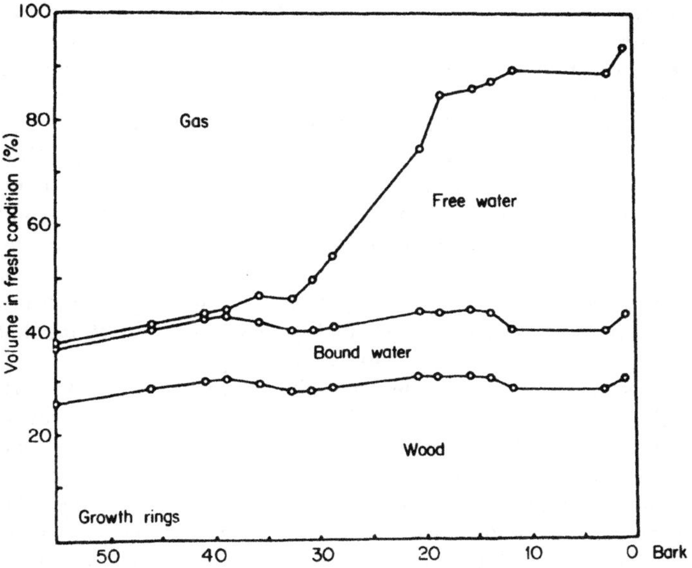

Fig. 6.1. Distribution of gas and water in the xylem of a spruce trunk (Trendelenburg 1939)

develop along a gradient from outer to inner xylem or changes may be abrupt, particularly when there is a distinct heartwood, e.g. in many conifers, where a relatively dry transition zone marks the boundary between dry heartwood and moist sapwood. Differences are often less clear-cut in angiosperms than in conifers, and, in some species, higher moisture contents exist in heartwood than in sapwood (Table 6.1; Skaar 1972, 1988).

The gas phase in wood is related reciprocally to free water. Thus, as wood dries the gaseous phase increases. Hence, due to the fact that under high water regimes it is the gaseous phase which largely influences fungal development (Sect. 6.3.2), a perhaps more useful way of considering differences between tissues is in terms of the percent volume of water, gases, and wood substance. The change in proportions of these components with age is illustrated in Fig. 6.1 for a spruce (*Picea*) trunk, from which it can be seen that an extensive gaseous phase develops in older tissues. Here follows a discussion of development of a gaseous phase in trees, central issues being development of internal positive or negative hydrostatic pressure, size and distribution of conducting pathways, and patterns of gas entry.

Water columns in wood are alternately subject to the imposition and relaxation of hydrostatic tension (Zimmermann 1971, 1983), which probably partly accounts for the marked seasonal differences in moisture content of sapwood, especially in angiosperms; commonly in spring moisture content is high, then it declines during summer and rises again following leaf abscission in autumn (Gibbs 1958). Cavitation (the introduction of gas-filled space into water columns under tension) is considered (Milburn 1979, Jones 1986) to be a common means by which gases enter the wood cylinder of standing dicotyledonous trees. Also, cavitation may be reversed by redissolution of gases when hydrostatic tension is lowered following absorption of water from external sources, and (at least in the lower part of the stem) when positive hydrostatic pressures develop during the spring. Gas arises from a variety of sources, including the freezing of xylem sap, since the formation of ice brings gases out of solution (Scholander 1958). Nucleation (the drawing of gases through pores in cell walls) may also contribute to ingress of gas, although this has not been substantiated for intact wood (Milburn 1979). Wounding is another major cause of access of gases into the wood cylinder (see Sect. 6.2.3). In intact xylem, irrespective of the cause of gaseous phase, air is not thought to pass readily from one conducting element to another, due to the presence of pit tori and small pit membrane apertures that resist passage of an air-water interface due to surface tension effects.

The gaseous phase in xylem may be similar to that of ambient air in vessels, which are directly linked and close to the external environment. However, O_2 levels are depleted and CO_2 levels elevated as a result of microbial and plant cell respiration. Carbon dioxide concentrations of 10–20% are common (e.g. Hintikka and Korhonen 1970), and rigorous studies employing mass spectrometry have demonstrated concentrations approaching 100% in living stems of *Acacia* (Carrodus and Triffett 1975). Other gases formed by microorganisms may be present (Sect. 6.2.4).

6.2.3 Effect of Wounding on Microclimate

The first consequence of mechanical injury extending beneath the vascular cambium is to expose xylem to the air which, unless the wood is submerged in water, results in drying. The extent of drying will depend on many factors, but particularly on whether the water columns within the xylem are subject to positive or negative pressure at the time of severing (Zimmermann 1983, Leben 1985). If the wound is deep and the water columns under tension (i.e., negative pressure), water will be immediately withdrawn from severed vascular elements and air will enter. Air can potentially enter as far as the connecting pit membranes of adjacent water-filled elements, but further passage of the menisci will be prevented by surface tension or pit aspiration. Lateral flow of water, from severed elements in inner regions, toward intact elements in outer regions, may occur. Also, intact elements above (and below) severed elements may be fed by intact lateral connections. Thus, air-accessible tissue takes the form of two inwardly tapering columns, one proximal and one distal to the wound, the latter being somewhat longer due to the upward/outward flow of water (Fig. 6.2). The extent of these dry columns will also depend upon the degree of tension and its duration in water columns, and on the size and distribution of conducting elements. Thus, because of their long vessels, angiosperm tissues are potentially likely to become aerated over greater distances than conifers (Sect. 6.2.1.1).

Shallow wounds and bark death also lead to variable patterns of aeration. Small wounds near the base of the tree may result in only slight peripheral aeration, although, higher up the tree, wedge-shaped patterns of aeration may

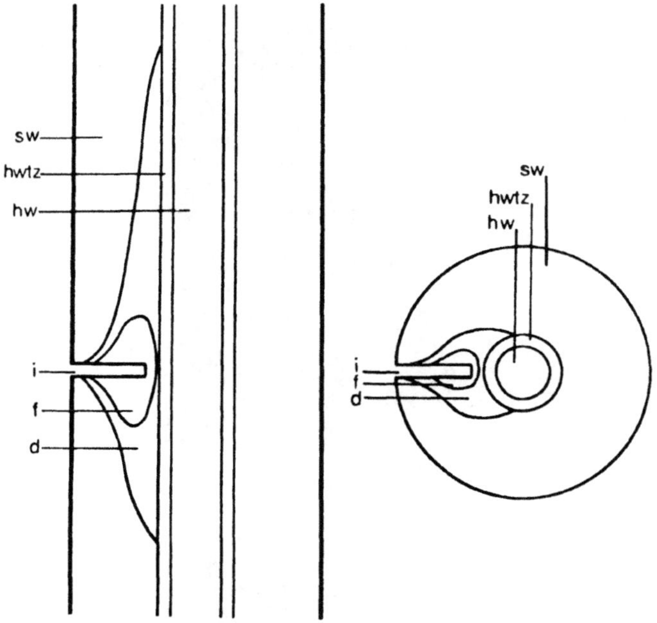

Fig. 6.2. Diagram of longitudinal and transverse sections through a conifer trunk being colonized (*f*) by *Heterobasidion annosum* which was spreading from a dowel inoculation (*i*). *d* Dry zone; *sw* sapwood; *hw* heartwood; *hwtz* heartwood transition zone. (Coutts 1976)

occur as a result of changes in moisture status of these tissues. If bark death/loss is extensive, there is a tendency for the whole wood cylinder extant at the time of damage to become aerated and dried. Regardless of the extent of wounding, aerated tissues eventually become sealed from adjacent xylem by the formation of tyloses and or gums, resins, and suberized layers (Sect. 6.2.1.1; Chaps. 7 and 8). Wood that forms subsequent to wounding remains functional as a result of these suberized barriers (Fig. 6.3). These reactions have sometimes been considered to be a defense response to prevent fungal invasion and, depending on their location, have been given a variety of names (see Sect. 6.3.1 for further discussion).

Patterns of air entry and drying are influenced also by the helical orientation of conducting elements in successive annual rings and their degree of lateral connection. In ring porous trees, e.g. oak, water conduction occurs mainly via the large earlywood vessels of the current year. Thus, the result of water stress at an individual root would be predominantly through an axial strip in the outermost growth ring, although large wounds will have the effects already described. By contrast, in conifers and diffuse porous angiosperms with smaller vessels, where water conduction is through more than one growth ring, aeration may follow more complex patterns, and in arborescent monocotyledons

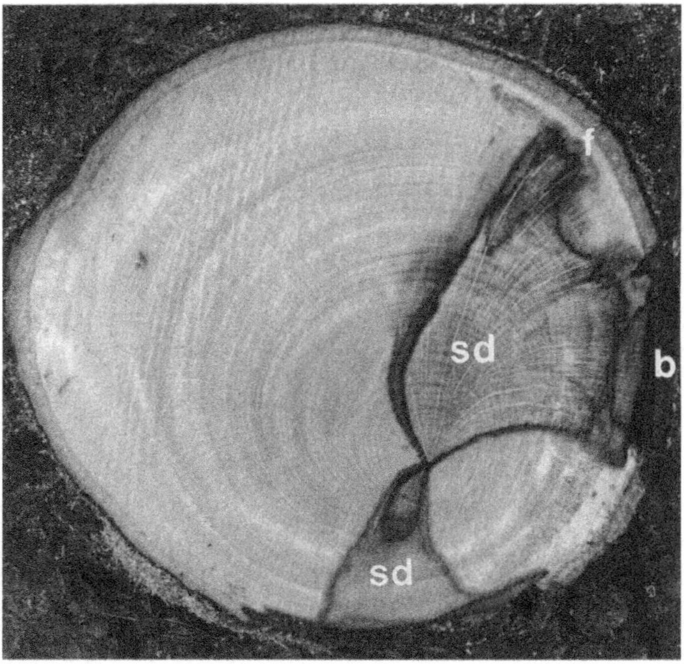

Fig. 6.3. Transverse section through an attached beech (*Fagus sylvatica*) branch (approx. 30 cm diameter) showing regions of stain and decay (*sd*) in the xylem behind dead bark (*b*). Xylem which formed subsequent to wounding remains functional (*f*)

extremely complex patterns of air ingress are likely due to the intricate arrangement and interconnections between the conducting pathways.

6.2.4 Effect of Microbial Colonization on Microclimate

Microbial colonization inevitably has the potential for altering the microenvironment within standing trees as either a direct or an indirect result of metabolic activities. Wood decay fungi cause an elevation in CO_2 and reduction in O_2 concentrations as a result of respiration; water content is increased due to breakdown of complex carbon compounds, and temperature may be elevated slightly in the vicinity of hyphae, although heat is usually rapidly dissipated. Also, extractives and other antifungal compounds produced as a result of wounding (Sect. 6.2.1.3; Chaps. 4, 7, 8, and 9) may be modified, detoxified, or degraded by some fungi. Degradation of the wood cell wall alters the chemical microenvironment and indirectly affects water relations and gaseous regime: porosity is increased, hence gaseous diffusion and wetting and drying are enhanced; also, the water relations of the woody cell walls are altered (Skaar 1972, 1988, Boddy 1986). Some specific effects are considered below.

6.2.4.1 Heartwood

Microbial colonization can have a particularly dramatic effect on the water and gaseous regime in heartwood when the condition known as wetwood develops. "Wetwood" describes any water-saturated, nonliving wood in living trees. It has been found associated with mechanical wounds, branch stubs, butt rots, frost cracks, insect attacks, and mistletoe cankers, and can also occur in intact, apparently healthy trees (Hartley et al. 1961, Etheridge and Morin 1962, Wilcox 1968). In some species of firs, e.g. *Abies grandis* and *A. concolor*, wetwood is the normal condition of the heartwood. Its etiology is probably diverse; wetwood often contains dense bacterial populations and a causal role has sometimes been attributed to them (Carter 1945, Zeikus and Ward 1974, Preece 1977). Such populations contain obligate anaerobes (indicating the highly anoxic conditions) whose activities can result in the formation of methane, e.g. in cottonwood (*Populus deltoides*) (Zeikus and Ward 1974). Further evidence for the hypoxic conditions is provided by the very low rate of O_2 diffusion (on average $35\,ng\,cm^{-2}\,min^{-1}$) in *Abies concolor*, which is similar to that found in waterlogged soils (Armstrong 1967, Worrall and Parmeter 1983).

In *Abies* spp., there is evidence of physiological activity being partly responsible for wetwood formation. The formation of a dry zone, possibly homologous with the heartwood transition zone (Sect. 6.2.2.2), has been implicated (Bauch et al. 1975, Coutts and Rishbeth 1977, Worrall and Parmeter 1982). Negative osmotic potentials, which are probably due to selective accumulation of potasium and calcium ions, and organic acids, have been demonstrated in wetwood relative to sapwood (Worrall and Parmeter 1982). Increased solute concentration would provide a mechanism whereby water would accumulate in tissues, thus contributing to the formation of wetwood.

6.2.4.2 Sapwood

In a number of cases, fungi are responsible directly for drying and aerating tissues or for extending air access that initially was caused by wounding. Canker-forming fungi, bark pathogens, and vascular-wilt fungi are examples of the former (Chaps. 2 and 13; Sect. 6.3.2), whereas the activities of *Heterobasidion annosum* provide a clear example of the latter. Inoculation of *H. annosum* into *Abies grandis* and several other conifer species resulted in the formation of extensive dry zones in advance of infection (Coutts 1976). *H. annosum* had grown to an equal extent (3–5 cm) upwards and downwards by 2 months after inoculation, with maximal extension in the inner sapwood. Tapered dry zones extended considerably in advance of infection, and, like mycelial spread, this was maximal in inner sapwood, but the dry zone progressed to a greater extent upwards than downwards (Fig. 6.2). Colonization and formation of dry zones was similar when *H. annosum* was inoculated into freshly felled logs of *Pinus* species, except that dry zones were of equal extent in both directions in the felled logs. This indicates that the asymmetry is a feature of the standing tree, and is not surprising in view of the earlier discussion of drying following wounding (Sect. 6.2.3).

Further experiments with the logs demonstrated that dry zones formed at least partly as a result of water loss from damaged tracheids by hydrostatic tension, and due to fungal lysis of bordered pit tori extending the range over which water was withdrawn. That an additional factor was also involved was indicated by the fact that drying extended beyond the region of lysis of pit tori; an active role of slowly dying ray parenchyma was indicated – a feature which is reminiscent of the formation of dry heartwood transition zones (Sect. 6.2.2.2).

Drying of large volumes of xylem can result from the pathogenic activities of fungi. Root death, for example as a result of colonization by pathogenic *Armillaria* spp. or *Heterobasidion annosum*, may result in water stress in xylem far removed from the source of infection, e.g. in branches in the canopy, due to the inability of the roots to absorb water. Bark death, resulting from the activities of canker-forming fungi, e.g. *Endothia parasitica* on chestnut and *Dichomera saubinetii* on sycamore maple (*Acer pseudoplatanus*), results in drying of underlying tissues (Sect. 6.3.2). Infection by vascular wilt pathogens, e.g. *Ceratocystis fagacearum* (the oak wilt fungus) and *Ophiostoma ulmi* (the cause of Dutch elm disease), results in loss of sapwood function due to blockage of water conduits by the spreading fungus and by host responses (Chap. 13). Further details and additional examples are provided by Zimmermann (1983).

6.3 Effect of Microenvironment on Development of Fungi in Xylem

6.3.1 Xylem Anatomy – a Major Determinant of Fungal Colonization Patterns

Following damage, colonization is typically localized within the vicinity of wounds (Fig. 6.4). The theory of compartmentalization of decay in trees (CODIT) attempts to explain this by suggesting that living sapwood tissues actively respond to injury and fungal colonization, and, by doing so, directly wall themselves off from further fungal ingress. Both chemical barriers in the form of allelopaths and physical barriers, including gums, tyloses, and suberized layers, are produced (Hart and Johnson 1970, McGinnes et al. 1977, Moore 1978, Tippett and Shigo 1980, 1981, Pearce and Rutherford 1981, Pearce and Holloway 1984, Pearce and Woodward 1986, see also reviews in this volume). Two different terminologies have been used to describe the formation of these physico-chemical barriers and their purported roles.

In one terminology, the barriers to colonization have been described as "walls" of which there are four different types, termed walls 1, 2, 3, and 4 in order of increasing resistance to fungi (Shigo and Marx 1977, Shigo 1979, 1984, Shortle 1979) (Fig. 6.5). Wall 1 consists of gums and tyloses which occlude the axial elements and limit axial spread. Wall 2 is a physical barrier consisting of dense wood layers of growth rings, which limits inward radial spread, but since it is laid down by normal cambial activity it cannot be regarded as induced. Wall 3 is the rays which impose a living (albeit discontinuous) barrier to tangential spread. Wall 4 divides newly formed wood from that extant at the time of

Fig. 6.4. Patterns of stain and decay developing in an oak (*Quercus robur*) branch inoculated with the basidiomycete, *Vuilleminia comedens*. Decay (*d*) is localized within the vicinity of the inoculated wound. Inoculum (*i*) consisted of a 1-cm diameter precolonized disk of oak wood inserted to a depth of 1–2 mm beneath the bark surface

injury, and is a modified (often suberized) growth ring formed by the vascular cambium.

The second descriptive terminology distinguishes between "barrier zones" and "reaction" zones (Shain 1967, 1971, 1979). The former are typically very abrupt demarcation zones with altered anatomy, apparently analagous to wall 4, and the latter are more diffuse colored demarcation regions abutting living sapwood present at the time of injury (Fig. 6.6).

The CODIT concept attempts to explain the spatial distribution of stain and

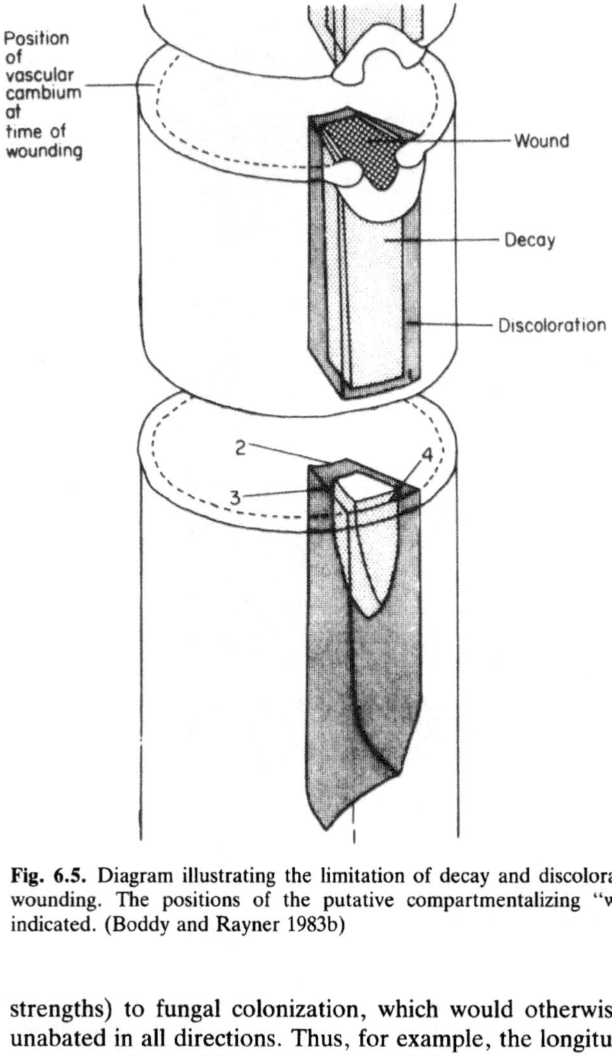

Fig. 6.5. Diagram illustrating the limitation of decay and discoloration in sapwood following wounding. The positions of the putative compartmentalizing "walls" *1*, *2*, *3*, and *4* are indicated. (Boddy and Rayner 1983b)

strengths) to fungal colonization, which would otherwise supposedly continue unabated in all directions. Thus, for example, the longitudinally extended shape of decay columns (Figs. 6.2, 6.4, and 6.5) is supposed to result from the relative weakness of wall 1. The model is based on the transgression of barriers which are actively produced and/or maintained.

However, the positioning and "relative strength" of walls 1, 2, and 3, and hence much of the shape of the decay column, can be explained by purely anatomical considerations and the facilitation of access of fungal hyphae and fluids (Boddy and Rayner 1983b, Rayner and Boddy 1988). Thus, the supposed "weakness" of wall 1 is related to ease of axial access facilitated by the predominantly elongated lumina of xylem elements, including vessels, tracheids, fibers, and axial parenchyma (Sect. 6.2.1.1). Radial access will be enhanced in damaged tissues by the nutrient-enriched, dead ray parenchyma cells, but not by

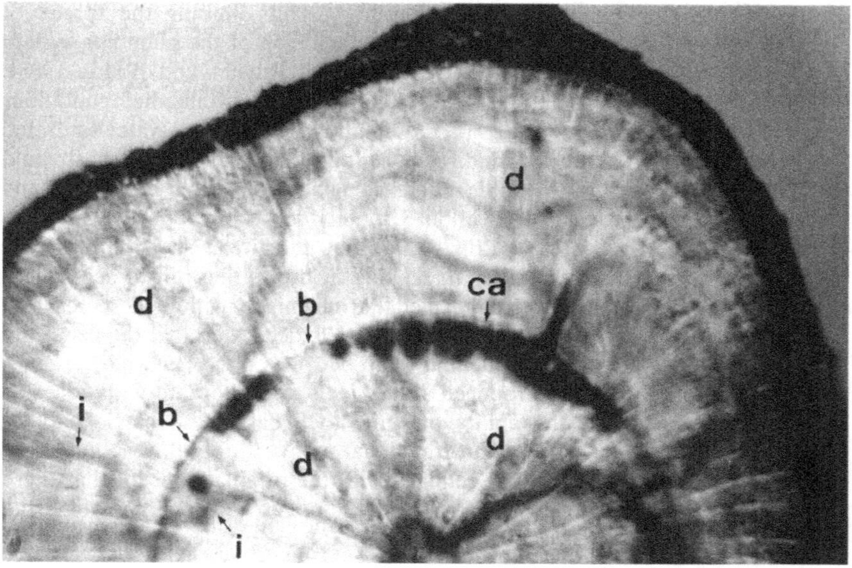

Fig. 6.6. Transverse section through a beech (*Fagus sylvatica*) branch, which had been subject to water stress over a period of several years, illustrating formation and breaching of "barrier" zones. Cankers (*ca*), which were formed several years previously, are associated with a barrier zone (*b*). The subsequently formed wood has, in this case, been accessed by decay fungi (*d*). Different decay columns are separated by interaction zone lines (*i*) produced by the fungi, and which directly traverse the barrier zone. (Rayner 1986)

living cells. Radial access, however, will be hindered by the closely packed xylem elements in the latewood of each annual ring. Thus, the implied ability of wall 2 to prevent inward but not outward spread is understandable. Tangential spread, supposedly impeded by wall 3, is restricted by the lack of any system of tangentially elongated lacunae in the wood, apart from, in a few cases, trauma-induced resin- or gum-filled canals. Like outward radial passage, spread must be effected via numerous woody cell walls. Thus, anatomical considerations adequately explain the shape of the longtudinally tapering wedge-shaped columns of stain and decay that are found not only in standing trees but also in felled trees and processed timber (Corbett 1963, 1965, Rayner and Boddy 1988).

It is not, however, possible to explain from these considerations alone the ultimate confinement of decay within columns, the limitation of colonization associated with the formation of barrier and reaction zones, and the tendency for preferential axial spread in inner sapwood. Once it is realized that wood anatomy forms similar routes of access for water and gases as it does for hyphae (Sect. 6.2.2.2) further questions are raised. In particular, it becomes apparent that there is/has been a problem in distinguishing between cause and effect. Thus, the CODIT concept has envisaged a direct causal relationship between the occurrence of barrier and reaction zones and the failure of fungi to invade as barriers keeping fungi out, but rather as sealants limiting the spread of aeration and drying, i.e., they maintain the functioning of the plumbing system of trees (Boddy and Rayner 1983b, Rayner 1986, Rayner and Boddy 1988). These zones then mark the divide between microenvironmental conditions favorable (i.e., aerated tissues) and those inhospitable (i.e., water-saturated tissues) to fungi, and hence only indirectly contain fungal spread. Thus, it could be argued that water is the major defense against invasion of xylem, and evidence for this is provided in Section 6.3.2. Evidence in support of the idea that barrier zones, etc. mark the divide between functional and nonfunctional tissue, rather than having a major, *direct* role in preventing fungal access into functional xylem, is provided by the fact that chemicals within these reaction and barrier zones vary considerably in their fungitoxicity, and that the zones are readily breached when the wood behind them becomes dried (Fig. 6.6).

Reaction zones are often produced successively and alternate with zones of decay, further indicating the ability of fungi to bypass these barriers under suitable conditions. This also demonstrates that many, if not all, reaction zones are static and fairly durable. This conflicts with the original concept that these zones were sites of dynamic interplay between fungus and tree, by which the fungus degraded the zone from within, while on the other side, healthy sapwood was converted to a reaction zone via a transition zone (Shain 1967, 1979, Pearce 1987). While reaction zones can impede fungal growth, they cannot entirely restrict it. The same arguments apply to the barrier zone, although this zone is supposedly much stronger than reaction zones.

6.3.2 Water and Aeration – Major Determinants of Mycelial Growth in Xylem

It is well known that high water contents inhibit growth and decomposition by wood decay fungi; indeed, wood is sometimes stored under water to reduce the decay rate (Cartwright and Findlay 1958, Boutelje and Kiessling 1964, Boddy 1986). Inhibition of fungal growth and germination, and of wood decomposition, probably results largely from lack of O_2 and buildup of CO_2 because gaseous diffusion is slow (Sect. 6.2.2.2) (Boddy 1986). Also, enzymes may not remain in the vicinity of hyphae in a water stream, and available nutrients will often be low (Sect. 6.2.1.2). There is now a large body of direct and indirect laboratory and field evidence indicating that high water contents and adverse gaseous regimes in standing trees do prevent fungal development. Some examples are discussed below.

6.3.2.1 Evidence from Laboratory Studies

In a study on the development of fungal communities in attached oak (*Quercus*) branches, it was found that the primary colonizers (wood-rotting Basidiomycotina) frequently formed longitudinally extensive (sometimes greater than 4 m) individuals in less than one growing season, as evidenced by the fact that no additional annual rings were formed in uncolonized regions (Boddy and

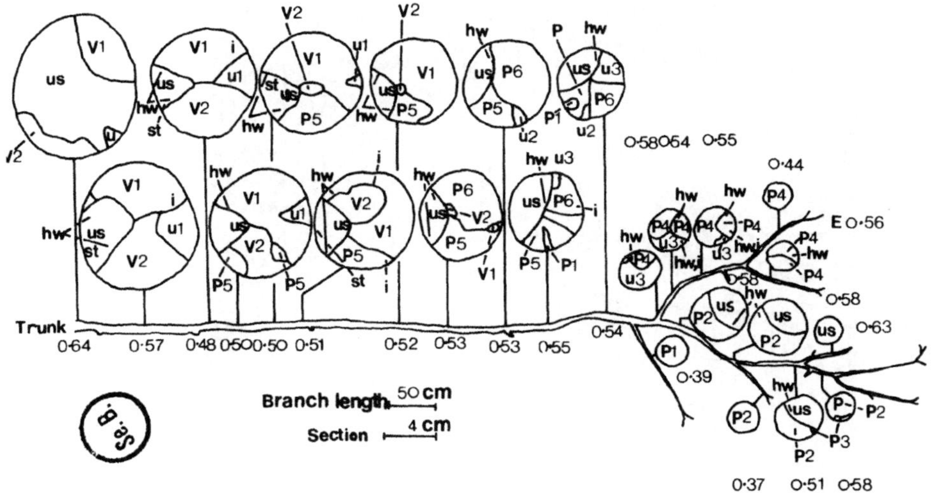

Fig. 6.7. Decay community structure in an attached oak (*Quercus robur*) branch. Note the extensive individuals of *Vuilleminia comedens* (two individuals, *V1* and *2*) and *Peniophora quercina* (six individuals P1–6) at least some of which developed in less than one growing season, as indicated by no additional growth rings in the functional sapwood (*us*). *u1–3* Unidentified basidiomycetes; *hw* heartwood wings; *i* interaction zone lines demarcating different individual basidiomycetes. Other *figures* represent the density of the wood ($g\,cm^{-3}$); density of undecayed wood is approx. $0.65\,g\,cm^{-3}$ (Boddy and Rayner 1983c)

Rayner 1982, 1983a,c) (Figs. 6.7 and 6.8). Since it was difficult to envisage that such rapid spread could be achieved by mycelial extension, based on laboratory growth rate data, some other means of colonization presumably occurred. It was therefore suggested that the phenomenon could be explained by the presence of latent, genetically identical fungal propagules (endophytes) distributed sparsely but extensively in columns of sapwood, which were only able to develop overtly as mycelia once the high moisture content (and/or associated gaseous regime and alterations within living cells) in functional sapwood had been alleviated. Such drying may result following death of associated photosynthetic tissues, death or damage of appropriate parts of the root system, drought, etc.

This hypothesis was examined experimentally by cutting freshly felled, healthy beech (*Fagus sylvatica*) branches into 10-cm lengths and subjecting them to five different drying regimes, ranging between a regime which maintained water saturation to rapid drying which allowed the water content to drop from $0.45\,g\,cm^{-3}$ to less than $0.1\,g\,cm^{-3}$ in 40 days (Chapela and Boddy 1988b). Other treatments included autoclaving, debarking, or inoculation with conidia or ascospores of *Biscogniauxia nummularia* and *Hypoxylon fragiforme* (which are common early colonizers of attached branches), respectively. Lengths were sampled at various times during the 92-day experiment. With the autoclaved sections only typical contaminants grew in from the ends of wood. Inoculations with conidia and ascospores were unsuccessful.

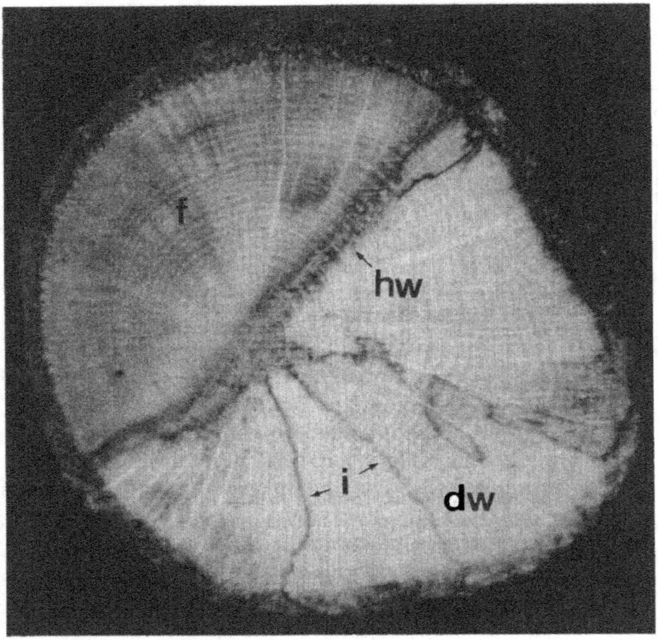

Fig. 6.8. Transverse section through a partly living, attached oak (*Quercus robur*) branch (approx. 10 cm diameter). Functional sapwood (*f*) has no more growth rings than decayed wood (*dw*). *hw* Heartwood wings; *i* interaction zone lines

Maintaining water saturation prevented fungal development in the wood, but under the other regimes active mycelia, of fungi commonly found as early colonizers of attached branches in the field, were isolated within 14 days after the start of drying. The more rapid the drying regime, the more rapid was initial fungal development in the wood, but under the most rapid drying regimes growth ceased before the end of the experiment, since the wood soon became too dry (Fig. 6.9). Extent and rate of drying differentially affected the development of different fungal species. *B. nummularia* and *H. fragiforme* were the most common species and increased in abundance with time (i.e., with drying), but later declined as drying continued beyond a crucial level (Fig. 6.10). Under a moderate drying regime (dropping from $0.45\,g\,cm^{-3}$ to $0.35\,g\,cm^{-3}$ after 40 days) maximum abundance (81%) occurred after 42 days (water content = $0.31\,g\,cm^{-3}$) and then declined to 30% after 90 days (water content about $0.2\,g\,cm^{-3}$). Maximum abundance was lower and decline more rapid under faster drying regimes, such that under the most rapid drying regime ($<0.1\,g\,cm^{-3}$ after 40 days) *B. nummularia* and/or *H. fragiforme* were absent after 46 days. Under the most rapid drying regime, the basidiomycete *Coniophora puteana* did not develop at all, but it did develop under the moderate regime, although only

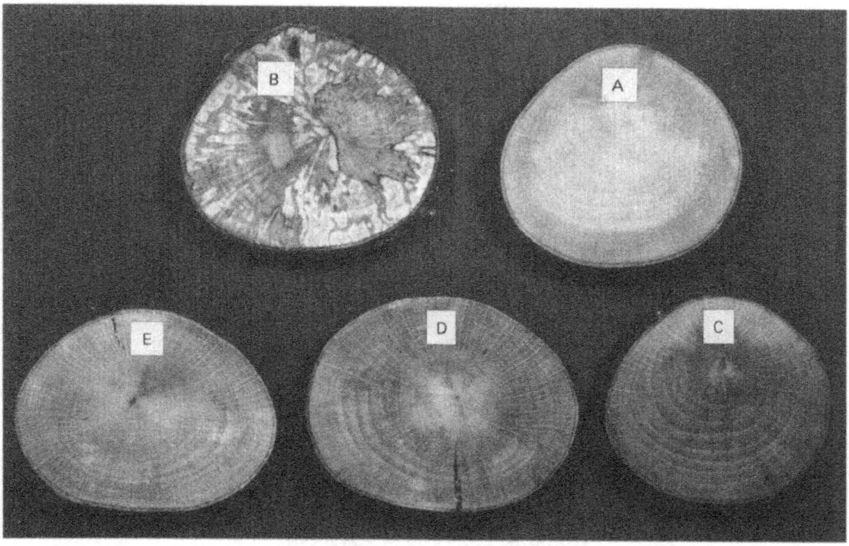

Fig. 6.9. Transverse sections of apparently healthy beach branch units cut after 12-week incubation under various drying regimes. *A* Maintained under water saturation; *B–E* increasingly faster drying regimes. Note extensive regions of decay due to *Hypoxylon fragiforme* and *Biscogniauxia nummularia* developing from latent propagules within the functional sapwood under gentle drying regime (*B*), absence of staining and decay under water saturation (*A*), and light staining and mottling under rapid drying regimes (*C–E*) (Chapela and Boddy 1988b)

after *B. nummularia* and *H. fragiforme* had begun to decline, and reached a maximum after 48 days (water content about $0.25\,g\,cm^{-3}$).

Most colonizers developed inwards from the bark, thus there was less fungal development in sections that had been debarked. However, some species, notably *B.nummularia* and *H.fragiforme*, also grew out from pockets within the wood (Fig. 6.11), each pocket containing a genetically different individual, determined by performing somatic compatibility pairing tests (Rayner *et al.* 1984). These pockets were prevalent in certain annual rings but not in others (Fig. 6.11a), suggesting the occurrence of some year to year variation in infection courts. These patterns did not occur in controls, neither were attempts at spore inoculation successful.

An essentially similar development has been shown to occur in American beech (*Fagus grandifolia*), trembling aspen (*Populus tremuloides*) (Chapela 1989), and hazel (*Corylus avellana*) (J.E. Hirst and L. Boddy unpublished, A.D.M. Rayner unpublished), although different fungal species are involved. Further, in hazel, genetically identical Basidiomycotina isolates have been found developing in different sections of the same branch, i.e., the situation postulated to account for rapid development of longitudinally extensive individuals in oak. Unfortunately, similar experiments have not yet been performed with oak.

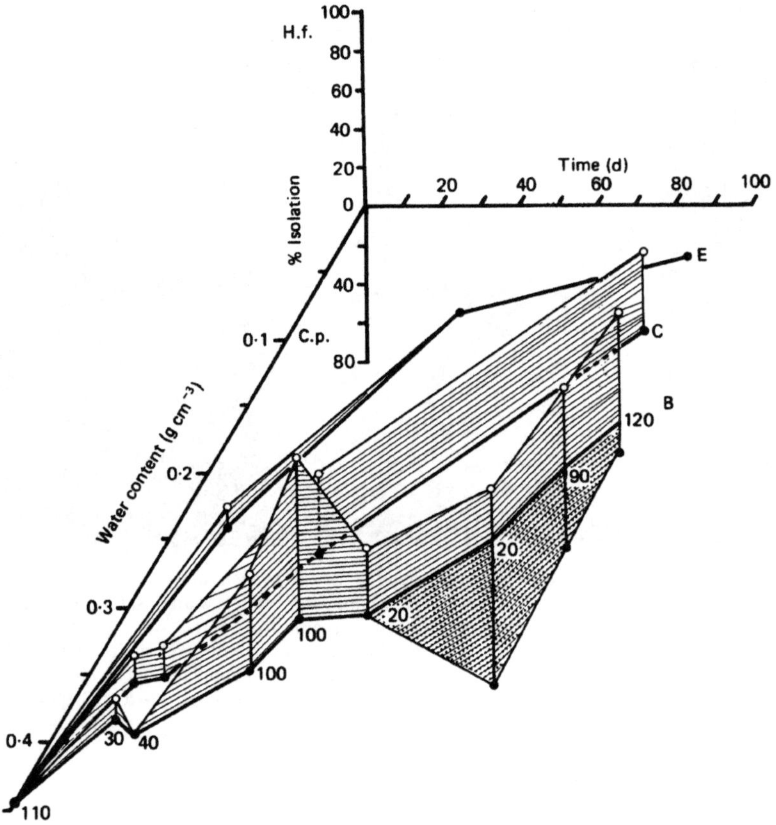

Fig. 6.10. Differential development of *Hypoxylon fragiforme/Biscogniauxia nummularia* (*H.f.*) and *Coniophora puteana* (*C.p.*), in sections from apparently healthy beech (*Fagus sylvatica*) branches, under three contrasting drying regimes (*B*, gentle drying regime; *C* and *E*, more rapid drying regimes). For each regime (on the horizontal plane) abundance (expressed as per-cent of chips yielding isolates) is plotted along the vertical axis (*H. fragiforme* upwards, *C. puteana* downwards). The *numbers* within the figure indicate the number of chips upon which the percent isolation is based (Chapela and Boddy 1988b)

Similar latency experiments have been carried out with ash (*Fraxinus excelsior*), beech (*Fagus sylvatica*), and oak (*Quercus robur*) twigs (Griffith and Boddy 1991a). In these studies, twigs (<1 cm diameter) were cut into 3-cm lengths; in some samples bark was left intact, in others bark was completely removed, and others were autoclaved as controls. The twigs were then incubated under a moderate drying regime for 9 weeks, during which time twigs were removed for making isolations and for determination of moisture content. Nonautoclaved twigs with bark were all colonized by 3 weeks, and species composition did not alter up to the end of the experiment. Autoclaved twigs and twigs with no bark were usually uncolonized, apart from an occasional external

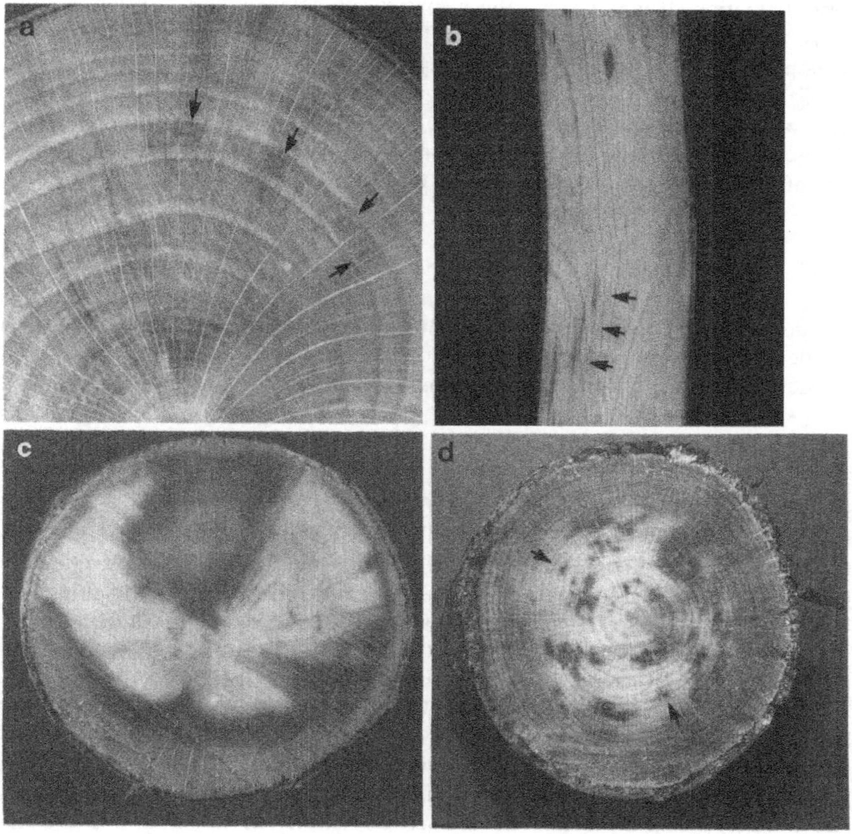

Fig. 6.11. Transverse (**a,c,d**) and longitudinal (**b**) sections from apparently healthy beech (*Fagus sylvativc*) branches, 4 weeks (**c,d**), 6 weeks (**b**), and 12 weeks (**a**) after incubation under a gentle drying regime, showing development of stain and discrete pockets of decay caused by *Hypoxylon fragiforme* (*arrows*) (Chapela and Boddy 1988b)

contaminant, implying that colonization during drying must have been effected from the bark. There was, however, one exception: *Cytospora ambiens* developed from within sapwood of oak.

Daldinia concentrica, *Phomopsis* spp., and *Xylaria* spp. were commonly isolated from all three tree species. However, interestingly, while *Phomopsis* spp. were found as early colonizers of attached twigs in the field, neither *D. concentrica* nor *Xylaria* spp. were (Griffith and Boddy 1988, 1991a), although *D. concentrica* is the major primary colonizer of attached ash branches and is also sometimes found in attached beech branches (Boddy et al. 1985, 1987). These differences between species developing within twigs in the laboratory, compared with the natural sequence of events in the field, may reflect the fact that the rate latter situation. It may be that these species can colonize under certain conditions, for example if twigs are broken off during storms but remain caught in the canopy where they will be subject to rapid drying.

Taken together, these experiments clearly demonstrate that water content, and/or closely related variables, are major determinants of the development of fungi in sapwood, and that early colonizers derive from endophyte inoculum latent in the bark and wood of healthy twigs and branches.

6.3.2.2 Field Evidence

Evidence for the controlling role of water in the field comes from observations of development patterns in tissues which have not been interfered with and in those whose water balance has been manipulated by girdling and leaf removal, and sometimes where fungi have been inoculated or woody tissues exposed to natural aerial inoculum.

6.3.2.2.1 Colonization of Sapwood Disrupted by Wounding

Although functional sapwood is an inhospitable environment for fungal growth, due to its high water content, low nutrient status, and possible responsiveness to infection, these stresses may by alleviated by disruption of sapwood function. Disruption by large-scale wounding is considered in this section. Disruption of sapwood by death of adjacent bark, which exposes tissues to colonization and allows aeration and drying, is discussed in the following section.

The effects of major wounds on fungal colonization of standing trees have been studied extensively, even to the point where the impression is gained that these are the sole routes for fungi into sapwood (Shigo 1966a,b, 1967, 1979, Shigo and Marx 1977, Mercer 1982). This, of course, is by no means the case. Localization of decay within the immediate vicinity of the wound and some hypotheses to account for this have been described in Section 6.3.1. Here, the interplay between wounding, aeration, and drying of xylem, and fungal colonization is considered.

Wounding results in the severence of part of the trees' water conduction system, and, as already explained in Section 6.2.2.2, sap flows outwards, and little air enters when water in the conduits is under positive hydrostatic pressures. However, when water columns are under tension, drying and air ingress are greater. Which of these situations occurs, depends on the physiology of the tree and on soil water status, atmospheric humidity etc., and hence varies with time of year. Also, extent of aeration and drying differs proximally and distally to a wound (Sect. 6.2.2.2). These factors all theoretically influence patterns of fungal colonization of xylem, and have indeed been shown to do so. Leben (1985) studied the formation of discolored wood columns and bark dieback in red maple (*Acer rubrum*) trunks wounded by making horizontal chain-saw cuts 1-cm wide, 4- to 6-cm deep and 10- to 16-cm-long. Wounds inflicted in September to November gave rise to much longer columns than those made in March to May (Table 6.2). Further, discoloration extended to greater distances upwards than downwards, and was correlated positively with

Table 6.2. Extent of bark dieback and wood discoloration after infliction of chain-saw wounds in spring and autumn in the stems of 93 red maples (*Acer rubrum*) (Leben 1985)

Season of wounding	Bark dieback (cm)		Length of the discolored wood column (cm)	
	Above wound	Below wound	Above wound	Below wound
Spring	1.44	1.75	6.80	3.24
Autumn	5.12	4.66	13.35	7.95

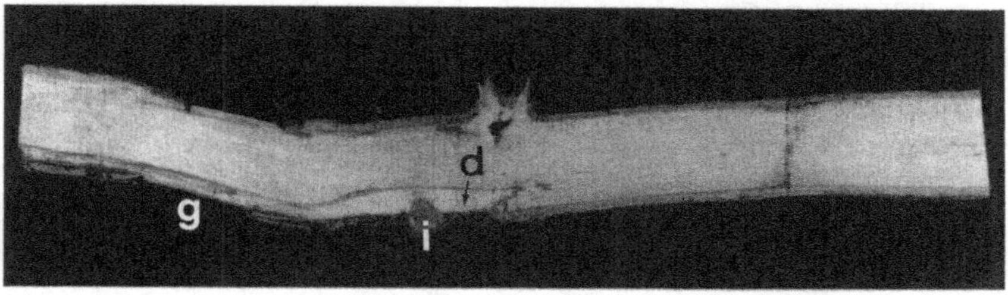

Fig. 6.12. Longitudinal section through an attached oak branch 6 months after inoculation with the basidiomycete *Vuilleminia comedens*. Spread of the decay column (**d**) from the inoculum plug (*i*) into the proximal portion of the branch is markedly restricted, while distal spread towards a girdled region (*g*) appears to have been facilitated in the outermost wood (Boddy and Rayner 1983b)

the area of bark dieback, which occurred in triangular patches above and below wounds. Likewise, discoloration extended further from wounds made in *Populus deltoides* in the dormant season relative to those made in the growing season (Shain and Miller 1988). These observations of fungal development correlate well with expected patterns of water distribution.

Similar evidence was obtained in experiments where primary colonizing wood decay fungi were inoculated into functional, attached oak (*Quercus robur*) branches, by inserting precolonized wood cylinders, 1 cm in diameter and 5–7 mm thick, into close-fitting holes. The time of year when inoculations were made and the proximity to artificial girdles resulted in differences in the extent of establishment (Boddy and Rayner 1984). *Exidia glandulosa*, *Phellinus ferreus*, and *Schizopora parodoxa* either failed to become established or growth was only very limited; *Peniophora guercina*, *Stereum gausapatum*, and *Vulleminia comedens* usually became established within 6 months but failed to extend further after this time (Fig. 6.4). In the absence of girdles, decay columns extended to 9 cm at most, usually much less; *Stereum gausapatum* colonized significantly more extensively following inoculation in August than in May or October to December, and *Vuilleminia comedens* usually failed to establish following inoculation in May. Spread of inoculated fungi toward girdled regions generally was far more extensive than away from these regions or from inocula not in the vicinity of girdles (Fig. 6.12).

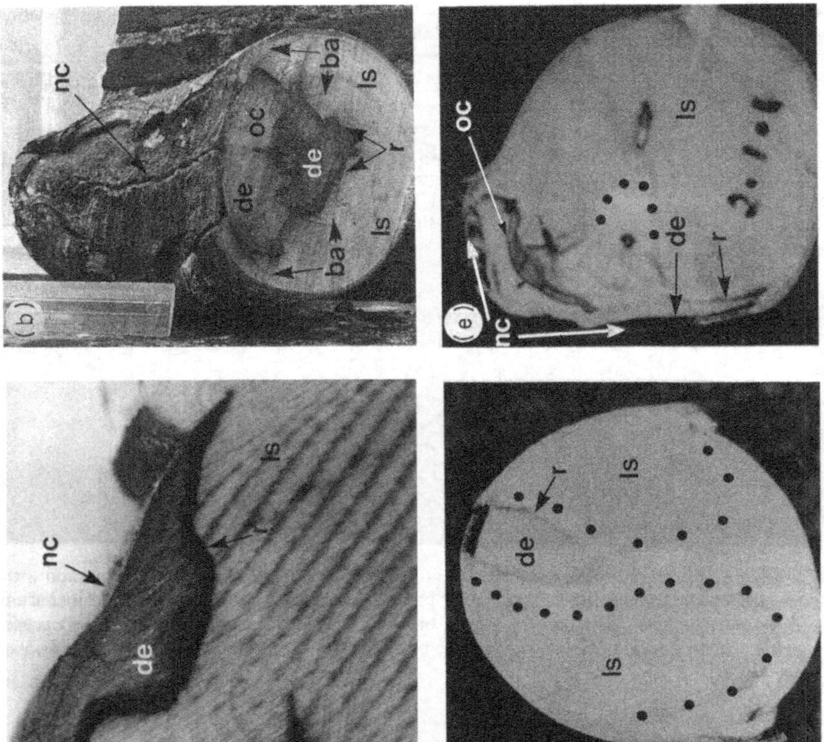
(a)
nc
de
nc
de
r
ls
oc
r
ls
(b)
nc
oc
de
de
ba
ba
ls
ls
r
(c)
de
r
ls
(d)
de
r
ls
ls
(e)
oc
nc
de
r
ls

6.3.2.2.2 Disruption of Sapwood Function Due to Bark Death

Shallow wounds and bark death, either in localized patches (cankers – see also Chaps. 3 and 14) or in continuous lengths, give rise to characteristic patterns of stain and decay in the underlying xylem. For example, the imperfect ascomycete *Dichomera saubinetii* forms diamond-shaped necrotic areas of bark on the main trunk of sycamore maple (*Acer pseudoplatanus*) following drought. Infection frequency increases with height, so that bark cover may be entirely lost at the tops of trees (Bevercombe 1980, Bevercombe and Rayner 1980). The wood beneath the cankers was relatively dry, discolored, and/or decayed and was often delimited by a greenish or brownish-black "reaction zone" (Fig. 6.13a). The position of these colonized regions and of the reaction zones was correlated with the expected extent of aeration of the tissues (Fig. 6.13c–e). Thus, in the lower trunk, colonization was usually limited to the outermost annual rings. In the upper and lower regions of severely infected trees, reaction zones formed along radii, delimiting wedge-shaped regions of discoloration and decay.

If death or removal of bark is more extensive, there is a tendency for the whole wood cylinder extant at the time of damage to become dried, aerated, and colonized by fungi. This is particularly well illustrated by strip cankers that formed on beech (*Fagus sylvatica*) in the UK after the 1976 and 1977 drought (Lonsdale 1983). These form tongues of dead bark which spiral up the tree, reflecting the underlying pattern of xylem, for many meters from ground level often up to a dead branch or stub. The "causative agents" appear to be the ascomycetes *Biscogniauxia nummularia* and *Eutypa spinosa*, and, along with other fungi, these cause decay of large volumes of underlying tissues (S.J. Hendry, L. Boddy and D. Lonsdale, unpublished) (Fig. 6.14).

The sooty bark fungus, *Cryptostroma corticale*, may also produce strip cankers, and was detected forming elongated stromata on sycamore maple (*Acer pseudoplatanus*) in the UK after the 1976/1977 droughts (Young 1978, Bevercombe 1980). Moreover, there is experimental evidence that spread of this fungus in the xylem is promoted by water stress (Dickenson and Wheeler 1981).

Fig. 6.13a–e. Development of discoloration and decay, reaction and barrier zones associated with diamond-canker formation, and variation in moisture in sycamore maple (*Acer pseudoplatanus*). **a** Part of a transverse section through a living tree where new cankers (*nc*) have developed over an old canker (*oc*). A dark 'reaction zone" (*r*) marks the junction between decayed (*de*) and functional sapwood (*ls*). **b** Development of a new canker by necrosis of callus tissue over an old canker. Decay is contained within radially aligned reaction zones, and barrier zones (*ba*) are associated with two episodes of bark death (**a** and **b** from Bevercombe and Rayner 1980). **c** Transverse section through a cankered region in the lower trunk. Drying and colonization is limited to an outer peripheral zone (*dotted line* indicates junction of moist sapwood and visibly dry wood). **d** Section through a large and much smaller canker on the upper part of a trunk. A region of drying (whose limits are indicated by *dotted line*) has developed behind the large and small canker. A wedge-shaped zone of decay (*de*) has developed behind the small canker and is delimited by a reaction zone (*r*). **e** Transverse section through a multiple cankered region. Drying and colonization has progressed in a wedge-shaped zone to the center of the trunk (**c–e** from Rayner 1986)

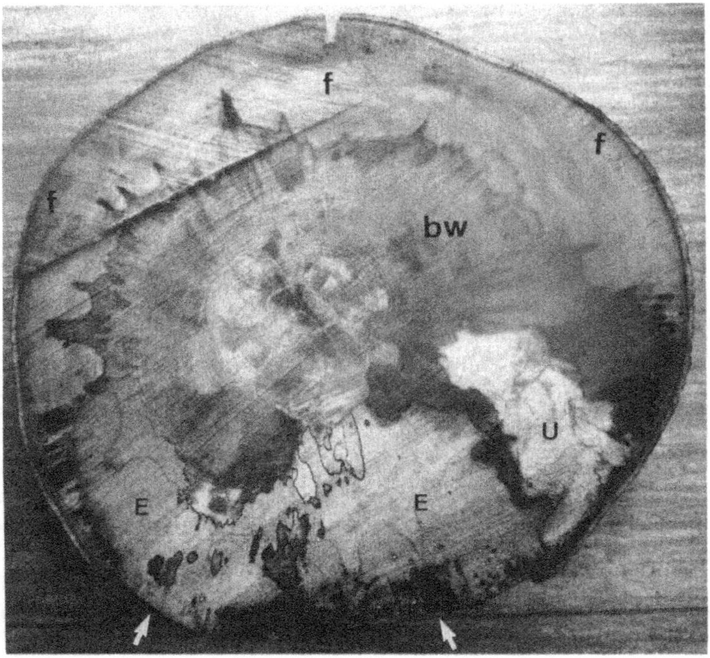

Fig. 6.14. Transverse section through a strip-cankered beech (*Fagus sylvatica*) trunk, 55 cm diameter. Decay has developed behind the strip canker (*arrows*) caused by *Eutypa spinosa* (*E*); *Ustulina deusta* (*U*) had also colonized. Stained central regions contain bacterial wetwood (*bw*). *f* Functional sapwood (Photograph courtesy of S.J. Hendry)

6.3.2.2.3 Distribution of Water and Decay in Non-wounded Attached Branches

In attached branches of oak (*Quercus* spp.), ash (*Fraxinus excelsior*), beech (*Fagus sylvatica*), and hazel (*Corylus avellana*), with no obvious signs of major wounding or bark pathogens, fungi often develop rapidly, from latent propagules within wood and bark, presumably when the high water content stress in xylem has been alleviated (Sect. 6.3.2.1). There is again a close correlation between development of decay in xylem and bark death, although whether one was the cause of the other or a combination of both is unclear (Boddy and Rayner 1982, 1983a,b,c, 1984, Chapela and Boddy 1988a,b, L. Boddy and S.J. Hendry, unpublished). In oak, wedge-shaped decay columns form which are demarcated from functional sapwood by dark-colored "reaction zones" extending radially from the center of the branch, or from central heartwood, to the junction between living and dead cambium (Figs. 6.7, 6.8, and 6.15). In these zones, the vessels are occluded by tyloses with deposited gums, and appear morphologically similar to true heartwood, hence they were termed "heartwood wings". Their location suggests that their formation may be triggered by air access and necrosis of ray parenchyma in a similar manner to the formation of true heartwood via a relatively dry transition zone. In the relatively narrow

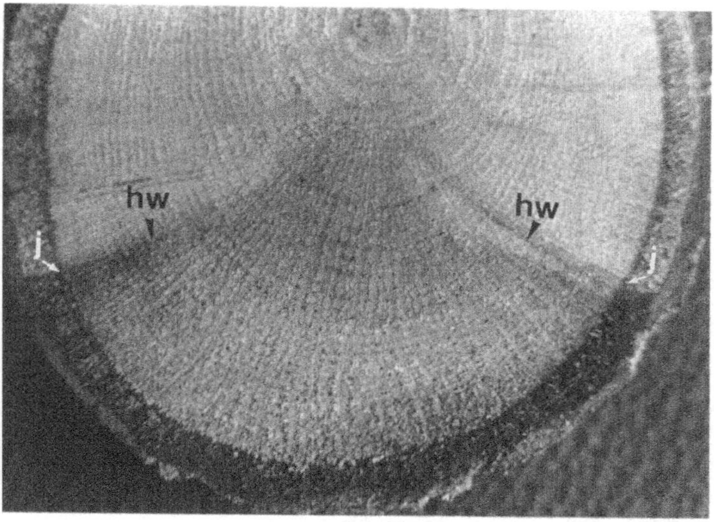

Fig. 6.15. Transverse section (about 10 cm diameter) through an attached oak (*Quercus robur*) branch where dark reaction zones, termed heartwood wings (*hw*), are in the process of forming. Heartwood wings extend from the center of the branch to the junction between living and dead cambium (*j*). Wood and bark between the wings is noticeably stained

poles of coppiced hazel (*Corylus avellana*) and in ash (*Fraxinus excelsior*) branches, the whole cross-sectional area usually becomes colonized, although occasionally in ash, colonization and bark death are confined to long segments. In beech, the pattern is rather complicated. Dieback often occurs from the end of the branches (see below), but sometimes long sectors of decay are formed, particularly by *Vuilleminia comedens*, in a manner analogous to that in oak. In large beech branches, it is common to find several foci of bark death, with associated sectors of stained and decayed xylem beneath (Fig. 6.3), commonly occupied by the basidiomycete *Oudemansiella mucida*.

In attached beech branches, the patterns of stain, decay, and dieback can be explained in terms of water, although to date no measurements of water potential have been made to support these ideas (Chapela and Boddy 1988a). The typical course of events is depicted schematically in Fig. 6.16. Main branches can be considered to consist of "fields" of sapwood which supply actively transpiring side branches with water. Such fields, while active in water conduction, would preclude the development of fungal mycelia, but outside of these fields decay could develop (Fig. 6.16a). If a side branch died in the vicinity of the front of fungal colonization, then the field which formerly supplied it with water would begin to dry out and become available for colonization (Fig. 6.16b). These regions would soon become colonized by fungi latent within bark and sapwood and also to some extent by fungi in the old colonization front. Subsequently, secondary colonizers would grow in from decayed regions (Fig. 6.16c).

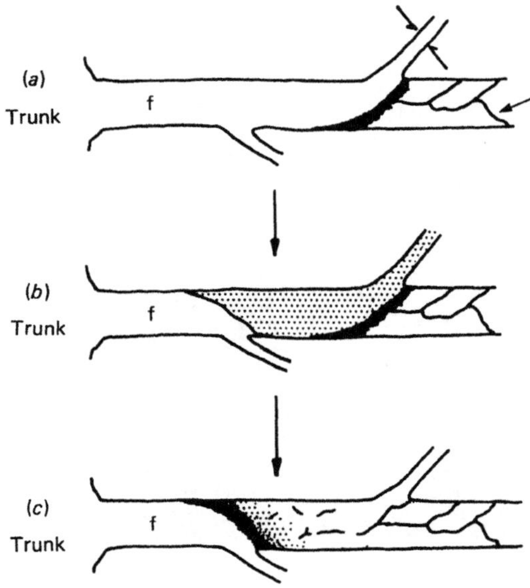

Fig. 6.16a–c. Schematic representation of proposed course of colonization of a primary, attached beech (*Fagus sylvatica*) branch. **a** Partly living branch. *Dark shading* represents stained region which sepataes functional sapwood (*f*) from decayed distal regions containing zone lines (*zl*). **b** The most distal side branch has died, causing wood behind it to dry, thus making it available for colonization (*stippled region*). **c** Colonization of drier wood occurs but does not extend into sapwood which is functional in transporting water to side branches (Chapela and Boddy 1988a)

6.3.2.2.4 Water and Fungal Development in Attached Twigs

With regard to twigs, the effect of removal of a high water content on subsequent development of decay communities has been examined in detail in ash (Griffith and Boddy 1990, 1991b). Branches giving rise to systems of twigs were girdled, where their diamater was between 2–3 cm, just after the trees had flushed (June), and leaves distal to the girdle were removed, as were any that developed subsequently. The experiment was performed in two different years. Twigs were removed from these stressed systems and from nonstressed, healthy systems for up to 17 months after girdling. The moisture content distribution of artificially stressed and nonstressed twigs altered with time after girdling (Fig. 6.17). Primary colonizers such as *Fusarium lateritium* and *Phomopsis platanoidis* began to establish, principally from bark, after about 1 month and were quite well established after about 3–4 months (Fig. 6.18). However, after 3 months, there was no significant difference ($P > 0.05$) in gross water content of stressed and nonstressed twigs. This does not, of course, mean that there was no difference nor does it preclude the possibility of considerable drying in localized regions. After 6 months, stressed twigs were significantly drier than nonstressed, and, interestingly, it was not until there were distinct differences in water

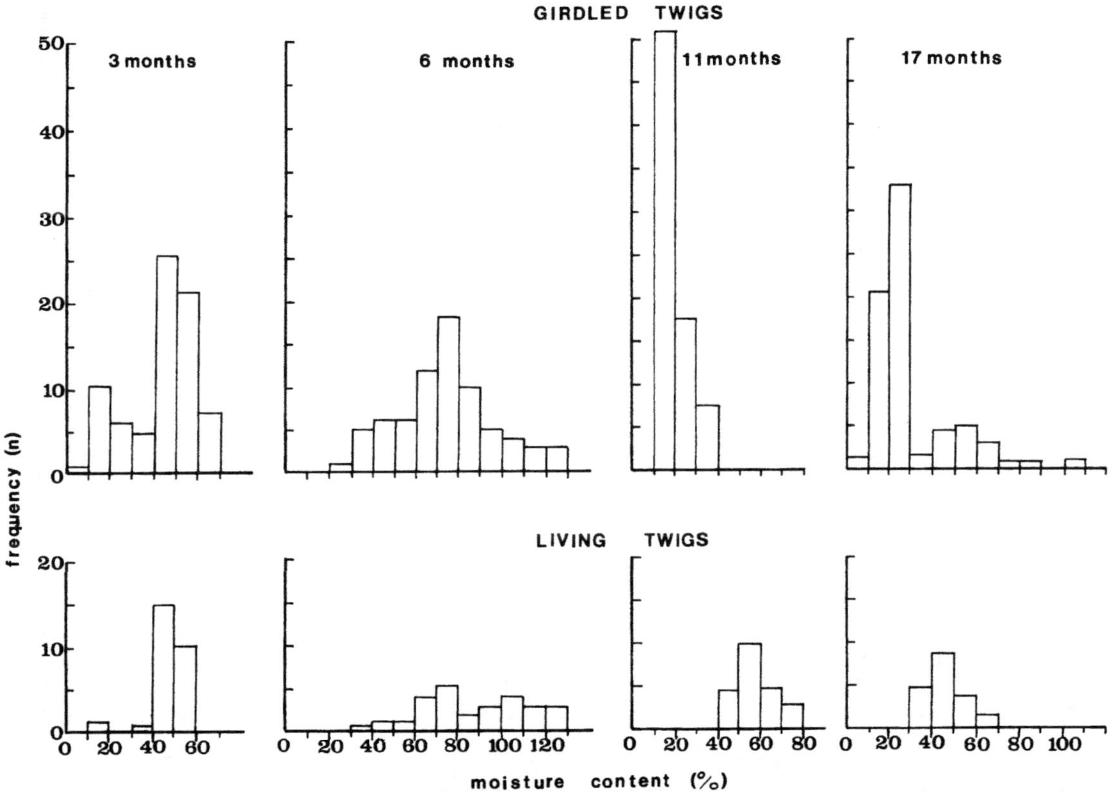

Fig. 6.17. Moisture content (% oven dry weight) distributions of 1-cm portions of wood removed from girdled (n = 75) and living (n = 25) ash (*Fraxinus excelsior*) twigs at various times after girdling in June 1986 (Griffith and Boddy 1990)

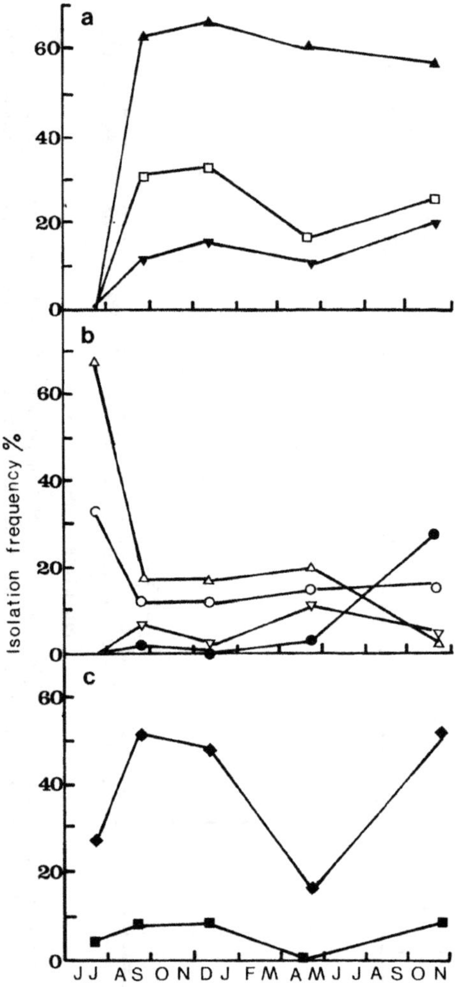

Fig. 6.18a–c. Change in the percent isolation frequency of fungi from living and girdled ash (*Fraxinus excelsior*) twigs with time. **a** and **b** girdled twigs, **c** living twigs. Twigs were girdled on 9–11 June 1986. ▲, *Phomopsis platanoidis*; □, Sp. 12; ▼, *Fusarium lateritium*; △, nothing isolated; ○ and ◆ incidental species; ▽, *Libertella fraxinea* – secondary colonizer; ●, *Peniophora lycii* – secondary colonizer; ■, primary colonizers combined (Griffith and Boddy 1990)

content that scondary colonizers, arriving from external sources, began to become established.

In healthy twigs, there was evidence of superficial colonization of wood located close to the bark, by fungi which were not commonly isolated from early stages of colonization of dead attached twigs, e.g. *Aureobasidium pullulans* and

Phoma macrostoma. Superficial colonization appeared to be seasonal, being maximal during the winter, perhaps associated with the fact that some parts of the twigs had lower moisture contents at this time (Fig. 6.17).

6.3.2.2.5 Water Stress Predisposition to Dissease

Further evidence that water is a major determinant of fungal colonization is provided by the preponderance of reports correlating soil water-stress with predisposition of plants to fungal pathogens causing root rots, stem cankers, and even vascular wilts (e.g. Schoeneweiss 1986). It must, however, be noted that water stress also results in changes in host metabolism.

With regard to wood decay fungi, penetration of *Heterobasidion annosum* into roots of 12-year-old *Pinus taeda* 6 months after inoculation was enhanced significantly by induced drought conditions, although, after 12 months, fungus growth had ceased (Towers and Stambaugh 1968). Also, root infection was increased significantly in 2-year-old seedlings when the soil was subjected to temporary wilting point depletion cycles.

Bark diseases and stem cankers commonly develop more rapidly and extensively under water stress (Bier 1964, Schoeneweiss 1978, 1981, 1986, Lonsdale 1983). The observations of McPartland and Schoeneweiss (1984) on the canker-forming fungus *Botryosphaeria dothidea* in vessels of nonstressed and drought-stressed stems of *Betula alba* are of particular interest with regard to xylem defense. In drought-stressed stems, hyphae were thick, branched, rectilinear, spread extensively through vessels, and were covered by a two-layered sheath when appressed to vessel walls. By contrast, in unstressed stems, hyphae were thin, contorted, highly vacuolated, restricted to vessels within 5 mm of the site of inoculation, and a high proportion of hyphal tips had burst. There was no evidence that resistance resulted from morphological barriers; it was more likely caused by the high water content of xylem or of host chemicals within the sap. With regard to the latter, chitinase and β-1,3-glucanse have been found in the phloem of various angiosperms, and these were capable of lysing cell walls of *Armillaria mellea* (Wargo 1975).

6.3.3 Extractives and pH – Major Determinants of Fungi-Colonizing Heartwood

The species of fungi which colonize heartwood are by and large different from those found in sapwood of standing trees. While the gaseous regime of heartwood is often ameliorated compared with sapwood, there is no evidence to suggest that the tolerances of heartrot fungi to high CO_2 and low O_2 are any less than for the majority of saprot fungi (Highley et al. 1983). The extractives found in many tree species, however, provide a highly selective environment (Sect. 6.2.1.3).

Extractives can confer durability to heartwood, but emphasis has been placed on the durability which they confer on timber in service (Scheffer and Cowling 1966), yet other factors may be relevant in the standing tree. Whereas large quantities of polyphenols, monoterpenes, etc. are undoubtedly toxic/inhibitory to the vast majority of fungi (Rudman and DaCosta 1961, Scheffer and Cowling

1966, Cobb et al. 1968, Hintikka 1970, DeGroot 1972, Väisälä 1974, 1978, Schuck 1977) there are distinct differences in tolerance to extractives between species. For example, many fungi which grow in coniferous wood are able to grow in atmospheres saturated with a variety of terpenes, but species common on angiosperms are sensitive to small amounts, correlating with the fact that in coniferous wood, but not angiospermous wood, the gas phase is probably saturated with terpene vapors (Hintikka 1970). Similarly, there is wide variation in the tolerance to phenols and other aromatics (Rudman 1963, Hintikka 1971, Popoff et al. 1975). For example, of 46 wood-rotting fungi tested on agar, *Gloeophyllum sepiarium*, *Phellinus pini*, and *Stereum sanguinolentum* were particularly tolerant of ferulic acid, whereas *Fomitopsis pinicola, Lentinus lepideus*, and *P. pini* were most tolerant of p-cresol (Hintikka 1971). *Deadalea quercina*, *Fomitopsis pinicola*, *Galerina mutabilis*, *Laetiporus sulphureus*, *Phellinus igniarius*, and *Piptoporus betulinus* were very tolerant, which correlates with the fact that *D. quercina* and *L. sulphureus* are common decayers of oak (*Quercus*) heartwood, which is rich in tannins (Hintikka 1971).

Some fungi are, therefore, tolerant of particular extractives or, indeed, of the generally fungitoxic substances found in their host, e.g. *Tyromyces amarus* on incense cedar (*Calocedrus decurrens*) (Wilcox 1970) and *Stereum sanguinolentum* on balsam fir (*Abies balsamea*). Moreover, some extractives may act as carbon sources, positively stimulate overall growth, and result in positive chemotropic growth of certain fungi, although little is known about such roles. Examples include stimulation of growth of *S. sanguinolentum* by volatile organic acids from *Pinus sylvestris* (Glasare 1970), and by hexanal and α-pinene plus hexanal (which are evolved by dried *Pinus* wood), although growth was inhibited by monoterpenes (which are found in large quantities in *Pinus* wood) (Flodin and Fries 1978). Also, some species of *Coniophora*, *Peniophora*, *Stereum*, and *Polyporus* showed increased biomass production when exposed to the vapors of some of the following: Δ-3-carene, α-pinene, β-pinene, limonene, longifolene, or a mixture of cuparene and thujopsene (Fries 1973). *Ganoderma applanatum* showed preferential growth toward nonanal (Fries 1961). Extractives from elm (*Ulmus* spp.) heartwood stimulated the growth of *Lyophyllum ulmarium*, *Pleurotus cornucopiae*, and *Rigidoporus ulmarius*, which are selective for this tree, but had no effect on fungi not selective for this tree (Rayner and Hedges 1982). With regard to use of extractives as nutritional sources, *Fistulina hepatica*, which is selective for *Castanea sativa* and *Quercus*, can utilize oak tannin as a sole source of carbon (Cartwright 1937). The pH is also probably a selective factor for some fungi; *F. hepatica* grows optimally at low pH, which again correlates with its selectivity for oak which has a pH of less than 4 (Cartwright and Findlay 1958).

Selection pressures are also likely to be imposed in the establishment phase as well as the growth phase, although this appears to have received little if any attention. The ability to germinate on an exposed heartwood surface is obviously likely to be critical and will be influenced by a wide range of abiotic and biotic factors. Also, heartrot fungi may initially become established in other tissues by active pathogenic mechanisms, via wounds or from latent propagules in sapwood, e.g. *Stereum gausapatum* (Sect. 6.3.2.1) where other selective pressures may be important.

Clearly then, extractives and pH considerably affect fungal growth although these effects are differential. These aspects of microenvironment, therefore, not only militate against the vast majority of fungal colonists, but also result in selectivity of heartrot fungi for different tree taxa, genotypes, and indeed different parts of the tree. Heartrot fungi display a broad spectrum of selectivity, depending presumably on their tolerance of specific microenvironmental factors and variations in these conditions. Some fungi are virtually specific for particular tree genera, e.g. *Fistulina hepatica* on oak (*Quercus*) and occasionally *Castanea* (which is closely related), *Phellinus pomaceus* on *Prunus*, *P. tremulae* on *Populus*, and *Rigidoporus ulmarius* on *Ulmus* in Europe (Philips and Burdekin 1982, Rayner and Boddy 1988). At the other end of the spectrum are species which exhibit little or no specificity, e.g. *Armillaria bulbosa* which can attack a wide range of angiospermous trees (Rishbeth 1982, 1985), and *Heterobasidion annosum* which can attack a wide range of coniferous and occasionally angiospermous trees, although there does appear to be host specialization within populations of *H. annosum* (Worrall et al. 1983). Between these extremes are fungi which exhibit strong selectivity for a particular tree species but which may occasionally occur on others [e.g. *Grifola frondosa*, *Inonotus dryadeus*, and *Stereum gausapatum* on *Quercus*, *Fomitopsis officinalis* on larch (*Larix*) and *Phellinus chrysoloma* on firs (*Abies*)] and those which are selective for two or more unrelated species, e.g. *Laetiporus sulphureus* which grows in the heartwood of *Castanea*, *Prunus*, *Quercus*, *Salix*, and *Taxus*. However, there is apparently no evidence as to whether or not populations on these different hosts are capable of interbreeding.

References

Allen SE, Grimshaw HM, Parkinson JA, Quarmby C 1974 Chemical analysis of ecological materials. Blackwell, Oxford

Armstrong W 1967 The relationship between oxidation-reduction potentials and oxygen-diffusion levels in some waterlogged organic soils. J Soil Sci 18:27–34

Basham HG, Cowling EB 1975 Distribution of essential elements in forest trees and their role in wood deterioration. In: Becker G, Liese W (eds) Organismen und Holz. Duncker & Humbolt, Berlin, 155–165

Bauch J, Höll W, Endeward R 1975 Some aspects of wetwood formation in fir. Holzforschung 29:198–205

Bevercombe GP 1980 Diseases affecting sycamore bark. PhD Thesis, Univ Exeter

Bevercombe GP, Rayner ADM 1980 Diamond-bark diseases of sycamore in Britain. New Phytol 86:379–392

Bier JE 1964 The relation of some bark factors to canker susceptibility. Phytopathology 54:272–275

Boddy L 1986 Water and decomposition processes in terrestrial ecosystems. In: Ayres PG, Boddy L (eds) Water, fungi and plants. Univ Press, Cambridge, 375–398

Boddy L, Bardsley DW, Gibbon, OM 1987 Fungal communities in attached ash branches. New Phytol 107:143–154

Boddy L, Gibbon OM, Grundy MA 1985 Ecology of Daldinia concentrica: effect of abiotic variables on mycelial extension and interspecific interactions. Trans Br Mycol Soc 78:337–351

Boddy L, Rayner ADM 1982 Population structure, inter-mycelial interactions and infection biology of Stereum gausapatum. Trans Br Mycol Soc 78:337–351

Boddy L, Rayner ADM 1983a Mycelial interactions, morphogenesis and ecology of Phlebia radiata and Phlebia rufa from oak. Trans Br Mycol Soc 80:437–448
Boddy L, Rayner ADM 1983b Origins of decay in living deciduous trees: the role of moisture content and re-appraisal of the expanded concept of tree decay. New Phytol 94:623–641
Boddy L, Rayner ADM 1983c Ecological roles of basidiomycetes forming decay communities in attached oak branches. New Phytol 93:77–88
Boddy L, Rayner ADM 1984 Internal spread of fungi inoculated into attached oak branches. New Phytol 98:155–164
Boutelje JB, Kiessling H 1964 On water-stored oak timber and its decay by fungi and bacteria. Arch Mikrobiol 49:305–314
Buchanan MA 1963 Extraneous components of wood. In: Browning BL (ed) The chemistry of wood. Interscience, New York London, 313–367
Carrodus BB, Triffett ACK 1975 Analysis of composition of respiratory gases in woody stems by mass spectrometry. New Phytol 74:243–246
Carter JC 1945 Wetwood of elms. Illinois Nat Hist Surv Bull 23:405–448
Cartwright K St G 1937 A re-investigation into the cause of "brown oak", Fistulina hepatica (Huds) Fr. Trans Br Mycol Soc 21:68–83
Cartwright K St G, Findlay WPK 1958 Decay of timber and its prevention, 2nd edn. HMSO, London
Chapela IH 1989 Fungi in healthy stems and branches of American beech and aspen: a comparative study. New Phytol 113:65–75
Chapela IH, Boddy L 1988a Fungal colonization of attached beech branches. I. Early states of development of fungal communities. New Phytol 110:39–45
Chapela IH, Boddy L 1988b Fungal colonization of attached beech branches. II. Spatial and temporal organization of communities arising from latent invaders in bark and functional sapwood, under different moisture regimes. New Phytol 110:47–57
Cobb FW Jr, Krstic M, Zavarin E, Barber HW Jr 1968 Inhibitory effects of volatile oleoresin components on Fomes annosus and four Ceratocystis species. Phytopathology 58:1327–1335
Corbett NH 1963 Anatomical, ecological and physiological studies on microfungi associated with decayed wood. PhD Thesis, Univ London
Corbett NH 1965 Micro-morphological studies on the degradation of lignified cell walls by Ascomycetes and Fungi imperfecti. J Inst Wood Sci 14:18–29
Coutts MP 1976 The formation of dry zones in the sapwood of conifers. I. Induction of drying in standing trees and logs by Fomes annosus and extracts of infected wood. Eur J For Path 6:372–381
Coutts MP, Rishbeth J 1977 The formation of wetwood in grand fir. Eur J For Pathol 7:13–22
DeGroot RC 1972 Growth of wood-inhabiting fungi in saturated atmospheres of monoterpenoids. Mycologia 64:863–870
Dickenson S, Wheeler BEJ 1981 Effect of temperature and water stress in sycamore on growth of Cryptostroma corticale. Trans Br Mycol Soc 76:181–185
Etheridge DE, Morin LA 1962 Wetwood formation in balsam fir. Can J Bot 40:1336–1345
Fengel D, Wegener G 1984 Wood chemistry, ultrastructure and reactions. Walter de Gruyter, Berlin New York
Flodin K, Fries N 1978 Studies on volatile compounds from Pinus sylvestris and their effect on wood-decomposing fungi. II. Effects of some volatile compounds on fungal growth. Eur J For Pathol 8:300–310
Fries N 1961 The growth-promoting activity of some aliphatic aldehydes on fungi. Svensk Bot Tidskr 55:1–18
Fries N 1973 The growth-promoting activity of terpenoids on wood-decomposing fungi. Eur J For Pathol 3:169–180
Gibbs RD 1958 Patterns in the seasonal water content of trees. In: Thimann KV (ed) The physiology of forest tress. Ronald, New York, 43–69
Glasare P 1970 Volatile compounds from Pinus silvestris stimulating the growth of wood-rotting fungi. Arch Mikrobiol 72:333–343
Good HM, Basham JT, Kadzielawa SD 1968 Respiratory activity of fungal associations in zones of heart rot and stain in sugar maple. Can J Bot 46:27–36

Good HM, Murray PM, Dale HM 1955 Studies on heartwood formation and staining in sugar maple, Acer saccharum Marsh. Can J Bot 33:31–41
Gray VR 1958 The acidity of wood. J Inst Wood Sci 1:58–64
Greenidge KNH 1952 An approach to the study of vessel length in hardwood species. Am J Bot 39:570–574
Griffith GS, Boddy L 1988 Fungal communities in attached ash (Fraxinus excelsior) twigs. Trans Br Mycol Soc 91:599–606
Griffith GS, Boddy L 1990 Fungal decomposition of attached angiosperm twigs. I Decay community development in ash, beech and oak. New Phytol 116:407–415
Griffith GS, Boddy L 1991a Fungal decomposition of attached angiosperm twigs. II. Moisture relations of ash twigs. New Phytol 117:251–257
Griffith GS, Boddy L 1991b Fungal decomposition of attached angiosperm twigs. III. Effect of water potential and temperature on fungal growth, survival and decay of wood. New Phytol 117:259–269
Hart HH 1968 Morphological and chemical differences between sapwood, discolored sapwood, and heartwood in black locust and osage orange. For Sci 14:334–338
Hart JA, Johnson KC 1970 Production of decay-resistant sapwood in response to injury. Wood Sci Technol 4:267–272
Hart JH, Wardell JF, Hemingway RW 1975 Formation of oleoresin and lignans in sapwood of white spruce in response to wounding. Phytopathology 65:412–417
Hartley C, Davidson RW, Crandall BS 1961 Wetwood, bacteria and increased pH in tress. US For Serv, For Prod Lab Rep 2215:1–34
Highley TL, Bar-Lev SS, Kirk TK, Larsen MJ 1983 Influence of O_2 and CO_2 on wood decay by heartrot and saprot fungi. Phytopathology 73:630–633
Hillis WE 1962 (ed) Wood extractives. Academic Press, New York London
Hillis WE 1971 Distribution, properties and formation of some wood extractives. Wood Sci Technol 5:272–289
Hillis WE 1972 Formation and properties of some wood extractives. Phytochemistry 11:1207–1218
Hillis WE 1977 Secondary changes in wood. In: Loewus FA, Runeckles VE (eds) The structure, biosynthesis and degradation of wood. Plenum, New York, 247–309
Hillis WE 1987 Heartwood and tree exudates. Springer, Berlin Heidelberg New York, 268 pp
Hintikka V 1970 Selective effect of terpenes on wood-decomposing Hymenomycetes. Karstenia 11:28–32
Hintikka V 1971 Tolerance of some wood-decomposing basidiomycetes to aromatic compounds related to lignin degradation. Karstenia 12:46–52
Hintikka V, Korhonen K 1970 Effects of carbon dioxide on the growth of lignicolous and soil-inhabiting Hymenomycetes. Comm Inst For Fenn 62(2):1–22
Jones HG 1986 Movement of liquid water and the effects of fungal infection. In: Ayres PG, Boddy L (eds) Water, fungi and plants. Univ Press, Cambridge, 223–239
Jorgensen E 1961 The formation of pinosylvin and its monomethyl ether in the sapwood of Pinus resinosa. Can J Bot 39:1765
King B, Oxley TA, Long KD 1974 Soluble nitrogen in wood and its redistribution on drying. Mat Org 9:241–254
Leben C 1985 Wound occlusion and discoloration columns in red maple. New Phytol 99:485–490
Lonsdale D 1983 Some aspects of the pathology of environmentally stressed trees. Yearb Int Dendrol Soc 1982:90–97
McGinnes EA, Phelps JE, Szopa PS, Shigo AL 1977 Wood anatomy after tree injury – a pictorial study. Univ Missouri-Colombia Coll Agr Exp Stn Res Bull 1025
McPartland JM, Schoeneweiss DF 1984 Hyphal morphology of Botryosphaeria dothidea in vessels of unstressed and drought-stressed stems of Betula alba. Phytopathology 70:1221–1224
Mercer PC 1982 Basidiomycete decay of standing trees. In: Frankland JC, Hedger JN, Swift MJ (eds) Decomposer basidiomycetes: their biology and ecology. Univ Press, Cambridge, 143–160

Merrill W, Cowling EB 1966 Role of nitrogen in wood deterioration: amounts and distribution of nitrogen in tree stems. Can J Bot 44:1555–1580
Milburn JA 1979 Water flow in plants. Longman, London
Moore KE 1978 Barrier-zone formation in wounded stems of sweetgum. Can J For Res 8:389–397
Mullick DB 1977 The non-specific nature of defense in bark and wood during wounding, insect and pathogen attack. In: Loewus FA, Runeckles VC (eds) The structure, biosynthesis and degradation of wood. Plenum, New York, 395–441
Packman DF 1960 The acidity of wood. Holzforschung 14:178–183
Pearce RB 1987 Antimicrobial defences in woody plants. In: Pegg GF, Ayres PG (eds) Fungal infection of plants. Univ Press, Cambridge, 219–238
Pearce RB, Holloway PJ 1984 Suberin in the sapwood of oak (Quercus robur L): its composition from a compartmentalization barrier and its occurrence in tyloses in undecayed wood. Physiol Plant Pathol 24:71–81
Pearce RB, Rutherford J 1981 A wound-associated suberized barrier to the spread of decay in the sapwood of oak (Quercus robur L.). Physiol Plant Pathol 19:359–369
Pearce RB, Woodward S 1986 Compartmentalization and reaction zone barriers at the margin of decayed sapwood in Acer saccharinum L. Physiol Mol Plant Pathol 29:197–216
Phillips DA, Burdekin DA 1982 Diseases of forest and ornamental trees. MacMillan, London Basingstoke
Platt WD, Cowling EB, Hodges CS 1965 Comparative resistance of coniferous root wood and stem wood to decay by isolates of Fomes annosus. Phytopathology 55:1347–1353
Popoff T, Theander O, Johansson M 1975 Changes in sapwood of roots of Norway spruce, attacked by Fomes annosus. II. Physiol Plant 34:347–356
Preece TF 1977 Watermark disease of the cricket bat willow. For Commiss Leaflet 20:1–9
Rayner ADM 1986 Water and the origins of decay in trees. In: Ayres PG, Boddy L (eds) Water, fungi and plants. Univ Press, Cambridge, 321–341
Rayner ADM, Boddy L 1988 Fungal decomposition of wood: its biology and ecology. John Wiley & Sons, Chichester
Rayner ADM, Coates D, Ainsworth AM, Adams TJH, Williams END, Todd NK 1984 The biological consequences of the individualistic mycelium. In: Jennings DH, Rayner ADM (eds) The ecology and physiology of the fungal mycelium. Univ Press, Cambridge, 509–540
Rayner ADM, Hedges MJ 1982 Observations on the specificity and ecological role of basidiomycetes colonizing dead elm wood. Trans Br Mycol Soc 78:370–373
Rayner ADM, Webber JF 1984 Interspecific mycelial interactions – an overview. In: Jennings DH, Rayner ADM (eds) The ecology and physiology of the fungal mycelium. Univ Press, Cambridge, 383–417
Rishbeth J 1982 Species of Armillaria in southern England. Plant Pathol 31:9–17
Rishbeth J 1985 Armillaria: resources and hosts. In: Moore D, Casselton LA, Wood DA, Frankland JC (eds) Developmental biology of higher fungi. Univ Press, Cambridge, 87–101
Rowe JW (ed) 1989 Natural products of woody plants I and II. Chemicals extraneous to the lignocellulosic cell wall. Springer, Berlin Heidelberg New York, 1243 pp
Rudman P 1963 The causes of natural durability in timber. XI. Some tests on the fungi toxicity of wood extractives and related compounds. Holzforschung 17:54–57
Rudman P 1964 The causes of natural durability in timber. Pt XVI. The causes of variation in decay resistance of jarrah (Eucalyptus marginata Sm.). Holzforschung 18:172–177
Rudman P 1965 The causes of natural durability in timber, pt XIX. Ageing of eucalypt heartwoods and its effect on decay resitance. Holzforschung 19:190–195
Rudman P 1966 The causes of variations in the natural durability of wood: inherent factors and ageing and their effects on resistance to biological attack. Mat Org Suppl 6 1:151–162
Rudman P, DaCosta EWB 1961 The causes of natural durability in timber. IV. Variation in the role of toxic extractives in the resistance of durable eucalypts to decay. Holzforschung 15:10–15
Scheffer TC, Cowling EB 1966 Natural resistance of wood to microbial deterioration. Annu Rev Phytopathol 4:147–170
Schoeneweiss DF 1978 Water stress as a prediaposing factor in plant disease. In: Kozlowski TT (ed) Water deficits and plant growth, vol 5. Academic Press, New York London, 61–99

Schoeneweiss DF 1981 The role of environmental stress in diseases of woody plants. Plant Dis 65:308–314
Schoeneweiss DF 1986 Water stress predisposition to disease – an overview. In: Ayres PG, Boddy L (eds) Water, fungi and plants. Univ Press, Cambridge, 157–174
Scholander PF 1958 The rise in sap in lianas. In: Thimann KV (ed) The physiology of forest trees. Ronald, New York, 3–17
Schuck HJ 1977 The effect of monoterpenes on the mycelial growth of Fomes annosus (Fr.) Cooke. Eur J For Pathol 7:374–384
Shain L 1967 Resistance of sapwood in stems of loblolly pine to infection by Fomes annosus. Phytopathology 57:1034–1045
Shain L 1971 The response of sapwood of Norway spruce to infection by Fomes annosus. Phytopathology 61:301–307
Shain L 1979 Dynamic responses of differentiated sapwood to injury and infection. Phytopathology 69:1143–1147
Shain L, Miller JB 1988 Ethylene production by excised sapwood of clonal eastern cottonwood and the compartmentalization and closure of seasonal wounds. Phytopathology 78:1261–1265
Shigo AL 1966 Decay and discoloration following logging wounds on northern hardwoods. US For Serv Res Pap NE-47
Shigo AL 1967a The early stages of discoloration and decay in living hardwoods in northeastern United States: a consideration of wound-initiated discoloration and heartwood. IUFRO Congr Proc 9:117–133
Shigo AL 1967b Succession of organisms in discoloration and decay of wood. Int Rev For Res II:237–299
Shigo AL 1979 Tree decay – an expanded concept. US Dep Agr For Inf Bull 419
Shigo AL 1984 Compartmentalization: a conceptual framework for understanding how trees grow and defend themselves. Annu Rev Phytopathol 22:189–214
Shigo AL, Marx HG 1977 Compartmentalization of decay in trees. US Dep Agr For Ser Agric Inf Bull 405
Shigo AL, Shortle WC 1979 Compartmentalization of discolored wood in heartwood of red oak. Phytopathology 69:710–711
Shortle WC 1979 Mechanisms of compartmentalization of decay in living trees. Phytopathology 69:1147–1151
Skaar C 1972 Water in wood. Univ Press, Syracuse
Skaar C 1988 Wood-water relations. Springer, Berlin Heidelberg New York, 283 pp
Swift MJ 1977 The role of fungi and animals in the immobilization and release of nutrient elements from decomposing branch-wood. In: Lohm U, Persson T (eds) Soil organisms as components of ecosystems. Swed Nat Sci Res Counc, Stockholm, Ecol Bull 25:193–202
Tattar TA, Shortle WC, Rich AE 1971 Sequences of microorganisms in constituents associated with discoloration and decay of sugar maples infected with Fomes connatus. Phytopathology 61:556–558
Tippett JT, Shigo AL 1980 Barrier zone anatomy of red pine roots invaded by Heterobasidion annosum. Can J For Res 10:224–232
Tippett JT, Shigo AL 1981 Barriers to decay in conifer roots. Eur J For Pathol 11:51–59
Towers B, Stambaugh WJ 1968 The influence of induced soil moisture stress upon Fomes annosus root rot of loblolly pine. Phytopathology 58:269–272
Trendelenburg R 1939 Das Holz als Rohstoff. Lehmann, Munich Berlin
Väisälä L 1974 Effects of terpene compounds on the growth of wood-decomposing fungi. Ann Bot Fenn 11:275–778
Väisälä L 1978 Effects of terpene compounds on the growth and peroxidase activity of Phellinus pini. Ann Bot Fenn 11:275–278
Wargo PM 1975 Lysis of the cell wall of Armillaria mellea by enzymes from forest trees. Physiol Plant Pathol 5:99–105
Wilcox WW 1968 Some physical and mechanical properties of wetwood in white fir. For Prod J 18:27–31
Wilcox WW 1970 Tolerance of Polyporus amarus to extractives from incense cedar heartwood. Phytopathology 60:919–923

Worrall JJ, Parmeter JR 1982 Formation and properties of wetwood in white fir. Phytopathology 72:1209–1212

Worrall JJ, Parmeter JR 1983 Inhibition of wood-decay fungi by wetwood of white fir. Phytopathology 73:1140–1145

Worrall JJ, Parmeter JR Jr, Cobb FW Jr 1983 Host specialization of Heterobasidion annosum. Phytopathology 73:304–307

Young CWT 1978 Sooty bark disease of sycamore. Arbor Leaflet 3. HMSO, London

Zeikus JG, Ward JC 1974 Methane formation in living trees: a microbial origin. Science 184:1181–1183

Zimmermann MH 1971 Tansport in xylem. In: Zimmermann MH, Brown CL (eds) Trees structure and function. Springer, Berlin Heidelberg New York, 169–220

Zimmermann MH 1983 Xylem structure and the ascent of sap. Springer, Berlin Heidelberg New York, 143 pp

Chapter 7

Biochemistry and Molecular Biology of Defense Reactions in the Xylem of Angiosperm Trees

L.C. Duchesne, M. Hubbes, and R.S. Jeng

7.1 Introduction

The xylem of woody angiosperms comprises the great majority of the biomass on earth (Lieth and Whittaker 1975, Woodwell et al. 1978, Duchesne and Larson 1989). Understandably, a wide array of living organisms have evolved mechanisms to acquire and metabolize the different components of tree xylem. On the other hand, the xylem of trees is critical for mechanical support, food reserve, and sap conduction, and therefore trees have evolved several mechanisms that allow them to preserve the integrity of their woody tissues.

A clear understanding of these defense mechanisms is of critical importance to the development of trees with enhanced resistance to pathogens or to the elaboration of silvicultural practices which optimize the trees' natural means of disease resistance. However, it is interesting to note that in the past very little attention has been given to this field in comparison to disease resistance of crops. Consequently, a great deal more knowledge is available on the mechanisms of resistance of crops compared with those of trees. One explanation for this is the difficulty associated with improving resistance to pathogens in tree species. However, recent advances in micropropagation and genetic engineering of trees (Klimaszewska 1989, Bekkaoui et al. 1990) will soon make possible the use of these techniques on a routine basis. This will be particularly useful for practices that involve growing trees on a shorter rotation, such as in energy plantations or in orchards (Hubbes 1990).

In general, we recognize that resistance of trees to pathogens is based on their ability to confine pathogens within the smallest area possible (Shigo 1984). This is achieved through the concerted action of active and passive defense mechanisms. Passive defense mechanisms encompass means of protection that are already in place before infection by pathogens, whereas active defense mechanisms are those means of protection triggered by the actions of pathogens.

7.2 Passive Defense Mechanisms

7.2.1 Lignin Biosynthesis

Passive defense mechanisms comprise chemical, anatomical, and morphological factors which are under genetic control and are not activated by the arrival of pathogens. Normal wood is constituted of approximately 20–35% lignin (Cote

1976). It is recognized that lignin provides woody tissues with protection against pathogens because of its resistance to biodegradation. In recent years, the mechanisms of lignin biosynthesis in plants have stimulated a great deal of research because of the undesirability of this substance to the paper industry. Although the biochemistry of lignin biosynthesis through the condensation of phenylpropane units in plants is well characterized (Higuchi 1990, Lewis and Yamamoto 1990), the processes involved at the onset of lignin biosynthesis are poorly understood. Some authors have even suggested that in the future it may be possible to engineer trees that do not produce lignin (Eriksson 1990, Higuchi 1990), and therefore would be of great value to the paper industry. It would be interesting to evaluate the resistance of these trees to wood decay fungi.

7.2.2 Heartwood Formation

The most striking passive defense mechanism of trees is the formation of heartwood, which shows enhanced resistance to wood decay fungi and enhanced mechanical properties (Bamber 1987, Hillis 1987). There are over 10000 secondary metabolites reported from plants including simple terpenes, sesquiterpenes, diterpenes, triterpenes/steroids, tetraterpenes, polyketides, polyacetylenes, flavonoids, phenylpropanoids, amines, alkaloids, and various other smaller groups (Wink 1988, Rowe 1989). A large number of chemicals from these different groups are accumulated in trees during heartwood formation. However, it is impossible to review all the mechanisms involved in the synthesis of these substances during heartwood formation because of the chemical diversity involved in tree secondary metabolites (Bell 1981, Callow 1983, Rowe 1989). Instead, we will concentrate on the events occurring at the sapwood-heartwood interface.

7.2.3 Heartwood-Sapwood Interface

The transformation of sapwood into heartwood is concomitant to the death of ray parenchyma cells, the loss of starch reserves, enhanced biosynthesis of both soluble and insoluble phenols (Hillis 1962, 1987, Sasaya 1987) and probably a transient increase in respiration before cell death. This is associated with an increase in succinate dehydrogenase and phosphatase at the heartwood-sapwood interface of some tree species (Rao et al. 1987). The resulting dead tissues are more resistant to pathogen invasion through chemical inhibition provided by extractives such as polyphenols or tropolones and through physical impedance of the advance of pathogens by the presence of highly polymerized substances such as lignin. The free radicals of oxidized phenols may function as nonspecific inhibitors indiscriminately affecting many enzymes of both fungal and plant origin (Smith 1982). Nevertheless, the presence of both soluble and nonsoluble phenolics in heartwood constitute an effective barrier against pathogens. The regulatory mechanisms which control the transformation of sapwood into heartwood are poorly understood. Factors that affect this phenomenon include mineral nutrition and light and water availability (Hillis 1962, 1977, 1987, Grand et al. 1982). Growth regulators may participate in both heartwood formation and the expression of active defense mechanisms (Yang and Hoffman 1984, Leguay et al. 1988). This is particularly evident for diseases that lead to tylosis formation since tyloses are the result of altered wall plasticity, which in turn is modulated by auxin metabolism (Beckman 1971). One practical implication, however, is that trees kept under optimal growth conditions are more efficient at producing heartwood.

7.2.4 Tree Nutrient Composition

The composition of xylem sap has been presumed to influence resistance to Dutch elm disease (Elgersma 1969, Singh and Smalley 1969). However, this hypothesis was later dismissed because results were not reproducible. Nevertheless, this kind of hypothesis may still be applicable to other diseases since it is evident that mineral nutrition of trees is critical to their health and may act as contributing factors in disease development (Manion 1985). Particularly, mineral nutrition may influence both passive and active defense mechanisms.

7.2.5 Host-Parasite Interactions

7.2.5.1 Viral Infections

Numerous viruses are known to attack trees (Nienhaus and Castello 1989). Following endocytosis, all viruses seem to require attachment to host ribosomes prior to multiplication. Another preformed biochemical mechanism of defense include the lack of attachment sites (Schlosser 1980). For attachment of viral mRNA to occur there should be affinity between the plant ribosomes and the former. Disease is prevented in cases where attachment does not take place (van den Elzen et al. 1989). This kind of phenomenon has yet to be reported for trees.

7.2.5.2 Toxins

Initial studies on toxins have demonstrated that pathogens are capable of producing metabolites in culture that cause symtoms similar to those found when natural infections take place (Goodman et al. 1986). Since then, a large number of putative fungal and bacterial toxins have been investigated. To date over 150 toxins are believed to be associated with pathogenesis (Scheffer 1983, Van Alfen 1989). Those toxins that exhibit similar host specificity are termed host-specific toxins, whereas those not solely responsible for pathogenicity of a pathogen are called nonhost specific toxins.

Ophiostoma ulmi synthesizes two known toxins, a high-molecular weight peptidorhamnomannan (Strobel et al. 1978) and the more intensively investigated cerato-ulmin, a 7.6-kD peptide (Otani et al. 1990). It was demonstrated that cerato-ulmin acts as a host-specific toxin (Otani et al. 1986). The present evidence suggests that its site of action is located on the vessel walls (Takai and Hiratsuka 1984). Although the exact site of action of cerato-ulmin on vessel walls is unknown, we speculate that nonhost resistance to DED is associated with a lack of suitable cell wall receptors for the toxin. Studies on crop plants have presented circumstantial evidence that the chemical structure of the action site of the toxin is under genetic control (Van Alfen 1989). The presence or absence of cellular receptors is, therefore, a passive defense mechanisms against tree diseases.

7.2.6 Interaction with Symbiotic Microorganisms

In nature, healthy trees are surrounded by soil microorganisms. Their tissues also carry numerous organisms, including mycorrhizal endophytic fungi. Recent evidence suggests that a tightly controlled relationship exists between trees and these microorganisms. The ectomycorrhizal fungus *Paxillus involutus* protects seedlings of *Pinus resinosa* against root rot caused by *Fusarium oxysporum* f. sp. *pini* (Duchesne et al. 1988a). The protective effect of the ectomycorrhizal fungus is associated with the synthesis of antifungal compounds by *P. involutus* (Duchesne et al. 1988b). Moreover, the synthesis of antifungal chemicals by *P. involutus* is stimulated by root exudates of *P. resinosa* (Duchesne et al. 1988b, 1989b). When the ability of three additional pine species to stimulate antibiosis in *P. involutus* was tested, it was shown that only *P. resinosa* root exudates have a stimulatory effect (Duchesne et al. 1989a). Further investigation indicated that de novo protein synthesis by *P. involutus* is associated with the protective effect of this fungus against *F. oxysporum* (Duchesne 1989). From these observations it was concluded that there is a genetic determinant provided by trees in disease protection against microorganisms. This phenomenon constitutes another preformed mechanism of disease resistance.

7.3 Active Defense Mechanisms

In comparison to diseases of crops, very little attention has been given to active defense mechanisms in the xylem of trees. In general, we recognize chemical and anatomical defense reactions. Chemical defense reactions include the synthesis of phytoalexins and proteins, whereas anatomical defense reactions include barrier formation, tylosis formation, cell wall lignification, and suberization.

7.3.1 Phytoalexins

Phytoalexins are low-molecular weight fungitoxic compounds accumulated by plants in response to several stress agents, notably in response to infection by microorganisms (Paxton 1980). Phytoalexin accumulation is associated with disease resistance, since fast accumulation of large amounts of phytoalexins is observed in incompatible host-pathogen interactions, whereas delayed accumulation and/or accumulation of small amounts of phytoalexins is characteristic of compatible host-pathogen interactions (Kuc 1976, Keen 1981, Deverall 1982, Mansfield 1982, Darvill and Albersheim 1984). Phytoalexin accumulation is one of the most extensively studied plant defense reactions. To date, over 200 putative phytoalexins have been isolated from over 20 plant families (Kuc 1976, Coxon 1982, Ingham 1982). Phytoalexins cause nonspecific oxidative stresses on living cells (Nebert et al. 1990).

A very limited number of chemicals have been investigated in trees with regard to phytoalexin accumulation. Interestingly, it is recognized that chemicals normally accumulated in the heartwood of trees can also accumulate in sapwood in a manner analogous to phytoalexin accumulation (Shain 1967, Hartmann and Nienhaus 1974, Langcake et al. 1979, Hart 1981, Coxon 1982, Burden and Kemp 1983, 1984, Kemp and Burden 1984). The best-known example of accumulation of phytoalexins in angiosperm trees is that of mansonone accumulation in elms.

7.3.2 Mansonone Accumulation in Elms

Elm heartwood accumulates small concentrations of fungitoxic sesquiterpene phenolics (Fig. 7.1) which were designated mansonones after their first reported origin, the west African tree, *Mansonia altissima* (Marini-Bettolo et al. 1965, Galeffi et al. 1968, Overeem and Elgersma 1970, Rowe et al. 1972, Hubbes and Jeng 1981, Dumas et al. 1983, 1984, Burden and Kemp 1984). In addition to heartwood formation, it was also shown that mansonone accumulation in elms is triggered by a number of factors, such as wounding and fungal infection (Overeem and Elgersma 1970, Elgersma and Overeem 1971, Dumas et al. 1983, Jeng et al. 1983, Burden and Kemp 1984).

To date, evidence indicates that the accumulation of mansonones is ubiquitous in elms. Mansonone C (Fig. 7.1) was observed in the heartwood of 13 elm species, including *Ulmus americana* and *U. pumila* (Rowe et al. 1972). Mansonones E and C were observed in the heartwood of *U. laciniata* (Suzuki et al. 1971) and in the sapwood of *U. parvifolia* (Yasuda et al. 1973). Mansonones E and F were shown to accumulate in *U. pumila* seedlings (Duchesne et al. 1985). Mansonones A, C, D, E, F, and G accumulate in *U. americana* (Dumas et al. 1983). *Ulmus glabra* synthesizes mansonones E and F in response to infection by *Ophiostoma ulmi* (Burden and Kemp 1984).

The accumulation of these compounds correlated with resistance to DED when seedlings of the susceptible *U. americana* and the resistant *U. pumila* were inoculated with either an aggressive or a nonaggressive isolate of *O. ulmi* (Duchesne et al. 1985, 1986, 1990, Proctor and Smalley 1988). However, several questions remain to be answered to ascertain the role of mansonone accumulation in resistance to DED. Particularly, it remains to be demonstrated that they accumulate in sufficient amounts in the areas infected by the pathogen (Elgersma

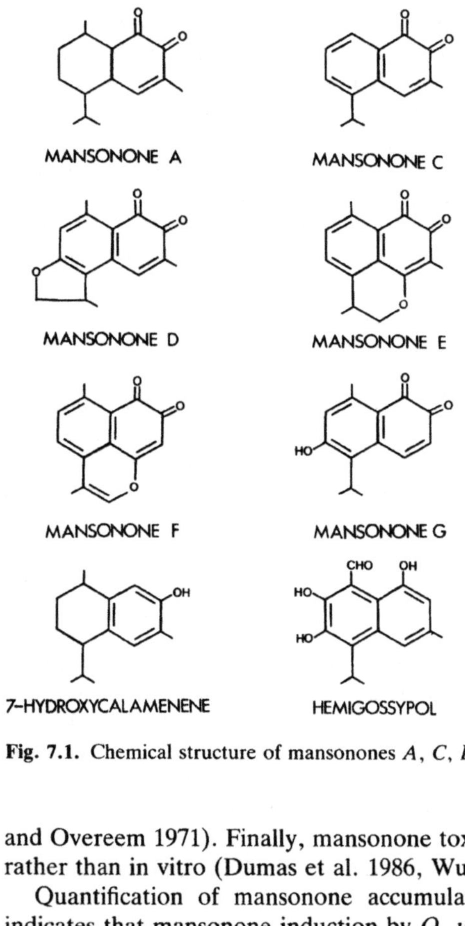

Fig. 7.1. Chemical structure of mansonones *A*, *C*, *D*, *E*, *F*, and *G*

and Overeem 1971). Finally, mansonone toxicity should be demonstrated in situ rather than in vitro (Dumas et al. 1986, Wu et al. 1989).

Quantification of mansonone accumulation in seedlings of *U. americana* indicates that mansonone induction by *O. ulmi* is dependent on isolate aggressiveness, since nonaggressive isolates stimulate more mansonone accumulation than do aggressive isolates (Duchesne et al. 1990). Co-inoculation of *U. americana* with both an aggressive and a nonaggressive isolate suggests that the aggressive isolates of *O. ulmi* have a suppressive influence on mansonone accumulation in this species (Fig. 7.2a), since inoculation with both an aggressive and a nonaggressive isolate leads to less mansonone accumulation than does inoculation with a nonaggressive isolate alone. These results provide further support to the hypothesis that aggressiveness in *O. ulmi* is linked to mansonone elicitation and suppression in elms (Fig. 7.3) (Duchesne et al. 1985, 1986, Yang et al. 1989). The evidence presented to date shows that mansonone turnover leads to a rapid degradation of mansonones in elms (Fig. 7.2a,b) (Duchesne et al. 1985, 1986). Therefore, resistance to DED must also be associated with other means of localization of the fungus to account for long-term compartmentalization of the pathogen within the vascular system of elms (Duchesne 1988). A recent investigation on nonhost resistance to DED corroborates these observa-

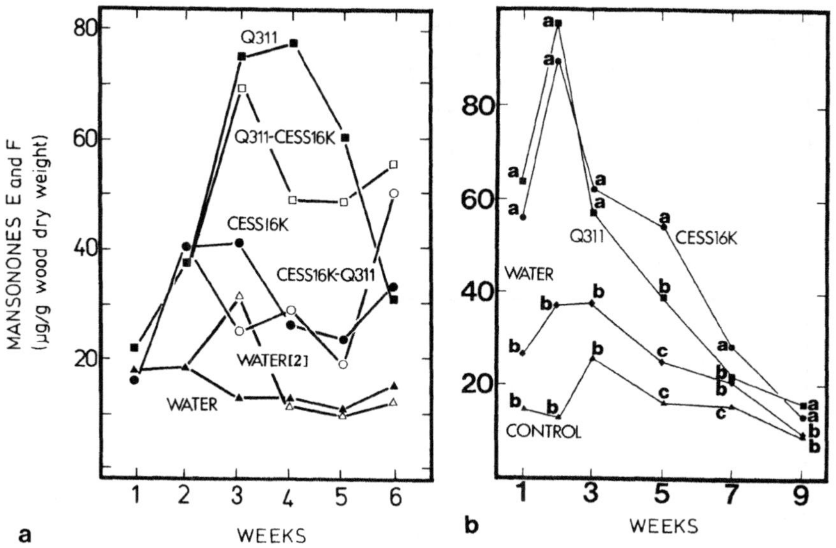

Fig. 7.2. **a** Mansonone accumulation in seedlings of *U. americana* in response to infection by an aggressive (*CESS16K*) and a nonaggressive (*Q311*) isolate of *Ophiostoma ulmi* or by both an aggressive and a nonaggressive isolate of *Ophiostoma ulmi* (*Q311-CESS16K*, *CESS16K-Q311*). Controls consist of seedlings inoculated with autoclaved distilled water. **b** Mansonone accumulation in seedlings of *U. pumila* in response to infection by an aggressive (*CESS16K*) and a nonaggressive (*Q311*) isolate of *Ophiostoma ulmi*. Controls consist of seedlings inoculated with autoclaved distilled water

tions and suggests that chemical inhibition may play a transient role early in the infection process before the onset of other mechanisms of defense (Rioux and Ouellette 1989).

7.3.3 Elicitation of Defense Reactions

In host-pathogen interactions the triggering of defense reactions is mediated through the presence of elicitors which are pathogen molecules derived from the pathogen culture fluid or their cell walls (Templeton and Lamb 1988). These molecules interact with the host at the plasma membrane level (Doke et al. 1987) and alter membrane potentials along with calcium and protein kinase mediated pathways (Dixon and Lamb 1990). Furthermore, virulent pathogens may also suppress phytoalexin accumulation through phytoalexin degradation or the use of suppressor molecules that shortcut the normal elicitation process (Deverall 1988).

Recent analyses have shown that cell wall fragments synthesized by *O. ulmi* in liquid culture can stimulate the production of mansonones in callus cultures of *U. americana* and *U. pumila* (Yang et al. 1989). The role of these molecules in

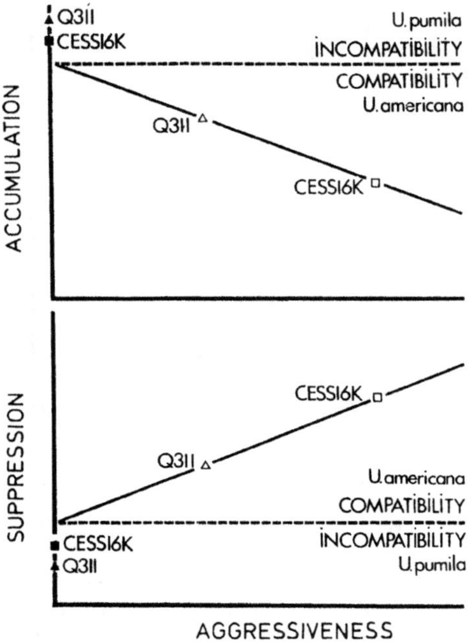

Fig. 7.3. Conceptual model showing the relationship between *O. ulmi* isolate aggressiveness, mansonone induction, and mansonone suppression in *U. americana*. Isolates with low aggressiveness are postulated to stimulate mansonone accumulation to a greater extent than more aggressive isolates. In turn, aggressive isolates are postulated to suppress mansonone accumulation

DED has yet to be demonstrated in situ since elicitor production may be different in vitro and in situ (De Wit et al. 1988).

The exact biochemical mechanisms involved in mansonone synthesis along with their regulation remains to be investigated. Gossipol, a phytoalexin of cotton (Al-Mousavi et al. 1983, Brinkeroff et al. 1984), is structurally similar to the mansonones. The synthesis of gossipol and related phytoalexins has been ascribed to the acetate-mevalonate pathway (Bell and Stipanovic 1978, Heinstein et al. 1979).

7.3.4 Protein Synthesis

Several enzymes normally present in healthy plants exhibit stimulated synthesis following mechanical injuries, UV irradiation, and pathogenic infection (Wingender et al. 1989). Most of these factors can cause increases in the respiration rate in elms (Jorgensen and Kondo 1967) and increases in number and activity of mitochondria specifically at different catabolic levels such as cytochrome oxidase, glucose-6-P-dehydrogenase, 6-phosphogluconate dehydrogenase, and invertase (Goodman et al. 1986). Most injuries can also stimulate enzymes associated with secondary metabolism such as peroxidase in *Pinus elliottii* (Lesney 1990) and polyphenoloxidase (Edreva et al. 1989). Others have linked increased peroxidase activity with enhanced lignification and resistance to pathogens (Grand et al. 1982, Moerschbacher et al. 1990). Injuries also stimulate the synthesis of phenylalanine ammonia-lyase and chalcone synthetase, and, concentrations of soluble secondary metabolites, including some phytoalexins (Bolwell et al. 1985). All of these processes are associated with enhanced RNA synthesis.

In many plant species resistance to pathogens is associated with de novo synthesis of numerous proteins such as hydroxyproline-rich glycoproteins (Chen and Varner 1985, Wilson and Fry 1986; Collinge and Slusarenko 1987, Corbin et al. 1987) protease inhibitors (Pena-Cortes et al. 1988, Ryan 1990) fatty acid synthetase (Willemot and Stumpf 1967), chitinases (Hedrick et al. 1988), pathogenesis-related proteins (Van Loon 1985, Legrand et al. 1987), and glucanases (Kauffmann et al. 1987). However, very few of these proteins have been reported in trees. One exception to this is the report that mechanical wounding of poplar leaves leads to de novo synthesis of proteins that are homologous to tobacco chitinases (Parsons et al. 1989). It has not been determined whether chitinase, as well as other induced proteins, can be detected in the xylem of woody angiosperms.

7.4 Future Research Emphasis

Comparison of the body of knowledge available on the biochemical mechanisms of resistance to tree diseases with the knowledge available on the same mechanisms in crop plants indicates clearly that the former is in its infancy. Fortunately, however, recent advances in tree biotechnology will soon allow the genetic manipulation of commercial woody species. Therefore, it will be important in the future to isolate tree genes that are useful in pest resistance. Particularly, it will be necessary to isolate a large number of genes in order to give horizontal or multigenic resistance to tree diseases. This is emphasized by the fact that trees have to resist numerous attacks from the same pathogens in the course of a forest, orchard, or urban rotation.

It is comparatively more difficult to investigate defense reactions of trees than those of crop plants. One reason for this is the slow growth of trees. This problem can be circumvented partly by the use of tissue cultures to study defense reactions (Diner and Karnosky 1987), and probe the mechanisms of secondary metabolism (Ellis 1984). Tissue culture techniques, along with further testing of results on whole plant systems, should favor an increase in the body of knowledge of tree defense reactions. Following this, it will be important to investigate the course of transcriptional activity of these factors to facilitate protection of trees against diseases. The application of genetic engineering and molecular biology tools to tree pathology should contribute greatly to this field of research.

7.5 Conclusions

Academic reasons compel researchers to present plant defense reactions in a synthetic manner that often misleads readers. It is of paramount importance to understand that any single mechanism of defense cannot be solely responsible for resistance to a single pathogen, although specific plant and pathogen genes may be involved in "gene-for-gene" resistance (Ellingboe 1981). It is more appropriate to envision disease resistance in trees as the result of several defense mechanisms that act in a complementary manner. These include a large array of passive defense mechanisms which are preestablished before pathogen infections. These will likely be the first line of defense to be encountered by pathogens. While various defense reactions are activated after infections, the timing of their arrival at the infection site is a critical determinant in the outcome of disease resistance. It is expected that a thorough understanding of the mechanisms involved in the biosynthesis of both passive and active defense reactions will help develop trees with enhanced disease resistance.

References

Al-Mousavi AH, Richardson PE, Essenberg M, Johnson WM 1983 Specificity on the envelopment of bacteria and other particles in cotton cotyledons. Phytopathology 73:4484–4489

Bamber RK 1987 Sapwood and heartwood. Tech Publ For Com NSW 2, Aust, 7 pp

Beckman CH 1971 The plasticizing of plant cell walls and tyloses formation. Physiol Plant Pathol 1:1–10

Bekkaoui F, Datla RSS, Pilon M, Tautorus TE, Crosby WL, Dunstan DI 1990 The effects of promoter on transient expression in conifer cell lines. Theor Appl Genet 79:353–359

Bell AA 1981 Biochemical mechanisms of disease resistance. Annu Rev Plant Physiol 32:21–81

Bell AA, Stipanovic RD 1978 Biochemistry of disease and pest resistance in cotton. Mycopathologia 65:91–106

Bolwell GP, Robbins MP, Dixon RA 1985 Metabolic changes in elicitor-treated bean cells. Enzymatic responses associated with rapid changes in cell wall components. Eur J Biochem 148:571–578

Brinkeroff LA, Verhalen LM, Johnson WM, Essenberg M, Richardson PE 1984 Development of immunity to bacterial blight of cotton and its implications for other diseases. Plant Dis 68:168–173

Burden RS, Kemp MS 1983 (−)-7-Hydroycalamenene, a phytoalexin from *Tillia europea*. Phytochemistry 22:1039–1040

Burden RS, Kemp MS 1984 Sesquiterpene phytoelexins from *Ulmus glabra*. Phytochemistry 23:383–385

Callow JA 1983 Biochemical plant pathology. John Wiley & Sons, New York, 484 pp

Chen J, Varner JE 1985 An extracellular matrix protein in plants: characterization of a genomic clone for carrot extensin. EMBO J 4:2145–2151

Collinge DB, Slusarenko AJ 1987 Plant gene expression in response to pathogens. Plant Mol Biol 9:389–410

Corbin DR, Sauer N, Lamb CJ 1987 Differential regulation of hydroxyproline-rich glycoprotein gene family in wounded and infected plants. Mol Cell Biol 7:4337–4344

Cote WA 1976 Wood ultrastructure in relation to chemical composition. In: Loewus FA, Runeckles VC (eds) The structure, biosynthesis and degradation of wood. Recent advances in biochemistry, vol 11. Plenum, New York, 1–44

Coxon DT 1982 Phytoalexins from other plant families. In: Bailey JA, Mansfield JW (eds) Phytoalexins. Blackie, London, 106–132

Darvill AG, Albersheim P 1984 Phytoelexins and their elicitation – a defense against microbial infection in plants. Annu Rev Plant Physiol 35:243–275

Deverall BJ 1982 Introduction. In: Bailey JA, Mansfield JW (eds) Phytoalexins. Blackie, London, 1–20

Deverall BJ 1988 Molecular bases of fungal parasitism. Aust J Plant Pathol 18:79–84

De Wit PJGM, Toma INJ, Joosten MHAJ 1988 Race specific elicitors and pathogenicity factors in the *Cladosporium fulvum*-tomato interaction. In: Keen NT, Kosuge T, Walling LL (eds) Physiology and biochemistry of plant-microbial interactions. Am Soc Plant Physiol, New York, 184 pp

Diner AM, Karnosky DF 1987 Tissue culture application to forest pathology and pest control. In: Bonga JM, Durzan DJ (eds) Cell and tissue culture in forestry, vol 2. Specific principles and methods: growth and developments. Nijhoff, Boston, 351–373

Dixon RA, Lamb CJ 1990 Molecular communication in interactions between plants and microbial pathogens. Annu Rev Plant Physiol Plant Mol Biol 41:339–367

Doke N, Chai HB, Kawaguchi A 1987 Biochemical basis of triggering and suppression of hypersensitive cell response. In: Nishimura S (ed) Molecular determinants of plant diseases. Springer, Berlin Heidelberg New York, 235–251

Duchesne LC 1988 Resistance mechanisms to Dutch elm disease: a review. Can Nat 115:163–167

Duchesne LC 1989 Protein synthesis in Pinus resinosa and the ectomycorrhizal fungus Paxillus involutus prior to ectomycorrhiza formation. Trees 3:73–77

Duchesne LC, Hubbes M, Jeng RS 1986 Mansonone E and F accumulation in Ulmus pumila, resistant to Dutch elm disease. Can J For Res 16:410–412

Duchesne LC, Jeng RS, Hubbes M 1985 Accumulation of phytoalexins in Ulmus americana in response to infection by a non-aggressive and an aggressive strain of Ophiostoma ulmi. Can J Bot 63:678–680

Duchesne LC, Jeng RS, Hubbes M, Sticklen MB 1990 The accumulation of mansonones E and F in seedling and callus cultures of elms in response to infection by Ophiostoma ulmi. Trees 4:187–190

Duchesne LC, Campbell SE, Koehler H, Peterson RL, Ellis BE 1989b Pine species influence suppression of Fusarium root rot by the ectomycorrhizal fungus Paxillus involutus. Symbiosis 7:139–148

Duchesne LC, Larson DW 1989 Cellulose and the evolution of plant life. BioScience 39:238–241

Duchesne LC, Peterson RL, Ellis BE 1988a Interaction between the ectomycorrhizal fungus Paxillus involutus and Pinus resinosa induces resistance to Fusarium oxysporum. Can J Bot 66:558–562

Duchesne LC, Peterson RL, Ellis BE 1988b Pine root exudate stimulates antibiotic synthesis by the ectomycorrhizal fungus Paxillus involutus. New Phytol 106:558–562

Duchesne LC, Peterson RL, Ellis BE 1989a The time course of disease suppression and antibiosis by the ectomycorrhizal fungus Paxillus involutus. New Phytol 111:693–698

Dumas MT, Strunz GM, Hubbes M, Jeng RS 1983 Isolation and identification of six mansonones from Ulmus americana infected by Ceratocystis ulmi. Experientia 39:1089–1090

Dumas MT, Strunz GM, Hubbes M, Jeng RS 1984 Inhibition of Ceratocystis ulmi by mansonones A, C, D, E, F, and G isolated from Ulmus americana. Eur J For Pathol 4:217–222

Edreva AM, Georgieva ID, Cholakova NI 1989 Pathogenic and non-pathogenic stress effects on peroxidases in leaves of tobacco. Environ Exp Bot 3:365–377

Elgersma DM 1969 Resistance mechanisms to Dutch elm disease. Meded Phytopathol Lab Willie Commelin Scholten. Baarn 77, 84 pp

Elgersma DM, Overeem JC 1971 The relation of mansonones to resistance against Dutch elm disease and their accumulation as induced by several agents. Neth J Plant Pathol 79:218–220

Ellingboe AH 1981 Changing concepts in host-pathogen genetics. Annu Rev Phytopathol 19:125–143
Ellis BE 1984 Probing secondary metabolism in plant cell cultures. Can J Bot 62:2912–2917
Eriksson KEL 1990 Biotechnology in the pulp and paper industry. Wood Sci Technol 24:79–101
Galeffi C, Casinovo CG, Monache EMD, Mirini-Bettolo GB 1968 A new method for the isolation of chloroform-soluble constituents of Mansonia wood. Ann Ist Super Sanata 4:305–316
Goodman RN, Kiraly Z, Wood KR 1986 The biochemistry and physiology of plant disease. Univ Missouri Press, Columbia, 433 pp
Grand C, Boudet AM, Ranjeva R 1982 Natural variations and controlled changes in lignification process. Holzforschung 36:217–224
Hart JH 1981 Role of phytostilbenes in decay and disease resistance. Annu Rev Phytopathol 19:437–458
Hartmann G, Nienhaus F 1974 The isolation of xanthoxylin from the bark of Phytophthora and Hendersonula-infected Citrus limon and its fungitoxic effect. Phytopathol Z 81:97–113
Hedrick SA, Bell JN, Boller T, Lamb CJ 1988 Chitinase cDNA cloning and mRNA induction by fungal elicitor, wounding and infection. Plant Physiol 86:182–186
Heinstein PF, Widmaier R, Wegner P, Howe J 1979 Biosynthesis of gossypol. Rev Adv Phytochem 12:313–333
Higuchi T 1990 Lignin biochemistry: biosynthesis and biodegradation. Wood Sci Technol 24:23–63
Hillis WE 1962 The distribution and formation of polyphenols within the tree. In: Hillis WE (ed) Wood extractives. Academic Press, New York London, 60–132
Hillis WE 1977 Secondary changes in wood. Recent Adv Phytochem 2:247–309
Hillis WE 1987 Heartwood and tree exudates. Springer, Berlin Heidelberg New York, 268 pp
Hubbes M 1990 Development of biotechnology programmes for energy forestry. Biomass 22:75–89
Hubbes M, Jeng RS 1981 Aggressiveness of Ceratocystis ulmi strains and resistance in Ulmus americana. Eur J For Pathol 11:257–254
Ingham JL 1982 Phytoalexins from the Leguminosae. In: Bailey JA, Mansfield JW (eds) Phytoalexins. Blackie, London, 21–82
Jeng RS, Alfenas AC, Hubbes M, Dumas MT 1983 Presence and accumulation of fungitoxic substances against Ceratocystis ulmi in Ulmus americana: possible relation to induced resistance. Eur J For Pathol 3:239–244
Jorgensen E, Kondo ES 1967 Measurement of gas exchange in woody plants. Nature (London) 213:413–414
Kauffmann S, Legrand M, Geoffroy P, Lamb CJ 1987 Biological function of pathogenesis-related proteins-4 PR proteins of tobacco have 1,3-beta-glucanase activity. EMBO J 6:3209–3212
Keen NT 1981 Evaluation of the role of phytoalexins. In: Staples RC (ed) Plant disease control. John Wiley & Sons, New York, 155–177
Kemp MS, Burden RS 1984 Isolation and structure of y-pyrufuran, a third induced antifungal benzofuran from the wood of Pyrus communis L. infected with Chondrostereum purpureum (Pers. ex Fr.) Pouzar. J Chem Soc Perkin Trans I:1441–1443
Kemp MS, Burden RS, Loeffler RST 1983 Isolation, structure determination, and total synthesis of the dibenzofurans a- and b- pyrufuran, new phytoalexins from the wood of Pyrus communis. J Chem Soc Perkin Trans I:2267–2272
Klimaszewska K 1989 Plantlet development from immature zygotic embryos of hybrid larch through somatic embryogenesis. Plant Sci 63:95–103
Kuc J 1976 Phytoalexins. In: Heitefuss RH, Williams PH (eds) Encyclopedia of plant physiology, vol 4. Springer, Berlin Heidelberg New York, 632–652
Langcake P, Cornford CA, Pryce RJ 1979 Identification of pterostilbene as phytoalexin from Vitis vinifera leaves. Phytochemistry 18:1025–1027
Legrand M, Kauffmann S, Geoffroy P, Fritig B 1987 Biological function of pathogenesis-related proteins: four tobacco pathogenesis-related proteins are chitinases. Proc Natl Acad Sci USA 84:6750–6754

Leguay JJ, Piecoup M, Puckett J, Jouanneau JP 1988 Common responses of cultured soybean cells to 2,4-D starvation and fungal elicitor treatment. Plant Cell Rep 7:19–22

Lesney MS 1990 Effect of elicitor on extracellular peroxidase activity on suspension-cultured slash pine (Pinus elliottii Engelm.). Plant Cell Tissue Organ Cult 20:173–175

Lewis NG, Yamamoto E 1990 Lignin: occurrence, biogenesis and degradation. Annu Rev Plant Physiol Plant Mol Biol 41:455–496

Lieth H, Whittaker RH 1975 The primary production of the biosphere. Springer, Berlin Heidelberg New York, 350 pp

Manion PD 1985 Tree disease concepts. Prentice-Hall, Englewood Cliffs, 399 pp

Mansfield JW 1982 The role of phytoalexins in disease resistance. In: Bailey JA, Mansfield JW (eds) Phytoalexins. Blackie, London, 106–132

Marini-Bettolo GB, Casinovo CG, Galeffi C 1965 A new class of quinones: sesquiterpenoid quinones of Mansonia altissima Chev. Tetrahedron Lett 52:4852–4864

Moerschbacher BM, Noll U, Gorrichon L, Reisener HJ 1990 Specific inhibition of lignification breaks hypersensitive resistance of wheat to stem rust. Physiol Plant 93:465–470

Nebert DW, Peterson DD, Fornace AJ 1990 Cellular responses to oxidative stress: the [Ah] gene battery as a paradigm. Environ Health Perspect 88:13–25

Nienhaus F, Castello JD 1989 Viruses in forest trees. Annu Rev Phytopathol 27:165–186

Otani H, Takai S, Richards WC, Kohmoto K, Nishimura S 1986 Effect of cerato-ulmin (CU) on electrolyte loss from elm leaf cells. Ann Phytopathol Soc Jpn 52:133 (Abstr)

Overeem JC, Elgersma DM 1970 Accumulation of mansonones E and F in *Ulmus hollandica* infected with Ceratocystis ulmi. Neth J Plant Pathol 9:1949–1952

Parsons JP, Bradshaw HD Jr, Gordon MP 1989 Systemic accumulation of specific mRNAs in response to wounding in poplar trees. Proc Natl Acad Sci USA 86:7895–7899

Paxton JD 1980 A new working definition of the term phytoalexin. Plant Dis 64:734

Pena-Cortes H, Sanchez-Serrano J, Rocha-Sosa M, Willmitzer L 1988 Systemic induction of proteinase inhibitor II gene expression in potato plants by wounding. Planta 174:84–89

Proctor RH, Smalley EB 1988 Localized accumulation of mansonones E and F in elms following inoculation with Ophiostoma ulmi. Can J Plant Pathol 10:371

Rao TA, Suresh PV, Sherieff AN 1987 Little known attributes of some mangrove plants of the Karnataka coast. Curr Sci India 56:725–727

Rioux D, Ouellette GB 1989 Light microscope observations of histological changes induced by Ophiostoma ulmi in various nonhost trees and shrubs. Can J Bot 67:2335–2351

Rowe JW 1989 Natural products of woody plants, vols 1 and 2. Springer, Berlin Heidelberg New York, 1243 pp

Rowe JW, Seikel MK, Roy DN, Jorgensen E 1972 Chemotaxonomy of Ulmus. Phytochemistry 11:2513–2517

Ryan CA 1990 Protease inhibitors in plants: genes for improving defenses against insects and pathogens. Annu Rev Phytopathol 28:425–449

Sasaya T 1987 Distribution and accumulation of extractives in tree trunks. I. Distribution of flavonoids in Katsura, Cercidiphyllum japonicum Sieb. et Lucc. wood. Res Bull College Exp For Hokkaido Univ 44:381–393

Scheffer RP 1983 Toxins as chemical determinants of plant diseases. In: Daly JM, Deverall BJ (eds) Toxins in plant pathogenesis. Academic Press, New York London, 1–40

Schlosser EW 1980 Preformed internal chemical defenses. In: Horsfall JG, Cowling EB (eds) Plant disease, vol 5. Academic Press, New York London, 161–177

Shain L 1967 Resistance of sapwood in stems of loblolly pine to infection by Fomes annosus. Phytopathology 57:1034–1045

Shigo AL 1984 Compartmentalization: a conceptual framework for understanding how trees grow and defend themselves. Annu Rev Phytopathol 22:189–214

Singh D, Smalley EB 1969 Nitrogenous and carbohydrates compounds in the xylem sap of Ulmacea species varying in resistance to Dutch elm disease. Can J Bot 47:335–339

Smith DA 1982 Toxicity of phytoalexins. In: Bailey JA, Mansfield JW (eds) Phytoalexins. Blackie, London, 218–252

Strobel GA, Van Alfen N, Hapner KD, McNeil M, Albersheim P 1978 Some phytotoxic glycopeptides from Ceratocystis ulmi, the Dutch elm disease pathogen. Biochim Biophys Acta 538:60–75

Suzuki H, Yasuda S, Hanzawa M 1971 The extractives of Ohyonire, *Ulmus laciniata* Mayr. The structure of lacinilene A and B, two new compounds from the heartwood. Mokuzai Gakkaishi 17:221–222

Takai S, Hiratsuka Y 1984 Scanning electron microscope observations of internal symptoms of white elm following *Ceratocystis ulmi* and cerato-ulmin treatment. Can J Bot 63:1365–1371

Templeton MD, Lamb CJ 1988 Elicitors and defence gene activation. Plant Cell Env iron 11:395–401

Van Alfen NK 1989 Reassessment of plant wilt toxins. Annu Rev Phytopathol 27:533–550

Van den Elzen PJM, Huisman MJ, Willink DPL, Jongedijk E, Hoekema A, Cornelissen BJC 1989 Engineering virus resistance in agricultural crops. Plant Mol Biol 13:337–346

Van Loon LC 1985 Effect of viral pathogenicity-related proteins on virus resistance in tobacco plants. Plant Mol Biol 4:111–116

Willemot C, Stumpf PK 1967 Fat metabolism in higher plants. Development of fatty acid synthetase as a function of protein synthesis in aging potato tuber slices. Plant Physiol 42:391–397

Wilson LG, Fry JC 1986 Extensin – a major cell wall glycoprotein. Plant Cell Environ 9:239–260

Wingender R, Rohrig H, Horicke C, Wing D, Schell J 1989 Differential regulation of soybean chalcone synthase genes in plant defence, symbiosis and upon environmental stimuli. Mol Gen Genet 218:315–322

Wink M 1988 Plant breeding: importance of plant secondary metabolites for protection against pathogens and herbivores. Theor Appl Genet 75:225–233

Woodwell GM, Whittaker RH, Reiners WA, Likens GE, Delwiche C, Botkin DB 1978 The biota and the world carbon budget. Science 199:141–146

Wu WD, Jeng RS, Hubbes M 1989 Toxic effects of elm phytoalexin mansonones on Ophiostoma ulmi, the causal agent of Dutch elm disease. Eur J For Pathol 9:343–357

Yang D, Jeng RS, Hubbes M 1989 Mansonone accumulation in elm callus cultures induced by elicitors of Ophiostoma ulmi, and general properties of elicitors. Can J Bot 67:3490–3497

Yang SF, Hoffman NE 1984 Ethylene biosynthesis and its regulation in higher plants. Annu Rev Plant Physiol 35:155–189

Yasuda S, Kinoshita Y, Hanzawa M 1973 Phenolic constituents of Akinire, Ulmus parviflora Jacq. Mokuzai Gakkaishi 19:593–598

Chapter 8

Biochemistry of Gymnosperm Xylem Responses to Fungal Invasion

T. Yamada

8.1 Introduction

Mechanisms limiting the spread of wood discoloration and decay in living trees have been discussed from several different viewpoints. The representative models are compartmentalization of decay in trees (CODIT) (Shigo and Marx 1977), reaction zone formation (Shain 1967, 1971), and an alternative view based on microenvironmental conditions (Boddy and Rayner 1983) (Chap. 6).

In this chapter, responses against fungal invasion of gymnosperm xylem formed before infection, particularly the production of inhibitory substances and histological barriers in the genera *Cryptomeria*, *Picea*, and *Pinus*, are described in terms of the reaction zone medel. Histological considerations, which are essential for understanding defense responses, are also discussed in combination with their biochemical features. Barrier zone responses are neglected in this chapter (Chap. 4).

The localized reactions of conifers essentially resemble the hypersensitive reactions of herbaceous plants. Phytoalexins, fungal toxic substances produced de novo following wounding or infection, contribute greatly to the localized, induced resistance mechanisms of plants (Lieutier and Berryman 1988a). Pinosylvin and pinosylvin monomethylether have been reported as induced inhibitory substances in the early studies of phytoalexins (Jorgensen 1961). However, much of our understanding of resistance mechanisms in gymnosperms depends on speculations based on results of studies with herbaceous plants. Although the responses described below are considered basically common in plants, several characters are specific to conifer xylem.

8.2 Reaction Zone and Transition Zone

Discoloration or decay of sapwood results in the formation of a reaction zone around discolored or decayed wood in response to injury or infection (Fig. 8.1). The original concept of the reaction zone is a necrotic region where phenols accumulate (Shain 1967). The transition zone, where most parenchyma cells are living, is formed contiguous to the sound area and encircles the reaction zone. Reaction zone formation is a nonspecific response in trees which are wounded or infected by various agents. The reaction zone is not the sole barrier responsible for the confinement of fungi in differentiated wood. The transition zone also contributes to their confinement in some instances. The term reaction zone

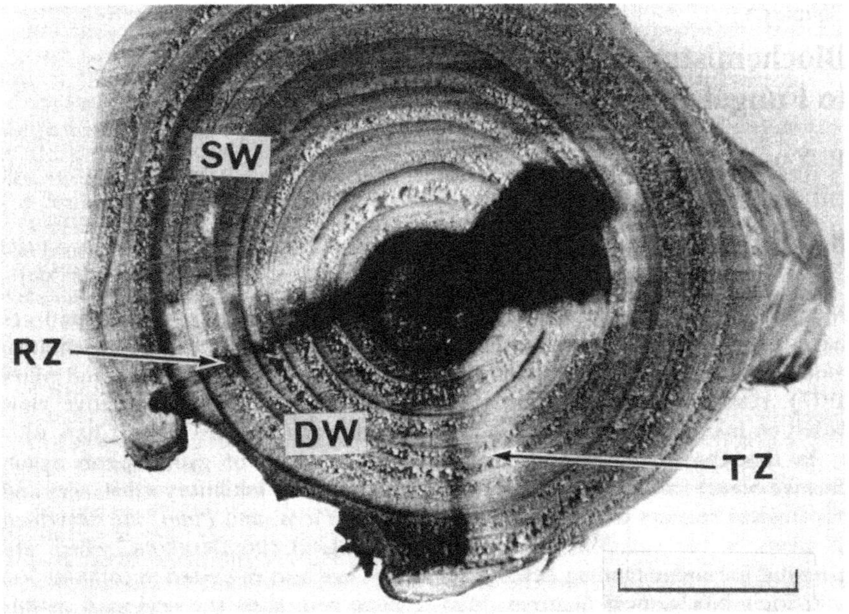

Fig. 8.1. Transverse view of *Cryptomeria japonica* infected with fungi following attack by the sugi bark borer. Sound sapwood (*SW*) is stained with acid fuchsin, indicating water conductivity. The dye does not penetrate the transition zone (*TZ*), reaction zone (*RZ*), or disclored sapwood (*DW*). Bar = 10 mm

barrier is used to describe an active defense barrier formed in differentiated wood and includes the reaction zone and the transition zone.

Chemical substances are the most important factor in defense mechanisms in the reaction zone barrier, but this barrier is not permanently impenetrable to fungi. Inhibitory extracts such as phenols and terpenoids may be detoxified by fungi (Loman 1970a, Popoff et al. 1975, Prior 1976) and, as the fungus slowly penetrates this tissue, the host continues to respond by producing additional protective layers (Shain 1967). Although this may be true when fungal hyphae grow outward from inner colonized wood, it is likely that the reaction zone barrier located in outer sapwood is more effective and durable when fungi invade through wounds. Furthermore, continuing active defense responses are assured by nutrients supplied to transition zone ray parenchyma from the cambium and phloem region.

A reaction zone was formed in advance of the pathogen *Heterobasidion annosum* (*Fomes annosus*), and nonhymenomycetous fungi were isolated occasionally from reaction zones (Shain 1967, 1971). The pathogen, however, invaded tissue impregnated with stilbenes, and then ceased growth in the pine-*Amylostereum* system described by Shain and Hillis (1972). Fungi invaded the reaction zone (outer discolored wood) of *Cryptomeria japonica* (sugi), but not the transition zone (intermediate wood) (Yamada et al. 1988) (Fig. 8.2).

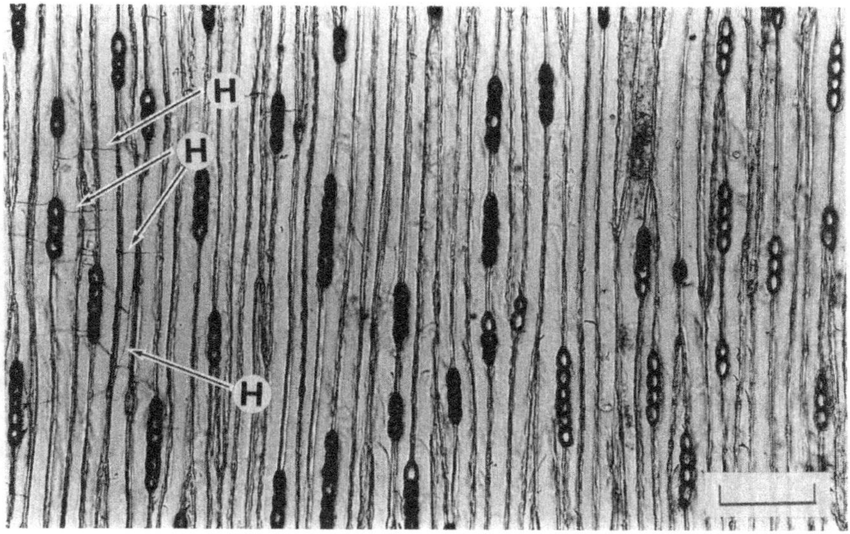

Fig. 8.2. Tangential section of sound and discolored sapwood boundary of *Cryptomeria japonica* colonized by fungi, stained for NAD diaphorase activity. Fungal hyphae (*H*) are abundant in reaction zone (*left*), but do not penetrate into the transition zone (*right*). Bar = 100 μm

The transition zone is a nonconducting, dry tissue where metabolic activity such as NAD diaphorase activity is greatly elevated (Shain 1971, Yamada et al. 1988) (Fig. 8.3). Ethylene production in the sapwood, which is implicated in triggering the synthesis of phenolic compounds, is attributable to host activity in the transition zone (Shain and Hillis 1972, 1973).

It has been reported that inhibitory substances accumulated not only in the reaction zone where parenchyma cells were not living, but in the transition zone in some cases. Stenlid and Johansson (1987) found phenol accumulation in the transition zone of *Picea abies* (Norway spruce) roots inoculated with *H. annosum*. No single zone which corresponded to the reaction zone in its original sense was observed in *Cryptomeria japonica*. In *C. japonica*, parenchyma cells of the inner transition zone, in which phenols accumulated, were still alive except for occasional dead cells (Yamada et al. 1988). Both living and dead cells have been observed in the inner transition zone (inner white zone) formed around heartwood of *C. japonica* (Nobuchi and Harada 1983), and the concentration of the phenol sequirin-C was maximum in the inner transition zone (Nobuchi et al. 1985). Survival rates of ray parenchyma cells from the outer part of the sapwood to the heartwood vary depending on tree species (Nobuchi et al. 1979, Yamamoto 1982). Whether parenchyma of phenol-enriched tissue is alive or dead may depend on tree species as well as the stage of reaction zone formation or season. All these observations are consistent with the concept that phytoalexins, which are toxic to host plant cells, accumulate in dead tissue. Even

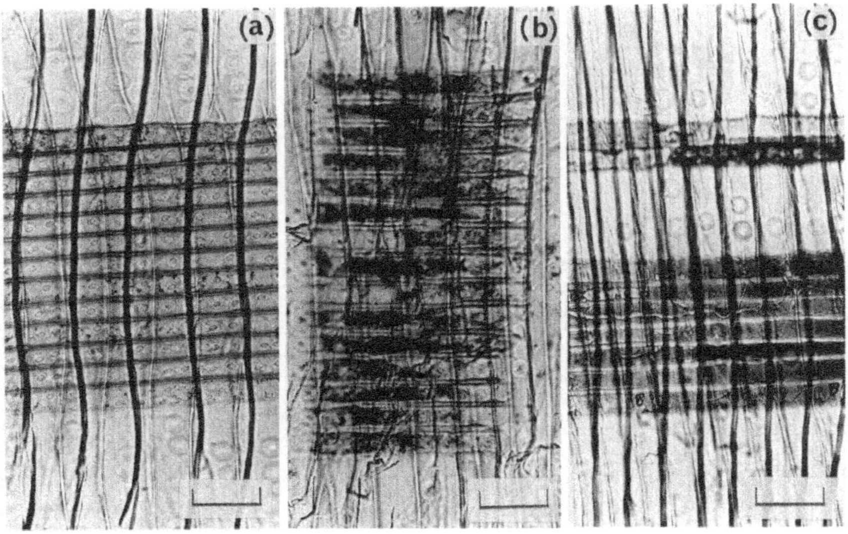

Fig. 8.3a–c. NAD diaphorase activity in xylem ray parenchyma of *Cryptomeria japonica* infected with fungi. Blue diformazan precipitation indicates NAD diaphorase activity. Radial section of **a** sound sapwood, **b** transition zone, and **c** discolored sapwood. No NAD diaphorase activity is observed in discolored sapwood parenchyma. Bar = 50 μm

in the phenol-enriched transition zone, phenols do not accumulate in living parenchyma cells but in tracheid lumina (Yamada et al. 1988).

Negative hydrostatic pressure in functioning sapwood, combined with possible pressure relaxation in discolored and decayed wood, induces the formation of a nonconducting dry zone (transition zone) (Coutts 1976). Exudation of deposits from parenchyma cells into tracheids, combined with hydrostatic pressure, could also be associated with pit aspiration or gas emboli. Pits become aspirated and emboli form in tracheids adjacent to plugged tracheids in *Pseudotsuga menziesii* (Douglas-fir) infected with *Ceratocystis wageneri* (Hessburg and Hansen 1987). Coutts (1977) postulated that altered metabolism contributed to the extended dry zone formation.

8.3 Inhibitory Substances

Heartwood phenols and their derivatives are involved not only in natural durability of heartwood as preformed agents, but also in postinfectional responses of sapwood. A large number of substances, such as terpenes (mono-, sesqui-, di-), flavonoids, phenylpropanoids, resin, and fatty acids, are involved in active defense mechanisms (Rowe 1989).

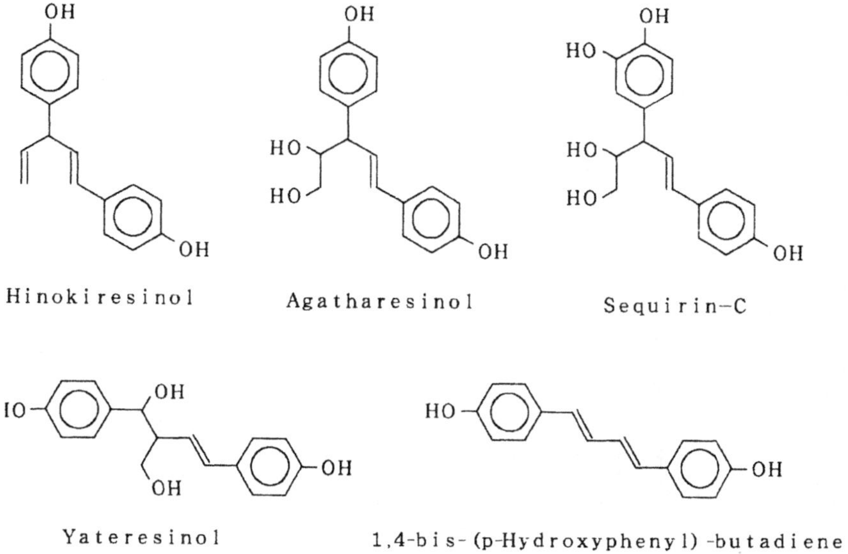

'ig. 8.4. Structures of norlignans

;.3.1 Phenolic Compounds

'henolic compounds play a large role and are critical to the formation of lefense mechanisms in conifer xylem. Defense responses of several conifer ;enera, *Cryptomeria*, *Picea*, and *Pinus* in relation to phenols, some of which are onsidered phytoalexins, are described here.

Norlignans have C_6-C_3-C_2-C_6 structures (Fig. 8.4), and are found in several oniferous families that lack resin canals in the xylem, e.g. Taxodiaceae, 'upressaceae, and Araucariaceae. Norlignans, as well as lignans, are derived rom two phenylpropanoid precursors.

In *Cryptomeria japonica*, norlignans which occur in very small amounts in ound sapwood are produced in the process of heartwood formation. They ccumulate also in the reaction zone barrier which surrounds the discolored apwood (Takahashi and Ogiyama 1985a,b, Yamada et al. 1988) (Fig. 8.5).

The ratio of each norlignan to total norlignan in the reaction zone barrier iffers from that in the heartwood. For example, heartwood contains little inokiresinol, yateresinol, and 1,4-bis-(p-hydroxyphenyl)-butadiene which are bundant in the reaction zone barrier (Takahashi and Ogiyama 1985a,b, 1986). 'he outermost (newly formed) sapwood has a high capability to produce inokiresinol (Takahashi and Ogiyama 1985a). Among norlignans produced in ıe reaction zone barrier, hinokiresinol and a few minor components have high ıhibitory activity against fungi (Yamada et al. 1988) (Table 8.1). These facts ıdicate that younger tissue has a greater inhibitory activity against fungal ıvasion, explaining the typical shape of discoloration with greater lateral and

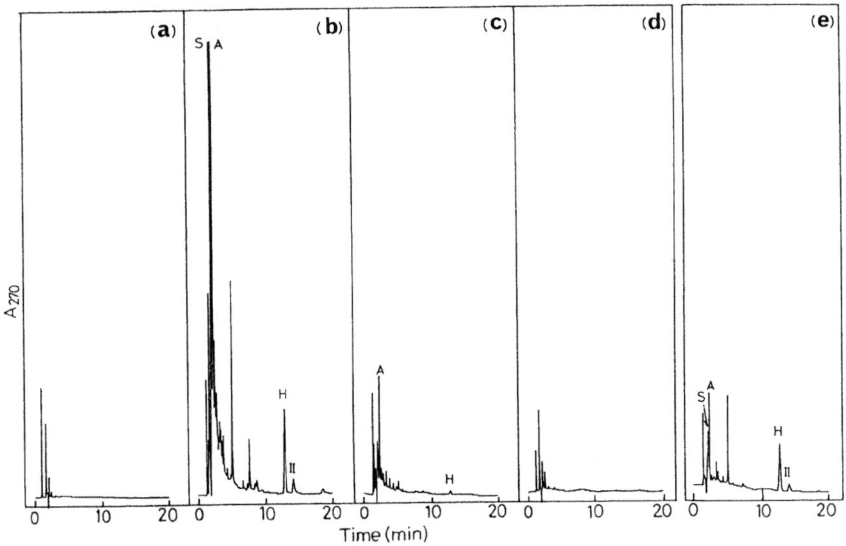

Fig. 8.5a–e. HPLC analysis of wood extract from *Cryptomeria japonica* naturally infected with fungi following bark borer attack. **a** Sound sapwood; **b** inner transition zone; **c** reaction zone; **d** discolored sapwood; **e** deposits in the inner transition zone. Peak *S*, sequirin-C; *A*, agatharesinol; *H*, hinokiresinol; *II*, unknown (Yamada et al. 1988)

Table 8.1. Relative inhibitory activity of extractives from the inner transition zone of *Cryptomeria japonica* against fungi isolated from discolored sapwood[a] (Yamada et al. 1988)

Compound	Fungi tested		
	Fusarium oxysporum	*Fusarium solani*	*Macrophoma sugi*
I	+++	+++	++
II	+	+	+
Hinokiresinol	+++	+++	+++
Agatharesinol	±	±	±
Sequirin-C	±	±	±

[a] Assessed on the size of the inhibition zone on TLC plates; −, no inhibition; ±, slight inhibition visible; +, ++, +++, increasingly large areas of inhibition.

longitudinal extension in inner sapwood. Residual norlignans present in the reaction zone may contribute to fungal confinement.

Other than norlignans, several terpenes and their derivatives accumulate in heartwood and probably in the reaction zone barrier of *C. japonica*. Ferruginol, a phenolic diterpene, is toxic to hyphae of *Lentinus lepideus* and *L. edodes* (Rudman 1965, Nakajima et al. 1980). Ferruginol and other diterpene phenols, as well as norlignans, may contribute to the delimitation of fungi in conifer xylem.

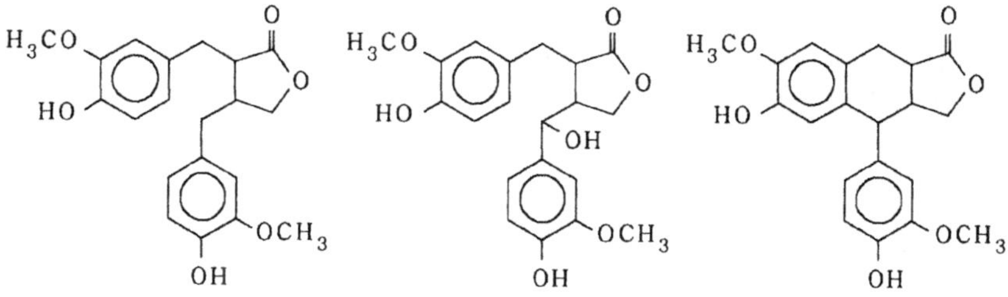

Fig. 8.6. Structures of lignans

Lignans are C_6-C_3-C_3-C_6 compounds (Fig. 8.6), and are distributed in Cupressaceae, Ginkgoaceae, Pinaceae, Podocarpaceae, Taxaceae, and several angiosperms. Hydroxymatairesinol, matairesinol, liovil, and conidendrin were identified in sound heartwood and reaction zones of *Picea abies* (Shain and Hillis 1971). Striking increases in hydroxymatairesinol in the reaction zone was observed, and the ratio of hydroxymatairesinol to total lignan in the reaction zone was higher than that in heartwood. Sound sapwood and wood in advanced stages of decay contained negligible quantities of lignans. The transition zone between sapwood and the reaction zone had a lignan content intermediate between that of sapwood and the reaction zone. Several other lignans were found in sapwood of *P. abies* attacked by *Heterobasidion annosum* (Popoff et al. 1975).

Hydroxymatairesinol was significantly more inhibitory to *H. annosum* than was matairesinol or conidendrin (Shain and Hillis 1971). It was thus concluded that hydroxymatairesinol in association with alkalinity in the reaction zone contributed to the resistance of the sapwood to *H. annosum*. Residual reaction zone components from the past, and present in incipiently decayed wood, probably contribute to its decay resistance (Shain 1971). Extensive central stem rot of *P. abies* caused by *H. annosum* was attributed to the lack of fungitoxic extractives in heartwood and the absence of oleoresin impregnation in inner sapwood (Gibbs 1968). Limited fungal penetration into living sapwood may be explained by the production of lignans.

Shain and Hillis (1972) found large quantities of phenols and relatively less resin soaking in *Sirex* lesions of *Pinus radiata* (Monterey pine) trees. They concluded that resin and phenol accumulations were triggered by different processes, and suggested that phenols, but not resin, participate in the resistance mechanisms.

General occurrence and antifungal activity of lignans, especially of hydroxymatairesinol, are controversial. No significant quantitative or qualitative differences were found in lignans between sound wood and decay-resistant wounded wood of *Picea glauca* (white spruce) (Hart et al. 1975). Although some of the phenolics accumulated in the reaction zone of *P. abies* roots attacked by *H. annosum* inhibited the growth of several wood-degrading fungi, the

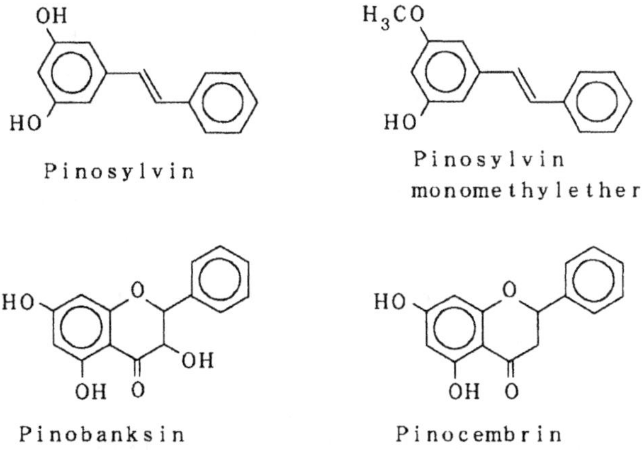

Fig. 8.7. Structures of stilbenes and flavanones

predominant lignan hydroxymatairesinol had no effect on these fungi (Popoff et al. 1975). The investigators assumed that the synergistic action of other minor components contributed to the prevention of fungal growth.

Stilbenes are C_6-C_2-C_6 compounds (Fig. 8.7) derived from a phenylpropanoid precursor which links with three acetate units (von Rudloff and Jorgensen 1963). Stilbenes, which occur naturally in a number of herbaceous and woody plants, are found in *Pinus* spp. Details on the role of stilbenes as inhibitory compounds have been reviewed (Hart and Shrimpton 1979, Hart 1981). Flavanones (C_6-C_3-C_6 compounds, Fig. 8.7) are widely distributed in the plant kingdom. Several flavanones such as pinobanksin, pinocembrin, and strobobanksin are known to occur in pines.

Sound sapwood of pine trees contains a very small amount of polyphenols (Shain 1967, Hillis and Inoue 1968). Abiotic stresses, e.g. desiccation, high humidity, ethylene, and UV, as well as biotic ones, are triggering agents that induce pinosylvins (pinosylvin and pinosylvin monomethylether) production (Hart 1981). Pinosylvins are reported to accumulate in wounded wood (Higuchi et al. 1967) and in the reaction zone (Shain 1967, Prior 1976) of many pine species, in response to fungal infection of the sapwood (Rishbeth 1972, Shrimpton 1973). Pinobanksin and pinocembrin also accumulate in wounded wood or reaction zones (Higuchi et al. 1967, Shain 1967, Shrimpton 1973).

Pinosylvin monomethylether extracted from *Pinus strobus* (eastern white pine) sapwood inhibited the growth of *Ceratocystis olivacea* (Rudman 1965) and *Armillaria ostoyae* (Mwangi et al. 1990). It has also been reported that pinosylvins significantly reduced the growth rate of *Heterobasidion annosum* (Shain 1967, Gibbs 1972). In addition, a negative correlation was shown between the extent of infection and pinosylvin concentration (Prior 1976). The restriction of the spread of the fungus appears to be due to the formation of stilbenes in pine sapwood after infection (Shain 1967, Hillis and Inoue 1968, Prior 1976).

Several problems that arise in assaying extracts of conifer xylem have been summarized by Hart (1981). The presence of cellulose or other woody materials can greatly reduce the toxicity of the extractives (Loman 1970b). However, Shain (1967) concluded that pinosylvins probably contributed to the resistance from the fact that extracted wood of the reaction zone and incipiently decayed region lost their resistance to decay.

The most conspicuous difference between the phenol content of the reaction zone and that of heartwood is the ratio among phenols, i.e., pinosylvin monomethylether:pinosylvin, and sequirin-C:hinokiresinol (Shain 1967, Hillis and Inoue 1968, Takahashi and Ogiyama 1985a,b, 1986). Since pinosylvin is more toxic to *H. annosum* than pinosylvin monomethylether, the tendency for a lower pinosylvin monomethylether:pinosylvin ratio in the reaction zone may be of some consequence for the defense mechanisms (Shain 1967). Similar consideration is applicable to the sequirin-C:hinokiresinol ratio in *C. japonica*. These alterations in the composition of phenols are common.

8.3.2 Terpenoids and Resin

The xylem of conifers, particularly of pines, produces large amounts of terpenes, which are constituents of oleoresin. It is characteristic in many, but not all, conifers that the resin is located in a resin canal system. However, only some genera in the Pinaceae have well-developed resin canals. Some genera lack resin ducts, and some develop only traumatic resin canals. Even in trees that lack resin canals, terpenoids are the quantitatively and qualitatively important components of heartwood or discolored wood.

Heavy resin impregnation after infection or wounding is a typical response in the xylem of pine, but not in other conifers. Fungal infection induced terpene accumulation in the xylem of pines (Shrimpton 1973, Wong and Berryman 1977, Cheniclet 1987). Resin acids also increase (Croteau et al. 1987), but at a lower rate than terpenes (Shrimpton 1973).

Volatile terpenes from conifers are mostly fungitoxic (Rudman 1962, Cobb et al. 1968, Gibbs 1972, Rishbeth 1972, Flodin and Fries 1978). The inhibitory effects vary widely among the fungi or substances tested (Bridges 1987). Nonvolatile terpenes and resin acids also generally inhibit fungal growth (Rudman 1965, Hartmann et al. 1981).

Resin production is assumed to be an important defense mechanism in conifer xylem, particularly of pines (Gibbs 1968, Hart et al. 1975). The role of resin in resistance, however, is still not fully understood. Resin impregnation was suggested to act as a nontoxic waterproofing layer which prevented fungal penetration (Verrall 1938, Rishbeth 1972, Hart et al. 1975, Prior 1976). Schuck (1980) indicated that the physical barrier depended on resin acids, whereas the monoterpenes of the resin acted as toxic chemical agents.

8.4 Site of Reactions and Inhibitory Substances

8.4.1 Origin of Resin

The site of synthesis of excess resin, which is conspicuous in the reaction zone in *Pinus* spp., is uncertain. It has been assumed that the excess resin accounting for impregnation, particularly in pines, originated from the resin canals (Shain 1967, Wong and Berryman 1977), because of the well-developed resin canal system and the large ratio of resin acids to fatty acids. It is, however, possible that metabolic changes in transition zone parenchyma (Shain 1971) cause excess production of resin by adjacent parenchyma.

Several observations support the view that resin is produced in ray parenchyma after injury or infection. The living tissues of *Pinus resinosa* (red pine) infected with *Heterobasidion annosum* were heavily resin-soaked along the rays (Shigo 1975). In paraquat-treated *Pinus elliottii* (slash pine), all affected xylem parenchyma cells, including those not directly associated with resin ducts, begin to synthesize excessive oleoresin and secrete it through pits into the lumina of contiguous tracheids (Birchem and Brown 1979). Lieutier and Berryman (1988a,b) also observed that resin-filled tracheids were usually located along rays and were not necessarily associated with resin canals, and tracheids close to resin ducts were often devoid of resin in pines inoculated with *Ceratocystis clavigera* or treated with chitosan. From their observations, they hypothesized that resin in tracheids originated from parenchyma.

Phenolic substances are also produced by parenchyma cells, and are excreted into adjacent conductive cells (Shain 1967, 1971, Gibbs 1968). Furthermore, in several cases described above, "resin" probably includes both terpenoids and phenols.

In coniferous species that lack resin canals in the xylem (e.g. *Cryptomeria* spp. and *Chamaecyparis* spp.), compartmentalization of infected tissue is observed similarly as in those which have a developed resin canal system (Yamada and Okuda 1987, Yamada et al. 1988). Defense mechanisms depending on oleoresin from the resin canal system can be negligible in the xylem of these trees. Phenols and terpenoids accumulated in the inner transition zone of *Cryptomeria japonica* indicate that they originate from parenchyma.

Two sorts of resin, primary and secondary, were distinguished on the basis of their origin for *Pinus contorta* var. *latifolia* (lodgepole pine) in response to attack by *Dendroctonus ponderosae* and blue stain fungi (Reid et al. 1967). They hypothesized that secondary resin originated from living parenchyma cells of xylem ray and phloem, and that it might therefore be parenchyma resin. Primary resin provides a rapid response to infection, since it is preformed. Reid et al. (1967) observed a primary oleoresin flow immediately after the beetle attack, which ceased after a few days, followed by a secondary resin flow that soaked the sapwood. They pointed out the lack of secondary resinosis in successfully attacked trees, and concluded that secondary resinosis was the most important factor in resistance to attack.

Changes in the percent composition of resin acids after infection or injury have been observed, although all components found in the defense responses were normal constituents of heartwood (Shrimpton 1973, Hart et al. 1975). There were quantitative and qualitative differences in the monoterpene composition of preformed xylem resin compared with those extracted from the hypersensitive tissue of *Pinus taeda* (loblolly pine) inoculated with *Ceratocystis minor* (Paine et al. 1987). Composition of the secondary resin is essentially the same as the primary constitutive resin, except for the ratio resin acids/terpenes (Cheniclet 1987). *Picea abies* wood reacts to injuries with an alteration of the terpene fraction (Schuck 1980, 1982). These findings suggested that the accumulation of resins was not due simply to movement of resins from the preformed system.

8.4.2 Tracheid Occlusion

It is supposed that parenchyma can be a defense barrier by itself in the wood of certain angiosperms (Shigo and Marx 1977). Because ray parenchyma of most gymnosperms, however, are single or consist of a few stranded cells that are short in height, fungal hyphae can easily bypass the ray parenchyma.

Suberization occurred at the reaction zone in *Tsuga canadensis* but not *Pinus strobus* (Biggs 1987). Discontinuous distribution of xylem ray and axial parenchyma in the wood formed before wounding suggests the limited efficiency of suberization in the reaction zone barrier (Yamada et al. 1988). Biggs (1987) reported that continuous boundaries were formed by suberization, tyloses, and plugging with insoluble material in woody plants. Suberization, however, seemed to be insufficient for compartmentalization in conifer xylem because of the limited extent of suberin formation after wounding.

Wound-initiated tyloses, as well as tylosoids in resin canals, are observed in tracheids of conifer xylem with thin-walled epithelial cells such as pines. Tyloses, developed from ray parenchyma, protruded into the lumina of adjacent tracheids in *Pinus monophylla* (single-leaf piñon pine), *P. ponderosa* (ponderosa pine) (Smith 1967), and *Pseudotsuga menziesii* (Hessburg and Hansen 1987) infected with *Ceratocystis wageneri*. Tyloses and tylosoids appear to play a partial role in confining the pathogen in these species.

Induced tracheid occlusion with resinous material has been observed in many conifer species, and may contribute to defense against fungi. Because droplet deposition within tracheids occurs at the normal sapwood-heartwood boundary (Nobuchi and Harada 1983), induced tracheid occlusion should be considered a nonspecific response.

Blockage of pit pairs by resinous compounds accumulating in the lesions of *Abies grandis* (grand fir) inoculated with *Trichosporium symbioticum* may explain the limited circumferential and radial spread of the pathogen (Wong and Berryman 1977). Plugging of tracheids by a resinous substance has been found in pine sapwood as the response to infection or treatment with chitosan (Hessburg and Hansen 1987, Lieutier and Berryman 1988a,b). Plugging was absent from the advancing margin of stain (Hessburg and Hansen 1987) and occurred in a large number of the conductive cells both in phloem and sapwood (i.e., sieve cells and tracheids). The barrier zone tracheids of *P. resinosa* in-

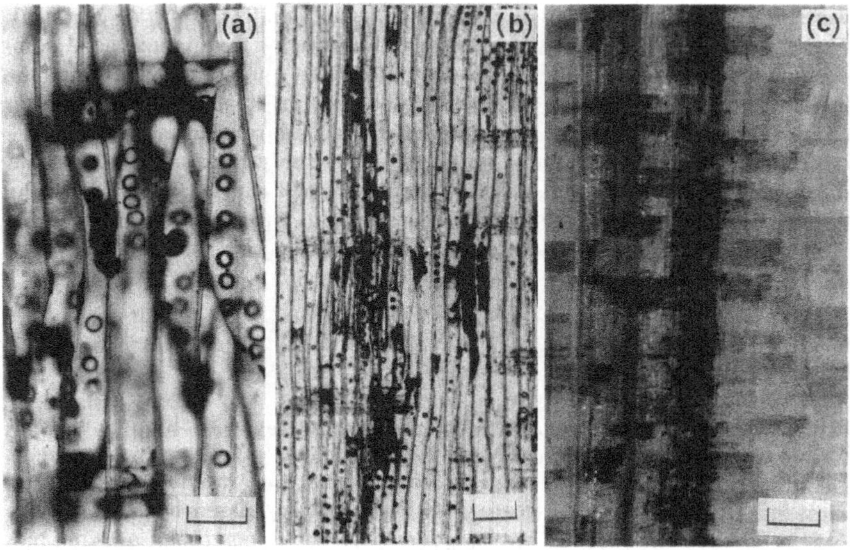

Fig. 8.8a–c. Droplet deposition in tracheid lumina in the inner transition zone of *Cryptomeria japonica*. **a** Droplet deposition in tracheid lumina and pits, stained with nile blue; bar = 50 μm. **b** Tracheid occlusion with viscous droplets, stained with nitroblue tetrazolium (NBT); bar = 100 μm. **c** Continuous barrier formed by deposits, stained with NBT; bar = 200 μm

fected with *Heterobasidion annosum* were also often resin impregnated (Tippett and Shigo 1980).

Shain (1967) indicated cytochemically that oleoresin and phenols were distributed in all xylarly elements of the reaction zone. Oil droplet-like deposits were found to occlude tracheid lumina of the inner transition zone circumscribing the reaction zone of *Cryptomeria japonica* (Yamada et al. 1988) (Fig. 8.8). Histochemical observation indicated that these deposits contained lipids and phenols.

Only a few chemical analyses have been conducted on the deposits occluding conifer tracheids. The deposits in tracheid lumina of *Tsuga heterophylla* (western hemlock) heartwood contained the lingans matairesinol, hydroxymatairesinol, and conidendrin, often in quantities sufficient to plug tracheids (Krahmer et al. 1970). The deposits lined tracheid walls and often encrusted the bordered pits. Lignans were also present in the cell walls. Deposits from tracheid lumina of the inner transition zone circumscribing discolored sapwood of *C. japonica*, contained antifungal norlignans such as hinokiresinol (Yamada et al. 1988) (Fig. 8.5). Every longitudinal tracheid is attached to a parenchyma cell. The deposits containing inhibitory substances block hyphal passage through tracheid lumina and pits, and form spatially continuous reaction zone barriers in sapwood of *C. japonica*. Accumulation of phenols and terpenoids in the reaction zone barrier and occluded tracheids suggest that similar defense mechanisms to *C. japonica* may operate in other conifer species.

8.5 Microenvironmental Factors

It is suggested that the development of decay in living trees can be explained in terms of wood moisture status (Boddy and Rayner 1983, Leben 1985) (Chap. 6). According to their theory, resistance of sapwood to fungal invasion is attributed to the high moisture content and insufficient amount of oxygen assured by the waterproofing layer (Boddy and Rayner 1983). They consider that microenvironmental conditions in water-filled xylem are inhospitable to fungi and that conditions in an aerated tissue are more favorable. Although low O_2 and high CO_2 concentrations are expected in the transition zone, *Heterobasidion annosum* grows equally well under aerobic and microaerophilic conditions (Gundersen 1961).

No fungal penetration was observed in the transition zone of *Cryptomeria japonica*, although the inner transition zone was not water-conducting and contained less water than that of sound sapwood (Yamada et al. 1988). This indicates that a high moisture content cannot directly prevent fungal spreading. It is unlikely that microenvironmental factors play a major role in resistance, because in many cases conducting sapwood is separated from the pathogen by a nonconducting reaction zone or transition zone where the moisture content is rather low (Shain 1971, Coutts 1976, Yamada et al. 1988).

Resin-impregnated wood has been suggested to act as a nontoxic waterproofing layer which prevented fungal penetration (Verrall 1938, Rishbeth 1972, Hart et al. 1975). However, it appears impossible that resin impregnation per se constitutes the water sealant which divides water-filled and aerated tissue, although the reaction zone is thought to serve as a waterproofing layer because of the low moisture content of the transition zone between the resin-impregnated region and sound sapwood. Functional changes resulting in barriers to water movement, e.g. pit aspiration, may play a role in tree defenses other than as barriers to fungi.

The characteristic shape of discolored wood and hence that of reaction zones are explained as the result of increased penetration by the fungus with increasing distance from the cambium and greater penetration longitudinally than transversely (Shain 1967). Rayner and Boddy (1988) interpret the positioning and "relative strength" of CODIT walls 1–3 according to purely anatomical considerations. Wong and Berryman (1977) also argued that long, tapered tracheids explained the greater vertical lesion expansion. The patterns of distribution of fungi colonizing from wounds and the corresponding position of reaction zones or barrier zones relate to the likely patterns of aeration and drying following injury (Rayner and Boddy 1988).

Lesions showed a rapid expansion phase followed by reduced expansion and stabilization in fungus-inoculated *Abies grandis* (Wong and Berryman 1977). Such an initial rapid expansion of lesions is often observed. Leben (1985) suggested that hydrostatic tension in the wood was a major determinant of the volume of the discolored wood column. The extent of initial rapid expansion of lesions may be determined by hydrostatic tension when a tree is infected by fungi via wounds.

Considering the water distribution in sapwood, the greater colonization of inner sapwood, explained as a result of less water and better aeration by Boddy and Rayner (1983), could be caused by less viability of parenchyma and hence a less active response, particularly in gymnosperms. The shape of discoloration can be explained as well by the active host responses.

Depletion of readily utilizable substances limits fungal spread. Shrimpton (1973) found a decrease in sugar levels that accompanied the increase in other extractives in the sapwood of *Pinus contorta* var. *latifolia* attacked by the bark beetle and associated fungi. The effective inhibition of fungal colonization could be due to lower sugar levels and the presence of growth inhibitors (Shrimpton 1973, Worrall and Harrington 1988). Wong and Berryman (1977) suggested that early fungal confinement in resistant *Abies grandis* was due to degenerative metabolism within the lesion.

8.6 Elicitation of Reactions

Phytoalexins play a important role in localized induced resistance. Their production by host plant cells is induced by low-molecular weight substances termed elicitors. They are usually derived from cell walls of pathogens or host plants. The elicitors responsible for systemic resistance originate from plant cell walls and diffuse throughout the plant. Several compounds, including polysaccharides, glycoproteins, peptides, or fatty acids are known as elicitors of defense reactions in plants (Darvill and Albersheim 1984).

The mechanisms of elicitation of defensive reactions in conifers have been investigated only recently. Chitosan and proteinase inhibitor-inducing factor (PIIF) have been used as model elicitors of conifers in several works. Chitosan is a fungal cell wall-derived elicitor of the hypersensitive reaction, and PIIF is an elicitor of systemic resistance derived from plant cell walls. Lieutier and Berryman (1988a) showed that chitosan and fungal inoculations elicited very similar reactions in pines; they were well-developed hypersensitive reactions, suggesting that similar mechanisms of response were induced.

Lieutier and Berryman (1988a,b) suggested that PIIF was involved only in the systemic reactions of conifers, because it induced no hypersensitive reaction, in spite of the fact that PIIF can induce localized reactions as well as systemic ones in some herbaceous plants. They hypothesized that wounds alone could initiate the systemic reaction, but not the localized one. Chitosan or PIIF, however, as well as *Ceratocystis clavigera* inoculation, induced monoterpene accumulation in *Pinus contorta* var. *latifolia* phloem (Miller et al. 1986). Croteau et al. (1987) also found that wounding, treatment with PIIF or chitosan, or *C. clavigera* inoculation induced elevated levels of monoterpenes and diterpene resin acids in *P. contorta* var. *latifolia*. These observations indicate the induction of localized response by PIIF or wounding, although at a lower intensity of response.

Necrosis of parenchyma cells did not result from direct hyphal attack, because death of xylem parenchyma cells occurs in advance of actual hyphal penetration (Shain 1967, Yamada et al. 1988). Furthermore, living host cells were responsive, though not invaded by fungi (Hessburg and Hansen 1987). Metabolic changes occur and phenols accumulate in the inner transition zone, where parenchyma cells are not in direct contact with fungal hyphae (Shain 1971, Yamada et al. 1988). These observations suggest that metabolites or signals capable of inducing cell death are produced in infected tissue and translocated to noninfected tissue. It is possible that elicitors derived from plant cells by injury contribute to the formation of the reaction zone barrier because of its nonspecific induction.

8.7 Perspectives

It is necessary to evaluate the models proposed to describe the patterns of wood discoloration and decay and therefore the mechanisms of restriction, and then to understand host reactions with an integrated view, supported by chemical and histological analyses. It is especially important to search for unity among the dynamic responses of conifers to injury and infection and ultimately develop a new understanding of tree defense mechanisms.

Antimicrobial microenvironment (for example, partial pressure of O_2 or of CO_2, pH, and accessibility to or availability of nutrients) works synergistically together with active responses. The reaction zone barrier is an inhibitory field where these factors interact. Cell walls of parenchyma and tracheids and, in part, fungi themselves, constitute such fields in the wood. The discolored region inhabited by fungi appears as hostile to fungi as the reaction and transition zones except for the lack of inhibitory substances in the first. Therefore, dynamic responses are considered the most important mechanisms contributing to defense in conifer xylem, although microenvironmental factors should not be ignored completely.

Phytoalexin production is a typical dynamic response and is induced nonspecifically by abiotic stresses as well as biotic agents. Elicitors are not only produced in host-parasite interactions, but autoregulated by the host under the influence of abiotic stresses. Mechanisms for the induction of defense responses by simple wounding may be explained in terms of elicitors.

Tracheid occlusion by substances secreted from adjoining parenchyma and containing phytoalexins is important in conifer xylem for blocking hyphal growth. Mechanisms causing dry zone formation are still poorly understood. Relationships between dry zone formation and the production of inhibitory substances have not been investigated. Future studies are required to elucidate these relationships and the elicitation process of defense responses.

References

Biggs AR 1987 Occurrence and location of suberin in wound reaction zones in xylem of 17 tree species. Phytopathology 77:718–725

Birchem R, Brown CL 1979 Ultrastructure of paraquat-treated slash pine (Pinus elliottii Engelm.). Am J Bot 66:1208–1218

Boddy L, Rayner ADM 1983 Origins of decay in living deciduous trees: the role of moisture content and a re-appraisal of the expanded concept of tree decay. New Phytol 94:623–641
Bridges JR 1987 Effects of terpenoid compound on growth of symbiotic fungi associated with southern pine beetle. Phytopathology 77:83–85
Cheniclet C 1987 Effects of wounding and fungus inoculation on terpene producing systems of maritime pine. J Exp Bot 38:1557–1572
Cobb FW Jr, Krstic M, Zavarin E, Barber HW Jr 1968 Inhibitory effects of volatile oleoresin components on Fomes annosus and four Ceratocystis species. Phytopathology 58:1327–1335
Coutts MP 1976 The formation of dry zones in the sapwood of conifers. I. Induction of drying in standing trees and logs by Fomes annosus and extracts of infected wood. Eur J For Pathol 6:372–381
Coutts MP 1977 The formation of dry zones in the sapwood of conifers. II. The role of living cells in the release of water. Eur J For Pathol 7:6–12
Croteau R, Gurkewitz S, Johnson MA, Fisk HJ 1987 Biochemistry of oleoresinosis. Monoterpene and diterpene biosynthesis in lodgepole pine saplings infected with Ceratocystis clavigera or treated with carbohydrate elicitors. Plant Physiol 85:1123–1128
Darvill AG, Albersheim P 1984 Phytoalexins and their elicitors. A defense against microbial infection in plants. Annu Rev Plant Physiol 35:243–275
Flodin K, Fries N 1978 Studies on volatile compounds from Pinus silvestris and their effect on wood-decomposing fungi. II. Effects of some volatile compounds on fungal growth. Eur J For Pathol 8:300–310
Gibbs JN 1968 Resin and the resistance of conifers to Fomes annosus. Ann Bot 32:649–665
Gibbs JN 1972 Tolerance of Fomes annosus isolates to pine oleoresins and pinosylvins. Eur J For Pathol 2:147–151
Gundersen K 1961 Growth of Fomes annosus under reduced oxygen pressure and the effect of carbon dioxide. Nature (London) 190:649–650
Hart JH 1981 Role of phytostilbenes in decay and disease resistance. Annu Rev Phytopathol 19:437–458
Hart JH, Shrimpton DM 1979 Role of stilbenes in resistance of wood to decay. Phytopathology 69:1138–1143
Hart JH, Wardell JF, Hemingway RW 1975 Formation of oleoresin and lignans in sapwood of white spruce in response to wounding. Phytopathology 65:412–417
Hartmann E, Renz B, Jung JA 1981 Untersuchungen über Bakterien- und Pilzhemmstoffe in höheren Pflanzen. Isolierung, Identifizierung und Wirkungsspektrum von zwei Harzsäuren aus Fichtenrinden. Phytopathol Z 101:31–42
Hessburg PF, Hansen EM 1987 Pathological anatomy of black stain root disease of Douglas-fir. Can J Bot 65:962–971
Higuchi T, Shimada M, Watanabe K 1967 Studies on the mechanism of heartwood formation. VI. On the artificial heartwood of Cryptomeria japonica and Pinus densiflora. Mokuzai Gakkaishi 13:274–279
Hillis WE, Inoue T 1968 The formation of polyphenols in trees. IV. The polyphenols formed in Pinus radiata after Sirex attack. Phytochemistry 7:13–22
Jorgensen E 1961 The formation of pinosylvin and its monomethyl ether in the sapwood of Pinus resinosa Ait. Can J Bot 39:1765–1772
Krahmer RL, Hemingway RW, Hillis WE 1970 The cellular distribution of lignans in Tsuga heterophylla wood. Wood Sci Technol 4:122–139
Leben C 1985 Wound occlusion and discolouration columns in red maple. New Phytol 99:485–490
Lieutier F, Berryman AA 1988a Elicitation of defensive reactions in conifers. In: Mattson WJ, Levieux J, Bernard-Dagan C (eds) Mechanisms of woody plant defenses against insects. Search for pattern. Springer, Berlin Heidelberg New York, 313–319
Lieutier F, Berryman AA 1988b Preliminary histological investigations of the defense reactions of three pines to Ceratocystis clavigera and two chemical elicitors. Can J For Res 18:1243–1247
Loman AA 1970a The effect of heartwood fungi of Pinus contorta var. latifolia on pinosylvin, pinosylvinmonomethyl ether, pinobanksin, and pinocembrin. Can J Bot 48:737–747

Loman AA 1970b Bioassays of fungi isolated from Pinus contorta var. latifolia with pinosylvin, pinosylvinmonomethyl ether, pinobanksin, and pinocembrin. Can J Bot 48:1303–1308
Miller RH, Berryman AA, Ryan CA 1986 Biotic elicitors of defense reactions in lodgepole pine. Phytochemistry 25:611–612
Mwangi LM, Lin D, Hubbes M 1990 Chemical factors in Pinus strobus inhibitory to Armillaria ostoyae. Eur J For Pathol 20:8–14
Nakajima K, Yoshimoto T, Fukuzumi T 1980 Substances inhibiting growth of shiitake mycelium in sugi wood (Cryptomeria japonica D. Don). Mokuzai Gakkaishi 26:698–702
Nobuchi T, Harada H 1983 Physiological features of the "white zone" of sugi (Cryptomeria japonica D. Don) – cytological structure and moisture content. Mokuzai Gakkaishi 29:824–832
Nobuchi T, Matsuno H, Harada H 1985 Radial distribution of heartwood phenols and the related cytological structure in the fresh wood of sugi (Cryptomeria japonica D. Don). Mokuzai Gakkaishi 31:711–718
Nobuchi T, Takahara S, Harada H 1979 Studies on the survival rate of ray parenchyma cells with ageing process in coniferous secondary xylem. Bull Kyoto Univ For 51:239–246
Paine TD, Blanche CA, Nebeker TE, Stephen FM 1987 Composition of loblolly pine resin defenses: comparison of monoterpenes from induced lesion and sapwood resin. Can J For Res 17:1202–1206
Pearce RB, Woodward S 1986 Compartmentalization and reaction zone barriers at the margin of decayed sapwood in Acer saccharinum L. Physiol Mol Plant Pathol 29:197–216
Popoff T, Theander O, Johansson M 1975 Changes in sapwood of roots of Norway spruce, attacked by Fomes annosus. Part II. Organic chemical constituents and their biological effects. Physiol Plant 34:347–356
Prior C 1976 Resistance by Corsican pine to attack by Heterobasidion annosum. Ann Bot 40:261–279
Rayner ADM, Boddy L 1988 Fungal decomposition of wood. Its biology and ecology. John Wiley & Sons, New York Brisbane, 587 pp
Reid RW, Whitney HS, Watson JA 1967 Reactions of lodgepole pine to attack by Dendroctonus ponderosae Hopkins and blue stain fungi. Can J Bot 45:1115–1126
Rishbeth J 1972 Resistance to fungal pathogens of tree roots. Proc R Soc London Ser B 181:333–351
Rowe JW (ed) 1989 Natural products of woody plants. Chemicals extraneous to the lignocellulosic cell wall. Springer, Berlin Heidelberg New York, 1243 pp
Rudman P 1962 The causes of natural durability in timber. IX. The antifungal activity of heartwood extractives in a wood substrate. Holzforschung 16:74–77
Rudman P 1965 The causes of natural durability in timber, pt.XVIII. Further notes on the fungi toxicity of wood extractives. Holzforschung 19:57–58
Schuck HJ 1980 Monoterpenes and resistance of conifers to fungi. In: Proc Worksh Genet Host-parasite Interact For, 169–175
Schuck HJ 1982 The chemical composition of the monoterpene fraction in wounded wood of Picea abies and its significance for the resistance against wound infecting fungi. Eur J For Pathol 12:175–181
Shain L 1967 Resistance of sapwood in stems of loblolly pine to infection by Fomes annosus. Phytopathology 57:1034–1045
Shain L 1971 The response of sapwood of Norway spruce to infection by Fomes annosus. Phytopathology 61:301–307
Shain L, Hillis WE 1971 Phenolic extractives in Norway spruce and their effects on Fomes annosus. Phytopathology 61:841–845
Shain L, Hillis WE 1972 Ethylene production in Pinus radiata in response to Sirex-Amylostereum attack. Phytopathology 62:1407–1409
Shain L, Hillis WE 1973 Ethylene production in xylem of Pinus radiata in relation to heartwood formation. Can J Bot 51:1331–1335
Shigo AL 1975 Compartmentalization of decay associated with Fomes annosus in trunks of Pinus resinosa. Phytopathology 65:1038–1039
Shigo AL, Marx HG 1977 Compartmentalization of decay in trees. US Dep Agr For Serv Agr Inf Bull 405, 73 pp

Shrimpton DM 1973 Extractives associated with wound response of lodgepole pine attacked by the mountain pine beetle and associated microorganisms. Can J Bot 51:527–534

Smith RS Jr 1967 Verticicladiella root disease of pines. Phytopathology 57:935–938

Stenlid J, Johansson M 1987 Infection of roots of Norway spruce (Picea abies) by Heterobasidion annosum. II. Early changes in phenolic content and toxicity. Eur J For Pathol 17:217–226

Takahashi K, Ogiyama K 1985a Phenols of discolored sugi (Cryptomeria japonica D. Don) sapwood. II. Norlignans of discolored sugi sapwood collected in the Kyushu region. Mokuzai Gakkaishi 31:28–38

Takahashi K, Ogiyama K 1985b Phenols of discolored sugi (Cryptomeria japonica D. Don) sapwood. III. Norlignans of "hachikami" and "shimi" woods. Mokuzai Gakkaishi 31:677–683

Takahashi K, Ogiyama K 1986 Phenols of discolored sugi (Cryptomeria japonica D. Don) sapwood. IV. Norlignans of Ayasugi (cultivar) in the Kyushu region. Mokuzai Gakkaishi 32:457–461

Tippett JT, Shigo AL 1980 Barrier zone anatomy in red pine roots invaded by Heterobasidion annosum. Can J For Res 10:224–232

Verrall AF 1938 The probable mechanism of the protective action of resin in fire wounds on red pine. J For 36:1231–1233

von Rudloff E, Jorgensen E 1963 The biosynthesis of pinosylvin in the sapwood of Pinus resinosa Ait. Phytochemistry 2:297–304

Wong BL, Berryman AA 1977 Host resistance to the fir engraver beetle. 3. Lesion development and containment of infection by resistant Abies grandis inoculated with Trichosporium symbioticum. Can J Bot 55:2358–2365

Worrall JJ, Harrington TC 1988 Respirometric testing of decay resistance of discolored root wood. Phytopathology 78:676–682

Yamada T, Okuda K 1987 Wood discoloration of hinoki and sugi living trees inoculated with Amylostereum sp. symbiotic to the Japanese horntail (Urocerus japonicus Smith). Trans 98th Annu Meet Jpn For Soc, 515–516

Yamada T, Tamura H, Mineo K 1988 The response of sugi (Cryptomeria japonica D. Don) sapwood to fungal invasion following attack by the sugi bark borer. Physiol Mol Plant Pathol 33:429–442

Yamamoto K 1982 Yearly and seasonal process of maturation of ray parenchyma cells in Pinus species. Res Bull Coll Exp For Hokkaido Univ 39:245–296

Chapter 9

Mechanisms of Defense in Gymnosperm Roots to Fungal Invasion

S. Woodward

9.1 Introduction

Root-infecting fungi cause major losses of timber in forestry worldwide, both through decay initiated following entry, and through the deaths of infected trees. This review will concentrate on responses to the two most studied of conifer root pathogens, *Heterobasidion annosum*, cause of Fomes root and butt rot, and *Armillaria*, formerly *A. mellea* but now divided into a number of distinct biological species, and also will include information on laminated root rot of conifers, caused by *Phellinus* (*Inonotus*) *weirii*.

Heterobasidion annosum is one of the most important pathogens in coniferous forests of temperate regions (Hodges 1969). In the UK, where many of the productive forests are still in the first or second rotations, *H. annosum* is the most common cause of root and butt rot in conifers, and is responsible for 90% of losses due to rot in Scotland (Low and Gladman 1962). Losses caused by *Armillaria* in UK forestry are much lower than those resulting from infection by *H. annosum*, although different species of *Armillaria* cause severe damage and kill trees in many other regions of the world. In interior coastal regions of northwestern North America, *Armillaria* may account for over 25% of deaths in trees of all ages (Wargo and Shaw 1985).

A further root disease of great importance in the coniferous forests of northwestern regions of North America is Poria or laminated root rot caused by *Phellinus weirii*. The pathogen causes estimated annual losses of more than 1 million m^3 of Douglas-fir (*Pseudotsuga menziesii*) timber in Washington and Oregon in the USA (Childs and Shea 1967). The same estimate has been reported for reductions in yield of *P. menziesii* in British Columbia (Bloomberg 1983).

Species of *Phytophthora* may affect mature trees in areas where the soil moisture content is high during warm periods. The principal species involved are *P. cinnamomi*, causing littleleaf disease of pines in the southeastern USA, and *P. lateralis*, which is responsible for root rot of Port Orford cedar (*Chamaecyparis lawsoniana*) in the Pacific Northwest of North America (Sinclair et al. 1987).

In addition to the important root disease-causing pathogens of conifers in forests, several relatively non-specific fungi, including *Cylindrocarpon destructans*, *Macrophomina phaseolina* (syn. *Sclerotium bataticola*), *Phytophthora* spp., *Pythium* spp., and *Helicobasidium purpureum*, may initiate disease on the root systems of transplants and tree seedlings in forest tree nurseries.

The presence of disorders or diseases of the root system in conifers becomes apparent with the onset of symptoms affecting the whole of the crown, a factor distinguishing these problems from those caused by many pests and pathogens in localized areas of the parts above ground. Loss of increment, coupled with yellowing of leaves, premature loss of needles and in some cases death of the host, are all symptomatic of root disorders. Severe decay in the root system also will render the tree more susceptible to wind throw.

With the possible exception of *Phytophthora* species, the fungi causing the most economically important root rots of conifers in the forest require a substantial food base from which to mount an attack on adjacent hosts. Symptoms of these diseases are typified by the formation of radiating infection foci in the forest, representing the circular spread of the pathogens from an original infected stump or tree. In temperate regions, *Armillaria* spp. colonize old stumps or roots of other trees and subsequently form rhizomorphs that radiate out from the infection site to invade other hosts. *Heterobasidion annosum* and *Phellinus weirii*, decay-causing pathogens with low competitive saprophytic abilities, pass from infected hosts to adjacent trees via root-to-root contacts. With these pathogens, infections are probably initiated by spores germinating on and colonizing the surfaces of fresh wounds or stumps of recently felled trees, although apparently, this event is relatively rare with *Armillaria* spp. (Rishbeth 1970).

With these heavy losses in timber production throughout the world, a detailed understanding of the nature of defense against root-infecting pathogens of conifers is essential for the development of suitable and effective control strategies.

9.2 Factors Predisposing Conifers to Attack by Root-Infecting Fungi

The susceptibility of conifers to attack by root-invading pathogens is influenced by many factors, including the characteristics of the planting site and the genetics of the host. As the effective operation of the defense mechanisms of the host depends on the overall vigor of the tree (Rishbeth 1972), environmental factors that may predispose trees to attack will be discussed briefly before consideration of the influence of host genetics on resistance.

9.2.1 Environmental Factors

The environment plays an important role in the development of root diseases in trees through predisposition of the hosts to attack and through its influence on the growth of the pathogens. Waterlogging and drought encourage the ingress of root-invading fungi by causing death of roots, opening up infection courts through which potentially damaging microorganisms may pass without making contact with the physical barriers of the periderm, or with active host defenses requiring the presence of live cells to be operative.

The relationship between host vigor and susceptibility to *Armillaria* has long been known. Day (1929) indicated that weakened trees, resulting from poor soil conditions or the presence of other injurious agents, are more susceptible to this root disease. This work has been confirmed by studies on the effects of light, nutrients, and soil pH on infection by *Armillaria* (Redfern 1978, Shields and Hobbs 1979, Singh 1983, Entry et al. 1986). Suppressed trees are significantly more susceptible to infection by *Armillaria* than those receiving full light (Redfern 1978). Such susceptibility is increased under conditions of reduced nutrient availibility (Entry et al. 1986). In a study of infections caused by *Armillaria* in the forests of northern Idaho, low soil pH coupled with low nitrogen was associated with decay in Douglas-fir (*Pseudotsuga menziesii*), whereas in grand fir (*Abies grandis*), decay was more prominent in trees on soil containing low levels of calcium and phosphorus, and high levels of potassium (Shields and Hobbs 1979).

Similar correlations have been found between light intensities and resistance to *H. annosum* (Rishbeth 1951, Gibbs 1967a). The roots of suppressed Scots pine (*Pinus sylvestris*) were readily colonized by *Heterobasidion* following mycelial inoculation, and showed no active resistance to the pathogen (Gibbs 1967). In contrast to reports of the beneficial effects of good nutrient supplies on the resistance of trees to root-invading pathogens, Alcubilla et al. (1987) found an increase in the susceptibility of Norway spruce to *Heterobasidion annosum*, with a concomitant reduction in the fungistatic effects of the inner bark, following application of fertilizer.

Wave-form types of forest dieback, seen in many natural stands, may also be explained in terms of changes in nutrient availibility throughout the lives of the trees. Studies in the USA on the role of light and nutrients on infection by *Phellinus weirii* demonstrated that vigorous young trees were more resistant to the pathogen (Waring et al. 1987). Following death of many trees in the forest due to attack by *P. weirii*, the increased nutrients released from decaying stumps and logs caused greater vigor, and consequently resistance to the pathogen, in the regeneration. Susceptibility reached critical levels once the stand attained an age of about 80 years. Laboratory-based studies on the susceptibility of mountain hemlock (*Tsuga mertensiana*) to *Phellinus* gave further evidence for the role of light intensity and nutrients, particularly nitrogen, in the wave-form dieback phenomenon (Matson and Waring 1984).

Soil pH has a great influence on the susceptibility of conifers to particular pathogens, both through an impact on the overall vigor of the trees, and through its effect on the microflora of the rhizosphere. A particularly good example of this effect is shown by the incidence and severity of *Heterobasidion annosum* in different areas of Thetford Forest in the UK. On sites with a soil pH of 6 or above, the susceptibility of pines to Fomes root rot increases markedly (Wallis 1960), whereas on sites with a lower pH, losses are much lower. This effect is a consequence of a reduction in the number of competing microorganisms, especially *Trichoderma*, in the rhizosphere on more alkaline sites, enabling more extensive and rapid growth of *H. annosum* on the root surface (Gibbs 1967b). The overall lowering of tree vigor also contributes to their greater susceptibility in these situations. Mortality due to *H. annosum* is more common in forests growing on dry sandy soils than on wetter sites (Rishbeth 1957, Jorgensen 1962, Maraite and Mayer 1966), and may be due to an increase in susceptibility as the soil dries out.

9.2.2 Other Infections/Pest Attacks

The possible effects of the presence of other pathogenic infections on the susceptibility of conifers to attack by root-infecting fungi have received relatively little attention. In Sitka spruce (*Picea sitchensis*), field observations in the UK suggested that infection of the root system by *Armillaria* was a prerequisite for subsequent infection by the butt-rotting basidiomycete *Phaeolus schweinitzii* (Barrett 1970). This conclusion was supported by the finding that *P. schweinitzii* readily colonized wood blocks previously colonized by *Armillaria*. *P. schweinitzii* is regarded as a relatively weak pathogen, and is less capable than *Armillaria* of degrading several of the constitutive defenses of the host, such as the antifungal stilbenes found in the inner bark of Sitka spruce (*Picea sitchensis*) (Woodward and Pearce 1988a).

The generally debilitating effects of pests and diseases of the crown may result in a greater susceptibility to root diseases (see Peace 1962), but the overall effects have been difficult to quantify. In a survey of the incidence of top dieback of Scots pine (*Pinus sylvestris*), caused by *Peridermium pini*, in Thetford Forest, East Anglia, UK, Greig (1988) found no correlation between the incidence of this disease and the presence of *Heterobasidion* in the roots and butt.

Unlike the negative effects of other pests and pathogens on the overall vigor of trees, the presence of mycorrhizae on the roots may increase resistance to root-invading pathogens (Marx 1973). This effect has been demonstrated in a number of interactions (see Marx 1972, Duchesne and Peterson 1989). The mechanism by which mycorrhizal infections protect plants against infection is unclear, although the possible roles of the mycelial barrier of the Hartig net in preventing fungal penetration (Marx and Davey 1969a,b), and the inhibition of pathogen growth through the production of antibiotic metabolites by the fungal symbionts, have been investigated in a number of conifer systems (Marx 1969, Trofast and Wickberg 1977, Garrido et al. 1982, Kope and Fortin 1989).

9.2.3 Wounding

Wounds, resulting from the actions of invertebrate and vertebrate animals and from routine management practices, provide important infection courts for many decay-causing fungi. Wounding provides the means through which fungi may bypass the defense mechanisms of the bark and, in deeper wounds, of the sapwood. A well-known example of this effect is that of the initial entry of *Heterobasidion annosum* into first rotation forests during thinning operations through freshly cut stumps (less than 2-weeks-old). Additional infection courts for root-invading pathogens result from the abrasion of roots during rocking in high winds (Redmond 1957, Stone 1977).

9.2.4 Host Age

With some root diseases of conifers, the susceptibity to attack changes as the tree ages. As the age of a stand increases, diseases such as *Heterobasidion* root rot spread over greater distances from the initial sites of infection. The potential resistance of individual trees may change with alterations in the nutrient status of the site (see Sect. 9.2.1), and in addition, the types of symptoms developed may be altered.

In pines, young trees are killed by *H. annosum*. However, marked increases in the resistance of the roots of older pines to attack by the fungus have been noted (Rishbeth 1951). In contrast, larch (*Larix* spp.) becomes more prone to infection as the crop ages (Gothe 1957). Older trees often are more prone to butt rot than mortality as a result of infection by *H. annosum*.

Resistance to killing by *Armillaria* is also known to increase with age of the host (Gibson 1960), but older trees become more prone to root rots and hence to windthrow. The effect of increasing resistance with age is a slowing down, and eventual halt in the spread of infection centers. The apparent increase in host resistance to killing with age may be a result of the rapidity with which stems of young trees are girdled by the pathogen (Phillips and Burdekin 1982). Alternatively, resistance factors present in the roots may alter with age (Rishbeth 1951), or during a rotation, and numbers of antagonistic micro-organisms present in the rhizosphere may increase sufficiently to prevent further rapid spread of the disease. With infections caused by *Armillaria*, a reduction in the inoculum potential, i.e. the food reserves on which the pathogen depends to sustain attacks on adjacent hosts, results in a reduced number of attacks with time (Garrett 1956).

As described in Section 9.2.1, the situation regarding the effects of aging on attack by *Phellinus weirii* contrasts with that described for *Heterobasidion annosum* and *Armillaria* spp.; the number of trees killed increasing once the stand reaches about 80 years. This effect is a result of the gradual alteration in the nutrient status of the site, rather than the overall resistance of the trees, as young seedlings of Douglas-fir (*Pseudotsuga menziesii*) are susceptible to infection in artificial inoculation experiments (see Sect. 9.2.5).

9.2.5 Genetics and Disease Escape

Several studies have been made on the relative resistance of different conifer species to attack by root-invading fungi. In vitro assays have tended to concentrate on the abilities of the pathogens to grow in wood blocks taken from stems or, occasionally, roots of potential hosts, or growth on media supplemented with sawdust or extractives (Platt et al. 1965, Scheffer and Cowling 1966, von Aguinagalde and Cerny 1974, Kaufmann and Wellendorf 1978,

Kaufmann et al. 1980). A few investigations have used seedlings as model systems for the study of resistance, but again, studies have concentrated on differences between, rather than within, host tree species (Kuhlman 1970).

The relative importance of host genetics in the resistance of individual conifer species to attack by root pathogens, therefore, has received little attention, although field observations suggest that several trees may "escape" an attack within a particular infection focus. This effect is undoubtedly influenced by the microenvironment and the overall vigor of the individual trees, but there is a high probability that genetic factors influence the response in some cases.

Tests for resistance of individual conifer species to *Heterobasidion* root rot have proved inconclusive. Von Weissenberg (1975) tested 1-year-old cuttings of four clones of *Picea abies* for resistance to *H. annosum* by inoculation into the lower stem through wounds made into the sapwood. Although some differences in growth of the pathogen in the host plants were noted, no absolute resistance was found in any clone. In a significantly larger test, using over 2000 Norway spruce trees from 31 clones, *H. annosum* gradually disappeared from all infected trees of each genotype over a 2-year period (Bazzigher 1986). The mechanisms responsible for the disappearance of *H. annosum* from infected plants were not investigated. Dimitri (1982) found considerable variations in the spread of *H. annosum* mycelium in inoculated Norway spruce (*Picea abies*), concluding that a broad-based genetic resistance (horizontal resistance) was operating in some individual trees.

In a research program instigated to search for Douglas-fir (*Pseudotsuga menziesii*) genotypes showing resistance to *Phellinus weirii*, Driver et al. (1982) found a small number of 2-year-old seedlings which survived inoculation with the pathogen. Relative resistance was measured by estimating the water stress of needles following exposure to *P. weirii*. Susceptible genotypes showed an increase in water stress from 16 days after inoculation, and most of these plants died quickly. The resistant seedlings, however, developed higher than normal water stress at the same time after inoculation, with this pattern being maintained over a period of 120 to 240 days before returning to control levels. Despite the significance of these results in terms of the possible development of resistant genotypes, the types of resistance mechanisms in operation in these genotypes were not investigated.

9.3 Physical and Chemical Barriers to Fungal Invasion

9.3.1 Resistance Mechanisms in Fine Roots

Little is known of the mechanisms of defense against fungal pathogens in the primary roots of conifers. Most of the available information on the responses of these tissues to fungi comes from studies on the formation of mycorrhizal interactions, which, although generally not considered as pathogenic and being largely of low specificity, do include incompatible reactions (Duddridge 1987). For example, the larch bolete, *Suillus grevillei*, forms associations with *Larix* species, the mycorrhizal roots forming a compact sheath with a well-formed Hartig net. With *Pinus sylvestris* and *Pseudotsuga menziesii*, however, although mycorrhiza were formed in vitro (Duddridge 1987), the symbionts were not completely compatible, the Hartig nets being irregular, or abnormalities developing at the host-fungus interface. In *Pinus nigra*, no Hartig net developed, and attempted penetrations of the host cortical cells resulted in cell death. Other responses characteristic of host defense reactions, including deposition of lignin in host cell walls, accumulation of phenolic compounds (Molina and Trappe 1982), and papilla formation (Nylund et al. 1982), have been noted in incompatible interactions between ectomycorrhizal symbionts.

The induction of chitinase, an enzyme believed to be involved in resistance to fungal attack in plants (Schlumbaum et al. 1986), in Norway spruce (*Picea abies*) seedling roots by cell wall preparations of *Amanita muscaria*, provides another example of induction of putative components of the defense response by a mycorrhizal fungus (Sauter and Hager 1989). Stimulation of chitinase and 1,3-β-glucanase activity has also been noted in seedling roots of Scots (*Pinus sylvestris*), Corsican (*P. nigra* var. *maritima*), and lodgepole (*P. contorta*) pines following challenge and infection with *Heterobasidion annosum* (Nsolomo and Woodward, unpublished).

Studies on the infection of *Pinus sylvestris* seedlings with the root pathogen *Cylindrocarpon destructans* have demonstrated the formation of papillae in the cortical cells (Bonello and Pearce, personal communication). The papillae comprised a pectic matrix impregnated with polyphenolic or lignin-like materials, as determined by staining with ruthenium red or molybdenum blue, and with phloroglucinol-HCl or Schiff's reagent, respectively, and contained high levels of calcium. In contrast to the situation found in many angiosperm-pathogen interactions (Aist 1976), in the *Pinus sylvestris-Cylindrocarpon* system, increased levels of silicon were not present in the papillae.

Two as yet uncharacterized antifungal compounds have been found in non-mycorrhizal primary roots of *Pinus sylvestris* seedlings (Bonello, personal communication). Following challenge with *C. destructans*, the levels of these compounds increase. Comparison with standards has shown that the active compounds are not the stilbenes pinosylvin or pinosylvin monomethylether, but may be related to abietic acid.

9.3.2 Resistance Mechanisms in Roots with Secondary Thickening

9.3.2.1 Resin Production

In some genera of the Pinaceae, such as *Picea*, *Pinus*, and *Larix*, resin ducts are present in all secondarily thickened tissues where an interconnecting system may be formed (see review in Gibbs 1968), the extent of this system varying between genera. Resin flow has been implicated in resistance to *Heterobasidion* (Gibbs 1968) and *Armillaria* (Rishbeth 1972) root rots, and is one of the initial visible symptoms of root disease in many conifers (see Sinclair et al. 1987).

Resin acids and volatile components (terpenes) are known to display direct antifungal activity against fungal pathogens (Cobb et al. 1968, Gibbs 1972,

Flodin 1979, Henriks et al. 1980) (Sect. 4.3.2). Reports of the relative effects of monoterpenes on inhibition of fungal growth have varied, possibly as a result of the different assay methods and strains of pathogens used in separate laboratories. However, several authors have described the inhibition of growth of *H. annosum* in saturated atmospheres of resin volatiles (Cobb et al. 1968, Gibbs 1972, Flodin 1979), suggesting a possible role for these compounds in resistance to fungal pathogens. In the vicinity of root wounds, it is conceivable that saturated atmospheres of resin volatiles would exist, causing reductions in growth of potential pathogens.

The conditions under which conifers were growing were shown to affect the abilities of different species to mobilize resins, and were correlated with resistance to *H. annosum* (Gibbs 1968). Lower yields of resins were obtained from trees growing on dry sites, where attacks by *H. annosum* were severe. In addition, in pines on alkaline soils, sites on which susceptibility is high, resin flow is intermittent, whereas copious flow is characteristic of more acidic soils. Further evidence for a role in resistance for resins was provided by the observation that suppressed trees gave very low yields when compared with subdominant and dominant individuals.

The work of Gibbs (1968) provides good circumstantial evidence of a role for these compounds in the resistance of conifers to attack by root-infecting fungi. The fact that resin components may be potent inhibitors of fungal growth under certain conditions (Sect. 4.3.2), coupled with the mechanical inhibition of hyphal growth by dried resin (Prior 1976), adds support to this conclusion.

9.3.2.2 Defense in Bark Tissues

In the absence of wounds or root grafts, the first obstacle that a potential root-invading pathogen encounters is the bark (Chap. 4). These tissues provide a formidable barrier to penetration, comprising the physically strong rhytidome, impregnated with suberin and lignin, protecting the underlying meristematic phellogen, and the phelloderm. As a result of the presence of suberin and lignin in the cell walls of the rhytidome, the outer layers of the bark are highly resistant to degradation, and very few pathogens are capable of penetrating this barrier. Internal to the phelloderm is the secondary bark cortex which contains high levels of inhibitory compounds, principally phenolics, which present a further barrier to the ingress of pathogens. In addition to these preformed barriers, the phellogen, phelloderm, and bark cortex are capable of redifferentiating to restore the integrity of the rhytidome following wounding (Woodward and Pearce 1988b). The resistance mechanisms which operate in conifer bark are discussed in detail in Chapter 4.

9.3.2.3 Defense in the Sapwood

The sapwood of trees consists of living tissues, comprising the functional xylem and associated companion cells, plus tangential ray parenchyma, and following wounding or attack by a pathogenic microorganism, the tissue is capable of initiating dynamic responses requiring the de novo establishment of metabolic pathways. This situation contrasts with that found in the heartwood, where in the majority of tree species, living cells are absent.

Responses of the xylem of conifer roots to wounding and to the ingress of pathogenic and decay-causing fungi resemble the non-specific reactions which occur in the stem wood of trees (Shigo and Marx 1977), where a series of compartmentalizing reactions occurs (see Chaps. 5,6,8, and 10). The theory of compartmentalization of decay in trees (CODIT) is based on the development of barrier zones, or walls within the xylem, which restrict the spread of decay to compartments delimited by defined anatomical boundaries visible when the affected trees are felled and cut into sections. Wall 1 forms the upper and lower boundaries of the compartment and may be relatively strong, wall 2 comprises the last cells formed in each growth ring, and wall 3 is composed of tangential ray cells and forms the sides of the compartment. Wall 4 is the strongest barrier, and is layed down only after wounding, separating the sapwood extant at the time of wounding from that formed subsequently.

An increase in the production of parenchyma cells and a concomitant accumulation of polyphenolics have been associated with the wounding response in conifer roots. In roots of red pine (*Pinus resinosa*) infected with *Heterobasidion annosum*, barrier zones were composed of tangential bands of resin ducts accompanied by axial parenchyma rich in polyphenols (Shigo 1979, Tippett and Shigo 1980). Similar responses were found in roots of balsam fir (*Abies balsamea*), red spruce (*Picea rubens*), tamarack (*Larix laricina*), and white pine (*Pinus strobus*) infected with *Armillaria*, or unidentified hymenomycetes (Tippett and Shigo 1981). Bands of resin ducts forming around wounds also constituted the major barrier zones forming around wounds made into roots of balsam fir (*Abies balsamea*) and eastern hemlock (*Tsuga canadensis*) (Tippett et al. 1982), emphasizing the non-specific nature of this response.

Shain (1979) describes the production of two types of tissue, the "transition zone" and the "reaction zone", produced in advance of fungal mycelium in the sapwood. These tissues separate functional sapwood from that infected by a pathogen, and may represent the early stages of barrier zone formation.

The transition zone, a tissue typically drier than adjacent, functional sapwood, has been noted in the response of conifer roots to infection by fungi (Coutts 1976a,b). In infections caused by *Heterobasidion annosum*, lysis by fungal enzymes of tori in the bordered pits of the xylem vessels resulted in the formation of the dry zone in Corsican pine (Coutts 1976a). Since this response could be induced by injecting 4 mM solutions of mercuric chloride into xylem vessels, a second mechanism for the induction of dry zones was postulated, involving the alteration of ray cell metabolism by diffusable toxins, resulting in the evolution of gas embolisms in adjacent tracheids (Coutts 1976b).

Reaction zones are usually visible as bands of discolored wood adjacent to the hyphal front in any infection of the sapwood. Extractives from the discolored area often contain antifungal activity, and formation of the transition zone in response to wounding has been likened to the formation of heartwood, a tissue that also may contain high levels of the same inhibitory substances as those present in the transition zone (Kuc and Shain 1977, Shain 1979). In species such as Corsican pine and slash pine (*Pinus taeda*), the stilbene derivatives, pinosylvin and pinosylvin monomethylether, which are active against many pathogenic fungi in vitro, including *H. annosum* (Hart 1981) and *A. ostoyae* (Mwangi et al. 1990), are formed in the reaction zones (Shain 1967, Prior 1976), whereas the equivalent tissue in Norway spruce contains high levels of lignans, the most active of which against *H. annosum* is hydroxymatairesinol (Shain and Hillis 1971). Prior (1976) demonstrated a significant correlation between the extent of infection by *H. annosum*, and the concentrations of pinosylvin and pinosylvin monomethylether in reaction zones in the roots of Corsican pine, although no absolute measurements of stilbene concentrations were made. It is difficult to draw absolute conclusions regarding the possible roles of the pinosylvins in resistance of the roots as both Shain (1967) and Prior (1976) used oven-dried material, which was then extracted using a Soxhlet apparatus. Under conditions such as these, stilbenes are likely to be degraded (Hart 1981). However, even with extractions made under these adverse conditions, the levels of pinosylvin and pinosylvin monomethylether recovered were probably high enough to significantly reduce the growth of *H. annosum*.

Detailed studies have been carried out in Sweden on the sapwood response of Norway spruce roots to wounding or challenge and infection with *Heterobasidion annosum* (Johansson and Theander 1974, Johansson and Stenlid 1985, Stenlid and Johansson 1987). The importance of dry zone formation was also stressed in this work (Stenlid and Johansson 1987). Reactions in the ray cells of excised roots, following wounding, included the rapid accumulation of phenolics, a decrease in starch levels and decreases in the activities of catalase, esterases and peroxidase. Inoculation of wounds with washed conidiospores of *H. annosum* caused oxidation of the phenols, decreases in starch levels, and induced the activity of β-glucosidase and acid phosphatase in the ray cells in advance of the growing mycelium, but caused reductions in the activities of peroxidases, dehydrogenases and esterases (Johansson and Stenlid 1985). A dry transition zone, characterized by a decreased starch content and increased levels of phenolic compounds, formed in advance of the growing mycelium. Extracts of the reaction zone were found to buffer strongly in the pH range of 6–9 (Johansson and Theander 1974), a finding confirming the earlier work on the same host-pathogen interaction by Shain (1971). This more recent work, however, has concentrated on responses in the roots, rather than stems, and it is therefore of greater relevance in terms of resistance early in the infection process.

The efficacy of barrier and reaction zones in preventing further spread of aggressive root-invading pathogens is the subject of some debate. Occurring late in the host-pathogen interaction, the discolored bands of cells within the wood may represent the sites of action of fungally produced enzymes in advance of the mycelium. In some angiosperm trees, apparently breached reaction zones, where the decay-causing fungus has grown through or around the barrier, have been noted (Pearce and Woodward 1986). Boddy and Rayner (1983) put forward a further hypothesis to explain the restriction of decay in angiosperms, indicating that a high moisture content and low carbon dioxide levels in the sapwood may prevent establishment and growth of many fungi (Chap. 6). Any stress applied to the host tree which prevents normal sapwood function will encourage the ingress of decay. Although the arguments of Boddy and Rayner were applied to angiosperms, several of their concepts also hold true for gymnosperms. Studies of the two-dimensional dynamics of the progress of decay fungi through wood of living trees may enable a more critical assessment to be made of the role of the various compartmentalization barriers in resistance.

9.3.2.4 Defense in the Heartwood

The heartwoods of conifers contain many compounds, principally phenolics, which show activity against fungal growth in in vitro assays (Kuc and Shain 1977). In common with the work on barrier zone formation, most of the available information comes from studies on extractives from stem heartwood, the timber product used for construction. It is likely, however, that many of the same, or similar compounds will occur in the root heartwood of the same species.

The most fungitoxic compounds isolated from heartwood of conifers are the tropolones from members of the Cupressaceae, characterized by a seven-carbon ring, but the activity of these substances against root-invading pathogens has received little attention compared with fungi responsible for the deterioration of wood products prepared from species of cedar. Heartwood of pines contains the fungitoxic stilbenes pinosylvin and pinosylvin monomethylether, whereas spruce contains the lignans hydroxymatairesinol, matairesinol and conidenrin, as described above (Sect. 9.3.2.3) (Kuc and Shain 1977). Taxifolin (3,3′,4′,5,7-pentahydroxyflavanone) comprises some 0.2–1.5% of the dry weight of Douglas-fir heartwood (Gardner and Barton 1960), and is the most inhibitory compound present in these tissues, with levels of 0.45% totally inhibiting growth of *H. annosum* in malt agar assays (Kennedy 1956). In in vitro assays, the ED_{50} of taxifolin against *P. weirii* was 0.2% (Cserjesi 1969), suggesting that the compound may also restrict the growth of this pathogen.

Notwithstanding the antimicrobial effects of the extractives, the nature of the lignin-cellulose matrix of the heartwood contributes greatly to the microbial decay resistance of these tissues (Scheffer and Cowling 1966). Also of importance is the availability of nutrients, particularly nitrogen, which may be a limiting factor for the growth of many fungal species in timber.

9.4 Conclusions

Improvements in our understanding of the nature of resistance of conifer roots to fungal diseases are essential for the development of more effective control strategies. Clearly, the use of any genetically based resistance to root diseases that may be present within a population of trees will be difficult to exploit, not simply because of the time required in breeding for resistance and for progeny testing, but also due to the lack of knowledge of the effects of differing environments on the expression of resistance in trees.

Despite the overall lack of information on responses of conifer roots to attack by pathogenic microorganisms, it is clear from the available literature that defense mechanisms in conifer roots and stems show many similarities. Further detailed studies on initial responses to wounding and attack in the roots may yield information leading to improved control methods for these economically important pathogens.

In the past, studies on the resistance of conifers to root-invading fungi concentrated mainly on the formation of mechanical barriers within the woody tissues at a relatively advanced stage of infection, and on the preformed chemicals present in the secondary tissues that show an antifungal effect against certain pathogens and decay-causing fungi. Investigations on the relative growth of different pathogens in sterilized wood blocks have yielded some useful data on the susceptibility of standing timber to decay by these species (von Aguinagalde and Cerny 1974, Kaufmann and Wellendorf 1978), but information on resistance to initial infections is limited in forest pathology (Johansson and Unestam 1982, Johansson and Stenlid 1985, Woodward and Pearce 1988a). During the life span of a tree, there is a high probability that wounds will open infection courts through which pathogens may bypass the active defenses present in the inner bark and the sapwood. Prevention of losses in young trees, rather than decay in older individuals, however, may be more amenable to the types of genetic resistance utilized in crop plants. It is well known from studies on herbaceous crops that the events which mark a resistant response are activated very rapidly following the initial recognition of host and pathogen by the respective organisms (Yoshikawa 1983). More in-depth investigations into the biochemistry of the events leading to recognition and development of a susceptible or resistant response in conifer-root pathogen interactions are required to provide markers for resistance which may be exploited rapidly in selection and breeding programs.

Acknowledgements. Thanks are due to Dr. Ray Pearce, and Mr. Enrico Bonello of the Oxford Forestry Institute for their critical reviews of the manuscript, and for making available previously unpublished data.

References

Aist JR 1976 Papillae and related wound plugs of plant cells. Annu Rev Phytopathol 14:145–163

Alcubilla M, von Aufsess, Cerny G, Heibl R, Rehfuess KE 1987 Stickstoffdüngungsversuche in einer Fichtenwuchsstockung (Picea abies Karst.) auf devastierter Kalkmergel-Rendzina. II. Verlauf künsticher Inokulationen mit Heterobasidion annosum (Fr.) Bref. und Pilzhemmwirkung von Bast und Holz im Biotest. Forstwiss Clb 106:27–44

Barrett DK 1970 Armillaria mellea as a possible factor predisposing roots to infection by Polyporus schweinitzii. Trans Brit Mycol Soc 55:459–462

Bazzigher G 1986 Infection studies with Heterobasidion annosum on young trees of Picea abies. Eur J For Pathol 16:125–128

Bloomberg WJ 1983 A ground survey method for estimating loss caused by Phellinus weirii root rot. III. Stimulation of disease spread and impact. Can For Serv, Pac For Res Cent Rep BC-R-7

Boddy L, Rayner ADM 1983 Origins of decay in living deciduous trees: the role of moisture content and a re-appraisal of the expanded concept of tree decay. New Phytol 94:623–641

Childs TW, Shea KR 1967 Annual losses from disease in Pacific Northwest forest. US Dep Agr For Serv Bull PNW-20

Cobb FW, Kristic M, Zavarin E, Barber HW 1968 Inhibitory effects of volatile oleoresin components on Fomes annosus and four Ceratocystis species. Phytopathology 58:1327–1335

Coutts MP 1976a The formation of dry zones in the sapwood of conifers. I. Induction of drying in standing trees and logs by Fomes annosus and extracts of infected wood. Eur J For Pathol 6:372–381

Coutts MP 1976b The formation of dry zones in the sapwood of conifers. II. The role of living cells in the release of water. Eur J For Pathol 7:6–12

Cserjesi AJ 1969 Toxicity and biodegradation of dihydroquercetin. Can J Microbiol 15:1137–1140

Day WR 1929 Environment and disease. A discussion on the parasitism of Armillaria mellea Vahl. Fr. Forestry 3:94–103

Dimitri L 1982 Some host/parasite relationships between Norway spruce (Picea abies) and Fomes annosus. In: Heybroek HM, Stephan BR, von Weissenberg K (eds) Resistance to diseases and pests in forest trees. PUDOC, Wageningen, 220–267

Driver CH, Morse DR, Edmonds RL 1982 Detection of disease resistance in Douglas fir seedlings and variation in pathogenicity in Phellinus weirii by monitoring water stress after inoculation. In: Heybroek HM, Stephan BR, von Weissenberg K (eds) Resistance to diseases and pests in forest trees. PUDOC, Wageningen, 275–282

Duchesne LC, Peterson RL 1989 The future of ectomycorrhizal fungi as biological control agents. Phytoprotection 70:51–57

Duddridge JA 1987 Specificity and recognition in ectomycorrhizal associations. In: Pegg GF, Ayres PG (eds) Fungal infection of plants. Univ Press, Cambridge, 25–44

Entry JE, Martin NE, Cromack K 1986 Light and nutrient limitation in Pinus monticola: seedling susceptibility to Armillaria infection. For Ecol Manag 17:189–198

Flodin K 1979 Effects of monoterpenes on Fomes annosus (Fr.) Cooke and its phenol oxidase activity. Eur J For Pathol 9:1–6

Gardner JAF, Barton GM 1960 The distribution of dihydroquercetin in Douglas-fir and Western larch. For Prod J 10:171–173

Garrett SD 1956 Rhizomorph behaviour in Armillaria mellea (Vahl) Quél. II. Logistics of infection. Ann Bot 20:193–209

Garrido N, Becerra J, Martirocorena C, Oehrens E, Silva M, Horak E 1982 Antibiotic properties of ectomycorrhizae and saprophytic fungi growing on Pinus radiata. Mycopathology 77:93–98

Gibbs JN 1967a The role of host vigour in the susceptibility of pines to Fomes annosus. Ann Bot 31:803–815

Gibbs JN 1967b A study of the epiphytic growth habit of Fomes annosus. Ann Bot 31:755–774

Gibbs JN 1968 Resin and the resistance of conifers to Fomes annosus. Ann Bot 32:649–665

Gibbs JN 1972 Tolerance of Fomes annosus to pine oleoresins and pinosylvins. Eur J For Pathol 2:147–151

Gibson IAS 1960 Armillaria root rot in Kenya pine plantations. Emp For Rev 39:94–99

Gothe H 1957 Beobachtungen über Stockfäule in Schlitzer Lärchenbeständen. Forst Holz 12:70–74

Greig BJW 1988 History of Peridermium stem rust of Scots pine (Pinus sylvestris L.) in Thetford Forest, East Anglia. Forestry 60:193–202

Hart JH 1981 Role of phytostilbenes in decay and disease resistance. Annu Rev Phytopathol 19:437–458

Henriks ML, Ekman R, von Weissenberg K 1980 Bioassay of some resin and fatty acids with Fomes annosus. Acta Acad Aboen Ser B 39:1–7

Hodges CS 1969 Modes of infection and spread of Fomes annosus. Annu Rev Phytopathol 7:247–266

Johansson M, Stenlid J 1985 Infection of roots of Norway spruce (Picea abies) by Heterobasidion annosum. I. Initial reactions in sapwood by wounding and infection. Eur J For Pathol 15:32–45

Johansson M, Theander O 1974 Changes in sapwood of roots of Norway spruce attacked by Fomes annosus. I. Physiol Plant 30:218–225

Johansson M, Unestam T 1982 The search for resistance to Heterobasidion root rot in Norway spruce – old and new approaches in studies of infection biology. Eur J For Pathol 12:346–357

Jorgensen E 1962 Fomes root rot investigations in Ontario. In: Conf Stud Tour Fomes annosus, Scotland 1960. IUFRO, Firenze, 44–47

Kaufmann U, Wellendorf H 1978 Genetic variation in Picea abies (L.) Karst. and Fomes annosus (Fr.) Cke. shown by growth of the fungus on wood dust. Eur J For Pathol 8:226–239

Kaufmann U, Wellendorf H, Dimitri L 1980 Tree-to-tree variation on Picea abies in growth inhibition against Fomes annosus. Eur J For Pathol 10:209–218

Kennedy RW 1956 Fungicidal toxicity of certain extraneous components of Douglas-fir heartwood. For Prod J 6:80–84

Kope HH, Fortin JA 1989 Inhibition of phytopathogenic fungi in vitro by cell free culture media of ectomycorrhizal fungi. New Phytol 113:57–63

Kuc J, Shain L 1977 Antifungal compounds associated with disease resistance in plants. In: Siegel MR, Sisler HD (eds) Antifungal compounds. Dekker, New York, 497–535

Kuhlman EG 1970 Seedling inoculations with Fomes annosus show variation in virulence and in host susceptibility. Phytopathology 60:1743–1746

Low JD, Gladman RJ 1962 Butt-rot of conifers in Scotland. A brief account of its distribution and occurrence. In: Conf Stud Tour Fomes annosus, Scotland 1960. IUFRO, Firenze, 56–65

Maraite H, Mayer JA 1966 Incidence de quelques facteurs du milieu sur la pourriture rouge de l'épicea. Bull Soc R For Belg 73:493–509

Marx DH 1969 The influence of ectotrophic mycorrhizal fungi on the resistance of pine roots to pathological infections. II. Production, identification and biological activity of antibiotics produced by Leucopaxillus cerealis var, piceina. Phytopathology 59:411–417

Marx DH 1972 Ectomycorrhizae as biological deterents to pathogenic root infections. Annu Rev Phytopathol 10:429–454

Marx DH 1973 Mycorrhizae and feeder root diseases. In: Marks GC, Kozlowski TT (eds) Ectomycorrhizae. Academic Press, New York London, 351–382

Marx DH, Davey CB 1969a The influence of ectotrophic mycorrhizal fungi on the resistance of pine roots to pathological infections. III. Resistance of aseptically formed mycorrhizae to infection by Phytophthora cinnamomi. Phytopathology 59:549–558

Marx DH, Davey CB 1969b The influence of ectotrophic mycorrhizal fungi on the resistance of pine roots to pathological infections. IV. Resistance of naturally occurring mycorrhizae to infections by Phytophthora cinnamomi. Phytopathology 59:559–565

Matson PA, Waring RH 1984 Effects of nutrient and light limitation on mountain hemlock susceptibility to laminated root rot. Ecology 65:1517–1524

Molina R, Trappe JM 1982 Patterns of ectomycorrhizal host specificity and potential among Pacific northwest conifers. For Sci 28:423–458

Mwangi LM, Lin D, Hubbes M 1990 Chemical factors in Pinus strobus inhibitory to Armillaria ostoyae. Eur J For Pathol 20:8–14

Nylund J-E, Kasimir A, Strandberg-Arveby S 1982 Cell wall penetration and papilla formation in senescent cortical cells during ectomycorrhiza synthesis in vitro. Physiol Plant Pathol 21:71–73

Peace TR 1962 Pathology of trees and shrubs. Univ Press, Oxford

Pearce RB, Woodward S 1986 Compartmentalization and reaction zone barriers at the margin of decayed sapwood in Acer saccharinum L. Physiol Mol Plant Pathol 29:197–216

Phillips DH, Burdekin DA 1982 Diseases of Forest and ornamental trees. Macmillan, London & Basingstoke

Platt WD, Cowling EB, Hodges CS 1965 Comparative resistance of coniferous root wood and stem wood to decay by isolates of Fomes annosus. Phytopathology 55:1347–1353

Prior C 1976 Resistance by Corsican pine to attack by Heterobasidion annosum. Ann Bot 40:261–279
Redfern DB 1978 Infection by Armillaria mellea and some factors affecting host resistance and the severity of the disease. Forestry 51:122–135
Redmond DR 1957 Infection courts of butt-rotting fungi in balsam fir. For Sci 3:15–21
Rishbeth J 1951 Observations on the biology of Fomes annosus. III. Natural and experimental infection of pines, and some factors affecting severity of the disease. Ann Bot 15:221–246
Rishbeth J 1957 Some further observations on Fomes annosus Fr. Forestry 30:69–89
Rishbeth J 1970 The role of basidiospores in stump infection by Armillaria mellea. In: Tousson TA, Bega RV, Nelson PE (eds) Root diseases and soil-borne pathogens. Univ Cal Press, Berkeley, 141–146
Rishbeth J 1972 Resistance to fungal pathogens of tree roots. Proc R Soc London Ser B 181:333–351
Sauter M, Hager A 1989 The mycorrhizal fungus Amanita muscaria induces chitinase activity in roots and in suspension-cultured cells of its host Picea abies. Planta 179:61–66
Scheffer TC, Cowling EB 1966 Natural resistance of wood to microbial breakdown. Annu Rev Phytopathol 4:147–170
Schlumbaum A, Mauch F, Vögeli U, Boller T 1986 Plant chitinases are potent inhibitors of fungal growth. Nature (London) 324:365–367
Shain L 1971 The response of sapwood of Norway spruce to infection by Fomes annosus. Phytopathology 57:1034–1045
Shain L 1971 The response of Sapwood of Norway spruce to infection by Fomes annosus. Phytopathology 61:301–307
Shain L 1979 Dynamic responses of differentiated sapwood to injury and infection. Phytopathology 69:1143–1147
Shain L, Hillis WE 1971 Phenolic extracives in Norway spruce and their effects on Fomes annosus. Phytopathology 61:841–845
Shields WJ, Hobbs SD 1979 Soil nutrient levels and pH associated with Armillariella mellea on conifers in northern Idaho. Can J For Res 9:45–48
Shigo AL 1979 Compartmentalization of decay associated with Heterobasidion annosum in roots of Pinus resinosa. Eur J For Pathol 9:341–347
Shigo AL, Marx HG 1977 Compartmentalization of decay in trees. US Dep Agr For Serv Inf Bull 405, 73 pp
Sinclair WA, Lyon HH, Johnson WT 1987 Diseases of trees and shrubs. Cornell Univ Press, Comstock
Singh P 1983 Armillaria root rot: influence of soil nutrients and pH on the susceptibility of conifer species to the disease. Eur J For Pathol 13:92–101
Stenlid J, Johansson M 1987 Infection of roots of Norway spruce (Picea abies) by Heterobasidion annosum. II. Early changes in phenolic content and toxicity. Eur J For Pathol 17:217–226
Stone EL 1977 Abrasion of tree roots by rock during wind stress. For Sci 23:333–336
Tippett JT, Shigo AL 1980 Barrier zone anatomy of red pine roots invaded by Heterobasidion annosum. Can J For Res 10:224–232
Tippett JT, Shigo AL 1981 Barriers to decay in conifer roots. Eur J For Pathol 11:51–59
Tippett JT, Bogle AL, Shigo AL 1982 Response of balsam fir and hemlock roots to injuries. Eur J For Pathol 12:357–364
Trofast J, Wickberg B 1977 Mycorrhizin A and chloromycorrhizin A, two antibiotics from a mycorrhizal fungus of Monotropa hypopitys L. Tetrahedron 33:875–879
Von Aguinagalde I, Cerny G 1974 Beziehungen zwischen der Ausbreitung künstlicher Fomes annosus – Infektionen in den Stämmchen zweier Fichtenklone und einigen holzchemischen Eigenschaften. Eur J For Pathol 4:138–148
Von Weissenberg K 1975 Variation in relative resistance to spread of Fomes annosus in four clones of Picea abies. Eur J For Pathol 5:112–117
Wallis GW 1960 Survey of Fomes annosus in East Anglian pine plantations. Forestry 33:203–214
Wargo PM, Shaw CG III 1985 Armillaria root rot: the puzzle is being solved. Plant Dis 69:826–832

Waring RH, Cromack K, Matson PA, Boone RO, Stafford SG 1987 Responses to pathogen-induced disturbance: decomposition, nutrient availibility, and tree vigour. Forestry 60:219–227

Woodward S, Pearce RB 1988a The role of stilbenes in the resistance of Sitka spruce (Picea sitchensis (Bong.) Carr.) to entry of decay fungi. Physiol Mol Plant Pathol 33:127–149

Woodward S, Pearce RB 1988b Wound-associated responses in Sitka spruce root bark challenged with Phaeolus schweinitzii. Physiol Mol Plant Pathol 33:151–162

Yoshikawa M 1983 Macromolecules, recognition, and the triggering of resistance. In: Callow JA (ed) Biochemical plant pathology. John Wiley & Sons, Chichester New York, 267–298

Chapter 10

Defense of Angiosperm Roots Against Fungal Invasion

M.R. NICOLE J.P. GEIGER and D. NANDRIS

10.1 Introduction

The major root disorders currently described in forest pathology mainly concern root rot diseases and vascular diseases. Although different tree-fungus interaction studies have been performed in such pathosystems (Nicole et al. 1982, 1985, 1986a,b, Balanchette 1984, Rishbeth 1985, Geiger et al. 1986a,b, Nandris et al. 1987a, Blanchette and Abad 1988, Blanchette et al. 1989), precise observations regarding defense of roots in angiosperm trees triggered by fungal invasion have rarely been made. Thus, little is known about how the root system of these trees defends itself against fungal infection. However, Thomas (1934) and Sharples (1936) published information about anatomical responses of *Prunus* sp., *Juglans regia*, and *Hevea brasiliensis* infected with *Armillaria mellea* and *Ganoderma* sp. Ito (1949). Sakurai (1952, cited by Akai 1959) and Struckmeyer et al. (1954) also described histological reactions of apple trees, mulberry trees, and *Quercus ellipsoidales* to *Helicobasidium* sp., *Roselinia* sp., and *Endoconidiophora* sp. Rishbeth (1972), probably providing the first review of resistance to fungal pathogens of tree roots but with few references to angiosperms (*Quercus robur*, *Fagus sylvatica*, *Acer pseudoplatanus*, *Populus* sp., some genera of the *Oleaceae* familly, and certain tropical species). Today, research on this subject is still poorly documented in comparison with studies on gymnosperm root defense (Chap. 9).

Our objectives in this review are (1) to describe the main active reactions of root angiosperm tissues; (2) to present the responses of lignified and unlignified cells; (3) to relate previous microscopical observations to biochemical modifications of the tissues; and (4) to discuss the concept of defense mechanisms in roots of angiosperm trees. The major studies dealt with in this chapter are concerned with rubber tree root rot diseases at cellular (Nicole et al. 1983, 1986c) and molecular (Geiger et al. 1986c, 1989) levels both on young artificially infected seedlings (Nandris et al. 1983) and on adult trees naturally infected with *Rigidoporus lignosus* and *Phellinus noxius* (Geiger et al. 1976, 1977).

10.2 Histological Reactions of Roots

Phytopathological surveys of tree plantations showed that among several diseased trees, some seemed to survive infection (Nandris et al. 1988). In rubber tree plantations, careful examination of root systems infected with *Rigidoporus lignosus* and *Phellinus noxius* sometimes revealed modifications of their anatomy, characterized by root witches' broom or by hypertrophied lateral roots (Fig. 10.1). In 10-year-old plantations, a survey showed that among taproots of 223 dead trees, only 3.6% had developed visible root reactions. Similar observations were also reported for trees inoculated with *Armillaria* sp. and with *Ganoderma pseudoferreum* (Sharples 1936, Rishbeth 1985). Such tissue modifications strongly suggest great pertubations in root histogenesis with respect to primary and lignified roots.

10.2.1 Increasing Phellogen Activity and Formation of a Secondary Periderm

The formation of protective layers in bark arises from the cork cambium (the phellogen) and can provide strong mechanical, and perhaps biochemical resistance to infection or wounding (Chaps. 2 and 3). The increase in the phellogen activity also occurs in roots and the formation of a secondary bark have been observed in some fruit tree species and specifically in *Juglans nigra* attacked by *Armillaria* sp. (Thomas 1934) and in *Pyrus malus* infected with *Helicobasodium mompa* (Akai 1959). Mathre et al. (1966) observed that penetration of older cotton roots by *Thielaviopsis basicola* was walled off by the development of heavy cork cambium. The pathogen *Armillaria* also stimulated the cork cambium in *Populus tremuloides* and *Fagus grandiflora*, the roots of which were dissected to examine the spatial limit of this reaction (Shigo and Tippett 1981b). In roots of avocado (*Isopogon ceratophyllus*) the necrophylactic periderm (as defined by Mullick and Jensen 1973) was induced by *Phytophthora cinnamomi* to wall off infected tissues (Phillips et al. 1987). In rubber tree root rot diseases, *Hevea brasiliensis* initated such reactions in roots specially contaminated by *Phellinus noxius* and *Rigidoporus lignosus* (Nicole et al. 1986c), as shown in Fig. 10.2. The stimulation of the phellogen activity, common in nonlignified roots, was also described in adult roots of the rubber tree infected with *Rigidoporus lignosus* or *Ganoderma pseudoferreum* (Sharples 1936). In Fig. 10.3, reaction tissues of *Hevea brasiliensis* were differentiated against *Sphaerostilbe repens*, another tropical root parasite decaying cortical tissues only. In some cases, small nodes originating from the phellogen confer an abnormal configuration to the young phellem.

10.2.2 New Vascular Cambium and Meristematic Activity

Barrier zones are differentiated by living trees in response to different stresses (Chap. 5) in both the phloem and xylem. The formation of a new vascular cambium alters the wood anatomy and seems to be initiated from cells adjacent to the former dead cambium (Rishbeth 1972). In rubber trees, such a reaction was observed rarely and only on roots of young seedlings (Nicole et al. 1986c). The newly differentiated cambium (Figs. 10.4 and 10.5) formed the phloem, which contained laticifers, tannin cells, and sieve tubes, and xylem tissue mainly composed of numerous punctated cells (Fig. 10.6). The resulting barrier on

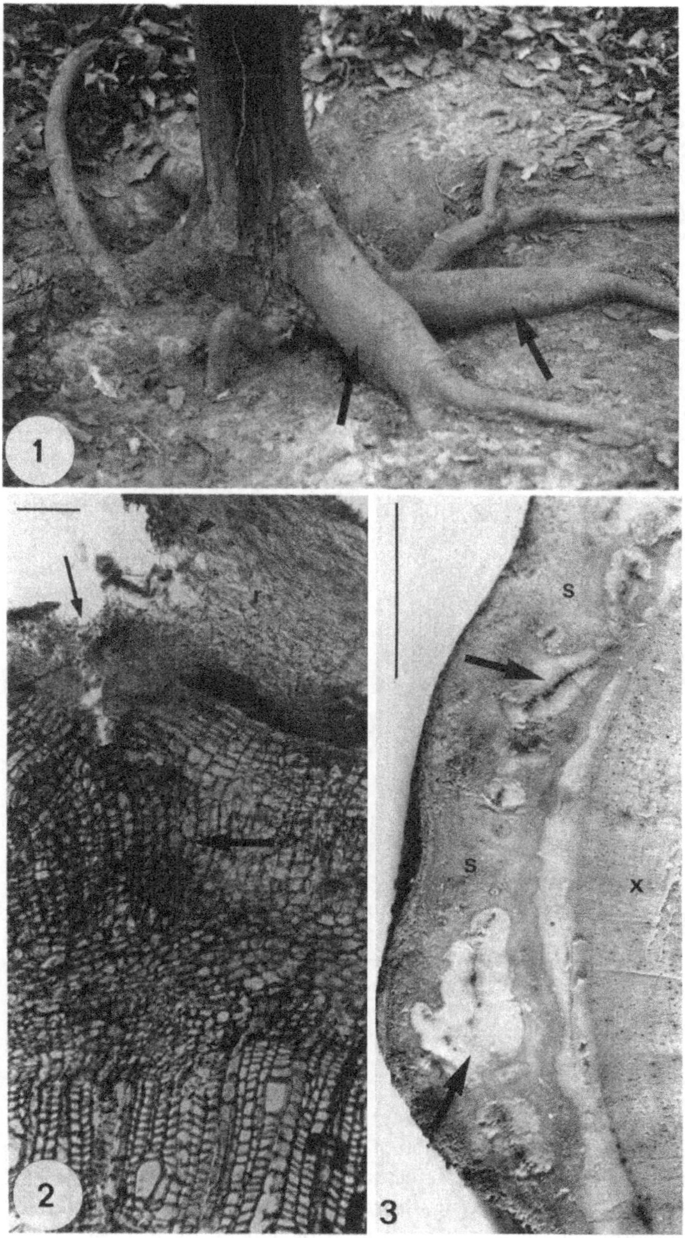

Fig. 10.1. The root system of a *Rigidoporus lignosus*-infected *Hevea brasiliensis* showing hypertrophied lateral roots (*arrows*) enabling the plant to survive root rot diseases

Fig. 10.2. Light photomicrograph of a cross-section through a young root infected with *Rigidoporus lignosus* rhizomorph (*r*) showing additional cell layers produced by phellogen stimulation (*large arrow*) under the penetration site (*small arrow*); *bar* = 12 μm

Fig. 10.3. Cross-section through an mature root infected with *Sphaerostilbe repens*. In the suberized reaction tissue (*s*), tertiar cambium activity (*arrows*) produced woody nodes to prevent fungal invasion of the healthy xylem (*x*); *bar* = 1 cm

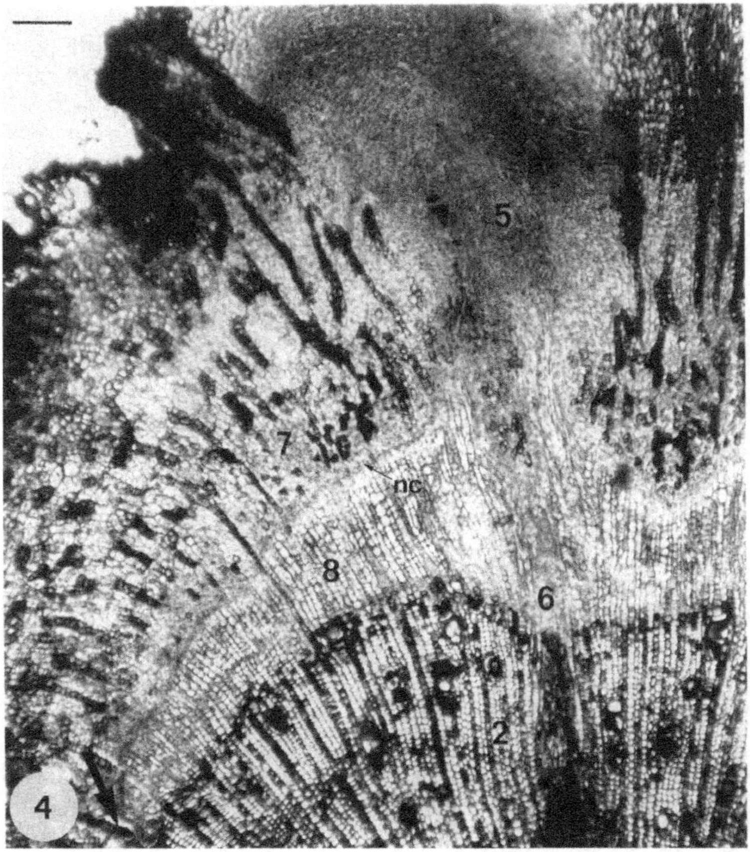

Fig. 10.4. Light photomicrograph of a cross-section through a young root infected with *Rigidoporus lignosus*. Meristematic tissue (*5*) is initiated from parenchymatous rays (*6*) of the decayed xylem (*2*). A new vascular cambium (*nc*), formed from the old one (*arrow*) which is probably killed, differentiated new xylem (*8*) and phloem (*7*); *bar* = 35 μm

small roots can be a complete ring, isolating the diseased xylem. Shigo and Tippett (1981a) also described a barrier zone formed by the vascular cambium in xylem of American elm root tissues infected with *Ceratocystis ulmi*. On the other hand, compartmentalization of pathogen in the secondary root phloem seems to be uncommon, although it was reported by Tippett et al. (1983, cited by Shigo 1984) on *Eucalyptus* sp. infected with *Phytophthora cinnamomi* or *Armillaria luteobubalina*.

Meristematic activity appears to be more frequent in trees, and Rishbeth (1985) has suggested that such activity occurs only when a parasite invades a root rather slowly. Our work agreed with this observation; the meristematic activity was considered indeed as a late reaction on rubber trees. On young seedlings this activity completely modified the anatomical organization of the

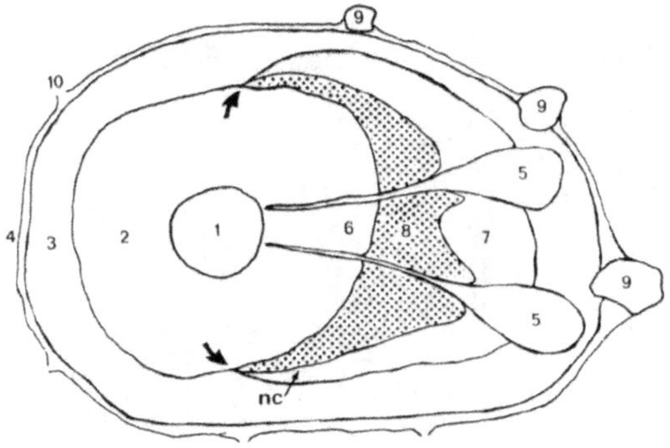

5

Fig. 10.5. Scheme showing anatomical reactions on a cross-section through a young root infected with *Rigidoporus lignosus*. *1* Medulla parenchyma; *2* decayed xylem; *3* infected phloem; *4* periderm; *5* meristematic tissue; *6* parenchymatous rays; *7* new phloem; *8* new xylem; *9* rhizomorphs; *10* lenticells. The new vascular cambium (*nc*) is formed from the old one (*arrows*)

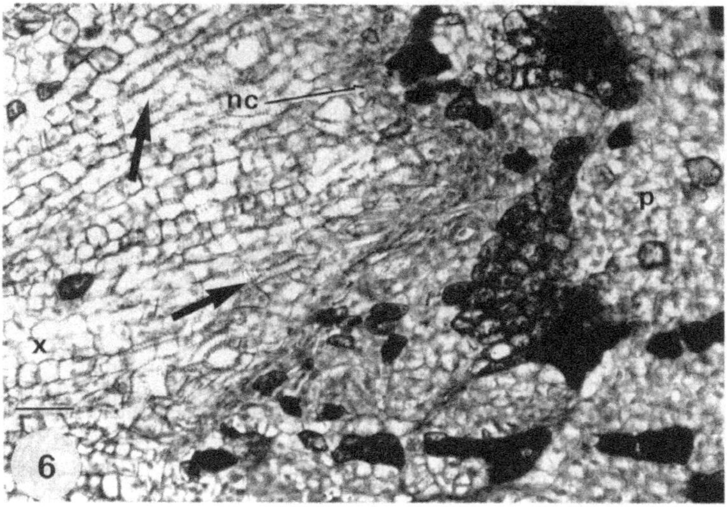

Fig. 10.6. Formation of new xylem (*x*) and new phloem (*p*) by the neo-cambium (*nc*). The new xylem is composed of numerous punctated cells (*large arrows*), typical of healing xylem; cross-section light photomicrograph through a young root infected with *Rigidoporus lignosus*; *bar* = 20 μm

root. The meristematic tissue arose from the parenchyma rays of the former xylem (Figs. 10.4 and 10.5) by dedifferentiation of cells into cellular masses initially composed of several cells (Fig. 10.7), and subsequently forming dense foci, whose organization is not unlike that of a meristem (Fig. 10.8). Such cells have a dense cytoplasm, a large nucleus, and a high nuclear/cytoplasmic ratio

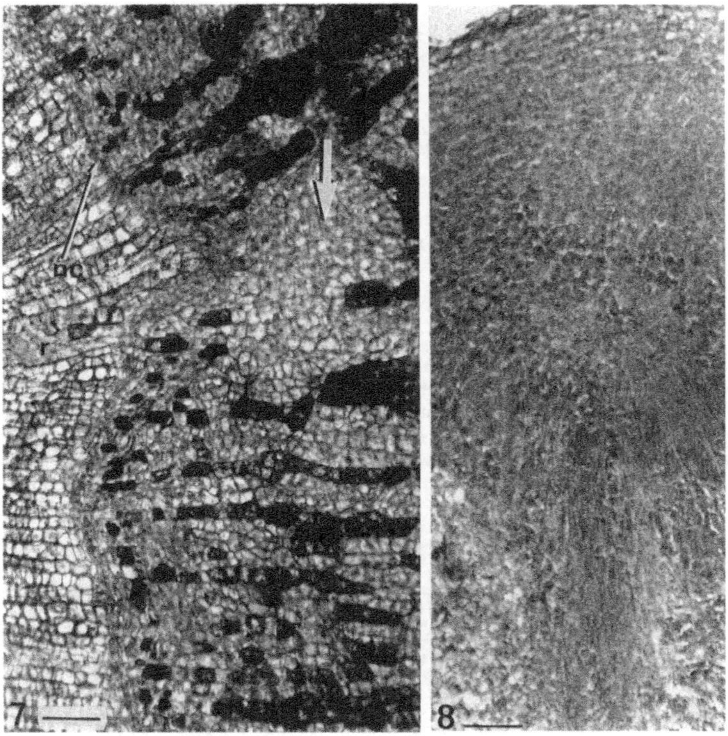

Figs. 10.7 and **10.8.** Dedifferentation of parenchymatous ray cells (*r*) results in the production of unorganized cellular masses (*arrow*) subsequently forming a meristematic tissue composed of cells with a dense cytoplasm and a large nucleus (*Fig. 10.8*); *nc* new cambium. Cross-section light photomicrographs through a young root infected with *Rigidoporus lignosus*; *bar* = 24 μm for both figures

and become generally impregnated with lignin and suberin as mentioned by Aist (1983).

10.2.3 Differentiation of New Roots

The formation of a lignified barrier, and of new lateral roots or taproots to prevent fungal invasion is the consequence of both new vascular cambium and meristematic activity. *Hevea brasiliensis* initiated these responses on young and adult trees infected with *Rigidoporus lignosus* and *Phellinus noxius* (Nicole et al. 1983, 1986c) or attacked by *Armillaria* sp. (Fox cited by Rishbeth 1972). Infected tissues were surrounded by the wall 4 barrier zone (Figs. 10.9, 10.10), as described by Pearce and Woodward (1986) on the butt of *Acer saccharinum* infected with *Ganoderma adspermum*. They were then replaced and compartmentalized by other roots (Fig. 10.11) to restrict the spread of decay. Differen-

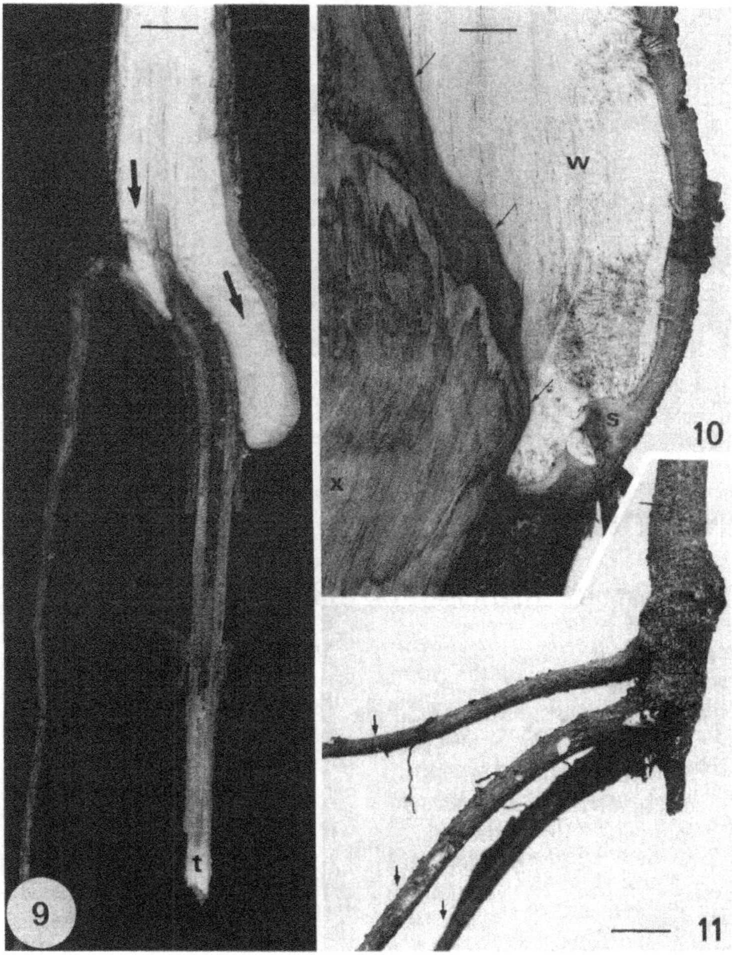

Fig. 10.9. Macrophotograph of a 2-month-old Hevea infected with *Rigidoporus lignosus*; longitudinal section through the root. The former taproot (*t*) is decayed by the fungus and compartmentalized by new tissues (*arrows*) differentiated from a new cambium; *bar* = 1.5 cm

Fig. 10.10. Macrophotograph of an adult Hevea taproot decayed by *Rigidoporus lignosus*; longitudinal section. The degraded xylem (*x*) and the reaction wood (*w*) are separated by a barrier zone (*arrows*) which prevent fungal spread into healthy tissues; *s* suberized tissue; *bar* = 1.5 cm

Fig. 10.11. Macrophotograph of a 2-month-old Hevea infected with *Rigidoporus lignosus* showing the complete decay of the former taproot and the differentiation of lateral roots (*arrows*) allowing the tree to survive; *bar* = 1 cm

tiation of new lateral roots often occurred after the former taproot had decayed, probably involving plant growth substances (see Sect. 10.4.3). Although there is no direct evidence, lignification and suberization in the barrier zone generally confer resistance to degradation of sapwood by fungi (Pearce and Rutherford 1981).

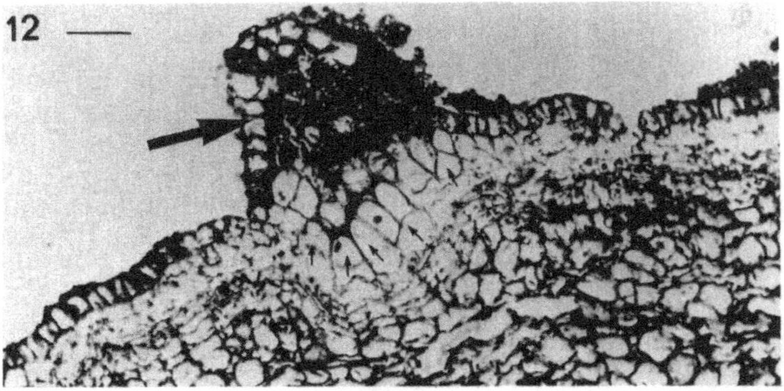

Fig. 10.12. Light photomicrograph of a cross-section through a root infected with *Phellinus noxius*. Hypertrophy of young phellem cells (*small arrows*) near cells colonized by fungal hyphae (*large arrow*); *bar* = 24 μm

10.3 Cellular Defenses

10.3.1 Cell Hyperplasia and Hypertrophy: The Hypersensitive-Like Response

A common response of *Hevea brasiliensis* to infection with *Phellinus noxius* concerns the first cell layers of the young root periderm. Hypertrophy and hyperplasia of the cells were observed during the penetration of hyphae into the root. The volume of some cells greatly increased near the penetration sites or around infected cells (Fig. 10.12), often in association with hyperplasia. Periclinal cell divisions also occurred in roots of avocado infected with *Phytophthora* sp., especially in the primary parenchyma, which acquired the appearance of an exophylactic periderm (Phillips et al. 1987).

The hypersensitive response, well known in foliar diseases, entails a reaction with an early tissue necrosis (Kiraly 1980). It merely constitutes a physical barrier, reducing the fungal penetration by surrounding the dead cells (Fig. 10.13) as specifically induced by *Phellinus noxius* on rubber tree roots. The hypersensitive necrosis probably results from recognition mechanisms, as yet unknown with root rot diseases, and might suggest the implication of fungal elicitors. Albersheim (1989) showed that oligosaccharide fragments of host cell walls, released by pathogen exo-enzymes, can kill plant cells. Enzymes secreted by *Phellinus noxius*, which assist hyphal penetration in roots (Geiger et al. 1986b), produce such components, which are able to elicit the hypersensitive-like response described in *Hevea* root.

10.3.2 Vascular Occluding Reactions

Gel, gum plugs, and tyloses are of widespread occurrence as root reactions in fungal wilt diseases (Beckman and Talboys 1981). Vascular gelation was

13

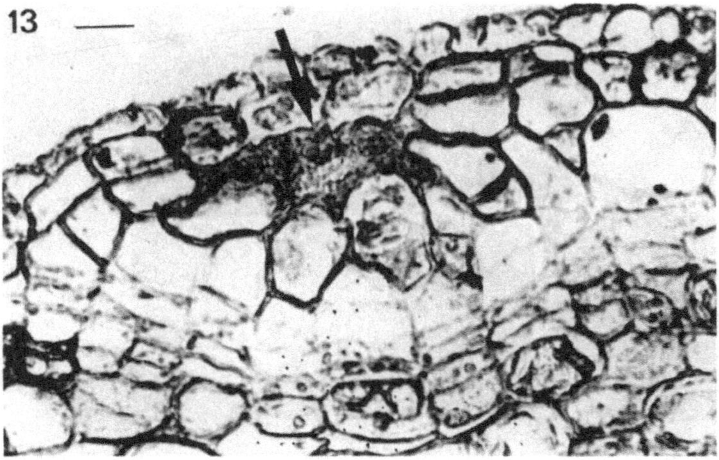

Fig. 10.13. Light photomicrograph of hypersensitive-like response in *Phellinus noxius*-infected root. Hypertrophied parenchyma cells develop around necrotic cells (*arrow*); *bar* = 8 μm

reported by VanderMolen et al. (1977) within vessel lumina in *Fusarium*-infected *Citrus* and *Verticillium*-infected cotton. These authors also described the pectic nature of this material. Tyloses, originating from the bulbing of adjacent parenchyma cells, are the more usual response in xylem vessels, for example in *Juglans regia* (Thomas 1934) and in *Quercus* sp. (Struckmeyer et al. 1954). However, formation of tyloses has rarely been described in root rot diseases, although they probably contribute to impede water transport, as observed on young rubber trees artificially infected with tropical root rot fungi. When gel, gums, and tyloses solidify they may act as physical barriers (Chap. 13).

10.3.3 Cell Wall Thickening

Cell wall thickening occurs very early in the infection process at the point of hyphal contact. In the apical region of the root of young infected rubber trees, the cells of the primary parenchyma undergo a spectacular thickening of the walls, which is orientated toward the parasite (Fig. 10.14) (Nicole et al. 1986c). Responding cells thus show enlarged walls, which can reach more than 2 μm in thickness. Positive tests enabled the identification of lignin in this new wall. In the cork barrier induced by *Phellinus noxius*, cell wall thickening becomes heavily suberized. Although suberization was mainly described in relation to wound-associated changes and postinfection (Pearce and Rutherford 1981), both lignification and suberization often take place in the barrier zone (Enebak and Blanchette 1989), the cell walls of which also contain unknown components (Ride 1983).

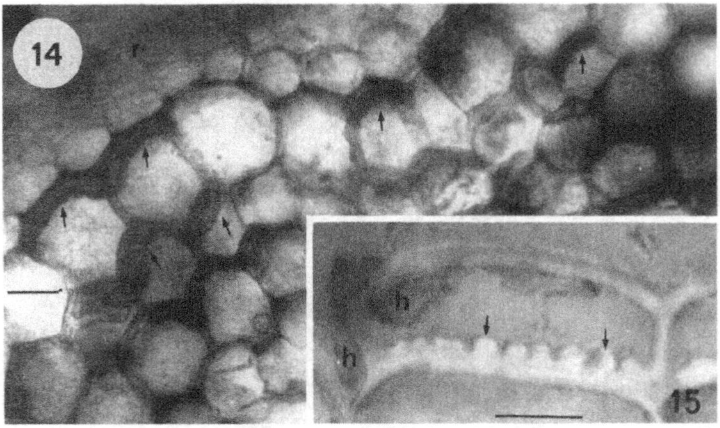

Fig. 10.14. Wall thickening of primary parenchyma cells (*arrows*) of a young root infected with *Rigidoporus lignosus*; histochemical tests (phloroglucinol/HCl; $KMnO_4$ oxidation) demonstrated the lignin nature of these walls; *r* rhizomorph. Light photomicrograph; *bar* = 8 μm

Fig. 10.15. Light photomicrograph of a cross-section through a root infected with *Rigidoporus lignosus*. Wall modification (*arrows*) of phellem cells colonized by fungal hyphae (*h*); UV illumination of sections after aniline blue staining reveals callose deposition along the wall; *bar* = 3 μm

10.3.4 Deposition of New Wall-Like Material

Wall appositions represent a general response of plants to infection by forming molecular barriers such as lignin, callose, and hydroxyproline-rich glycoprotein (Mazau and Esquerré-Tugayé 1986). In diseased woody plants, deposition of wall-like material was described in lignified and unlignified roots. Light microscopic observations of *Rigidoporus lignosus*-infected *Hevea brasiliensis* revealed that young, suberizing cell walls present a notched appearance (Fig. 10.15) with a fluorescent aspect when observed under UV light. Ultrastructural examination showed deposition of material along the cell wall between the plasmalemma and the wall itself, suggesting that these wall appositions could be induced by fungal progression inside the wall (Fig. 10.16). Electron microscopy also revealed that they are electron-dense, indicating the presence of other components than the callose shown by histochemical tests. The efficiency of these induced papillae in young phellem cells is probably low because *Rigidoporus lignosus* produces the enzymes (β-glucanases) able to degrade them (Nicole et al. 1986a). Coating material in vessels of the root xylem has also been described in infected *Hevea brasiliensis* (Nicole et al. 1986c). It appeared as a new, irregular, and lignified layer, located on the secondary cell wall of vessels, and was never observed in healthy roots. In trees, such a "protective layer" was also investigated in paravascular and ray parenchyma cells of elm infected with *Ceratocystis ulmi* (Ouellette 1981).

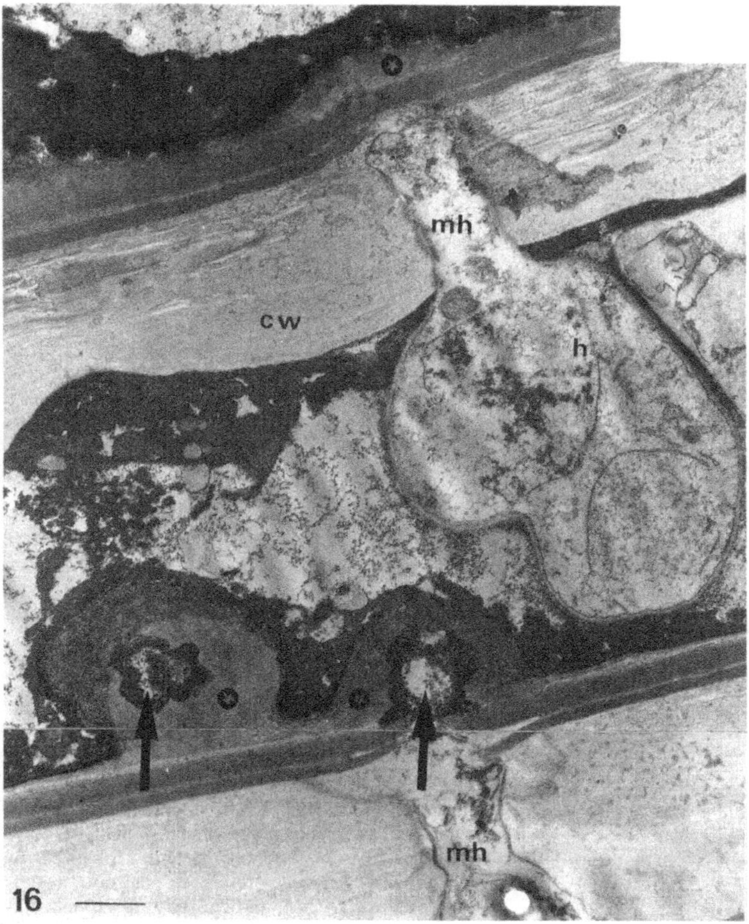

Fig. 10.16. Transmission electron micrograph of a cross-section through a *Rigidoporus lignosus*-infected root. Papillae (★) deposited along the cell wall in contact with fungal hyphae (*h*). The pathogen can degrade these wall appositions which become more electron-dense (*arrows*). *mh* Microhyphae; *cw* cell wall; glutaraldehyde/osmium tetroxide fixatives; *bar* = 0.5 μm

10.4 Biochemical Modifications of the Xylem

10.4.1 Lignification

Histological and cytological reactions of wood include various metabolic disturbances in response to stresses (Chaps. 5, 7, and 8). Among them, stimulation of the lignin biosynthesis pathway is of importance in resistance (Vance et al.

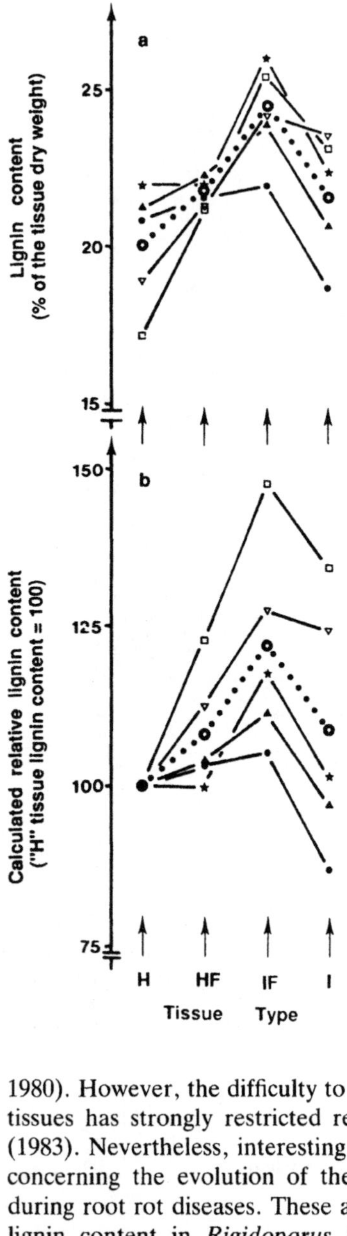

Fig. 10.17a,b. Lignin content recorded for different tissue types of rubber tree taproots. **a,b** Values expressed as a percentage of dry weight of tissue types in case of taproots partially colonized by *Rigidoporus lignosus*; for each taproot, results registered for *HF*, *IF*, and *I* tissues are expressed as a percentage of corresponding *H* tissue lignin content. Healthy tissue *H* and *I* infected tissue are located far from the fungal progression line; *HF* and *IF* are, respectively, healthy and infected tissues located close to the fungal progression line

1980). However, the difficulty to quantify induced lignin-like material in lignified tissues has strongly restricted research on the subject, as mentioned by Ride (1983). Nevertheless, interesting results were published by Geiger et al. (1986c) concerning the evolution of the lignin content of *Hevea brasiliensis* taproots during root rot diseases. These authors showed an increase of up to 50% of the lignin content in *Rigidoporus lignosus*-infected lignified tissues close to the fungal progression line (Fig. 10.17). On the other hand, accumulation of lignin in xylem infected with *Phellinus noxius* is rather heterogeneous. It would be interesting to test the monomer composition of the newly formed defense lignin of barrier zones to see whether it is different from the native xylem as demon-

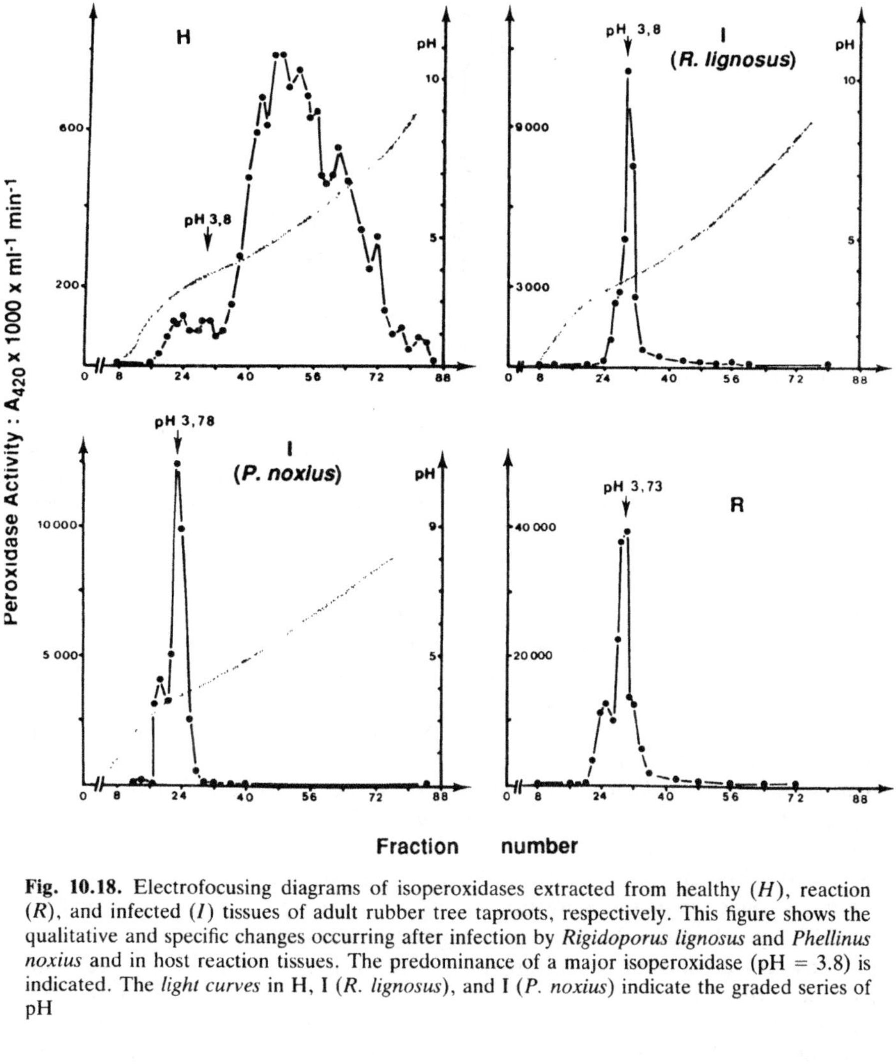

Fig. 10.18. Electrofocusing diagrams of isoperoxidases extracted from healthy (*H*), reaction (*R*), and infected (*I*) tissues of adult rubber tree taproots, respectively. This figure shows the qualitative and specific changes occurring after infection by *Rigidoporus lignosus* and *Phellinus noxius* and in host reaction tissues. The predominance of a major isoperoxidase (pH = 3.8) is indicated. The *light curves* in H, I (*R. lignosus*), and I (*P. noxius*) indicate the graded series of pH

strated by Asada and Matsumoto (1972) and Hammerschmidt et al. (1985) on herbaceous angiosperms.

10.4.2 The Peroxidase Reaction

Peroxidase activity has been the most studied plant reaction to infection, except in highly lignified tissues. Research carried out with rubber trees showed that healthy and infected tissues, both localized close to the fungal progression line, had a high peroxidase activity (Geiger et al. 1976). Figure 10.18 reports

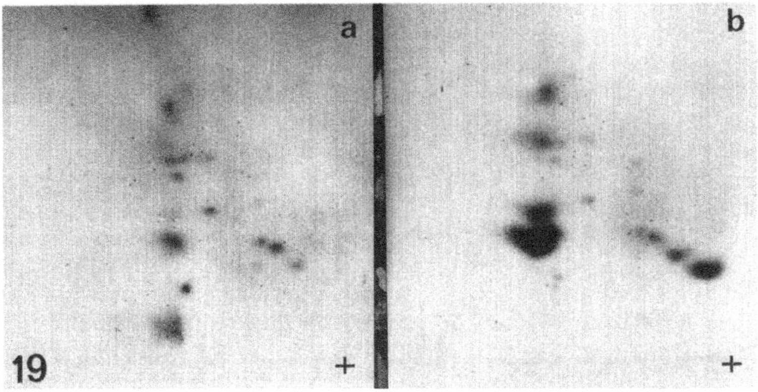

Fig. 10.19a,b. Trypsic map (fingerprint) of the peroxidase (P2a) purified from infected tissues of **a** *Rigidoporus lignosus*-infected taproots and **b** reaction zone tissues produced in *Sphaerostilbe repens*-infected taproots. This result with respect to other physicochemical characteristics reveals that the isoperoxidases produced in infected roots and in the reaction zone are identical

electrofocusing results, showing a considerable increase in one enzyme species in both parasitized and reaction tissues. The identified isoperoxidase, called P2a, was responsible for the major enzyme activity, which may reach a factor as great as 170 by comparison with healthy tissues. Polyacrylamide gel electrophoresis, amino acid composition, and trypsic fingerprints (Fig. 10.19) demonstrated that the physicochemical characteristics of the purified peroxidases were similar in *Rigidoporus lignosus*-infected taproots and in the uninfected reaction zone tissues induced by *Sphaerostilbe repens*. Thus, the increase of the P2a isoperoxidase activity was considered as a host reaction to infection (Geiger et al. 1989). This activity probably results in the stimulation of a de novo synthesis of the enzyme or in the activation of a preexisting apoenzyme inactivated in healthy tissue because the heme of the peroxidase is lacking.

Close correlations between peroxidase activity and lignification enhancement and host resistance (Vance et al. 1980) have often been established. In the rubber tree, P2a isoperoxidase can carry out the terminal step in lignin biosynthesis, i.e., the polymerization of p-coumaryl and coniferyl alcohols to form condensed products analogous to lignin (Geiger and Huguenin 1981, cited by Nicole et al. 1986c). This indicates that P2a activity enhancement may induce specific functions during the biosynthesis of lignin, like other peroxidases during lignification of trees (Harkin and Obst 1973). Although the peroxidase reaction is widespread in herbaceous plants, this research revealed that the root system of trees is able to initiate similar responses, specific in terms of its expression and the intensity of its stimulation. However, the induced peroxidase activity may also be involved with other biochemical processes occurring during root resistance, such as those involved in growth substance metabolism, synthesis of hydroxyproline-rich glycoprotein, and/or suberization.

10.4.3 Other Biochemical Responses

Biochemistry of tree defense reactions in the xylem of gymnosperms and angiosperms is described in Chapters 7 and 8, respectively. In roots of angiosperms, some reactions are of interest and concern, especially proteins, the phenolic pool, and hormonal stimulations.

Infection-related enzymes were differentiated in the phloem of roots of *Quercus rubra*, *Q. alba*, and *Acer saccharinum* infected with *Armillaria mellea*. It was demonstrated that chitinase and β-1,3 glucanase hydrolized the hyphal cell walls of this pathogen (Wargo 1975) and may account for the resistance of healthy tissues to fungal invasion.

In cotton (*Gossypium hirsutum*) catechin and gallocatechin, localized in the endodermis of young roots, played a role as chemical barriers to penetration and colonization of the root of vascular pathogens (*Verticillium dahliae*) by inhibiting sporulation and fungal pectinases (Mace and Howell 1974). In the same way, dihydroxyphenols, normally absent from the stele tissue of cotton roots, were synthesized by the xylem ray cells, whereas paravascular cells and the root cortex produced terpenoids (Bell 1981). Production of phytoalexins was also stimulated in root tissues of resistant cultivars of *Gossypium hirsutum*, as for example gossypol, a phenolic antioxidant, which was accumulated in response to infection by *Fusarium oxysporum* f. sp. *vasinfectum* and *Rhizoctonia* infection (Dallakyan et al. 1978, Bell 1981). Lastly, stilbenes were associated with resistance of avocado against *Phytophthora* root rot.

Several kinds of reactions and symptoms concern modifications of hormone metabolism (Pegg 1988). Ethylene is produced, for example, in roots of *Phytophthora*-infected peach trees (Ugalde and Taylor 1988). The role of ethylene was also stressed in hyperplasia and hypertrophy responses. Gibberellic acid and auxins promote the formation of tyloses and the differentiation of xylem (Pegg 1976). Generally, differentiation of vascular tissues is under hormonal control (Aloni 1987, Roberts et al. 1988). In addition, histological and biochemical reactions appear to be regulated by plant growth substances as reported by Hoque (1982) in his review on biochemical aspects of stress physiology of ligneous plants. The scheme outlined below (Fig. 10.20) summarizes the different pathways involved in the defense of the rubber tree that also may be under hormonal control. It reveals how complex the regulation of the root defense can be.

10.5 Defense in Angiosperm Roots Triggered by Fungi

10.5.1 The Time Sequence of Root Defense

Root reactions to fungal invasion depend greatly on plant aging, which changes their ability to defend themselves. Mature trees thus become less susceptible to the pathogens. Although natural physical barriers locally limit fungal colonization, they are often inefficient in stopping the infection with time. Thus, roots

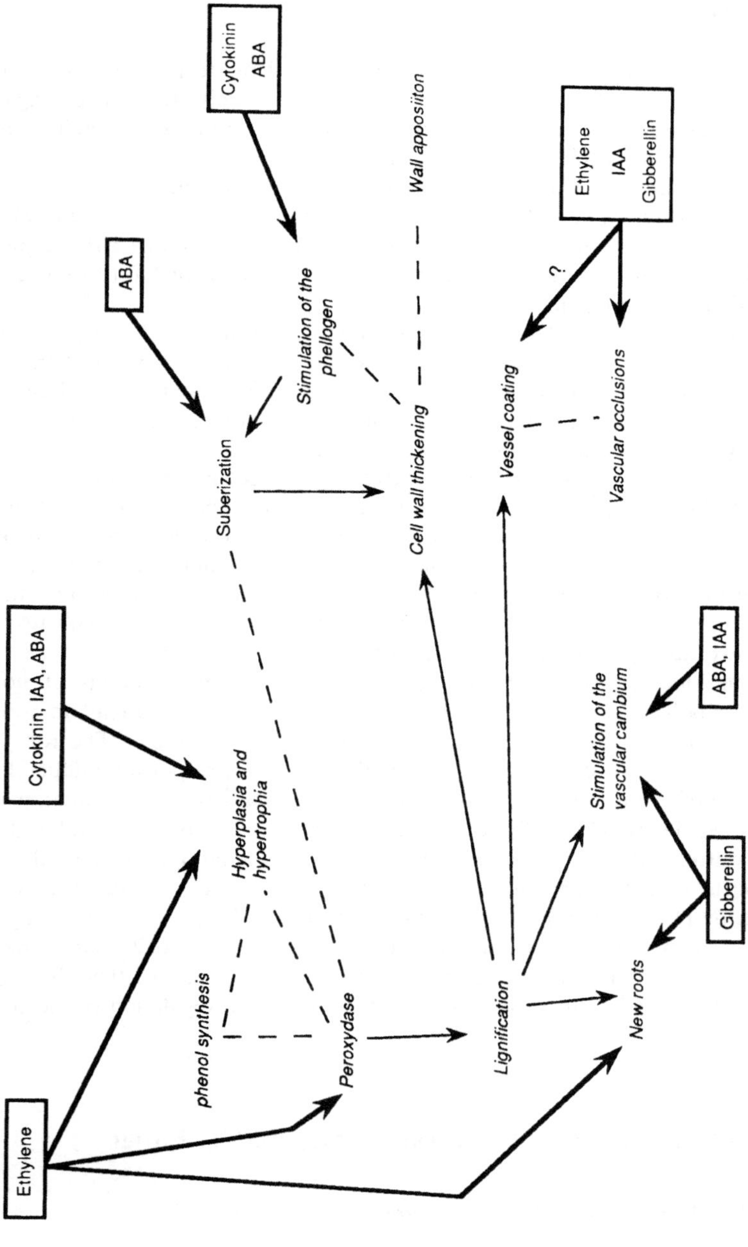

Fig. 10.20. Possible relations between root reactions and hormonal balance in rubber tree infected with root rot fungi (→ hormonal control; → connected responses; .... responses occurring simultaneously)

produce sequential reaction zones whose chemical components, such as lignins, terpenoids, or enzymes, help the plant to resist. Plant defense accordingly increases with maturity; this is likely to be the case with roots which are infected very early. Bell (1980) reported that "defense against root diseases increases rapidly and remains at a high level, even in the mature plant", as described in roots infected with *Fusarium*, *Rhizoctonia*, or *Phytophthora*. Older cotton roots were more resistant to *Thielavopsis* root rot because of cork cambium stimulation (Mathre et al. 1966).

Induced defense by roots of the rubber tree infected with *Rigidoporus lignosus* and *Phellinus noxius* illustrate well the sequence of events in root defense (Fig. 10.21). Among the responses observed within these root rot diseases, most were concerned with roots of yound seedlings (Nicole et al. 1986c). Some are developed as soon as the hyphae penetrate the root periderm (cell hypertrophia and hyperplasia, stimulation of peroxidase activity, and the hypersensitive-like response). Other responses, such as cork-phelloderm cambium stimulation, wall thickening, wall appositions, and cell wall lignification or suberization occur during the colonization of nonlignified tissues. Protective layers, peroxidase activity, plugging material, the stimulation of the vascular cambium, and the lignification of phloem or medulla parenchyma cells are induced when the fungi degrade the vascular tissues. However, the newly differentiated tissues that wall off the decayed roots occurred late and only in a few trees.

Early reactions suggest recognition mechanisms between host and pathogens. In root-fungi interactions, these mechanisms are probably common and nonspecific. In this respect, the rapid cell death of rubber tree periderm only occurs after close contact with *Phellinus noxius*. In foliar diseases, for example, immediate responses of cells appear within seconds or minutes after infection (or wonding), whereas the rapid reactions in root infections, and especially with root-rot diseases, occur over a period of several hours to some days. Rubber tree roots attacked with *Rigidoporus lignosus* initiated the first observable reactions only 5 days after artificial infection, and slow reactions such as root differentiation were produced several weeks later (Nicole et al. 1983). Thus, in infected roots the sequence of defense must not only be considered in time but also in space as a result of a precise situation (i.e., periderm penetration, xylem colonization, or middle lamella degradation).

10.5.2 Efficacy of Root Defense to Fungal Invasion

As postulated by Beckman and Talboys (1981), roots are known to be the most vulnerable plant organs, in particular young roots. Although root defense increases with plant maturity, efficacy of reactions often decrease with the frequency and the successful penetrations of the tissues by the parasites. Root-rotting fungi penetrate the root system in several locations (Nicole et al. 1987), and the ectotrophic infection habit of these fungi suggests that the more rapidly a pathogen progresses into the taproot, the less the plant can react (Garrett 1970).

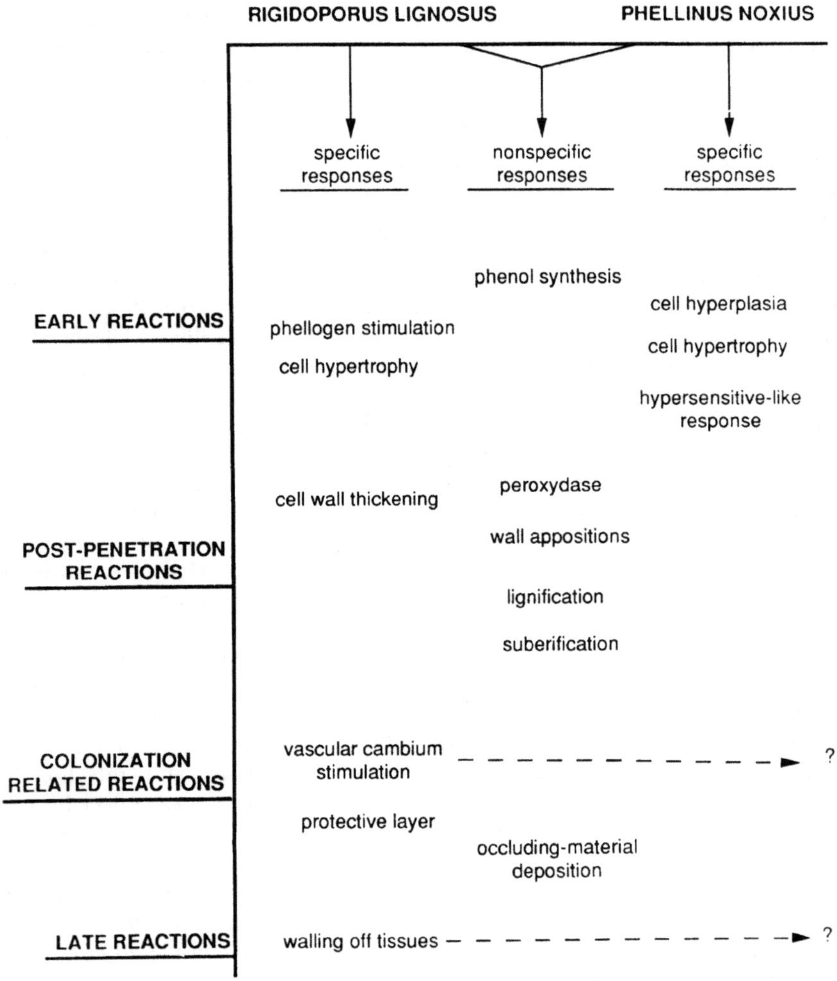

Fig. 10.21. Sequence of histological and biochemical reactions in *Hevea brasiliensis* roots following infections with *Rigidoporus lignosus* and *Phellinus noxius*. Rapid responses occur within a few days, late responses within several weeks after infection

When considered individually, none of the root responses described above are specific; moreover, none of them confers complete resistance to the whole tree; at best a balance is established between fungi and trees (Nicole et al. 1983). As Shigo (1984) stated, if environmental conditions do not favor the tree, the pathogen has an advantage. However, local efficacy was often reported for some of these reactions. Hypersensitivity has been suggested to precede the cessation of fungal growth (Ingram 1982), as observed in rubber tree roots infected with *Phellinus noxius*. The hypersensitive-like response of *Hevea brasiliensis* is always associated with hypertrophied cells that prevent local colonization of cortical cells. This suggests that cell death or damage occurring in a hypersensitive response can induce host defense responses (Hahn et al. 1989). Similar efficient associations have also been described; xylem hyperplasia and hypertrophy, for example, often accompanied tylosis formation in vascular diseases (Beckman and Talboys 1981). However, to our knowledge, few references have reported resistance of callose to woody root rot fungi. Callose deposition in the young phellem cells of rubber tree roots do not resist fungal hyphae, which possess enzymes able to degrade these wall appositions (Fig. 10.16) (Nicole et al. 1986a,c).

When all reactions are considered together, survival of trees to root infection often depends on the success of root regeneration. Both meristematic and cambial activities contribute to the differentiation of new tissues (including lignification, suberization, and polysaccharide pathway processes), which may play a more positive role in resistance mechanisms than responses at the cellular level. Necrophylactic periderm and barrier zones restrict fungal spread by compartmentalizing decayed tissues in both bark and vascular tissues. No fungal penetration into and beyond the neodifferentiated tissues was observed (Pearce and Woodward 1986). Indeed, microscopic observations and isolation tests performed on the xylem barrier zone of roots of rubber trees infected with *Rigidoporus lignosus*, never allowed identification of the fungus. Although it has not been investigated, it is possible that the monomer composition of this neoxylem was different, thus preventing lignin degradation by the fungus (Geiger et al. 1986a).

The CODIT model elaborated by Shigo and Marx (1977) was applied to angiosperm roots (Shigo and Tippet 1981b, Nicole et al. 1986c). Accordingly, responses such as vessel occlusion, wall apposition, and hyperplasia or hypertrophy correspond to the movable walls 1, 2, and 3, whereas protective tissues correspond to the stationary wall 4. However, defense by compartmentalization is not always successful (Shigo 1984), and the tree often dies when the system does not function. In rubber tree plantations or under greenhouse conditions, tree survival of root rot diseases depends mainly on the efficacy of the barriers decribed by the CODIT model.

10.5.3 Elicitation of Some Defense Reactions in Tree Roots

Attempts to trigger defense reactions of woody plants by fungal elicitors have been made rarely in comparison with elicitation of herbaceous plant defense (De Wit 1986, Mazau et al. 1987, Esquerré-Tugayé et al. 1990). However, some authors reported in young trees, different aspects of elicitation as gossypol stimulation by conidia cell walls of *Verticillium* in *Gossypium barbadense* (Bell 1981). Miller et al. (1986) showed that monoterpene synthesis was induced in stem phloem of lodgepole pine (*Pinus concorta*) after inoculation with *Ceratocystis clavigerum* or treatment with chitosan. Rapid callose synthesis was also elicited in *Pinus elliotii* hypocotyls after injection of a conidia suspension of the pitch canker fungus (*Fusarium subglutinans*) (Valluri and Soltes 1990).

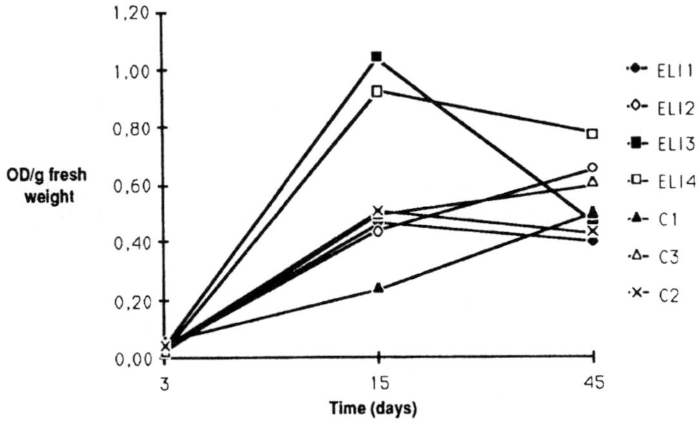

Fig. 10.22. Elicitation of the cinnamyl-alcohol dehydrogenase (CAD) of rubber tree roots. The CAD activity is expressed in OD/g fresh weight; *ELI1*, *2*, *3*, and *4* different *Rigidoporus lignosus* elicitors; *C1* healthy plants; *C2* plants wounded with a needle; *C3* plants injected with water

Recents works on root rot diseases of the rubber tree revealed that elicitation of some reactions also succeed on roots of lignified plants (Nicole et al. 1991). Cambium activity and callose depositions in parenchyma cells were induced by *Rigidoporus lignosus* elicitors. Lignification was also triggered by fungal extracts according to microscopic observation and by assaying cinnamy-lalcohol dehydrogenase (CAD), a specific enzyme for lignin synthesis (Fig. 10.22). Elicitation of these defense responses by fungal cell wall fragments contributes to the strengthening of the structural barriers of rubber tree roots in order to prevent the extension of decay.

10.5.4 Lignification in Wood: a Model in Root Defense

Many papers point out the role of lignification in plant defense (Vance et al. 1980, Touzé and Esquerré-Tuguayé 1982). Lignification is known to restrict fungal progress in numerous ways, but in trees little attention has been devoted to the importance of lignification in root defense. An analysis of data collected on *Hevea brasiliensis* infected with *Rigidoporus lignosus* and *Phellinus noxius* suggest that lignification is strongly involved in wood defense and contributes to the resistance of some trees to decay (Fig. 10.23). Even though no close relations were established between the different rubber tree root responses, cell wall thickening, xylem neosynthesis, enhancement of peroxidase activity, involvement of this peroxidase in lignin monomer polymerization, increase in the lignin content of infected tissues, and the induction of the CAD with fungal elicitors, there is evidence for the contribution of lignin synthesis to tree defense (Nicole et al. 1986c). Survival of trees attacked by soil fungi depends mainly on the

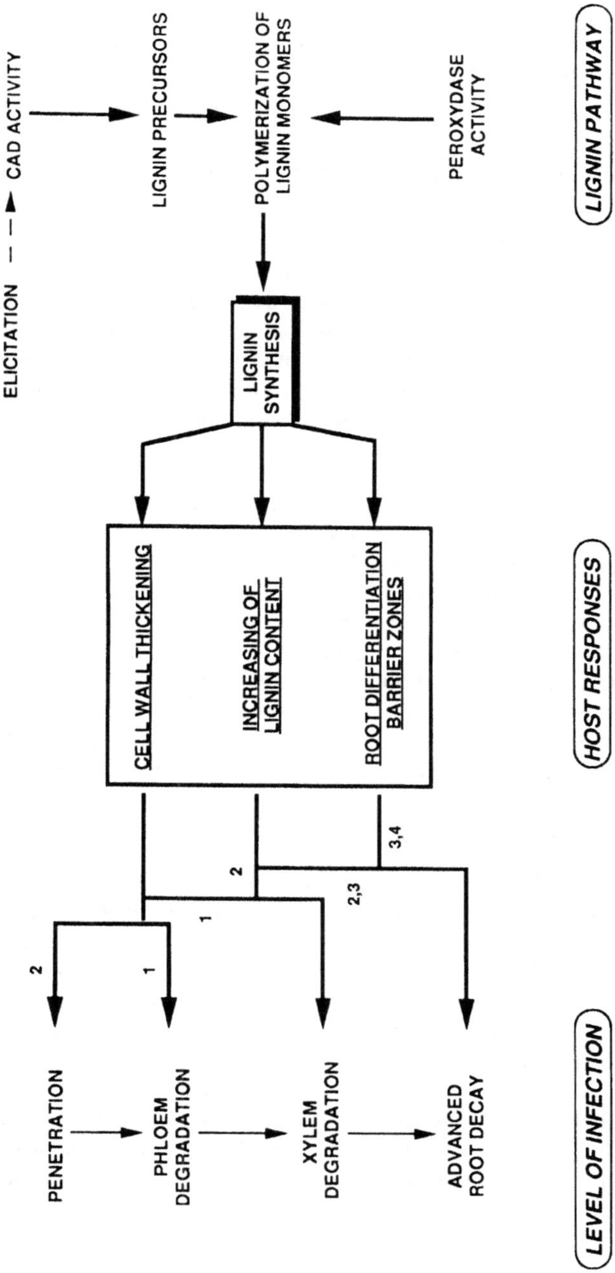

Fig. 10.23. Lignification as a defense reaction in roots of rubber trees infected with root rot fungi. *1* Possible local efficacy; *2* stop fungal spread; *3* stop decay extension; *4* induced tolerance or resistance of trees

efficacy of the wall 4 barrier of the CODIT model, often concerned with intense lignification. It is therefore likely that specific genes may be involved in the biosynthesis of defense lignin (Walter 1990).

10.6 Root Defense: Tolerance or Resistance to Fungal Invasion?

Tolerance and resistance are concepts that have been used to describe the host-parasite interaction. However, the definitions given by several authors are sometimes confusing and their use in tree infections, such as root rot diseases, might be debatable. Clarke (1986) suggested that we should establish tolerance scale ranging from highly tolerant to acutely intolerant plants; a similar continuum has been described for resistance. He proposed different forms of tolerance related to the parasite (tolerance to the parasite), to the disease (tolerance to disease), or to both (overall tolerance). Tolerance can be shown not only by the whole plant, but also at the cellular and/or field level (Mussell 1980), assuming that the physiological mechanisms of their reactions are known.

Identification of tolerance or resistance among trees infected with root-rotting fungi, if foliar symptoms do not occur, requires the uprooting of the plants to provide visual evidence of infection. The tolerance of the rubber tree to root-rotting fungi seems to be mainly determined by the severity of the disease. A given strain of *Rigidoporus lignosus* or *Phellinus noxius* causes damage which varies from tree to tree. In a population of artificially infected seedlings of *Hevea brasiliensis* clone GT1, only 5% survived the disease. Although the level of infection was identical with that in dead trees, infection was severely restricted in the surviving plants by host reactions, decayed roots being replaced by newly formed tissues.

More generally, in tree root diseases, it would be interesting:

1. To understand the role of tree vigor in regrowth of damaged root tissues. As stressed by Rishbeth (1972), survival of a tree often depends on the balance between regeneration and killing of roots. The level of infection-induced disturbances of the tree metabolism interferes strongly with the plant's ability to allocate resources to defend itself against root fungal infection (McLaughlin and Shriner 1980), thus determining the level of tolerance.
2. To know whether root tolerance varies with the variable aggressiveness of strains. Such a variation of pathogenicity is well known in root rot fungi (Raabe 1967, Prillinger and Molitoris 1979, Nandris et al. 1987b) as well as in vascular pathogens. Evidence of pathotypes of *Rigidoporus lignosus* and *Phellinus noxius* was demonstrated in isolates originating from various countries and even from the same plantation (Nicole et al. 1985).
3. To find out if tolerance, or resistance, of tree roots is of genetic origin. In tree breeding projects, resistance or tolerance to diseases is now being taken into account. This implies, first, the identification of resistance mechanisms and, second, the determination of the possible level of specificity. According to Carson and Carson (1989), without specificity, genetic control seems

impossible. However, in tropical tree plantations (i.e., the rubber tree), the genetic origin of the "clones" is almost unknown. Each population of a root rot fungus is considered as a mosaic of clones with different pathogenic potential (Redfern 1975, Nicole et al. 1985). In addition, fungi such as *Rigidoporus lignosus* are known to infect more than 100 species of trees. Thus, selection of resistant or tolerant trees should increase genetic diversity in order to maintain a low level of disease.

References

Aist JR 1983 Structural responses as resistance mechanisms. In: Bailey JA, Deverall BJ (eds) The dynamics of host defence. Academic Press, New York London, 33–70

Akai S 1959 Histology of defense in plants. In: Horsfall JG, Dimond AE (eds) Plant pathology. Academic Press, New York London, 391–434

Albersheim P 1989 Oligosaccharins are biochemical messengers. NATO ASI Ser L28:1–18

Aloni R 1987 Differentiation of vascular tissues. Annu Rev Plant Physiol 38:179–204

Asada Y, Matsumoto I 1972 The nature of lignin obtained from downy mildew-infected Japanese radish root. Phytopathol Z 73:208–214

Beckman CH, Talboys PW 1981 Anatomy of resistance. In: Mace ME, Bell AA, Beckman CH (eds) Fungal wilt diseases of plants. Academic Press, New York London, 487–521

Bell AA 1980 The time sequence of defense. In: Horsfall JG, Cowling EB (eds) Plant disease, an advanced treatise, vol 5. How plants defend themselves. Academic Press, New York London, 53–73

Bell AA 1981 Biochemical mechanisms of disease resistance. Annu Rev Plant Physiol 32:21–81

Blanchette RA 1984 Screening wood decayed by white-rot fungi for preferential lignin degradation. Appl Environ Microbiol 48:647–653

Blanchette RA, Abad AR 1988 Ultrastructural localization of hemicellulose in birch wood (Betula papyrifera) decayed by brown and white-rot fungi. Holzforschung 42:393–398

Blanchette RA, Abad AR, Farrell RL, Leathers TD 1989 Detection of lignin peroxydase and xylanase by immunocytochemical labelling in wood decayed by basidiomycetes. Appl Environ Microbiol 55:1457–1465

Carson SD, Carson MJ 1989 Breeding for resistance in forest trees – a quantitative genetic approach. Annu Rev Phytopathol 27:373–395

Clarke DD 1986 Tolerance of parasites and disease. In: Ingram DS, Williams PH (eds) Advances in plant pathology, vol 5. Academic Press, New York London, 162–197

Dallakyan GA, Markarova EN, Veselovskii VA, Tarusov BN 1978 The antioxidant content in root secretions and root tissues of cotton seedlings infected by wilt and exposed to thermal treatment. Sel'skokhoz Biol 13:593–597

De Wit PJGM 1986 Elicitation of active resistance mechanisms. NATO ASI Ser H1:149–169

Enebak SA, Blanchette RA 1989 Canker formation and decay in sugar maple and paper birch infected by Cerrana unicolor. Can J For Res 19:225–231

Esquerré-Tugayé MT, Fournier J, Mazau D, Pouenat ML, Rickauer M, Rumeau D 1990 Cellular and molecular approaches of defense in plants. In: Ranjeva R, Boudet AM (eds) Signal, perception and transduction in higher plants. Springer, Berlin Heidelberg New York, 237–280

Garrett SO 1970 Pathogenic root infecting fungi. Cambridge Univ Press, London, 294 pp

Geiger JP, Goujon M 1977 Etude de deux peroxydases différentes extraites des tissus racinaires d'Hévéa sains et parasités par Leptoporus lignosus. C R Acad Sci Paris Ser D 284:1053–1056

Geiger JP, Huguenin B, Nicole M, Nandris D 1986a Laccases of Rigidoporus lignosus and Phellinus noxius. II Effects of R. lignosus laccase L1 on thioglycolic lignin of Hevea brasiliensis. Appl Biochem Biotech 13:97–111

Geiger JP, Nandris D, Goujon M 1976 Activité des laccases et des peroxydases au sein des racines d'Hévéa attaquées par le pourridié blanc (Leptoporus lignosus). Physiol Vég 14:271–282
Geiger JP, Nicole M, Nandris D, Rio B 1986b Root rot diseases of Hevea brasiliensis. I. Physiological and biochemical aspects of root aggression. Eur J For Pathol 16:22–36
Geiger JP, Rio B, Nicole M, Nandris D 1986c Biodegradation of Hevea brasiliensis wood by Rigidoporus lignosus and Phellinus noxius. Eur J For Pathol 16:147–159
Geiger JP, Rio B, Nicole M, Nandris D 1989 Peroxidase production in tissues of the rubber tree following infection by root rot fungi. Physiol Mol Plant Pathol 34:241–256
Hahn MG, Bucheli P, Cervone F, Doares SH, O'Neill RA, Darvill A, Albersheim P 1989 Roles of cell wall constituents in plant-pathogen interactions. In: Kosuge T, Nester EW (eds) Plant-microbe interactions, molecular and genetic perspectives, vol 3. McGraw-Hill, New York, London, 131–181
Hammerschmidt R, Bonnen AM, Bergstrom GC, Baker K 1985 Association of epidermal lignification with nonhost resistance of cucurbits to fungi. Can J Bot 63:2393–2398
Harkin JH, Obst JR 1973 Lignification in trees: indication of exclusive peroxidase participation. Science 180:296–297
Hoque E 1982 Biochemical aspects of tree physiology of plants and some considerations of defense mechanisms in conifers. Eur J For Pathol 12:280–296
Ingram DS 1982 A structural view of active defence. In: Wood RKS (ed) Active defense mechanisms in plants. NATO ASI Ser A: 19–38
Kiraly Z 1980 Defenses triggered by the invader: hypersensitivity. In: Horsfall JG, Cowling EB (eds) Plant disease, an advanced treatise, vol 5. How plants defend themselves. Academic Press, New-York London, 201–224
Mace ME, Howell CR 1974 Histochemistry and identification of condensed tannin precursors in roots of cotton seedlings. Can J Bot 52:2432–2426
Mathre DE, Ravenscroft AV, Garber RH 1966 The role of Thielaviopsis basicola as a primary cause of yield reduction in cotton in California. Phytopathology 56:1213–1216
Mazau D, Esquerre-Tugayé MT 1986 Hydroxyproline-rich glycoprotein accumulation in the cell walls of plants infected by various pathogens. Physiol Mol Plant Pathol 29:147–15724
Mazau D, Rumeau D, Esquerre-Tugaye MT 1987 Molecular approaches to understanding cell surface interactions between plant fungal pathogens. Plant Physiol Biochem 25:337–343
Mc Laughlin SB, Shriner DS 1980 Allocation of resources to defense and repair. In: Horsfall JG, Cowling EB (eds) Plant disease, an advanced treatise, vol 5. How plants defend themselves. Academic Press, New York, London, 407–431
Miller RH, Berryman AA, Ryan CA 1986 Biotic elicitors of defense reactions in lodgepole pine. Phytochemistry 25:3611–612
Mullick DB, Jensen GD 1973 New concepts and terminology of coniferous periderms: necrophylactic and exophylactic periderms. Can J Bot 51:1459–1470
Mussell H 1980 Tolerance to disease. In: Horsfall JG, Cowling EB (eds) Plant disease, an advanced treatise, vol 5. How plants defend themselves. Academic Press, New York London, 39–52
Nandris D, Nicole M, Geiger JP 1983 Infections artificielles de jeunes plants d'Hevea brasiliensis par Rigidoporus lignosus et Phellinus noxius. Eur J For Pathol 13:65–76
Nandris D, Nicole M, Geiger JP 1987a Root rot diseases of rubber tree. Plant Dis 71:298–306
Nandris D, Nicole M, Geiger JP 1987b Variations virulence among Rigidoporus lignosus and Phellinus noxius isolates from West Africa. Eur J For Pathol 17:271–281
Nandris D, Nicole M, Geiger JP 1988 Root rot diseases of the rubber tree in the Ivory Coast. 1. Severity, dynamics, and characterization of epidemics. Can J For Res 18:1248–1254
Nicole M, Geiger JP, Nandris D 1986a Penetration and degradation of suberized cells of Hevea brasiliensis infected with root rot fungi. Physiol Mol Plant Pathol 28:181–185
Nicole M, Geiger JP, Nandris D 1986b Ultrastructure of laticifer modifications in Hevea brasiliensis infected with root rot fungi. J Phytopathology 116:259–268
Nicole M, Geiger JP, Nandris D 1986c Root rot diseases of Hevea brasiliensis. II. Some host reactions. Eur J For Pathol 16:37–55
Nicole M, Geiger JP, Nandris D 1987 Ultrastructural aspects of rubber tree root rot diseases. Eur J For Pathol 17:1–10

Nicole M, Nandris D, Geiger JP 1982 Interactions hôte-parasites entre Hevea brasiliensis et les agents de pourriture racinaire Rigidoporus lignosus et Phellinus noxius: étude physiopathologique comparée. Phytopathol Z 105:311–326
Nicole M, Nandris D, Geiger JP 1983 Cinétique de l'infection de plantes d'Hevea brasiliensis par Rigidoporus lignosus. Can J For Res 13:359–364
Nicole M, Nandris D, Geiger JP, Rio B 1985 Variability among African populations of Rigidoporus lignosus and Phellinus noxius. Eur J For Pathol 15:293–300
Nicole M, Toppan A, Geiger JP, Roby D, Nandris D, Rio B 1991 Defense responses of Hevea brasiliensis to elicitors from root rot fungi. Can J Bot 69:1819–1824
Ouellette GB 1981 Ultrastructural cell wall modification in secondary xylem of american elm surviving the acute stage of Dutch elm disease: fibres. Can J Bot 59:2425–2438
Pearce RB, Rutherford J 1981 A wound-associated suberized barrier to the spread of decay in the sapwood of oak (Quercus robur). Physiol Plant Pathol 19:359–369
Pearce RB, Woodward S 1986 Compartmentalization and reaction zone barriers at the margin of decayed sapwood in Acer saccharinum. Physiol Mol Plant Pathol 29:197–216
Pegg GF 1976 Endogenous auxins in healthy and diseased plants. In: Heitefuss R, Williams PH (eds) Physiological plant pathology Springer, Berlin Heidelberg New York, 560–581
Pegg GF 1988 The role of growth hormones in plants pathogenesis. In: 5th Int Congr Plant Pathol, Kyoto, Sec VI:221
Phillips D, Grant BR, Weste G 1987 Histological changes in the roots of an avocado cultivar, Duke 7, infected with Phytophthora cinnamomi. Phytopathology 77: 691–698
Prillinger H, Molitoris PH 1979 Genetic analisys in wood decaying fungi. I. Genetic variation and evidence for allopatric speciation in Pleurotus ostreatus using phenoloxydase zymograms and morphological criteria. Physiol Plant 46:265–277
Raabe RD 1967 Variation in pathogenicity and virulence in Armillaria mellea. Phytopathology 57:73–75
Redfern DB 1975 Influence of food base on rhizomorph growth and pathogenicity of Armillaria mellea isolates. In: Bruehel GW (ed) Biology and control of soilborne plant pathogens. Am Phytopathol Soc, St Paul, 69–73
Ride JP 1983 Cell walls and other structural barriers in defence. In: Callow JA (ed) Biochemical plant pathology. John Wiley & Sons, New York, 215–236
Rishbeth J 1972 Resistance to fungal pathogens of tree roots. Proc R Soc London Set B 181:333–351
Rishbeth J 1985 Infection cycle of Armillaria and host response. Eur J For Pathol 15:332–341
Roberts LW, Gahan PB, Aloni R 1988 Vascular differentiation and plant growth regulators. Springer, Berlin Heidelberg New York, 154 pp
Sharples ARC 1936 Diseases and pests of the rubber tree. MacMillan, New York London, 479 pp
Shigo AL 1984 Compartmentalization: a conceptual framework for understanding how trees grow and defend themselves Annu Rev Phytopathol 22:189–214
Shigo AL, Marx M 1977 Compartmentalization of decay in trees (CODIT). US Dep Agr Inf Bull 405:73 pp
Shigo AL, Tippett JT 1981a Compartmentalization of American elm tissues infected by Ceratocystis ulmi. Plant Dis 65:715–718
Shigo AL, Tippett JT 1981b Compartmentalization of decayed wood associated with Armillaria mellea in several tree species. US Dep Agr Res Pap NE-488:20 pp
Struckmeyer BE, Beckman CH, Kuntz JE, Riker AJ 1954 Plugging of vessels by tyloses and gums in wilting oaks. Phytopathology 14:148–153
Thomas HE 1934 Studies on Armillaria mellea, infection, parasitism, and host resistance. J Agr Res 48:187–218
Touzé A, Esquerré-Tugayé MT 1982 Defence mechanisms of plants against varietal non-specific pathogens. In: Wood RKS (ed) Active defense mechanisms in plants. NATO ASI Ser A:103–117
Ugalde D, Taylor P 1983 Ethylene production from roots and trunk of the peach tree in response to infection with Phytophthora cactorum. In: Proc 4th Int Congr Plant Pathol, Melbourne 781

Valluri JV, Soltes EDJ 1990 Callose formation during wound inoculated reaction of Pinus elliottii to Fusarium subglutinans. Phytochemistry 29:71–72
Vance CP, Kirk TK, Sherwood RT 1980 Lignification as a mechanism of disease resistance. Annu Rev Phytopathol 18:259–288
VanderMolen GE, Beckman CH, Rodehorst E 1977 Vascular gelation: a general response phenomenon following infection. Physiol Plant Pathol 11:95–100
Walter MH 1990 Regulation of lignification in defense. In: Boller T, Meins F (eds) Advances in plant gene research, vol 8. Genes involved in plant defense. Springer, Berlin Heidelberg New York (in press)
Wargo PM 1975 Lysis of the cell wall of Armillaria mellea by enzymes from forest trees. Physiol Plant Pathol 5:99–105

Chapter 11

Defense Mechanisms in Leaves and Fruit of Trees to Fungal Infection

J.E. ADASKAVEG

11.1 Introduction

Most biotic, foliar diseases of trees are caused by fungi. In order to establish infection, the fungal pathogen must overcome natural host defense mechanisms. Disease resistance is the ability of the host to prevent or resist disease. In foliar diseases of trees, host and nonhost resistance is dependent on the characteristics of the entire leaf including both biochemical and structural mechanisms as preexisting barriers or induced responses from host-pathogen interactions. The apoplast, or the extracellular matrix that includes cell wall, intercellular spaces, and xylem, is considered to have a central role in the interface of plants with the environment and in the defense of plants against pathogens (Bowles 1990). Studies on plant resistance have focused on either preinfectional host mechanisms that prevent and inhibit pathogen penetration or postinfectional host responses to disease development.

Preinfectional mechanisms have been referred to as preformed, passive, or static host defenses. Many studies on preinfectional barriers of plants that may prevent or delay penetration of fungi have been reviewed. These include physical barriers such as cell walls or cuticle (Akai 1959, Martin 1964, Martin and Juniper 1970, Overeem 1976, Royle 1976, Campbell et al. 1980, Kolattukudy and Köller 1983, Ride 1983) and biochemical compounds derived from shikimic acid, tricarboxylic acid, fatty acid, and mevalonic acid cycles (Goodman et al. 1967, Schlösser 1980, Stoessl 1983).

Host resistance to disease development reflects the dynamic interactions between the host and pathogen. Recognition of a pathogen elicits responses from individual host cells and tissues to provide defenses and thus, attempts to limit the infection. Although host responses originate in the protoplasm or symplast, many defense reactions occur in the apoplast (Bowles 1990). Active or postinfectional responses include physiological and biochemical changes that involve the production of defense compounds such as phytoalexins, β-1,3-glucanases, proteases, peroxidases, phenoloxidases, glycoproteins, lignin, and callose (Walker and Stahmann 1955, Kuć 1966, 1990, Goodman et al. 1967, Vance et al. 1980, Cruickshank 1980, Patil 1980, Mansfield 1982, 1983, Bowles 1990). These biochemical responses have been placed into three classes: (1) defense compounds that act directly on the pathogen (e.g., β-1,3-glucanases, peroxidases) or catalyze the synthesis of antimicrobial compounds such as oxidized phenolics, tannins, or *o*-quinones; (2) structural responses that repair, alter, or strengthen the cell wall; and (3) responses that occur during pathogenesis but have an unknown function. Structural changes may occur from responses of single cells as in the formation of cytoplasmic aggregates or papillae, or from responses of multiple cells that form lignified, suberized, or meristematic tissues as structural defense barriers (Akai 1959, Aist 1976, 1983, Beckman 1980, Heath 1980, Sherwood and Vance 1982, Ride 1983).

Distinctions between preinfectional biochemical barriers and postinfectional inhibitors that are formed in response to infection of the host are not always easily distinguished and may be interrelated (Stoessl 1983). Furthermore, structural responses are directly dependent on physiological changes and are often coincident as discussed by Aist (1983). These distinctions are made occasionally by researchers for the purpose of discussion. Heath (1981), in a review of resistance of plants to rust infection, characterized most preinfectional characters and structural changes of cells in response to infection as mechanisms of nonhost resistance; he characterized most postinfectional (posthaustorial) responses that inhibit the growth of the pathogen as mechanisms of host resistance. Preinfectional mechanisms or structural responses, however, may contribute to the basic resistance of the host and in many cases, distinctions may not be obvious. In this chapter, host defense mechanisms will be reviewed as preinfectional barriers and postinfectional responses to fungal infection. Many host responses (including recognition, gene activation, and protein synthesis) occur in the formation of compounds that may be ultimately responsible for or involved in host resistance. This discussion, however, will provide an overview of those mechanisms implicated in preventing or inhibiting development of disease in relation to leaves and fruit of trees and woody plants. Host defenses to wounds will not be discussed. In many cases, however, responses to disease may be recognized as functions of wound response and healing rather than specific host defenses to pathogenic fungi.

11.2 Preinfectional Morphological Barriers

Fungi enter leaves of trees and other plants either directly or indirectly. With direct entry, fungal pathogens must first penetrate the cuticle and then the cell walls. With indirect entry, fungi enter leaves through natural openings or wounds. Natural openings on tree leaves include stomata, hydathodes, and glands. Evidence for preinfectional barriers (morphological or physiological functions such as stomatal operation) have been correlated mainly with host resistance, but, several studies have indicated contributory roles of some of these defenses. Information available on leaf structures of trees influencing host resistance have included trichomes, stomata, cuticle, and cell walls.

11.2.1 Trichomes

Trichomes are unicellular or multicellular structures that extend from epidermal layers of many plants and plant tissues. Furthermore, many morphological forms of trichomes occur, while their functions vary from undetermined to glandular.

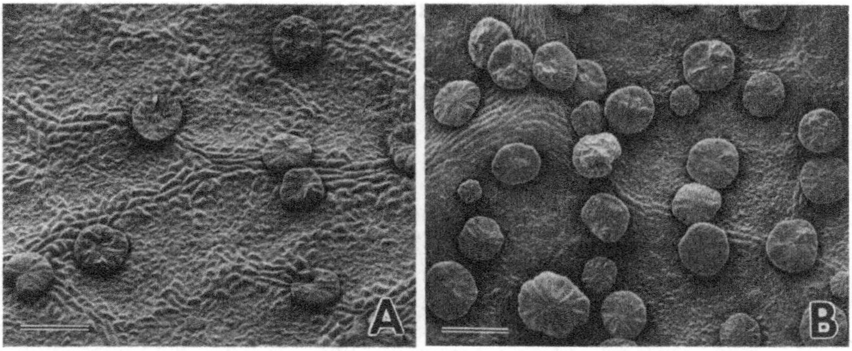

Fig. 11.1A,B. Micrographs of leaves of *Carya illinoensis* resistant and susceptible to scab infection caused by *Cladosporium caryigenum*. **A,B** Scanning electron micrographs (SEM) of abaxial surface of immature leaves with low (**A**) and high (**B**) glandular trichome densities in cultivars resistant (**A**) and susceptible (**B**) to scab. *Bars* = 100 μm (Wetzstein and Sparks 1983)

Trichomes influence air turbulence and wettability of the leaf surface that may affect spore impaction and microclimatic conditions influencing disease (Gay and Pearce 1984). In studies of disease resistance, trichomes have been associated with barriers or implicated as infection sites for fungal penetration. As barriers, glandular trichomes may release fungitoxic excretions (Hafiz 1952, Beckman et al. 1972) or prevent surface wetness. In studies of foliar and fruit diseases of trees, Johnstone (1931) noted that apple leaves resistant to *Venturia inaequalis* had more trichomes that prevented water from reaching the cuticle and thus functioned as a barrier rendering leaves less susceptible to disease. In other studies, bases of trichomes or trichome sockets may serve as infection sites in peaches and apricots for *Monilinia fructicola* (Curtis 1928, Hall 1971) and in pecan leaves for *Cladosporium caryigenum* (Latham and Rushing 1988). Smith (1936) indicated that *M. fructicola* infects peaches through trichome sockets and that infection time was reduced by one-half when trichomes were injured from brushing of peach fruit. Brown rot resistance in peaches was correlated with lower numbers of trichomes, shallow trichome sockets, as well as other characters of fruit surfaces (Adaskaveg et al. 1989). Lower numbers of trichomes may minimize potential infection sites, while shallow trichome sockets may delay fungal penetration into noncutinized cells. Wetzstein and Sparks (1983) studied anatomical differences in pecan leaves to scab infection caused by *C. caryigenum* and attributed resistance in some cultivars and older leaves of susceptible cultivars to lower densities of peltate trichomes (Fig. 11.1). Trichomes of the susceptible cultivars often abutted the abaxial surface of leaves where secretions were found adjacent to some trichomes. Wetzstein and Sparks suggested that higher trichome densities might increase relative humidity and free water retention and, therefore, contribute to a more conducive microclimate for disease. Furthermore, they indicated that secreted substances may promote germination of conidia and hyphal growth. Latham and Rushing (1988) studying modes of penetration of *C. caryigenum* in pecan leaves determined that only 16% of 2000 conidia evaluated were associated with trichomes. However, 82% of those conidia germinated and grew toward the base of trichomes where infection occurred.

11.2.2 Stomata

Stomata are specialized openings that function in regulation of transpiration, exchange of O_2 and CO_2, and release of volatile metabolites. Resistance and susceptibility of hosts to pathogens have been related to numbers, spatial arrangement, structure, and operation of stomata (Royle 1976). Many foliar pathogens enter directly through intact surfaces, but many others enter solely or alternatively through stomata. In white pine blister rust, caused by *Cronartium ribicola*, infection occurs through stomata (Patton and Johnson 1970). In species of pines susceptible to white pine blister rust, Spaulding (1925) observed that stomata occurred on the inner and outer surfaces of needles, while pines resistant to the fungus had stomata primarily on the inner surface of needles. Spaulding assumed that numbers of stomata differed between the two groups and suggested that needles with stomata on both surfaces would be more likely to be infected. Germ tubes of *Dothistroma pini* also infect needles of pines through stomata. However, Peterson and Walla (1978), studying development of *Dothistroma pini* in two species of pines, found no marked differences in numbers of stomata between resistant and susceptible pines or between tips and bases of needles despite a higher frequency of infection at the needle tips. In this study, germ tubes of the pathogen were observed to grow in a positive direction to stomata. Edwards and Bowling (1986) suggested that germ tubes of *Uromyces viciae-fabae* may grow toward stomata in response to pH gradients.

Some fungi penetrate directly or indirectly. In *Marssonina* leaf spot of *Populus*, germ tubes of *M. brunnea* enter primarily by direct penetration and not through stomata (Spiers and Hopcroft 1984). Not surprisingly, Spiers and Hopcroft (1984), studying susceptibility of poplar cultivars to *M. brunnea*, found no significant correlation between stomatal frequency and infection. Other leaf characteristics of poplar, however, were considered important in disease resistance (see Sect. 11.2.3).

Structural differences in stomata have been implicated as mechanisms of resistance to fungal infections. In studies on rust fungi, Wynn (1976) and Hoch et al. (1987) observed thigmotropic responses in growth and appressorial formation to topography of leaf and inert surfaces. Hoch et al. (1987), studying penetration of *Uromyces appendiculatus* on bean, determined that a simple ridge of $0.5\,\mu m$ (but not >1 or $<0.25\,\mu m$) served as a differentiation signal for the formation of appressoria and associated this ridge with a stomatal lip of the host plant. Other rust fungi may also respond to surface topography and thus, differences in leaf structure may function in host resistance. Studying differences in poplar species to foliar penetration by *Melampsora larici-populina*, Mlodzianowski et al. (1978) observed greater surface relief or sculpturing on the lower epidermis of susceptible clones of poplar. Siwecki and Werner (1980) and Siwecki et al. (1982) observed lower numbers of appressoria of the fungus on

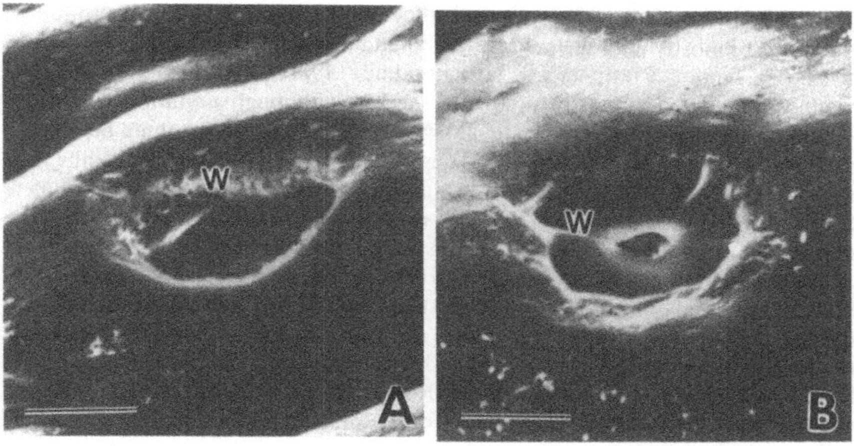

Fig. 11.2A,B. Transmission electron micrographs of stomata in needles of *Pinus radiata*. **A** Smooth cuticular wax of needle surface and granular texture of cuticular wax (**W**), inside stomatal pore, of an open stoma of young trees (5–10 years old). **B** Occluded stoma of mature tree (40 years old). Wax emanates from the rim of the pore and also from stomatal antechamber. Occluded stomata may contribute to resistance of pine needles to *Dothistroma pini*. *Bars* = 5 μm (Franich et al. 1977)

leaves of resistant clones and attributed this to structural differences in surface morphology of stomata. Differences in stomatal function and water relations of the plant also have been observed between resistant and susceptible clones (Siwecki and Przybyl 1981). Thigmotropic responses to surface topography of leaves, water relations of leaves, and hypersensitive reactions (see below) have all been related to disease resistance in leaves of *Populus*.

Other structural differences of stomata have been reported to mechanically inhibit indirect penetration by fungi. In studies of penetration of *Botrytis cinerea* in fruit of plum, Valleau (1915) suggested that parenchymatous plugs and corky cells in stomata may delay infection in resistant varieties. Franich et al. (1977) indicated that resistance in older needles of *Pinus radiata* to *Dothistroma pini* might be due to wax occlusions of stomatal pores (Fig. 11.2) and in a subsequent publication, Franich et al. (1978) indicated that changes in resin acids of waxes around stomata may prevent the waxes from being metabolized by the fungus. Furthermore, Patton and Spear (1980) observed complete occlusion of some epistomatal chambers by wax deposits and suggested that penetration by *Cronartium ribicola* in needles of pine might be prevented by this mechanical barrier.

11.2.3 Cuticle

The cuticle is a multilayered, extracellular structure that covers the epidermal cells of most plants. This structure consists of layers of pectin, cutin, and wax

(Kolattukudy and Köller 1983). Cutin is the main component and is an insoluble polyester of mostly unbranched, saturated and unsaturated aliphatic acid derivatives (C_{16} and C_{18} fatty acids) (Kolattukudy 1981). Quantity and chemical composition depends mostly on plant species, although variation can occur within species (Baker 1971). Cuticular waxes form the most outer layer, and are predominately long-chained alkanes that can vary both chemically and morphologically between species (Martin and Juniper 1970). Although cuticle production occurs throughout leaf expansion, thickness and quantity of the cuticle varies not only between plant species but within a species grown in different environments (Christensen and DeVay 1955, Martin 1964, Martin and Juniper 1970, Günthardt-Goerg 1986). Lee and Priestly (1924) noted that in species of *Ribes* and *Prunus*, the cuticle is thicker in leaves with high levels of nitrogen and thinner in leaves with high levels of calcium. Functions assessed to the cuticle include prevention of water loss, diffusion of light, protection from microwounds by abrading wind-blown dust, and inhibition of penetration by pathogens.

Evidence for direct mechanical penetration has been reviewed by Martin (1964), Martin and Juniper (1970), and Royle (1976). Previous researchers suggested that fungi penetrated cell walls by force and therefore they measured resistance of tissue to mechanical penetration. Much of this work, however, has not distinguished between effects of the cuticle and other physical factors influencing penetration, including shape and size of cells, thickness and composition of cuticle and cell walls, and turgor pressure. For a review of this topic, see Royle (1976).

One physical property that may influence fungal penetration is surface wettability. Fisher and Corke (1971) studied the wettability of red and green sides of apples in relation to their susceptibility to apple scab caused by *Venturia inaequalis*. Red sides of the apples differed from green sides in wettability, quantity of wax, and greater thickness of the cuticle. Of 100 apples inoculated on red and green sides of the same fruit, scab resulted on all red and only three green sides. Water repellency of waxes may also affect spore deposition. Heather (1967a) attributed resistance of young leaves of *Eucalyptus bicostata* primarily to a glaucous coating that inhibited deposition of spores of *Phaeoseptoria eucalypti* on the leaf surface.

Evidence of the cuticle functioning as a mechanical barrier to fungal penetration has been debated. In some studies, cuticle thickness has been correlated with disease resistance. In withertip of lime caused by *Gloeosporium limetticola*, Roberts and Martin (1963) found that disease resistance correlated with substantial increases in the quantity of waxes and cutin in older leaves. Marks et al. (1965), examining mechanisms of penetration by *Colletotrichum gloeosporioides* in a species of *Populus*, determined that cuticular resistance causes appressorial collapse and prevents infection by the fungus. This was considered to be one factor in leaf resistance. Spiers and Hopcroft (1984), studying another leaf spot of poplar caused by *Marssonina brunnea*, also attributed higher disease resistance in older than in younger leaves of several cultivars to stronger cutinization and cell wall ultrastructure (Fig. 11.3). In studies of resistance in conifer needles, Campbell (1972) considered that resistance to *Lophodermella sulcogena*

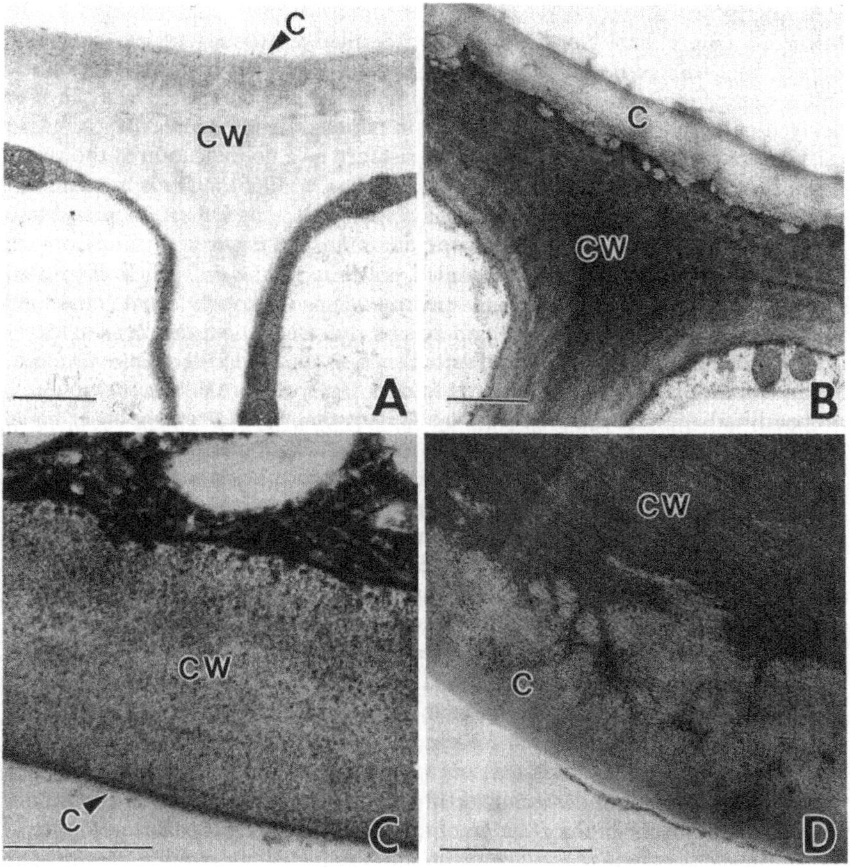

Fig. 11.3A–D. Transmission electron micrographs of vertical sections of epidermal cell walls of leaves of species of *Populus*. **A** Soft, expanding leaf of *Populus* sp., susceptible to *Marssonina brunnea*, showing poorly developed cuticle (*C*) and amorphous, electron-lucent cell wall (*CW*) with conspicuous microfibrils. **B** Mature leaf of *Populus* sp., resistant to *M. brunnea*, showing well-developed cuticle and striate, electron-dense cell wall with conspicuous microfibrils. *Bars* = 1 μm (Spiers and Hopcroft 1984). **C,D** Transmission electron micrographs of epidermal cell walls of leaves of *Carya illinoensis* resistant and susceptible to scab infection caused by *Cladosporium caryigenum*. **C** Immature, susceptible leaf with poorly developed cuticle. **D** Mature leaf resistant to scab showing well-developed cuticles and fibrous reticulate layers continuous with cell wall materials. *Bars* = 0.5 μm (Wetzstein and Sparks 1983)

in older needles of *Pinus nigra* var. *maritima* may be related to the development of a thicker cuticle and cell wall. Similarly, a thicker cuticle and epidermal wall on older portions of needles of *Larix occidentalis* was considered a mechanical barrier to penetration by *Hyphodermella laricis* (Cohen 1967). In studies of brown rot of fruit of sweet cherry, cuticle and wall thickness was correlated with longer incubation periods and lower incidence of infection caused by *M. fructicola* (Biggs and Northover 1989). Similarly, Adaskaveg et al. (1989) showed that fruit of peach cultivars resistant to brown rot had a thicker cuticle and a denser epidermis than susceptible cultivars and correlated resistance to delayed penetration and longer incubation periods for infection. Although correlations of cuticle thickness to disease resistance give no indication of thoroughness of coverage or number of infectibility sites (Kuć 1966), a thicker cuticle and epidermal wall may necessitate longer incubation periods for infection and thus allow time for other mechanisms to operate in host resistance.

Weinhold and English (1964) studied powdery mildew of peach caused by *Sphaerotheca pannosa* in different environments. Although cuticle and wall thickness correlated with leaf age and disease resistance, no differences in leaves of comparable age were observed between resistant and susceptible varieties. When resistant trees were transferred to darkness for 6 weeks osmotic potential dropped and they became susceptible. Penetration by the fungus occurred in both susceptible and resistant leaves, although in resistant leaves haustoria were rudimentary. Furthermore, they were able to induce resistance of susceptible leaves by floating them on certain concentrations of sucrose and thus they attributed resistance to increases in osmotic pressure of host cells. In a similar study, Vidhyasekaran and Cheeran (1972) related resistance to anthracnose caused by *Gloeosporium ampelophagum* in leaves of grape to increases in osmotic pressure. Resistance in rose to powdery mildew caused by *S. pannosa* (Mence and Hildebrandt 1966) and resistance in apple to apple scab caused by *Venturia inaequalis* (Valsangiacomo and Gressler 1988) could not be attributed to cuticle thickness since haustoria or hyphae, respectively, were found in both resistant and susceptible cultivars. Resistance of rose cultivars was attributed to internal factors, while resistance of apples has been attributed to reduced formation of stromata of *V. inaequalis*. In studies of grasses resistant to powdery mildew caused by *Erysiphe graminis*, Lupton (1956) considered several defense mechanisms operative. These included failure of hyphal penetration of cuticle or cell wall, degenerative haustoria, and hypersensitive reactions.

In general, the role of the cuticle in disease resistance has been comprehensively reviewed; it is not considered to be an efficient barrier to the penetration of fungi (Martin 1964, Martin and Juniper 1970, Royle 1976) under optimal conditions for disease. The cuticle may not be an absolute barrier to fungal penetration of leaves and fruit; however, it may contribute to host resistance as a quantitative barrier by delaying infection. Most likely, the cuticle serves in nonhost or basic resistance and as a barrier for fungi that penetrate through stomata or wounds. Although numerous correlations exist, conclusive evidence for many specific diseases is still needed. The role of the cuticle as a qualitative barrier has been demonstrated. In studies of infection of pea caused by *Fusarium solani* f. sp. *pisi*, disease was prevented when cutinase activity was inhibited using specific antibodies or inhibitors (Maiti and Kolattukudy 1979). Köller et al. (1982a,b,c) prevented disease by inactivating cutinase and showed that disease developed normally if the cuticle was mechanically removed. Dickman et al. (1982) showed similar results for *Colletotrichum gloeosporioides* on papaya. In this study, the addition of cutinase purified from

C. gloeosporioides allowed a wound pathogen (*Mycosphaerella* sp.) to penetrate and cause disease on papaya fruit with intact cuticles.

11.2.4 Cell Walls and Leaf Age

Primary cell walls are composed mostly of cellulose, hemicelluloses, and pectic substances (Darvill et al. 1980). Physical properties of cell walls functioning as mechanical barriers to infection in tree leaves have been investigated, but physical tests also indicate other properties of the leaf (see Sect. 11.2.3). Numerous reports have shown changes in resistance to disease with tree leaf age (Roberts and Martin 1963, Weinhold and English 1964, Mence and Hildebrandt 1966, Campbell 1972, Gibson 1972, Sharma et al. 1980, Wetzstein and Sparks 1983, Spiers and Hopcroft 1984, Gottwald 1985, Valsangiacomo and Gressler 1988). Factors attributed to increased disease resistance with leaf age include changes in the cuticle, as discussed above, and in the cell wall (Fig. 11.3). In ultrastructural studies, Spiers and Hopcroft (1984) showed that in a species of poplar, hyphae of *M. brunnea* penetrated electron-lucent primary cell walls of young, expanding leaves more readily than electron-dense secondary walls in mature leaves and as a result disease was expressed earlier. Wetzstein and Sparks (1983), using transmission electron microscopy, also noted differences between cell walls of resistant and susceptible cultivars of *Carya illinoensis* (Fig. 11.3). These differences may reflect biochemical changes in the cell wall and are discussed in Section 11.3.3.

Epidermal walls may be strengthened and made resistant to enzymatic action during plant development by the addition of calcium or silicon. A structural function of calcium (a divalent cation) in cross-linking pectin in cell walls of maturing bean plants has been proposed as a mechanism of disease resistance (Bateman et al. 1969). Garbaye and Pinon (1973) noted that leaves of poplar deficient in calcium and manganese were more susceptible to infection by *M. brunnea*. Calcium has been implicated in plant resistance of other diseases (Goodman et al. 1967), although it also is involved in the activation of some pectolytic enzymes of pathogens (Hancock and Miller 1965). Silicon is another element deposited in or on cell walls as amorphous hydrated silica ($SiO_2 \cdot xH_2O$) that has been implicated in resistance of rice to infection by *Pyricularia oryzae* (Volk et al. 1958). Postek (1981) also suggested that hydrated silica is involved in disease resistance in leaves of *Magnolia* spp. Silicon's proposed function includes strengthening of cell walls and reducing transpiration. Specific studies with calcium or silicon and their role in disease resistance of foliar diseases of trees have not been done.

Porosity is a physical property of cell walls that allows molecular sieving to occur and thus restricts enzymes and their degradation. The size of an enzyme can determine its location in the pathogen and its activity in host cell walls (Chang and Trevithick 1974). Cowling (1975) indicated that fungal cellulases are probably restricted in degradation of secondary cell walls since most of these enzymes are larger than the wall capillaries. A more detailed review of this subject is provided by Cooper (1983).

11.3 Preinfectional Biochemical Barriers

11.3.1 Unidentified Compounds

The inability of microorganisms to colonize plants often has been attributed to the presence of inhibitory compounds within host tissue (Ingham 1973). Preinfectional or constitutive compounds occur in healthy plants and include those compounds that are released from inactive compounds following injury of tissue during pathogen penetration.

11.3.1.1 Cuticular Exudations and Leachates

Plants may exude materials, such as sugars and amino acids, through the cuticle to the leaf surface. These compounds can stimulate or inhibit spore germination and hyphal growth of fungi (Brown 1922, Martin and Juniper 1970, Blakeman 1971, Blakeman and Atkinson 1981, Frossard 1981, Millar 1981, Ragazzi et al. 1985). Pero and Howard (1970) showed that diffusates from susceptible cultivars of *Juniperus virginiana* stimulated spore germination and germ tube growth of the pathogen *Phomopsis juniperovara*, while diffusates of a resistant cultivar had to be concentrated 10 to 40 times to have the same effect. Volatile hydrocarbons of pine needles have also been found to affect a foliar pathogen. In a study of resistance in foliage of *Pinus radiata* to *Dothistroma pini*, Franich et al. (1982) found that needles from young, susceptible trees produced twice as much volatile, monoterpene hydrocarbons than needles of older, resistant trees. Relative amounts of the 13 hydrocarbons found also varied with age. Mixtures of the hydrocarbons at 10, 100, and 300 ppm stimulated germination and cultural growth of the pathogen. Mycelial growth was inhibited only at 1000 ppm. Inhibition of fungal pathogens may also occur indirectly from enhanced growth of bacteria from exudates on the surface of larch needles (McBride 1971), from growth of yeasts on pine needles (Campbell 1972), or from growth of endophytes in conifer needles (Carroll et al. 1977).

Direct inhibition of pathogens may result from toxic materials exuded or leached onto the leaf cuticle. Leaf washings of *Populus* and other plants inhibited spore germination of several fungi (Kovács 1955, Kovács and Szeöke 1956). Topps and Wain (1957) examined washings of undamaged leaves of 12 species of woodland trees for activity against spore germination and germ tube length of *Botrytis cinerea*. Extracts of ten species were inhibitory; extracts of *Sambucus* and *Ligustrum* spp. were found to be the most fungitoxic. Washings tested directly were not inhibitory, however, Topps and Wain contended that greater concentrations would occur if water films were in contact with leaves for long periods of time. In a study using a thin-layer chromatography bioassay, a single compound from each leachate from leaves of *Betula pubescens* and *Tilia platyphyllos* inhibited growth of *Cladosporium herbarum* (Warren 1976). Other accounts of additional hosts are reviewed by Martin and Juniper (1970), Preece and Dickinson (1971), Blakeman (1981), and Blackeman and Atkinson (1981).

11.3.1.2 Cuticular Extracts

The wax fraction of the cuticle may also contain fungitoxic compounds. Martin et al. (1957) examined ether extracts of apple leaves resolved into acidic, oil, and true wax components for suppression of conidial germination of *Podosphaera leucotricha*. Although the wax fraction showed some activity, the acidic fraction was the most fungistatic. Other examples of fungitoxic extracts of leaf waxes of trees shown to inhibit spore germination include: oil and true wax fractions from leaves of *Ginkgo biloba* to *M. fructicola* (Johnston and Sproston 1965); wax components of lime leaves resistant to *G. limetticola* (Martin et al. 1966); a water-soluble fraction from leaves of *Eucalyptus* resistant to *P. eucalypti* (Heather 1967b); ether extracts of adaxial leaf surfaces to *Botrytis cinerea* (Atkinson and Blakeman 1982); and wax components from needles of some species of conifers to *Botrytis cinerea* (Schütt 1971). Specific fatty and resin acids are highly fungistatic to *Dothistroma pini*, a needle pathogen of *Pinus radiata* (Franich et al. 1983). A long-chain fatty acid, *w*-hydroxy fatty acids, and oxidized resin acids inhibited in vitro spore germination and mycelial growth of the fungus. In needles spray-extracted with acetone to remove epicuticular and stomatal pore fatty and resin acids, the average infection was approximately twice that of control plants. Franich et al. (1983) suggested that these compounds could be preinfectional barriers contributing to resistance of mature trees of *P. radiata*.

11.3.1.3 Leaf Extracts

Early studies on apple scab, caused by *V. inaequalis*, correlated host resistance to leaf extracts functioning as biochemical inhibitors of fungal germination (Wiltshire 1915, Johnstone 1931, Nusbaum and Keitt 1938). Gilliver (1947) found that 23% of 1915 plant extracts tested inhibited conidial germination of *V. inaequalis*. Additionally, Verrall (1934) observed more resin production in species of pines resistant to brown spot needle blight caused by *Mycosphaerella dearnessii*. Thus, in these early studies, constituents of tree leaves were found to contain compounds that inhibit pathogenic fungi.

11.3.2 Identified Antifungal Compounds

A number of preinfectional antimicrobial compounds have been identified and reported to have a role in plant disease resistance. The compounds are chemically diverse, generally occur in high concentrations, and may be converted into more potent active forms. Furthermore, these compounds have been reported from different plant tissues and from a number of species in many different plant families. The following sections contain specific examples by chemical groups of preinfectional compounds found in or on leaves of woody plants.

11.3.2.1 Phenolics and Related Compounds

Phenolics are a group of compounds that have an aromatic ring bearing a hydroxyl component. Many of these compounds, their oxidation products, and their derivatives (certain phytoalexins) have antimicrobial properties and are often present in host cell vacuoles as nontoxic glycosides (Mueller and Beckman 1976). Fourteen groups of phenolic compounds from plants have been recognized. These include simple phenols, flavonoids, phenolic quinones, lignin, tannins, as well as other compounds (Harborne 1964). Reviews of structures, synthesis, and functions have been provided by Pridham (1960), Harborne (1964), Ingham (1973), Misaghi (1982), and Vidhyasekaran (1988a). Phenolic compounds have been reported primarily in postinfectional responses to disease, although in a few studies they have been implicated as preinfectional barriers. In some plants, these compounds occur in high concentrations in healthy tissue and they may also increase in concentration in wounded or diseased tissue. Constituent phenolics, however, are not always related to disease resistance. Phenolics may occur in high concentrations in plant hairs including those of many tree species (Beckman et al. 1972) and in vacuoles of leaf epidermal cells as in *Carya illinoensis* (Wetzstein and Sparks 1983). In the latter study, large amounts of phenolics in leaf cells were observed and correlated to cultivars resistant to scab caused by *C. caryigenum* (Fig. 11.4). Furthermore, phenolics are often a constituent of leaf leachates on plant surfaces such as maple (Dix 1974, 1979) and poplar (Dormaar 1970). Phenolic compounds have also been found in leachates of *Syringa vulgaris* and found to inhibit spore germination and hyphal growth of *B. cinerea* and *Alternaria alternata* (Godfrey and Clements 1978). Some examples of phenolics that have been implicated in disease resistance are gallic acid, *o*-dihydroxy phenolics, phlorodzin/phloretin, oxidized phenolics, and flavonoids.

Gallic acid has long been known for its fungitoxic properties, especially in wood of trees, and is considered a preinfectional barrier. Dix (1974) collected gallic acid from dew on leaves of *Acer platanoides* and suggested that this compound may be one reason for low populations of *Cladosporium herbarum* on these leaves compared to other plants. In a later study, gallic acid inhibited all phylloplane fungi tested except *Aureobasidium pullulans*, while fungi that decompose leaf litter were not affected (Dix 1979). Fungi sensitive to gallic acid produced brown, pigmented reactions in agar containing gallic acid, while non-sensitive fungi produced no reaction. The mechanism of inhibition of gallic acid was suggested to be from the acidic oxidation of polyphenoloxidases which results in the formation of toxic compounds (Dix 1979). Irvine et al. (1978) demonstrated that leaf washings from *A. platanoides* also contained three other inhibitory compounds, but based on their calculations, gallic acid contributed to about half of the observed inhibition of phylloplane fungi. Furthermore,

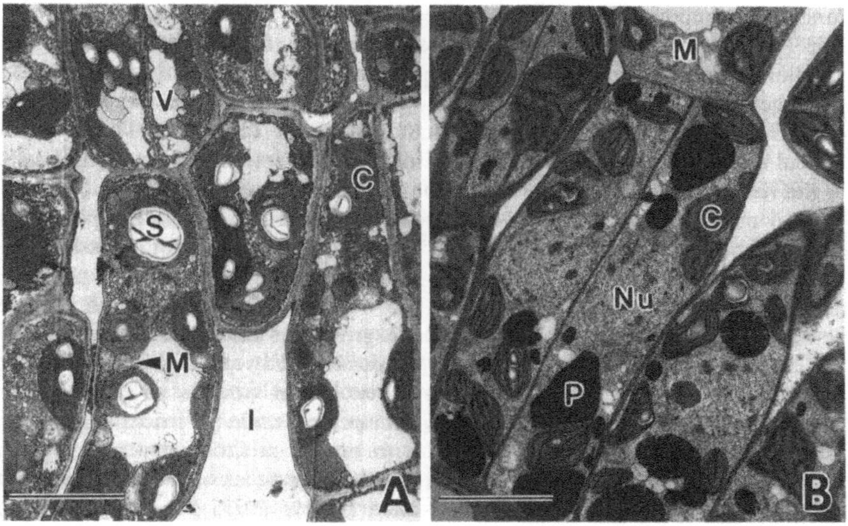

Fig. 11.4A,B. Transmission electron micrographs of palisade parenchyma in leaves of *Carya illinoensis*. **A** Susceptible cultivar with clear, unstained vacuoles. **B** Cultivar resistant to *Cladosporium caryigenum* is correlated with phenolic compounds condensed in vacuoles. *C* Chloroplast; *i* intercellular space; *P* phenolics; *m* mitochondria; *S* starch; *Nu* nucleus; *V* vacuole. *Bar* = 5 μm (Wetzstein and Sparks 1983)

inhibitory activity of washings decreased as the season progressed and this was correlated with increases in growth of fungi on leaf surfaces.

In grapevines, resistance to anthracnose caused by *Gloeosporium ampelophagum* occurs between cultivars, as well as in older leaves of susceptible varieties. Vidhyasekaran (1973, 1988a) correlated high concentrations of *o*-dihydroxy phenolics with disease resistance in a resistant variety and in older leaves of susceptible varieties and found that these compounds were inhibitory to the pathogen. No correlation was found with total phenolics. Disease has also been correlated with amino acids, ascorbic acid, and osmotic pressure in grape leaves (Vidhyasekaran and Cheeran 1972, Vidhyasekaran 1988a,b). In many host-pathogen interactions, ascorbic acid may accumulate and suppress symptom development. High concentrations of ascorbic acid (which are dependent on the sugar content of the host), however, may inhibit accumulation of quinones or oxidized phenolics and this may result in susceptibility of some grape varieties (Vidhyasekaran 1973, 1988b). Thus, resistance of grape leaves to anthracnose appears to involve high concentrations of *o*-dihydroxy phenolics.

In wounded or diseased apple leaves, the formation of the phenolic phloretin is a result of its liberation from its nonfungitoxic glycoside, phloridzin. Phloridzin is commonly found in apple leaves resistant or susceptible to apple scab caused by *V. inaequalis* (Williams 1960). In resistant leaves, however, cells around the point of fungal penetration immediately collapse. Phloridzin is hydrolyzed to phloretin and glucose by liberated β-glucosidase; phloretin is then oxidized to fungitoxic compounds by phenoloxidases (Noveroske et al. 1964a,b, Kuć 1966). Raa (1968) considered the intermediate *o*-quinones, formed in transformation reactions of phloridzin, as the fungitoxic agents. In susceptible leaves, cells do not react immediately, phenolic compounds are not formed from the conversion of phloridzin to phloretin, and the pathogen is not contained. Therefore, a critical aspect of this resistance mechanism is the hypersensitive reaction of cells in the resistant variety and the formation of toxic phenols.

Two mechanisms have been described for changes in constituent phenolics that may be involved in the resistance of apple fruit to brown rot caused by *M. fructigena*. In the first mechanism, the fungus enzymatically converts phenolic acids in the apple cultivar Edward VII to 4-hydroxybenzoic acid and related compounds. Since these compounds inhibit growth of the fungus in vitro, they are considered to be involved in disease resistance (Fawcett and Spencer 1967, 1968). In the second mechanism, host phenolics in wounds of cider apples (resistant to brown rot caused by *Monilinia* species) undergo oxidation by host polyphenoloxidase and these polymerize to brown oxidation products. These products are capable of precipitating extracellular enzymes of the pathogen and are considered to be involved in resistance (Byrde 1957, Byrde et al. 1960, Byrde and Willetts 1977). The roles of both of these mechanisms need to be further evaluated.

Other phenolics, such as p-coumaryl-quinic and chlorogenic acids, have been correlated with disease resistance in apple fruits to infection by *Botrytis cinerea, Penicillium expansum*, and *Alternaria* sp. (Ndubizu 1976). As concentrations of these phenolics decrease in maturing fruit, susceptibility increases. These phenolics have also been implicated in disease resistance of apple fruits to *Pezicula malicorticis* (Hulme and Edney 1960). Other studies, however, suggest that it is unlikely that preformed phenolics, including chlorogenic acid, are involved in maintaining latency of infection by this pathogen. Verhoeff (1974) and Noble and Drysdale (1983) suggested that the reduction of phenolics occurred early in the season and did not correlate with the typical development of bitter-rot in postharvest storage. Furthermore, the latter study indicated that chlorogenic acid, a major component of total phenolics, had little affect on mycelial growth of *P. malicorticis* at higher concentrations than that found in fruit tissue. Noble and Drysdale (1983) suggested that other mechanisms may be involved in the latent development of bitter rot of apple (see Sect. 11.5.2).

Several different flavonoids have been identified and implicated as pre-infectional mechanisms of disease resistance. Three of these phenol-related compounds were studied from species of *Vitis*, *Ribes*, and *Populus*. Immature grapes are highly resistant to *Botrytis cinerea*. Hill (1977) observed a correlation between increased concentrations of a cell wall bound inhibitor and resistance to disease. Later, Hill et al. (1981) identified proanthocyanidins from the skins but not the flesh of grapes and negatively correlated these compounds with fruit susceptibility. Proanthocyanidins are inhibitors of fungal macerating enzymes and are considered to be significant in suppressing the formation of spreading lesions of this pathogen. In *Ribes nigrum*, the flavone sakuranetin was found in adaxial but not abaxial leaf glands (Atkinson and Blakeman 1982). These leaf glands senesce early in the season and release their contents onto the leaf surface. Observations indicated that germination of conidia of *B. cinerea* is inhibited only on the adaxial leaf surface. In vitro tests indicated that sakuranetin inhibited germination, but effective concentrations were generally $\geq 500\,\mu g/ml$ and it was therefore concluded that other factors were also involved in disease resistance. Another flavone that may function in disease resistance is pinocembrin, a compound secreted by leaf glands of eastern cottonwood (Shain and Miller 1982). In this study, pinocembrin was found to be toxic to spores of the leaf pathogens *Melampsora medusae* ($16\,\mu g/ml$), while inhibitory to spore germination of *Marssonina brunnea* ($32\,\mu g/ml$). Concentrations of the flavanone on the leaf surface decreased with age. Thus, resistance of young leaves to infection by *M. medusae* was correlated with high concentrations of pinocembrin.

The preceding studies have mostly implicated phenolics to be involved in disease resistance. In other studies, however, the role of phenolics is less well understood. For example, the role of juglone in scab resistance in pecan is unclear. Juglone is a naphthoquinone isolated from leaves, nuts, and other plant tissues of members of the Juglandaceae. The compound is known mostly for its alleopathic effect on certain plants, although it also has been shown to be fungitoxic to *C. caryigenum* (Langhans et al. 1978). Juglone was suggested to be involved in disease resistance since concentrations inhibitory to the fungus occur at different growth stages of the plant and scab-resistant species such as walnut, hickory, and some cultivars of pecan have higher concentrations of juglone than pecan cultivars susceptible to the disease (Hedin et al. 1979). Currently, the compound is not considered to have a major role in scab resistance because its concentrations decrease during the growing season while resistance increases. Additionally, in susceptible cultivars, inhibitory levels occur in young susceptible leaves (Langhans et al. 1978). Isolated from fruit peels of tangerine, the flavones nobiletin and tangeritin were implicated in resistance to the fruit pathogen, *Deuterophoma tracheiphila*. In other species of citrus, however, no direct correlation was observed between resistance to this disease and concentrations of flavones in leaf tissues (Piatelli and Impellizzeri 1971). The roles of juglone and nobiletin and tangeritin, however, may be in nonhost or basic resistance.

11.3.2.2 Cyanogenic Glycosides

These compounds are present in many species of plants. Cyanogenic glycosides in species of trees include amygdalin (Rosaceae), acacipetalin (Mimosaceae), linamarin and methyllinamarin (Euphorbiaceae), proteacin (Proteaceae, i.e., *Macadamia integrifolia*), prunasin (Myrtaceae and Rosaceae), sambunigrin (Caprifoliaceae), and taxiphyllin (Taxaceae) (Ejólfsson 1970). These compounds are considered to be mostly present in cell vacuoles, however, information on their ultrastructural location is limited. Cyanogenic plants may liberate hydrogen cyanide (HCN) from mechanical, chemical, or biotic injuries (Wattenbarger et al. 1968, Ejólfsson 1970, Millar and Hemphill 1978). In disrupted cells, decompartmentalization occurs and hydrolytic enzymes release HCN from cyanogenic glycosides (Mansfield 1983). Cytochrome oxidase, a key enzyme of aerobic respiration, is rapidly inactivated by cyanide. Thus, release of HCN from cyanogenic glycosides can be considered a potential mechanism in disease resistance, although no correlation was found in race-specific resistance in flax (Trione 1960). Fry and Evans (1977) indicated that many plant pathogens were tolerant to HCN and considered that HCN may be involved in nonhost (basic) resistance. Of the fungi tested, all 11 pathogens of cyanogenic plants produced formamide hydrolyase, an enzyme that converts HCN to the nontoxic formamide. Only four fungi of the 14 pathogenic fungi of noncyanogenic plants and none of the six saprophytic fungi produced the enzyme. Cyanogenic glycosides may function in nonhost (nonspecific) resistance. In host resistance, however, these compounds may inhibit other defenses of the host, including phytoalexin production. Lieberei (1988) correlated a high capacity of HCN liberation with susceptibility of rubber tree leaves during infection by the HCN-tolerant pathogen *Microcyclus ulei*. Furthermore, HCN liberation was inhibitory to the accumulation of the phytoalexin scopoletin, an active defense mechanism in leaves of rubber trees (Lieberei et al. 1989).

11.3.3 Cell Wall Characteristics

The primary cell wall may function as a barrier to microorganisms not capable of producing cellulolytic or pectolytic enzymes and, like other preformed characters (Heath 1981), it is considered to be a possible factor in nonhost resistance (Cooper et al. 1981). During plant growth, biochemical changes occur in the primary cell wall (Lamport 1970). These changes may influence basic resistance of hosts to fungal pathogens. Proportions of monosaccharides in cell walls is dependent on tissues within the plant and on maturation of specific tissues (Jones and Albersheim 1972). Disease resistance of older tissues may be related to changes in walls that render them less susceptible to cell wall degrading enzymes as in the addition of calcium to cell walls (Bateman et al. 1969), or more susceptible to degradation, as in pectolytic changes in cell walls of ripening apple fruit (Knee et al. 1975, Knee 1978). Proteins in plant cell walls have also been studied and shown to function as nonselective inhibitors of *endo*-polygalacturonases from several species of fungi, while not inhibiting other cell wall degrading enzymes (Albersheim and Anderson 1971). In disease resistance, the roles of compositional changes in monosaccharides and occurrences of protein inhibitors of *endo*-polygalacturonase within cell walls have not been extensively investigated, especially in studies of foliar diseases of trees, although ultrastructural changes have been observed (Fig. 11.3).

Epidermal cells may also be protected by cutin and waxes or they can be strengthened by the addition of calcium, silicon, suberin, and occasionally lignin. Physical and chemical properties of the cuticle and cell wall were discussed in the preceding sections. Lignin describes a group of aromatic phenolic polymers that are formed from dehydrogenative polymerization of three hydroxycinnamyl alcohols (p-coumaryl, coniferyl, and sinapyl alcohols). The phenolic properties of lignin indicate its potential role in preinfectional, disease resistance of tissues. Within tree leaves, however, lignification is generally associated with a cessation of metabolic activity as in the formation of xylem tissue. Its role as a preinfectional barrier to fungal infection of tree leaves has not been evaluated. For general information on lignin as a preformed barrier, see Ride (1983) or in other plants, see Vance et al. (1980).

11.4 Postinfectional Hypersensitive Reactions

In many plant-pathogen interactions, a hypersensitive response is often the result of incompatibility (Tomiyama et al. 1979, Bowles 1990). This response is widespread among plants (Tomiyama et al. 1979) and phenotypically observed as a rapid, but limited necrosis of one or a few plant cells at the interaction site. Although this plant reaction may be a consequence of infection (Brown et al. 1966, Király et al. 1972), and may not be related to disease resistance as reviewed by Crute et al. (1985) and Vidhyasekaran (1988a), hypersensitivity may be involved in disease resistance of the plant depending on the specific host-pathogen interaction (Heath 1976, Ingham 1978, Király 1980). As a mechanism of resistance in the host, cell death has been related directly to or as an intermediate reaction in a complex series of events including the synthesis of phytoalexins, phenolics, lignin, and callose that results in the inhibition of the microorganism and disease resistance in the plant (Heath 1976, Ingham 1978, Vidhyasekaran 1988a, Bowles 1990).

Hypersensitive responses in tree leaves have been reported and reviewed by Kinloch (1982). In early studies of foliar diseased apple trees, hypersensitive reactions were reported in leaves responding to infection by *Gymnosporangium juniperivirginianae*, but the reactions were not considered responsible for host resistance since fungal cell death preceded host cell death (Nusbaum 1935). Nusbaum (1938) also reported hypersensitivity in leaves of apple to *V. inaequalis* and considered it part of host resistance. The importance of the hypersensitive response was discussed previously in relation to the release of phloretin and subsequent phenolic compounds from phloridzin (Sect. 11.3.2.1). Hypersensitive reactions have been reported for leaves of poplar to infection by *M. larici-populina* where it was associated with rapid accumulation of phenolic compounds (Siwecki et al. 1982). Hoff and McDonald (1971) described premature needle shed as a hypersensitive response in limiting development of *Cronartium ribicola* into stem tissue of *Pinus monticola*. Other defense mechanisms to blister rust have been described (Bingham et al. 1971) and include a necrotic reaction of host and fungal tissue at the needle fascicle (Hoff and McDonald 1971), foliar lesions in *P. armandii* (Hoff and McDonald 1975), and lesions consisting of two distinct reactions, yellow and red, in response to infection in *P. monticola* (McDonald and Hoff 1975). These authors attempted to relate host reactions to genetic mechanisms using either segregation ratios of necrotic fascicle reactions in self and full-sib families or relative frequencies of reaction classes (red and yellow) in inoculated plants, but conclusive evidence was not obtained. Apparently, these mechanisms are under polygenic control (Hoff and McDonald 1980). In *Pinus lambertiana*, however, a hypersensitive reaction in needle tissue to mycelium of the blister rust fungus is controlled by a single dominant gene

(Kinloch and Littlefield 1977). In this response, a rapid degeneration results as cytoplasm of affected cells degenerates and becomes dense with tannins and is devoid of organelles as in hypersensitive reactions of other hosts to rust infections (Onoe et al. 1987). Regardless of genetic control, hypersensitivity is part of a series of reactions that may be ultimately involved in disease resistance.

11.5 Postinfectional Biochemical Responses

In a defense situation, many coordinated responses occur in plant cells at or near infection sites of a pathogen. These host responses result in the accumulation of an array of biochemicals leading to: (1) modifications of the host cell wall; (2) direct antimicrobial activities; and (3) unknown functions that are correlated with defense responses. Enhanced production of proteins (Bowles 1990) and phenolic compounds via the shikimic acid pathway (Kosuge 1969, Beckman 1980, Vance et al. 1980) are responses to infection common to most plants. In general, biochemical responses may involve modifications of host cell walls by the addition of structural proteins (Bowles 1990) or by phenol-conjugated lignification or suberization that results in structural barriers delimiting the pathogen (Aist 1976, 1983). Vance et al. (1980) suggested that lignification may hinder fungal growth by physically strengthening host cell walls, prevent enzymatic dissolution of cell walls, and restrict diffusion of fungal enzymes and toxins, as well as inhibit diffusion of water and nutrients to the fungus from the host. Furthermore, host proteins (Bowles 1990) and phenolic compounds may be directly toxic to the pathogen, with the latter including phytoalexins (Vance et al. 1980, Aist 1983). Host proteins that may have antifungal activity include endohydrolases that have enzymatic activity on β1,3-β-D-glucan and chitin (components of cell walls of many pathogenic fungi), proteinase inhibitors, and possibly lectins. For general reviews see Boller (1987), Bowles (1990), and Ryan and An (1988). Phenolic compounds, their precursors, or their breakdown products (Sect. 11.3.2.1) may also inactivate fungal enzymes, toxins, elicitors, or membranes (Kosuge 1969, Sherwood and Vance 1976, Ride 1978, Vance et al. 1980, Pearce and Rutherford 1981, Hammerschmidt and Kuć 1982, Aist 1983). Additionally, hyphae may be "lignified' by these compounds and may be inactivated (Ride 1978, Vance et al. 1980). Thus, the processes of lignification and suberization may directly affect the pathogen or may result in structural barriers within the host that restrict the growth of the pathogen. The following sections review phytoalexins and other induced biochemicals considered as resistance mechanisms of tree leaves.

11.5.1 Phytoalexins

Phytoalexins are low-molecular-weight, antimicrobial compounds produced by plants in response to exposure or infection by microorganisms including pathogens and nonpathogens (Paxton 1981, Bailey and Mansfield 1982). These compounds are secondary metabolites (Stoessl 1982) and have been mostly identified as isoflavonoids, terpenoids, and polyacetylene compounds (Vidhyasekaran 1988a). Synthesis of phytoalexins has been discussed by Stoessl (1982) and later classified into three groups based on their formation (Stoessl 1983). These groups are: (1) degradation of normal plant constituents; (2) simple oxidations and hydroxylations of normal plant constituents affected by newly formed low specificity enzymes; and (3) complex synthesis involving high specificity enzymes. Compounds of the first two groups are important in the lignification processes and may function as fungitoxicants directly. The third group refers to compounds formed as a specific host response to a potential pathogen or its metabolites, they have in vitro activity, and they are involved in host resistance in situ (Cruickshank 1980, Stoessl 1983). Comprehensive reviews are provided by Kuć et al. (1976), Cruickshank (1977), and Stoessl (1983). In general, phytoalexins (including subgroups) inhibit spore germination, germ tube elongation, mycelial growth of fungal pathogens, and can be toxic to plant cells. Inhibitory affects of phytoalexins to microorganisms and plants are reviewed by Mansfield (1982), Misaghi (1982), Smith (1982), and Vidhyasekaran (1988a).

Phytoalexins are reported from species in 17 higher plant families, although most are reported from two families, the Solanaceae and the Leguminaceae (Ingham 1973, Deverall 1976, Coxon 1982, Stoessl 1983). Families with woody species reported to produce phytoalexins include Moraceae, Rutaceae, Vitaceae, probably Ginkgoaceae (Coxon 1982), and Euphorbiaceae (Giesemann et al. 1986). Phytoalexins reported thus far are chemically diverse. Mansfield (1983) summarized the chemical diversity of phytoalexins as follows: (1) members of plant families usually produce chemically similar types of phytoalexins, e.g., isoflavonoids from the Leguminaceae, terpenoids from the Solanaceae, and polyacetylenes from the Compositae (Ingham and Harborne 1976); and (2) most plants produce several closely related phytoalexins, although one or two may predominate.

There are several examples of compounds found in trees that were first classified as phytoalexins but are considered presently in other categories of resistance mechanisms or are unidentified. Two compounds, pinoresinol and hydroxymatairesinol, reported from wood of pines, were considered to be phytoalexins (Shain 1967), but they are currently believed to be preformed lignans that are capable of inhibiting fungal enzymes (Popoff et al. 1975, Johansson et al. 1976, Stoessl 1983). In other studies, Hare (1971) reported a phytoalexin in needles of shortleaf pine formed in response to fusiform rust infection, while Coxon (1982) indicated that a phytoalexin may be produced in leaves of *Ginkgo biloba* that prevents penetration, but not germination or growth of *Botrytis allii*. These compounds, however, have not been identified. Flores and Hubbes (1980) also reported a phytoalexin from stem tissue of a species of poplar (Salicaceae) and characterized it as a preformed phenolic glycoside.

Foliage of woody species in three plant families, Vitaceae (Langcake et al. 1977), Euphorbiaceae (Giesemann et al. 1986), and Moraceae (Takasugi et al. 1980a,b) has been reported to produce phytoalexins in response to infection and these will be discussed as examples of phytoalexins in leaves of woody plants. In grape leaves, Langcake and Pryce (1976) identified blue fluorescing viniferins, a series of stilbene phytoalexins, formed in response to infection by *B. cinerea*. The viniferins are oligomers of *trans*-resveratrol and at least four others have been identified as α-, β-, γ-, and ε-viniferin. Although these compounds are constitutive factors in lignified tissues of grapevines (Langcake and Pryce 1977), their synthesis can be induced as stress metabolites in grape leaves by ultraviolet radiation, various chemicals, and inoculation with *B. cinerea* or *Plasmopara viticola* (Langcake and Pryce 1976, 1977, Langcake 1981, Stein and Hoos 1984, Barlass et al. 1987). Langcake (1981) reported that viniferins differed in their fungitoxicity with α- and ε-viniferin being the most active. These phytoalexins inhibited spore germination, germ tube elongation, and mycelial growth of *B. cinerea* and other fungi at concentrations generally less than 100 μg/ml. In studies of downy mildew of grape, Dercks and Creasy (1989) found that the viniferin precursor resveratrol was more stable in aqueous sporangial suspensions of *P. viticola* and more inhibitory to germination of sporangia than ε-viniferin. Although Langcake and Pryce (1977) did not consider *trans*-resveratrol a phytoalexin, due to its low activity against the fungi they tested, Dercks and Creasy (1989) suggested that resveratrol may be more important than the viniferins in resistance to downy mildew of grape. Not all derivatives of *trans*-resveratrol are considered phytoalexins. Langcake et al. (1979) considered a methylated derivative of *trans*-resveratrol, *trans*-pterostilbene, a potential phytoalexin because of its high toxicity and its production in response to infection by *P. viticola*. Langcake (1981), however, indicated that concentrations of the compound were low and it probably did not contribute to resistance of the species of grape studied. Thus, in grape leaves, several stilbene phytoalexins have been implicated in disease resistance to at least two foliar pathogens.

In studies of disease resistance in leaves of *Hevea brasiliensis*, Tan and Low (1975) found at least two fungitoxic compounds produced in response to infection by *Colletotrichum gloeosporioides*. One of these compounds also produced a blue fluorescence under ultraviolet light similar to other phytoalexins (e.g., *trans*-resveratrol). Based on these observations, Giesemann et al. (1986) induced a blue fluorescing compound in leaves of species of *Hevea* inoculated with *Microcyclus ulei* and identified the compound as the phytoalexin scopoletin. The phytoalexin is chemically similar to scopolin, a glycoside found in large amounts in potatoes infected with *Phytophthora infestans*. Scopolin, however, is rarely observed in leaves of species of *Hevea*. Purified scopoletin inhibits spore germination of *M. ulei*, although greater inhibition is found in leachates from lower leaves of infected tissue, suggesting the possibility of other toxic compounds (Giesemann et al. 1986).

A number of phytoalexins have been reported from cortical and phloem tissue of shoots (Takasugi et al. 1980a, 1984a,b) and leaves of *Morus* sp. (Takasugi et al. 1980b). In leaves inoculated with *Fusarium solani* f. sp. *mori*, a compound designated chalcomoracin was identified as a major phytoalexin component. This compound completely inhibited spore germination of several fungi at concentrations of 10^{-4}–10^{-5} M. Apparently, chalcomoracin is derived from a rather complex group of compounds, the benzofurans (Takasugi et al. 1980b), and may be a reduction product of preformed compounds in leaves and other tissues of mulberry (Stoessl 1983).

11.5.2 Other Induced Compounds

Other fungitoxic compounds formed in response to fungal infection have been identified. As discussed in Section 11.3.2.1, mechanisms other than total phenolics have been considered to be involved in latent infections of apple by *P. malicorticis*. One mechanism, the induced production of benzoic acid, has been implicated as a factor that may be involved in resistance. Benzoic acid is capable of inhibiting mycelial growth of this pathogen at concentrations of 12 μg/ml and at a pH similar to that of apple fruit (Noble and Drysdale 1983). In studies of resistance of apple fruit to infection by *Nectria galligena*, Brown and Swinburne (1973) indicated an ED_{50} value of 5 μg/ml benzoic acid for inhibition of spore germination. Benzoic acid may also be involved in resistance of apples to this pathogen, since resistant cultivars produce two to three times the amount of this compound than susceptible cultivars (Brown and Swinburne 1973, Noble and Drysdale 1983).

Similarly, cells adjacent to lesions caused by *D. pini* in needles of *P. radiata* synthesized and accumulated benzoic acid (Franich et al. 1986). Benzoic acid was toxic to mesophyll cells of needles, fungistatic to *D. pini* in vitro, and prevented immediate colonization of tissue by *D. pini* in natural infections. Lesion size was proportional to the amount of benzoic acid isolated from dead tissue and a linear dose response relationship was obtained for benzoic acid injected into needles of *P. radiata*. Furthermore, the magnitude of response in needles to the toxin dothistromin produced by *D. pini* was correlated with observed field resistance to this pathogen among control cross-progeny comprising 50 full-sib families of *P. radiata*. Thus, benzoic acid was considered an inducible mechanism of resistance involved in direct inhibition of the pathogen, as well as being indirectly involved in lignin accumulation in tissue adjacent to infected areas (Franich et al. 1986). For additional information of lignin as a barrier, see Sections 11.5 and 11.6.3.

11.6 Postinfectional Structural Responses

Resistance to fungal penetration in many plants is expressed by the failure of infection hyphae to penetrate into or through a cell wall of a plant (Aist 1976, Ride 1978, Heath 1980). Active responses of the host include single or multiple cell responses that modify the structure of cell walls at the point of fungal attack. At the interaction site, cell wall modifications begin with the formation of cytoplasmic aggregates containing microorganelles involved in the secretion of materials into the cell wall (Aist 1976, 1983). Although this has not been studied extensively in foliar and fruit diseases of trees, it is considered a fundamental process in the cellular response. Specific wall modifications are characterized as appositions (papillae, callose deposits, collars, or encasements) that accumulate within living cells beneath infection hyphae (Aist 1976, 1983) or around haustoria (Mims and Glidewell 1978, Littlefield and Heath 1979, Tainter 1985). In grass species, depending on the specific plant-pathogen interaction, appositions may differ in morphology and color (Sherwood and Vance 1980). Other cellular host responses to infection include halos in cell walls at penetration sites and cytoplasmic striate inclusions. Halos may be sites of cell wall degrading activity by the pathogen or a host response to attempted penetration (Aist 1983), whereas Ride (1983) considered halos as part of papillae formation. Chemically, halos may contain silica, lignin, or callose as discussed by Sherwood and Vance (1976), Ride and Pearce (1979), Heath (1980), and Aist (1983). Striate inclusions have been suggested to consist of fungitoxic compounds that aggregated in the cytoplasm of the host (Aist 1983). In fungal diseases of fruit and foliage of trees, these responses have not been investigated extensively. For reviews, see Sherwood and Vance (1976), Aist (1983), and Ride (1983).

Plant tissues, involving several to many cells, may respond to infection by a pathogen. In species of trees and other plants, these responses include: lignification (Akai 1959, Friend 1976, Vance et al. 1980) and suberization (Akai 1959, Wood 1967, Kolattukudy 1980) of cell walls, stationary interfaces of needles (Williamson et al. 1976, Jewell 1990), and meristematic barriers in leaves and fruit (Akai 1959, Beckman 1980, Aist 1983). The roles of lignification in response to biotic infection and as a mechanism of plant resistance have been discussed in Section 11.4 and are reviewed by Glazener (1982) and Vance et al. (1980), while those of suberization are discussed by Wood (1967), Akai (1959), and Kolattukudy (1980). Furthermore, lignification and suberization may occur during the development of meristematic tissues in response to infection. Specific responses observed in leaves or fruit of trees are considered in the following sections.

11.6.1 Cell Wall Appositions

In many living plants, cells respond to attempted fungal penetration or other physical stresses at a specific point on the cell surface by producing cell wall appositions or papillae (Aist 1976). This structure is a deposit between the plasma membrane and adjacent cell wall and consists commonly of callose, although many contain other materials including lignin, other phenolic derivatives, cellulose, protein, and silicon (Heath 1980, Aist 1983). The composition, however, may depend on plant species, nature of the pathogen, type of cell infected, and age of the papilla (Ride 1983). Papillae have been studied mostly in grasses (Aist 1976, 1983, Heath 1980, Sherwood and Vance 1980, Stolzenburg et al. 1984). The function of papillae is considered to be in the physical blocking of fungal penetration (Ride and Pearce 1979, Sherwood and Vance 1980). Although the response occurs in both resistant and susceptible hosts, quantitative differences are positively correlated with resistant hosts (Crute et al. 1985). Stolzenburg et al. (1984) indicated that papillae formation in barley was not necessary for resistance to powdery mildew, although there was a strong link between papillae formation and an *ml-o* gene that conditions resistance. Using 2-deoxy-D-glucose, an inhibitor of callose formation, Bayles et al. (1990) demonstrated a 92 to 41% decrease in papillae frequency in the *ml-o* mutant barley and a 15 to 78% increase in penetration efficiency of *Erysiphe graminis* f.

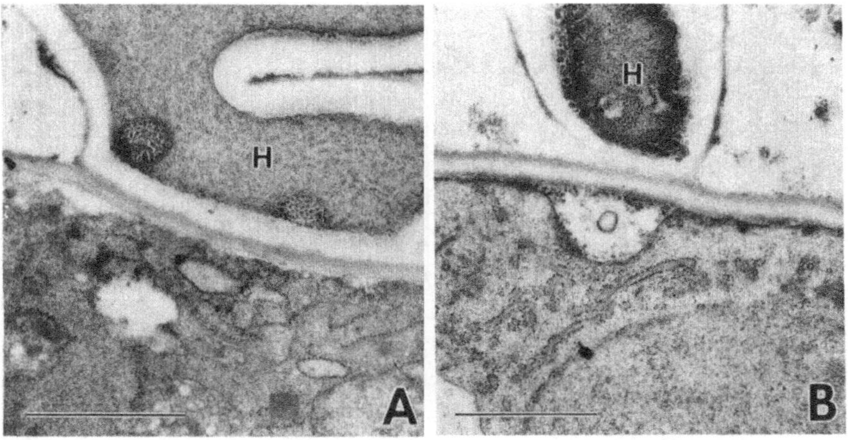

Fig. 11.5A,B. Transmission electron micrographs of cells (palisade and mesophyll) in leaves of species of *Populus* infected with *Marssonina brunnea*. **A** Cytoplasmic response of host cell to attempted hyphal (*H*) penetration. Increased endoplasmic reticulum and number of ribosomes in cytoplasm of host beneath hypha. **B** Electron-opaque deposit of cell wall surrounded host plasmalemma formed in response to attempted penetration of fungal hypha. *Bars* = 1 μm (Spiers and Hopcroft 1983)

sp. *hordei*. This suggests that in the *ml-o* resistant plants, papillae formation (enhanced by callose production) is the primary mechanism of resistance (Bayles et al. 1990).

Papillae or papilla-like structures have been observed in host-parasite interactions of foliar or fruit diseases of trees. Papilla-like structures occur directly beneath appressoria of *Colletotrichum gloeosporioides* on epidermal cells of resistant leaves of *Populus tremuloides* (Marks et al. 1965), and in cell walls beneath appressoria of several fungi attempting penetration of leaves of *Ginkgo biloba* (Adams et al. 1962). In the latter study, staining differences between appositions and cell walls were also observed. In leaves of *Populus* sp., Spiers and Hopcroft (1983) observed papillae-like structures in host cell walls beneath penetrating hyphae of species of *Marssonina* (Fig. 11.5). These structures impeded hyphal growth, but they did not prevent penetration and their role in disease resistance was considered minimal. In contrast, Heintz and Blaich (1990) observed that grape cells forming thick papillae in response to attempted penetration by *Uncinula necator* were not infected. In this study, the authors suggested that host cell penetration by *U. necator* may be blocked by papillae composed of carbohydrates, silica, and phenolic compounds.

In host responses to biotrophic fungi, collars around the neck or complete encasements of haustoria may occur. In rust fungi, collars and encasements of both monokaryotic (M-) and dikaryotic (D-) haustoria have also been observed (Littlefield and Heath 1979). These responses occur in older infections of susceptible plants as well as in resistant plants and are considered nonspecific responses to infection. Furthermore, these structures are observed commonly as

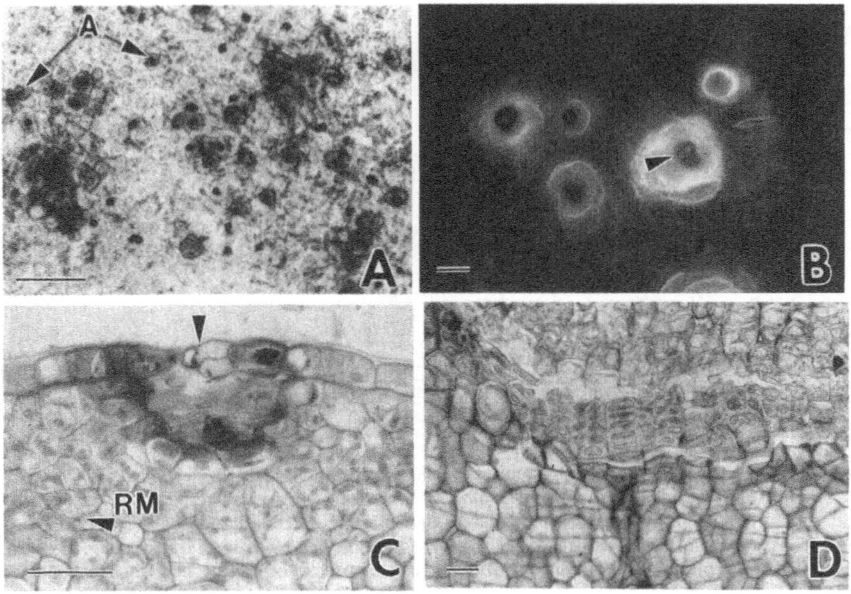

Fig. 11.6A–D. Bright-field micrographs of paraffin sections of fruit of *Carica papaya*. **A,B** Epidermal strip of detached, immature fruit showing appressoria (*A*, *arrows*) of *Colletotrichum gloeosporioides* associated with halo-like callose deposits stained in resorcinal blue. **A** *Bar* = 70 μm; **B** *bar* = 10 μm. **C** Cross-section of fungal lesion in epicarp tissue showing callose deposits and renewed meristematic (*RM*) activity, *arrow* indicates point of fungal penetration. *Bar* = 42 μm. **D** Cross-section of attached, immature fruit showing lignified derivatives of the wound phellogen 30 days after injury. *Bar* = 20 μm (Stanghellini and Aragaki 1966)

electron-lucent structures with membranous profiles and may be similar in composition to papillae-like deposits in containing callose (Littlefield and Heath 1979, Ride 1983). Reports of collars around necks of haustoria or encased haustoria in tissues of woody plants include pine needles infected by *Cronartium ribicola* (Tainter 1985) and grape leaves infected by *Uncinula necator* (Heintz and Blaich 1990). Specific functions of collars and encasements have been discussed in other plant-pathogen interactions (Littlefield and Heath 1979, Cohen et al. 1990), however, the exact role in host resistance is unclear.

On fruit of *Carica papaya*, Stanghellini and Aragaki (1966) observed callose deposits within numerous epidermal cells around sites of attempted appressorial penetration of the pathogen *C. gloeosporioides* and the nonpathogens *C. falcatum* and *Helminthosporium stenospilum* (Fig. 11.6). In a specific area of epidermal cells, 94.2% of the callose deposits were associated with appressoria. Only 50% of the total appressoria, however, had callose deposits (Stanghellini and Aragaki 1966), indicating the potential of the host response when artificially inoculated. The response was localized within individual cells, although many cells stained for callose due to the widespread attempted penetrations by the fungi studied. In most studies, papillae, papillae-like structures, or callose deposits were observed as nonspecific responses of the host since their formation could be induced by nonpathogens. Heath (1980) described papillae formation as widespread among plants and as a primary response of nonsusceptible hosts (nonsuscepts), although these structures may function in host resistance. Furthermore, this response may be a mechanism of nonhost (basic or nonspecific) resistance of plants and, along with other mechanisms of nonhost resistance, determines the host species range of a fungus (Heath 1981, 1987).

11.6.2 Stationary Interfaces

Tissue barriers reported to prevent or inhibit development of fungal pathogens include stationary interfaces composed of dead cells and intercellular matrices. Williamson et al. (1976), studying the histochemistry of needles of *Pinus nigra* var. *maritima* infected by *Lophodermella sulcigena*, reported that prior to hypertrophy and hyperplasia of cells in the host-parasite interface, a stationary interface is formed in response to infection. Initially, the fungus colonizes intercellular spaces of mesophyll, followed by endodermis, hypodermis, and epidermis. Later in the season, the stationary interface is formed and is composed of a fungal-free zone, dead mesophyll cells, and an intercellular matrix (Fig. 11.7). Although the composition of the matrix is unknown, histochemically it is similar to resin in healthy resin canals. Unlike resin, however, it is insoluble in alcohol. The matrix, in addition to other mechanisms such as tannin accumulation, may provide a barrier in needle tissue to prevent the pathogen from advancing into healthy tissue. Similarly, Jewell (1990) also reported a stationary interface comprised of dead cells and an intercellular matrix at lesion boundaries of needles of *Pinus palustris* infected by *Ploioderma hedgcockii*. No hypertrophy or hyperplasia occurred in or at the boundaries of diseased tissue and the fungus did not invade the endodermis in this host-pathogen interaction. In both studies, extensive growth of hyphae of the pathogens was observed in diseased tissue. Hyphae were not observed, however, in or beyond the matrix into healthy tissue, while lateral spread of disease appeared to be contained by the matrix (Fig. 11.7).

11.6.3 Meristematic Barriers

In gymnosperms and dicotyledonous plants, meristematic barriers may be formed from parenchymatous tissue in response to infection or wounding (Akai 1959, Beckman 1980, Aist 1983). Adjacent to injuries or infections, parenchyma cells may dedifferentiate, enlarge, and become meristematic, dividing, enlarging, and dividing again. Later, this new layer of tissue may differentiate into an infection barrier (abscission zone, corky layer, etc.) after lignification, suberization, or deposition of phenolic materials. In 1928, Cunningham reported on the delimitation of fungal infections from healthy tissue by a barrier or cicatrix in leaves of herbaceous plants and tree species. In tissues of *Prunus* species responding to infections by *Blumeriella jaapii* (Higgins 1914) or *Wilsonomyces carpophilus* (Samuel 1927, Akai 1959, Adaskaveg and Ogawa 1988), in leaves

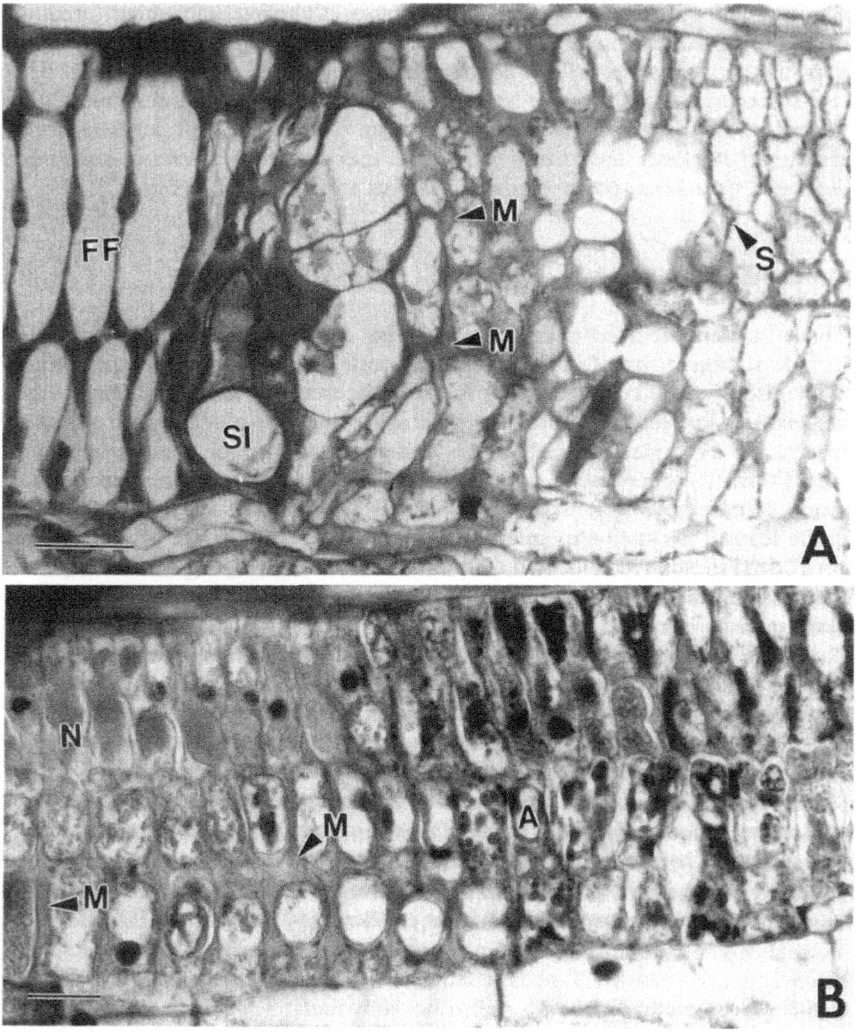

Fig. 11.7A,B. Bright-field micrographs of a longitudinal section of diseased needles of *Pinus* spp. **A** Methacrylate sections of *Pinus nigra* infected with *Lophodermella sulcigena*. Diseased and healthy tissue of a needle consisting of infected mesophyll, stationary interface (*SI*) with hypertrophied mesophyll cells, fungal-free zone (*FF*), intercellular matrix (*M*), and healthy cells with peripheral plastids and starch (*S*). *Bar* = 60 μm (Williamson et al. 1976). **B** Paraffin sections of *Pinus palustris* infected with *Ploioderma hedgcockii*. Boundary of affected (*A*) and nonaffected (*N*) mesophyll tissue separated by intercellular matrix (*M*). *Bar* = 25 μm (Jewell 1990)

of *Pyrus* infected by *Botryosphaeria* sp. (Hesler 1916), in grapes infected by *Elsinoe ampelina*, in citrus fruit infected with *Elsinoe fawcetii* (Gaumann 1950, Akai 1959) or *Diaporthe citri* (Bach and Wolf 1928), and in fruit of *Carica papaya* infected with *Colletotrichum gloeosporioides* (McWhorter 1927, Stanghellini and Aragaki 1966), cells adjacent to infections may respond to fungal pathogens by initiating meristematic barriers.

In shot hole disease of some *Prunus* species caused by *W. carpophilus*, tissues in stems and leaves produce meristematic barriers resulting in abscission layers (Samuel 1927, Akai 1959, Adaskaveg and Ogawa 1988). Epidermal, palisade, and spongy parenchyma cells, 10–20 cells away and in a circular arrangement around the infection are involved in the host response. Some of these cells become lignified, while cells adjacent to healthy tissue dedifferentiate, enlarge, and begin to divide to form a meristematic layer of isodiametric cells that occlude the intercellular spaces (Fig. 11.8). Cells of the meristematic layer have enlarged nuclei, fewer chloroplasts, and more protoplasm than healthy cells. As cells of this layer continue to enlarge, the cuticle breaks; while infected tissue, isolated from the healthy tissue, becomes necrotic, collapses, and separates from the new abscission layer. Some lignification and suberization of the abscission layer (healthy tissue) complete the process. According to Samuel (1927), the enlargement of abscission cells in moist atmospheres and the concurrent desiccation of the delimited, infected tissue cause this tissue to be discharged by physical forces.

In leaves of *Prunus domestica* infected by *B. jaapii*, a wound periderm consisting of phellogen and phelloderm is produced at the margin of the necrotic tissue (Higgins 1914, Cunningham 1928). The healing tissue consists of several layers of enlarged cells. Cells lying adjacent to necrotic tissue are densely granular, while, cells further away are devoid of cytoplasm. Both layers are lignified and suberized. The phelloderm and phellogen are adjacent to the healthy tissue with the former tissue filled with dense protoplasm. Cells in these tissues have few to no chloroplasts and are thick-walled (Cunningham 1928, Akai 1959).

Renewed meristematic activity of cells may occur in fruits of plants in response to fungal infections. In the epicarp of fruit, affected tissues are often suberized to form a corky layer or barrier and thus the tissue resembles a phellogen. In uninjured citrus fruit infected with *D. citri*, melanose spots are delimited by a band of cork layers formed by meristematic cells that cause the infected tissue to bulge outward. Cells of infected tissue turn brown and collapse, while the pathogen is contained within this tissue (Bach and Wolf 1928, Akai 1959). Similarly, lesions of peach fruit caused by *W. carpophilus* can be demarcated by a corky layer. When attached, immature fruit of papaya were wounded, Stanghellini and Aragaki (1966) observed meristematic activity within 48 h, periderm formation after 7 days, and lignified derivatives of the phellogen within 30 days (Fig. 11.6). In nonwounded, attached fruit inoculated with the pathogen, meristematic cells resembling phellogen or callose deposition were observed as a response to infection by the fungus, similar to that caused by wounding. No periderm formed where mitotic activity ceased: in wounded,

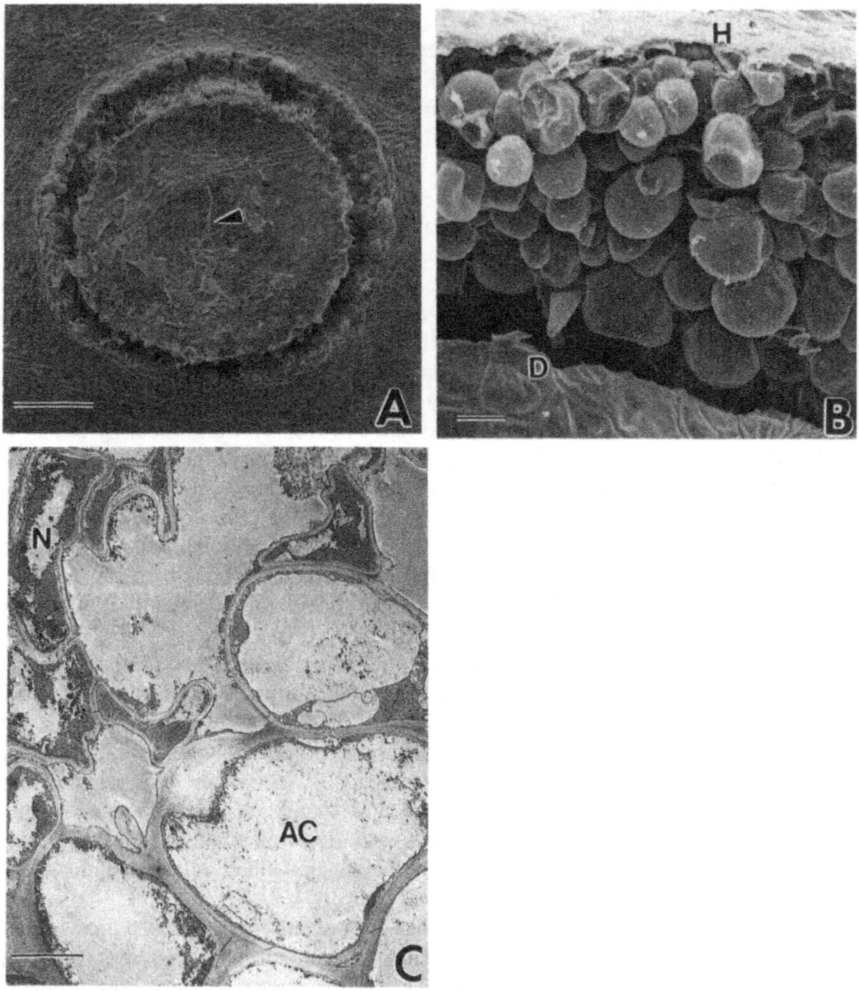

Fig. 11.8A–C. Micrographs of abscission zone formed around lesions caused by *Wilsonomyces carpophilus* in leaves of *Prunus dulcis*. **A,B** Scanning electron micrographs. **A** Shot hole lesion separated from healthy tissue by abscission layer cells. Note fungal spore (*arrow head*) in center. *Bar* = 150 μm. **B** Isodiametric, abscission cells separating healthy (*H*) from diseased (*D*) tissue. *Bar* = 20 μm. **C** Transmission electron micrograph of abscission cells (*AC*) with large vacuoles separating healthy from collapsed necrotic cells (*N*) with dense granular cytoplasm. *Bar* = 10 μm

detached, immature fruit or in attached or detached mature fruit (Stanghellini and Aragaki 1966).

Evidence for meristematic tissue functioning as structural barriers to proliferation of fungal infections is based on observations of restricted or delimited fungal growth to within compartmentalized tissue. Suggested mechanisms of structural responses are similar to those discussed for mechanisms of lignification in Section 11.5. In some host-pathogen interactions, advancement of fungal growth is checked prior to the meristematic response. Mullick (1977) indicated that in some plants, a pathogen is restricted by an autofluorescent, nonsuberized impermeable layer formed before the development of a meristematic layer. Growth of the fungus may be restricted by phenolic compounds present in the nonsuberized, impermeable layer (Mullick 1977). Thus, other structural barriers and biochemical responses may operate in concert with meristematic barriers in restricting fungal growth and disease.

11.7 Concluding Remarks

Plants are generally resistant to fungal infection. Many different mechanisms in plants are involved in resistance to pathogenic fungi and most are likely to be multipurpose and parasite nonspecific (Heath 1987). As indicated in the introduction to this chapter, resistance in plants has been studied and characterized mostly as pre- or postinfectional, structural or biochemical mechanisms of host defense. Although one mechanism may be identified as the primary character of resistance in a plant species, resistance of the host is probably dependent on many characteristics (specific and nonspecific) of the whole plant from preformed biochemicals or structures to postinfectional responses. Heath (1981, 1987) discussed mechanisms of host resistance as specific (host) or nonspecific (nonhost or basic). In specific resistance, mechanisms are postinfectional and seem to depend on interference in the host-pathogen relationship with responses such as phytoalexin production or by impairing metabolic pathways, while in nonspecific resistance, mechanisms include preformed structural and biochemical characteristics, as well as postinfectional responses, that inhibit or prevent further development of the pathogen (Heath 1981, 1987). Indeed, many nonspecific responses (biochemical or structural) to infection are basic mechanisms of the plant to mechanical or chemical injury (Aist 1976, Mullick 1977, Kuc 1978) and may be related directly to wound healing and not to host defense (Aist 1983). In this chapter, responses to fungal infections were considered as part of host defense, since injury is always associated with fungal pathogens of leaves.

Many different mechanisms, such as morphological structures, preformed biochemicals, postinfectional biochemicals affecting pathogen metabolism, or structural responses of the host affecting the cell walls of leaves and fruit of trees, have been implicated and studied as mechanisms of host resistance. The precise nature and importance of many of these mechanisms, however, have not been determined. Much of our information is observational or correlative and future experimentation should include studies to determine the role of morphological structures in basic resistance of the host in providing microclimates favorable for disease in the field; the quantity and toxicity of preformed chemicals inhibiting fungal pathogens at microsites of penetration; and the site, quantity, and timing of production of cell wall modifications or biochemicals directly inhibitory to fungal pathogens. Much of our information on mechanisms of disease resistance has been dependent on information obtained from herbaceous plants or grasses used as model systems. These model systems have advanced our understanding of the specific nature and function of structural and biochemical mechanisms. Although there are many advantages to such systems, including ease of manipulation and rapid growth of test plants, mechanisms in leaves and fruit of woody plants may differ and cannot always be directly related to other plant-pathogen interactions. Thus, future studies are needed to determine specific mechanisms of resistance in leaves and fruit of perennial, woody plants to fungal pathogens.

References

Adams PB, Sproston T, Tietz H, Major RT 1962 Studies on the disease resistance of Ginkgo biloba. Phytopathology 52:233–236

Adaskaveg JE, Feliciano AJ, Ogawa JM 1989 Comparative studies of resistance in peach genotypes to Monilinia fructicola. Phytopathology 79:1183–1184

Adaskaveg JE, Ogawa JM 1988 Ultrastructural study of shot hole disease of almond and the causal organism Stigmina carpophila. Phytopathology 78:1519

Aist JR 1976 Papillae and related wound plugs of plant cells. Annu Rev Phytopathol 14:145–163

Aist JR 1983 Structural responses as resistant mechanisms. In: Bailey JA, Deverall BJ (eds) The dynamics of host defence. Academic Press, New York London, 33–70

Akai S 1959 Histology of defense in plants. In: Horsfall JG, Dimond AE (eds) Plant pathology, vol 1. Academic Press, New York London, 435–467

Albersheim P, Anderson A 1971 Proteins from plant cell walls inhibit polygalacturonases secreted by plant pathogens. Proc Natl Acad Sci USA 68:1815–1819

Atkinson P, Blakeman JP 1982 Seasonal occurrence of an antimicrobial flavanone, sakuranetin, associated with glands on leaves of Ribes nigrum. New Phytol 92:63–74

Bach WJ, Wolf FA 1928 The isolation of the fungus that causes citrus melanose and the pathological anatomy of the host. J Agr Res 37:243–252

Bailey JA, Mansfield JW 1982 Phytoalexins. Wiley & Sons, New York, 334 pp

Baker EA 1971 Chemical and physical characteristics of cuticular membranes. In: Preece TF, Dickinson CH (eds) Ecology of leaf surface micro-organisms. Academic Press, New York London, 55–65

Barlass M, Miller RM, Douglas, TJ 1987 Development of methods for screening grapevines for resistance to infection by downy mildew. II. Resveratrol production. Am J Enol Viticult 38:65–68

Bateman DR, Van Etten HD, English PD, Nevins DJ, Albersheim P 1969 Susceptibility to enzymatic degradation of cell walls from bean plants resistant and susceptible to Rhizoctonia solani Kuhn. Plant Physiol 44:641–648

Bayles, CJ, Ghemawat MS, Aist JR 1990 Inhibition by 2-deoxy-D-glucose of callose formation, papilla deposition, and resistance to powdery mildew in and ml-o barley mutant. Physiol Mol Plant Pathol 36:63–72

Beckman CH 1980 Defenses triggered by the invader: physical defences. In: Horsfall JG, Cowling EB (eds) Plant disease: an advanced treatise, vol 5. Academic Press, New York London, 225–267

Beckman CH, Mueller WC, McHardy WE 1972 The localization of stored phenols in plant hairs. Physiol Plant Pathol 2:69–74

Biggs AR, Northover J 1989 Association fo sweet cherry epidermal characters with resistance to Monilinia fructicola. Hort Sci 24:126–127

Bingham RT, Hoff RJ, McDonald GI 1971 Disease resistance in forest trees. Annu Rev Phytopathol 9:433–452
Blakeman JP 1971 The chemical environment of the leaf surface in relation to growth of pathogenic fungi. In: Preece TF, Dickinson CH (eds) Ecology of leaf surface micro-organisms. Academic Press, New York London, 255–268
Blakeman JP 1981 Microbial ecology of the phylloplane. Academic Press, New York London, 502 pp
Blakeman JP, Atkinson P 1981 Antimicrobial substances associated with the aerial surfaces of plants. In: Blakeman JP (ed) Microbial ecology of the phylloplane. Academic Press, New York London, 245–263
Boller T 1987 Hydrolytic enzymes in plant disease resistance. In: Kosuge T, Nester EW (eds) Plant-microbe interactions: molecular and genetic perspectives, vol 2. MacMillan, New York, 385–413
Bowles DJ 1990 Defense-related proteins in higher plants. Annu Rev Biochem 59:873–907
Brown AE, Swinburne TR 1973 Factors affecting the accumulation of benzoic acid in Bramley's seedling apples infected with Nectria galligena. Physiol Plant Pathol 3:91–99
Brown JF, Shipton WA, White NH 1966 The relationship between hypersensitive tissue and resistance in wheat seedlings infected with Puccinia graminis tritici. Ann Appl Biol 58:279–290
Brown W 1922 Studies in the physiology of parasitism. VIII. On the exosmosis of nutrient substances from the host tissue into the infection drop. Ann Bot 34:101–119
Byrde RJW 1957 The varietal resistance of fruits to brown rot. II. The nature of resistance in some varieties of cider apple. J Hort Sci 32:227–238
Byrde RJW, Fielding AH, Williams AH 1960 The role of oxidized polyphenols in the varietal resistance of apples to brown rot. In: Bridham JB (ed) Phenolics in plants in health and disease. Pergamon, New York, 95–99
Byrde RJW, Willetts HJ 1977 The brown rot fungi of fruit: their biology and control. Pergamon, New York, 171 pp
Campbell CL, Huang J-S, Payne G 1980 Defense at the perimeter: the outer walls and gates. In: Horsfall JG, Cowling EB (eds) Plant disease: an advanced treatise, vol 5. Academic Press, New York London, 103–120
Campbell R 1972 Electron microscopy of the epidermis and cuticle of the needles of Pinus nigra var. maritima in relation to infection by Lophodermella sulcigena. Ann Bot 36:307–314
Carroll FE, Müller E, Sutton BC 1977 Preliminarty studies on the incidence of endophytes in some European conifers. Sydowia 29:87–103
Chang PLY, Trevithick JR 1974 How important is secretion of exo-enzymes through apical cell walls of fungi? Arch Microbiol 101:281–293
Christensen JJ, DeVay JE 1955 Adaptation of plant pathogen to host. Annu Rev Plant Physiol 6:367–392
Cohen LI 1967 The pathology of Hypodermella laricis on larch, Larix occidentalis. Am J Bot 54:118–124
Cohen Y, Eyal H, Hanania J 1990 Ultrastructure, autofluorescence, callose deposition and lignification in susceptible and resistant muskmelon leaves infected with the powdery mildew fungus Sphaerotheca fuliginea. Physiol Mol Plant Pathol 36:191–204
Cooper RM 1983 The mechanisms and significance of enzymic degradation of host cell walls by parasites. In: Callow JA (ed) Biochemical plant pathology. Wiley & Sons, New York, 101–135
Cooper RM, Wardman PA, Skeleton JEM 1981 The influence of cell walls from host and non-host plants on the production and activity host plants on the production and activity of polygalacturonide-degrading enzymes from fungal pathogens. Physiol Plant Pathol 18:239–255
Cowling EB 1975 Physical and chemical constraints in the hydrolysis of cellulose and lignocellulosic materials. In: Wilkie CR (ed) Biochemical and bioengineering symposium 5. Wiley & Sons, New York, 163–181
Coxon DT 1982 Phytoalexins from other plant families. In: Bailey JA, Mansfield JW (eds) Phytoalexins. Wiley & Sons, New York, 106–132

Cruickshank IAM 1977 A review of the role of phytoalexins in disease resistance mechanisms. Pontif Acad Sci Scr Varia 41:509–561
Cruickshank IAM 1980 Phytoalexins. Annu Rev Phytopathol 1:351–374
Crute IR, DeWitt PJGM, Wade M 1985 Mechanisms by which genetically controlled resistance and virulence influence host colonization by fungal and bacterial parasites. In: Fraser RSS (ed) Mechanisms of resistance to plant diseases. Nijhoff Junk, Boston, 197–309
Cunningham HS 1928 A study of the histologic changes induced in leaves by certain leaf-spotting fungi. Phytopathology 18:717–751
Curtis KM 1928 The morphological aspect of resistance to brown rot in stone fruit. Ann Bot 42:39–68
Darvill AG, McNeil M, Albersheim P, Delmer DP 1980 The primary walls of flowering plants. In: Tolbert NE (ed) The biochemistry of plants, vol 1. Academic Press, New York London, 91–162
Dercks W, Creasy LL 1989 The significance of stilbene phytoalexins in the Plasmopara viticola – grapevine interaction. Physiol Mol Plant Pathol 34:189–202
Deverall BJ 1976 Current perspectives in research on phytoalexins. In: Friend J, Threlfall DR (eds) Biochemical aspects of plant-parasite relationships. Academic Press, New York London, 207–223
Dickman MB, Patil SS, Kolattukudy PE 1982 Purification and characterization of an extracellular cutinolytic enzyme from Colletotrichum gloeosporioides on Carica papaya. Physiol Plant Pathol 20:333–337
Dix NJ 1974 Identification of a water-soluble fungal inhibitor in the leaves of Acer platanoides L. Ann Bot 38:505–514
Dix NJ 1979 Inhibition of fungi by gallic acid in relation to growth on leaves and litter. Trans Brit Mycol Soc 73:329–336
Dormaar JF 1970 Seasonal pattern of water-soluble constituents from leaves of Populus X "Northwest" (Hort.). J Soil Sci 21:105–110
Edwards MC, Bowling DJF 1986 The growth of rust germ tubes toward stomata in relation to pH gradients. Physiol Mol Plant Pathol 29:185–196
Ejólfsson R 1970 Recent advances in chemistry of cyanogenic glycosides. Fortschr Chem Org Naturst 28:74–108
Fawcett CH, Spencer DM 1967 Antifungal phenolic acids in apple fruits after infection with Sclerotinia fructigena. Ann Appl Biol 60:87–96
Fawcett CH, Spencer DM 1968 Sclerotinia fructigena infection and chlorogenic acid content in relation to antifungal compounds in apple fruits. Ann Appl Biol 61:245–253
Fisher RW, Corke ATK 1971 Infection of Yarlington Mill fruit by the apple scab fungus. Can J Plant Sci 51:535–542
Flores G, Hubbes M 1980 The nature and role of phytoalexin produced by aspen (Populus tremuloides Michx.). Eur J For Pathol 10:95–103
Franich RA, Carson MJ, Carson SD 1986 Synthesis and accumulation of benzoic acid in Pinus radiata needles in response to tissue injury by dothistromin, and correlation with resistance of P. radiata families to Dothistroma pini. Physiol Mol Plant Pathol 28:267–286
Franich RA, Gadgel PD, Shain L 1983 Fungistatic effects of Pinus radiata needle epicuticular fatty and resin acids on Dothistroma pini. Physiol Plant Pathol 23:183–195
Franich RA, Gaskin RE, Wells LG, Zabkiewicz, JA 1982 Effect of Pinus radiata needle monoterpenes on spore germination and mycelial growth of Dothistroma pini in vitro in relation to mature tree resistance. Physiol Plant Pathol 21:55–63
Franich RA, Wells LG, Barnett JR 1977 Variation with tree age of needle cuticle topography and stomatal structure in Pinus radiata D. Don. Ann Bot 41:621–626
Franich RA, Wells LG, Holland PT 1978 Epicuticular wax of Pinus radiata needles. Phytochemistry 17:1617–1623
Frossard R 1981 Affect of guttation fluids on growth of micro-organisms on leaves. In: Blakeman JP (ed) Microbial ecology of the phylloplane. Academic Press, New York London, 213–226
Fry WE, Evans PH 1977 Association of formamide hydro-lyase with fungal pathogenicity to cyanogenic plants. Phytopathology 67:1001–1006

Garbaye J, Pinon J 1973 Mineral nutrition and susceptibility to Marssonina brunnea of Populus "J-214": preliminary study of young cuttings under controlled conditions. Ann Sci For 30:423–431

Gäumann E 1950 Principles of plant infection, English edn. Hafner, New York, 543 pp

Gay J, Pearce R 1984 The structure of plant surfaces. In: Roberts DW, Aist JR (eds) Infection processes of fungi. Rockefeller Found, New York, 16–30

Gibson IAS 1972 Dothistroma blight of Pinus radiata. Annu Rev Phytopathol 10:51–72

Giesemann A, Biehl B, Lieberei R 1986 Identification of scopoletin as a phytoalexin of the rubber tree Hevea brasiliensis. J Phytopathol 117: 373–376

Gilliver K 1947 The effect of plant extracts on the germination of conidia of Venturia inaequalis. Ann Appl Biol 34:136–143

Glazener JA 1982 Accumulation of phenolic compounds in cells and formation of lignin-like polymers in cell walls of young tomato fruits after inoculation with Botrytis cinerea. Physiol Plant Pathol 20:11–25

Godfrey BES, Clements DM 1978 Effect of a lilac leaf leachate on germination of Alternaria alternata and Botrytis cinerea. Trans Brit Mycol Soc 70:163–165

Goodman RN Király Z, Zaitlin M 1967 The biochemistry and physiology of infectious plant disease. Van Nostrand, Princeton NJ, 354 pp

Gottwald TR 1985 Influence of temperature, leaf wetness period, leaf age, and spore concentration on infection of pecan leaves by conidia of Cladosporium caryigenum. Phytopathology 75:190–194

Günthardt-Goerg MS 1986 Epicuticular wax of needles of Pinus cembra, Pinus sylvestris and Picea abies. Eur J For Pathol 16:400–418

Hafiz A 1952 Basis of resistance in gram to Mycosphaerella blight. Phytopathology 42:422–424

Hall R 1971 Pathogenicity of Monilinia fructicola, pt 2. Penetration of peach leaf and fruit. Phytopathol Z 72:281–290

Hammerschmidt R, Kuć J 1982 Lignification as a mechanism for induced systemic resistance in cucumber. Physiol Plant Pathol 20:61–71

Hancock JG, Miller RL 1965 Relative importance of polygalacturonate transeliminase and other pectolytic enzymes in southern anthracnose, spring black stem, and Stemphylium leaf spot of alfalfa. Phytopathology 55:346–355

Harborne JB 1964 Biochemistry of phenolic compounds. Academic Press, New York London, 618 pp

Hare RC 1971 Physiology and biochemistry of resistance to pine rusts. In: Biology of rust resistance in forest trees. In: Proc NATO-IUFRO Adv Stud Inst, August 17–24, 1969. US For Serv Misc Publ 1221:465–478

Heath MC 1976 Hypersensitivity, the cause or the consequence of rust resistance. Phytopathology 66:935–936

Heath MC 1980 Reactions of nonsuscepts to fungal pathogens. Annu Rev Phytopathol 18:211–236

Heath MC 1981 Resistance of plants to rust infection. Phytopathology 71:971–974

Heath MC 1987 Evolution of plant resistance and susceptibility to fungal invaders. Can J Plant Pathol 9:389–397

Heather WA 1967a Susceptibility of juvenile leaves of Eucalyptus bicostata Maiden et al. to infection by Phaeoseptoria eucalypti (Hanf.) Walker. Aust J Biol Sci 20:769–775

Heather WA 1967b Leaf characteristics of Eucalyptus bicostata Maiden et al. seedlings affecting the deposition and germination of spores of Phaeoseptoria eucalypti (Hanf.) Walker. Aust J Biol Sci 20:1155–1160

Hedin PA, Langhans VE, Graves Jr CH 1979 Identification of juglone in pecan as a possible factor of resistance to Fusicladium effusum. J Agr Food Chem 27:92–94

Heintz C, Blaich R 1990 Ultrastructural and histochemical studies on interactions between Vitis vinifera L. and Uncinula necator (Schw.) Burr. New Phytol 115:107–117

Hesler LR 1916 Black rot, leaf spots, and canker of Pomaceous fruits. Cornell Agr Exp Stn Bull 379:53–148

Higgins BB 1914 Contribution to the life history and physiology of Cylindrosporium on stone fruits. Am J Bot 1:145–173

Hill G 1977 Frühphase der Pathogenese von Botrytis cinerea auf unterschiedlichen Entwicklungsstadien vegetativer und generativer Organe von Vitis vinifera L. Diss, Univ Giessen, FRG
Hill G, Stellwag-Kittler F, Huth G, Schlösser EW 1981 Resistance of grapes in different developmental stages to Botrytis cinerea. Phytopathol Z 102:328–338
Hoch HC, Staples RC, Whitehead B, Comeau J, Wolf Ed 1987 Signaling for growth orientation and cell differentiation by surface topography in Uromyces. Science 235:1659–1662
Hoff RJ, McDonald GI 1971 Resistance to Cronartium ribicola in Pinus monticola: short shoot fungicidal reaction. Can J Bot 49:1235–1239
Hoff RJ, McDonald GI 1975 Hypersensitive reaction in Pinus armandii caused by Cronartium ribicola. Can J For Res 5:146–148
Hoff RJ, McDonald GI 1980 Resistance to Cronartium ribicola in Pinus monticola: reduced needle-spot frequency. Can J Bot 58:574–577
Hulme AC, Edney KL 1960 Phenolic substances in the peel of Cox's Orange Pippin apples with reference to infection by G. perennans. In: Prideham JB (ed) Phenolics in plants in health and disease. Pergamon, New York, 87–94
Ingham JL 1973 Disease resistance in higher plants. Phytopathol Z 78:314–335
Ingham JL 1978 Cell death and resistance to biotrophs. Ann Appl Biol 89:291–295
Ingham JL, Harborne JB 1976 Phytoalexin induction as a new dynamic approach to the study of systematic relationships among higher plants. Nature (London) 260:241–243
Irvine JA, Dix NJ, Warren RC 1978 Inhibitory substances in Acer platanoides leaves: seasonal activity and effects on growth of phylloplane fungi. Trans Brit Mycol Soc 70:363–371
Jewell FF, Sr 1990 Histopathology of longleaf pine needles infected by Ploioderma hedgcockii (Dearn.) Darker. Eur J For Pathol 20:24–31
Johansson M, Popoff T, Theander O 1976 Effect of spruce root constituents on extracellular enzymes of Fomes annosus. Physiol Plant 37:275–282
Johnston HW, Sproston Jr T 1965 The inhibition of fungus infection pegs in Ginkgo biloba. Phytopathology 55:225–227
Johnstone KH 1931 Observations on the varietal resistance of the apple to scab (Venturia inaequalis, Aderh.) with special reference to its physiological aspects. J Pom Hortic Sci 9:30–52, 195–227
Jones TM, Albersheim P 1972 A gas chromatographic method for determination of aldose and uronic acid constituents of plant cell wall polysaccharides. Plant Physiol 49:926–936
Kinloch BB 1982 Mechanisms and inheritance of rust resistance in conifers. In: Heybroek HM, Stephan BR, von Weissenberg K (eds) Resistance to diseases and pests in forest trees. Proc Cent Agr Publ Doc (Pudoc), Wageningen, 119–129
Kinloch BB, Littlefield JL 1977 White pine blister rust: hypersensitive resistance in sugar pine. Can J Bot 55:1148–1155
Király Z 1980 Defenses triggered by the invader: hypersensitivity. In: Horsfall JG, Cowling EB (eds) Plant disease: an advanced treatise, vol 5. Academic Press, New York London, 201–224
Király Z, Barna B, Érsek T 1972 Hypersensitivity as a consequence, not the cause, of plant resistance to infection. Nature (London) 239:456–458
Knee M 1978 Properties of polygalacturonate and cell cohesion in apple fruit cortical tissue. Phytochemistry 17:1257–1260
Knee M, Fielding AH, Archer SA, Laborda F 1975 Enzymic analysis of cell wall structure in apple fruit cortical tissue. Phytochemistry 14:2213–2222
Kolattukudy PE 1980 Biopolyester membranes of plants: cutin and suberin. Science 208:990–1000
Kolattukudy PE 1981 Structure, biosynthesis, and biodegradation of cutin and suberin. Annu Rev Plant Physiol 32:539–567
Kolattukudy PE, Köller W 1983 Fungal penetration of the first line defensive barriers of plants. In: Callow JA (ed) Biochemical plant pathology, Wiley & Sons, New York, 79–100
Köller W, Allan CR, Kolattukudy PE 1982a Inhibition of cutinase and prevention of fungal penetration into plants by benomyl – a possible protective mode of action. Pest Biochem Physiol 18:15–25

Köller W, Allan CR, Kolattukudy PE 1982b Role of cutinase and cell wall degrading enzymes in infection of Pisum sativum by Fusarium solani f. sp. pisi. Physiol Plant Pathol 20: 47–60

Köller W, Allan CR, Kolattukudy PE 1982c Protection of Pisum sativum from Fusarium solani f. sp. pisi by inhibition of cutinase with organophosphorus pesticides. Phytopathology 72:1425–1430

Kosuge T 1969 The role of phenolics in host response to infection. Annu Rev Phytopathol 7:195–222

Kovács A 1955 Über die Ursachen der unterschiedlichen Resistenz der Zuckerrübensorten gegen Cercospora beticola Sacc. Phytopathol Z 24:283–298

Kovács A, Szeöke É 1956 Die phytopathologische Bedeutung der kutikulären Exkretion. Phytopathol Z 27:335–349

Kuć J 1966 Resistance of plants to infectious agents. Annu Rev Microbiol 20:337–370

Kuć J 1978 Changes in intermediary metabolism caused by disease. In: Horsfall JG, Cowling EB (eds) Plant disease: an advanced treatise, vol 3. Academic Press, New York London, 349–374

Kuć J 1990 A case for self defense in plants against disease. Phytoparasitica 18:3–8

Kuć J, Currier WW, Shih MJ 1976 Terpenoid phytoalexins. In: Friend J, Threlfall DR (eds) Biochemical aspects of plant-parasite relationships. Academic Press, New York London, 225–237

Lamport DTA 1970 Cell wall metabolism. Annu Rev Plant Physiol 21:235–269

Langcake P 1981 Disease resistance of Vitis spp. and the production of stress metabolites resveratrol, ε-viniferin, α-viniferin and pterostilbene. Physiol Plant Pathol 18:213–226

Langcake P, Cornford CA, Pryce RJ 1979 Identification of pterostilbene as a phytoalexin from Vitis vinifera leaves. Phytochemistry 18:1025–1027

Langcake P, Pryce RJ 1976 The production of resveratrol by Vitis vinifera and other members of the Vitaceae as a response to infection or injury. Physiol Plant Pathol 9:77–86

Langcake P, Pryce RJ 1977 The production of resveratrol and viniferins by grapevines in response to ultraviolet irradiation. Phytochemistry 16:1193–1196

Langhans VE, Hedin PA, Graves Jr CH 1978 Fungitoxic chemicals in pecan tissue. Plant Dis Rep 62:894–898

Latham AJ, Rushing AE 1988 Development of Cladosporium caryigenum in pecan leaves. Phytopathology 78:1104–1108

Lee B, Priestly JH 1924 The plant cuticle. I. Its structure, distribution, and function. Ann Bot 38:525–545

Lieberei R 1988 Relationship of cyanogenic capacity (HCN-c) of the rubber tree Hevea brasiliensis to susceptibility to Microcyclus ulei, the agent causing South American leaf blight. J Phytopathol 122:54–67

Lieberei R, Biehl B, Giesemann A, Junqueira NTV 1989 Cyanogenesis inhibits active defense reaction in plants. Plant Physiol 90:33–36

Littlefield LJ, Heath MC 1979 Ultrastructure of rust fungi. Academic Press, New York London, 277 pp

Lupton FGH 1956 Resistance mechanisms of Triticum and Aegilops and of amphidiploids between them to Erysiphe graminis DC. Trans Brit Mycol Soc 39:51–59

Maiti IB, Kolattukudy PE 1979 Prevention of fungal infection of plants by specific inhibition of cutinase. Science 205:507–508

Mansfield JW 1982 The role of phytoalexins in disease resistance. In: Bailey JA, Mansfield JW (eds) Phytoalexins. Wiley & Sons, New York, 253–288

Mansfield JW 1983 Antimicrobial compounds. In: Callow JA (ed) Biochemical plant pathology. Wiley & Sons, New York, 237–265

Marks GC, Berbee JG, Riker AJ 1965 Direct penetration of leaves of Populus tremuloides by Colletotrichum gloeosporioides. Phytopathology 55:408–412

Martin JT 1964 Role of cuticle in the defense against plant disease. Annu Rev Phytopathol 2:81–100

Martin JT, Baker EA, Byrde RJW 1966 The fungitoxicities of cuticular and cellular components of citrus lime leaves. Ann Appl Biol 57:491–500

Martin JT, Batt RF, Burchill RT 1957 Fungistatic properties of apple leaf wax. Nature (London) 180: 796–797

Martin JT, Juniper BE 1970 The cuticles of plants. St. Martin's, New York, 339 pp

McBride RP 1971 Micro-Organism interactions in the phyllosphere of larch. In: Preece TF, Dickinson CH (eds) Ecology of leaf surface micro-organisms. Academic Press, New York London, 545–555

McDonald GI, Hoff RJ 1975 Resistance to Cronartium ribicola in Pinus monticola: an analysis of needle spot types and frequencies. Can J Bot 53:2497–2505

McWhorter FA 1927 The diseases of Carica papaya. Thesis, Cornell Univ, 67 pp

Mence MJ, Hildebrandt AC 1966 Resistance to powdery mildew in rose. Ann Appl Biol 58:309–320

Millar CS 1981 Infection processes on conifer needles. In: Blakeman JP (ed) Microbial ecology of the phylloplane. Academic Press, New York London, 185–209

Millar RL, Hemphill R 1978 β-Glucosidases associated with cyanogenesis in Stemphylium leafspot of birdsfoot trefoil. Physiol Plant Pathol 13:259–270

Mims CW, Glidewell DC 1978 Some ultrastructural observations on the host-pathogen relationship within the telial gall of the rust fungus Gymnosporangium juniperi-virginianae. Bot Gaz 139:11–17

Misaghi IJ 1982 Physiology and biochemistry of plant-pathogen interactions. Plenum, New York, 287 pp

Mlodzianowski F, Werner A, Siwecki R 1978 Germination of Melampsora larici-populina uredospores on poplar leaves. Eur J For Pathol 8:119–125

Mueller WC, Beckman CH 1976 Ultrastructure and development of phenolic-storing cells in cotton roots. Can J Bot 54:2074–2082

Mullick DB 1977 The non-specific nature of defense in bark and wood during wounding, insect and pathogen attack. In: Loewus FA, Runeckles VC (eds) The structure, biosynthesis and degradation of wood, vol 11. Plenum, New York, 395–436

Ndubizu TOC 1976 Relations of phenolic inhibitors to resistance of immature apple fruits to rot. J Hortic Sci 51:311–319

Noble JP, Drysdale R 1983 The role of benzoic acid and phenolic compounds in latency in fruits of two apple cultivars infected with Pezicula malicorticis or Nectria galligena. Physiol Plant Pathol 23:207–216

Noveroske RL, Kuć J, Williams EB 1964a Oxidation of phloridzin and phloretin related to resistance of Malus to Venturia inaequalis. Phytopathology 54:92–97

Noveroske RL, Williams EB, Kuc J 1964b β-Glucosidase and phenol oxidase in apple leaves and their possible relation to resistance to Venturia inaequalis. Phytopathology 72:98–103

Nusbaum CJ 1935 A cytological study of the resistance of apple varieties to Gymnosporangium juniperi-virginianae. J Agr Res 51:573–596

Nusbaum CJ 1938 A cytological study of host-parasite relations of Venturia inaequalis on apple leaves. J Agr Res 56:595–618

Nusbaum CJ, Keitt GW 1938 A cytological study of host-parasite relations of Venturia inaequalis on apple leaves. J Agr Res 56:595–617

Onoe T, Tani T, Minagawa S, Sagawa H 1987 Ultrastructural changes of stomata in relation to specificity of rust fungi. In: Nishimura S, Vance CP, Doke N (eds) Molecular determinants of plant diseases. Jpn Sci Soc Press, Tokyo; and Springer, Berlin Heidelberg New York, 29–45

Overeem JC 1976 Pre-existing antimicrobial substances in plants and their role in disease resistance. In: Friend J, Threlfall DR (eds) Biochemical aspects of plant-parasite relationships. Academic Press, New York London, 195–206

Patil SS 1980 Defenses triggered by the invader: detoxifying the toxins. In: Horsfall JG, Cowling EB (eds) Plant disease: an advanced treatise, vol 5. Academic Press, New York London, 269–277

Patton RF, Johnson DW 1970 Mode of penetration of needles of eastern white pine by Cronartium ribicola. Phytopathology 60:977–982

Patton RF, Spear RN 1980 Stomatal influences on white pine blister rust infection. Phytopathol Medit 19:1–7

Paxton JD 1981 Phytoalexins – a working redefinition. Phytopathol Z 101:106–109
Pearce RB, Rutherford J 1981 A wound-associated suberized barrier to the spread of decay in the sapwood of oak (Quercus robor L.). Physiol Plant Pathol 19:359–369
Pero RW, Howard FL 1970 Activity of juniper diffusates on spores of Phomopsis juniperovora. Phytopathology 60:491–495
Peterson FW, Walla JA 1978 Development of Dothistroma pini upon and within needles of Austrian and ponderosa pines in eastern Nebraska. Phytopathology 68:1422–1430
Piattelli M, Impellizzeri G 1971 Fungistatic flavones in the leaves of Citrus species resistant and susceptible to Dueterophoma tracheiphila. Phytochemistry 10:2657–2659
Popoff T, Theander O, Johansson M 1975 Changes in sapwood of roots of Norway spruce, attacked by Fomes annosus, pt II. Organic chemical constituents and their biological effects. Physiol Plant 34:347–356
Postek MT 1981 The occurrence of silica in the leaves of Magnolia grandifolia L. Bot Gaz 142:124–134
Preece TF, Dickinson CH 1971 Ecology of leaf surface micro-organisms. Academic Press, New York London, 640 pp
Pridham JR 1960 The formation and possible function of phenolic glycosides. In: Pridham JR (ed) Phenolics in plants in health and disease. Pergamon, New York, 9–15
Raa J 1968 Polyphenols and natural resistance of apple leaves against Venturia inaequalis. Neth J Plant Pathol 74 (Suppl): 37–45
Ragazzi A, Raddi P, Dellavelle Fedi I 1985 Pre-infection behavior of Cronartium flaccidum basidiospores on the needle surface of three pine species. In: Barrows-Broaddus J, Powers HR (eds) Proc Rusts of hard pines working party Conf, 1984, Athens, Ga. Ga Center Cont Educ, Univ Ga, 201–206
Ride JP 1978 The role of cell wall alterations in resistance to fungi. Ann Appl Biol 89:302–306
Ride JP 1983 Cell walls and other structural barriers in defence. In: Callow JA (ed) Biochemical plant pathology. Wiley & Sons, New York, 215–236
Ride JP, Pearce RB 1979 Lignification and papilla formation at sites of attempted penetration of wheat leaves by non-pathogenic fungi. Physiol Plant Pathol 15:79–92
Roberts MF, Martin JT 1963 Withertip disease of limes (Citrus aurantifolia) in Zanzibar. III. The leaf cuticle in relation to infection by Gloeosporium limetticola Clausen. Ann Appl Biol 51:411–413
Royle DJ 1976 Structural features of resistance to plant diseases. In: Friend J, Threlfall DR (eds) Biochemical aspects of plant-parasite relationships. Academic Press, New York London, 161–193
Ryan CA, An G 1988 Molecular biology of wound-inducible proteinase inhibitors in plants. Plant Cell Environ 11:345–349
Samuel G 1927 On the shot-hole disease caused by Clasterosporium carpophilum and on the "shot-hole" effect. Ann Bot 41:375–404
Schlösser EW 1980 Preformed internal chemical defenses. In: Horsfall JG, Cowling EB (eds) Plant disease: an advanced treatise, vol 5. Academic Press, New York London, 161–177
Schütt P 1971 Untersuchungen über den Einfluß von Cuticularwachsen auf die Infektionsfähigkeit pathogener Pilze. Eur J For Pathol 1:32–50
Shain L 1967 Resistance of sapwood in stems of loblolly pine to infection by Fomes annosus. Phytopathology 57:1034–1045
Shain L, Miller JB 1982 Pinocembrin: an antifungal compound secreted by leaf glands of eastern cottonwood. Phytopathology 72:877–880
Sharma JK, Heather WA, Winer P 1980 Effect of leaf maturity and shoot age of clones of Populus species on susceptibility to Melampsora larici-populina. Phytopathology 70:548–554
Sherwood RT, Vance CP 1976 Histochemistry of papillae formed in reed canarygrass leaves in response to noninfecting pathogenic fungi. Phytopathology 66:503–510
Sherwood RT, Vance CP 1980 Resistance to fungal penetration in Gramineae. Phytopathology 70:273–279
Sherwood RT, Vance CP 1982 Initial events in the epidermal layer during penetration. In: Asada Y, Bushnell WR, Ouchi S, Vance CP (eds) Plant infection: the physiological and

biochemical basis. Jpn Sci Soc Press, Tokyo; and Springer, Berlin Heidelberg New York, 27–44

Siwecki R, Przybyl K 1981 Water relations in the leaves of poplar clones resistant and susceptible to Melampsora larici-populina. Eur J For Pathol 11:348–357

Siwecki R, Werner A 1980 Resistance mechanism involved in the penetration and colonization of poplar leaf tissues by Melampsora rust. Phytopathol Medit 19:27–29

Siwecki R, Werner A, Krzan Z, Mlodzianowski F 1982 Resistance mechanisms in interactions between poplars and rust. In: Heybroek HM, Sephan BR, von Weissenberg K (eds) Resistance to diseases and pests in forest trees. Proc Cent Agr Publ Doc (Pudoc), Wageningen, 130–142

Smith DA 1982 Toxicity of phytoalexins In: Bailey JA, Mansfield JW (eds) Phytoalexins. Wiley & Sons, New York, 106–132

Smith MA 1936 Infection studies with Sclerotinia fructicola on brushed and nonbrushed peaches. Phytopathology 26:1056–1060

Saulding P 1925 A partial explanation of the relative susceptibility of the white pines to the white pine blister rust (Cronartium ribicola, Fischer). Phytopathology 15:591–597

Spiers AG, Hopcroft DH 1983 Ultrastructural study of the pathogenesis of Marssonina species to poplars. Eur J For Pathol 13:414–427

Spiers AG, Hopcroft DH 1984 Influence of leaf age, leaf surface and frequency of stomata on the susceptibility of poplar cultivars to Marssonina brunnea. Eur J For Pathol 14:270–282

Stanghellini ME, Aragaki M 1966 Relation of periderm formation and callose deposition to anthracnose resistance in papaya fruit. Phytopathology 56:444–450

Stein U, Hoos G 1984 Induktions- und nachweismethoden für stilbene bei vitaceen. Vitis 23:179–194

Stoessl A 1982 Biosynthesis of phytoalexins In: Bailey JA, Mansfield JW (eds) Phytoalexins. Wiley & Sons, New York, 133–180

Stoessl A 1983 Secondary plant metabolites in preinfectional and postinfectional resistance. In: Bailey JA, Deverall BJ (eds) The dynamics of host defence. Academic Press, New York London, 71–122

Stolzenburg MC, Aist JR, Israel HW 1984 The role of papillae in resistance to powdery mildew conditioned by the ml-o gene in barley. II. Experimental evidence. Physiol Plant Pathol 25:347–361

Tainter FH 1985 The ultrastructure of germinating aeciospores, and intercellular and intracellular structures of Cronartium comandrae. In: Barrows-Broaddus J, Powers HR (eds) Proc Rusts of hard pines working party Conf, 1984, Athens, Ga. Ga Cent Cont Educ, Univ Ga, 155–178

Takasugi M, Anetai M, Masamune T, Shirata A, Takahashi K 1980a Broussonins A and B, new phytoalexins from diseased paper mulberry. Chem Lett 339–340

Takasugi M, Nagao S, Masamune T, Shirata A, Takahashi K 1980b Chalcomoracin, a natural diels-alder odduct from diseased mulberry. Chem Lett 1573–1576

Takasugi M, Niino N, Anetai M, Masamune T, Shirata A, Takahashi K 1984a Structure of two stress metabolites, spirobroussonin A and B, from diseased paper mulberry. Chem Lett 693–694

Takasugi M, Niino N, Nagao S, Anetai M, Masamune T, Shirata A, Takahashi K 1984b Eight minor phytoalexins from diseased paper mulberry. Chem Lett 689–692

Tan AM, Low FC 1975 Phytoalexin production by Hevea brasiliensis in response to infection by Colletotrichum gloeosporioides and its effect on other fungi. In: Int Rubb Conf, Kuala Lumpur, 217–227

Tomiyama K, Doke N, Nozue M, Ishiguri Y 1979 The hypersensitive response of resistant plants. In: Daly JM, Uritani I (eds) Recognition and specificity in plant host-parasite interactions. Jpn Sci Soc Press, Tokyo; and Univ Park Press, Baltimore, 69–84

Topps JH, Wain RL 1957 Fungistatic properties of leaf exudates. Nature (London) 179:652–653

Trione EJ 1960 The HCN content of flax in relation to flax wilt resistance. Phytopathology 50:482–486

Valleau WD 1915 Varietal resistance of plums to brown-rot. J Agr Res 5:365–395

Valsangiacomo C, Gressler C 1988 Role of the cuticular membrane in ontogenic and vf-resistance of apple leaves against Venturia inaequalis. Phytopathology 78:1066–1069

Vance CP, Kirk TK, Sherwood RT 1980 Lignification as a mechanism of disease resistance. Annu Rev Phytopathol 18:259–288

Verhoeff K 1974 Latent infections by fungi. Annu Rev Phytopathol 12:99–110

Verrall AF 1934 The resistance of saplings and certain seedlings of Pinus palustris to Septoria acicola. Phytopathology 24:1262–1264

Vidhyasekaran P 1973 Possible role of orthodihydroxy phenolics in grapevine anthracnose disease resistance. Indian J Exp Biol 11:473–475

Vidhyasekaran P 1988a Physiology of disease resistance in plants, vol 1. CRC, Boca Raton, 128 pp

Vidhyasekaran P 1988b Physiology of disease resistance in plants, vol 2. CRC, Boca Raton, 149 pp

Vidhyasekaran P, Cheeran A 1972 Osmotic pressure of grapevine leaves in relation to anthracnose disease incidence. Indian J Exp Biol 10:398–399

Volk RJ, Kahn RP, Weintraub RL 1958 Silicon content of the rice plant as a factor influencing its resistance to infection by the blast fungus, Piricularia oryzae. Phytopathology 48:179–186

Walker JC, Stahman MA 1955 Chemical nature of disease resistance in plants. Annu Rev Plant Physiol 6:351–366

Warren RC 1976 Buds and leaf microbes. In: Dickinson CH, Preece TF (eds) Microbiology of aerial plant surfaces. Academic Press, New York London, 361–374

Wattenbarger DW, Gray E, Rice JS, Reynolds JH 1968 Effects of frost and freezing on hydrocyanic acid potential of sorghum plants. Crop Sci 8:536–538

Wetzstein HY, Sparks D 1983 Anatomical indices of cultivar and age-related scab resistance and susceptibility in pecan leaves. J Am Soc Hortic Sci 108:210–218

Weinhold AR, English H 1964 Significance of morphological barriers and osmotic pressure in resistance of mature peach leaves to powdery mildew. Phytopathology 54:1409–1414

Williams AH 1960 The distribution of phenolic compounds in apple and pear leaves. In: Pridham JB (ed) Phenolics in plants in health and disease. Pergamon, New York, 3–7

Williamson B, Mitchell CP, Millar CS 1976 Histochemistry of Corsican pine needles infected by Lophodermella sulcigena (Rostr.) v. Höhn. Ann Bot 40:281–288

Wiltshire SP 1915 Infection and immunity studies on the apple and pear scab fungi (Venturia inaequalis and V. pirina). Ann Appl Biol 1:335–350

Wood R 1967 Physiological Plant Pathology. Blackwell, Oxford, 570 pp

Wynn WK 1967 Appressorium formation over stomates by the bean rust fungus: response to a surface contact stimulus. Phytopathology 66:136–146

Chapter 12

Anatomy of Resistant and Susceptible Pines to Fusiform Rust

F.F. JEWELL, SR. and D.C. JEWELL

12.1 Introduction

The host/pathogen relationships between species of *Pinus* and various gall-rust organisms have been studied rather extensively. Probably, some of the earliest work concerning gall rusts of pine was by Hartig (1874) and Wolff (1876). Another early paper reported observations on the effects of a gall rust, *Peridermium cerebrum* Pk., on jack pine, *P. banksiana* (Stewart 1916). Later, Colley (1918) published a detailed description of the relationship of *P. strobus*, Eastern white pine, and the blister rust pathogen, *Cronartium ribicola*. Dodge and Adams (1918) followed with a paper concerning *P. cerebrum* affecting pitch pine, *P. rigida*. Although these early papers described rather completely the morphology of the rust pathogens and the anatomy of the affected pine species, few, if any, observations concerning anatomical manifestations of host resistance to the rust pathogens were recorded.

An initial paper concerning resistance in pine to a gall-rust pathogen was published by Hutchinson (1935), which reported three reactions in *P. sylvestris*, Scots pine, to inoculation by a *Peridermium* sp. The reactions ranged from susceptibility (typical gall formation) to small swellings or necrotic areas of bark that eventually sloughed off. These results are quite similar to more recent observations of gall rust/pine relationships, and will be discussed later in this chapter. True (1938) and McKenzie (1942) reported instances of incompatibility among mycelium of gall-rust pathogens and host cells in certain pine samples. True (1938) also described a tissue barrier that prevented additional invasion of the host by the mycelium of the pathogen. These authors did not specify resistant traits among the pines studied, but their observations indicate that resistance was possibly expressed as an anatomical abnormality in the pine host. Possibly, the first reference to resistance in pine to a heteroecious *Cronartium* rust was for *P. strobus*, where a wound periderm was observed preventing host tissue invasion by *C. ribicola* (Struckmeyer and Riker 1951).

Histopathological research on Southern pine/gall-rust associations was initiated through study of the Southern fusiform rust pathogen, *Cronartium guercuum* (Berk.(Miyabe ex. Shirai f.sp. *fusiforme* (= *Cronartium fusiforme*) (Burdsall and Snow 1977) infecting loblolly pine. *P. taeda*, and slash pine, *P. elliottii* var. *elliottii*. These pine species are widely affected by this pathogen with severe losses occurring, particularly in plantations (Phelps 1973).

Interest in anatomical expressions of resistance in pine to fusiform rust was intensified in 1963 with a paper by Snow et al. reporting loblolly and slash pine seedlings recovering from infection by this rust. At that time, no reason for the recovery of the hosts was advanced. The first paper to pursue an explanation of recovery from fusiform rust infection was by Jewell and Snow (1972). These authors, working with slash pine seedlings inoculated with *Cronartium guercuum* f.sp. *fusiforme* and *C. guercuum* (Berk.)Miyabe ex. Shirai f.sp. *echinata* (= *C. guercumm*) (Burdsall and Snow 1977), concluded that anatomical resistance was expressed in certain slash pines, which gave these hosts the ability to recover from rust infection and develop or continue to grow normally, regardless of the rust pathogen involved in the original infections. Results of this research initiated rather intensive examination of rust resistance, particularly in slash pine and loblolly pine, expressed by the anatomy of the affected hosts.

12.2 Susceptible Host/Pathogen Relationship

The initial research on the pathological anatomy of fusiform rust galls was published in 1958 by Jackson and Parker. These authors reported on the anatomical characteristics of gall tissue in 7-year-old loblolly pines (*P. taeda*), as well as the morphology and mode of action of the pathogen in the host gall tissue. When compared to the normal (rust-free) loblolly material, gall tissue exhibited particular characteristics: irregularly shaped and shorter tracheids; higher xylem rays composed of an increased number of and larger parenchyma cells; increase in the number of vertical resin ducts; phloem rays similar to those of the xylem; and, with age, crushing of the phloem elements. Figure 12.1A,B exhibits, generally, anatomical differences in normal and fusiform rust gall tissue or slash pine. The pathogen in the host tissue was typified by intercellular hyphae throughout the gall tissue with haustoria evident in parenchyma-type cells of the host. The pathogen was not observed in the pith.

In 1962, Jewell et al. described the pathological anatomy of galls from 1-year-old slash pines (*Pinus elliottii* var. *elliottii*) previously inoculated with the fusiform rust pathogen (Jewell 1960). In general, the host gall tissue and the pathogen exhibited characteristics similar to those reported for loblolly pine by Jackson and Parker (1958). However, the 1962 paper (Jewell et al.) defined certain host and pathogen activities not previously reported for fusiform rust galls. Xylem rays in gall tissue lacked ray tracheids. In certain slash pine samples, the rust pathogen was observed in the pith. An increase in the number of resin ducts in gall tissue was considered proportional to the increase in volume of xylem in rust galls, thus not an actual increase in number. These authors commonly observed an unusual layer of vertical (wood) parenchyma cells, not associated with xylem rays, which consistently marked the inner (toward the pith) extent of the pathogen and corresponding abnormalities of the host tissue in the xylem. This layer of wood parenchyma was termed "reaction parenchyma." Similar parenchymatous strands in pine have been reported by Stewart (1916). Hyphae did not extend beyond the layer of wood parenchyma. Galls in which the pathogen was observed in the pith did not exhibit reaction parenchyma. Tannin development and cellular necrosis are additional features consistent with active fusiform rust galls (Walkinshaw 1978).

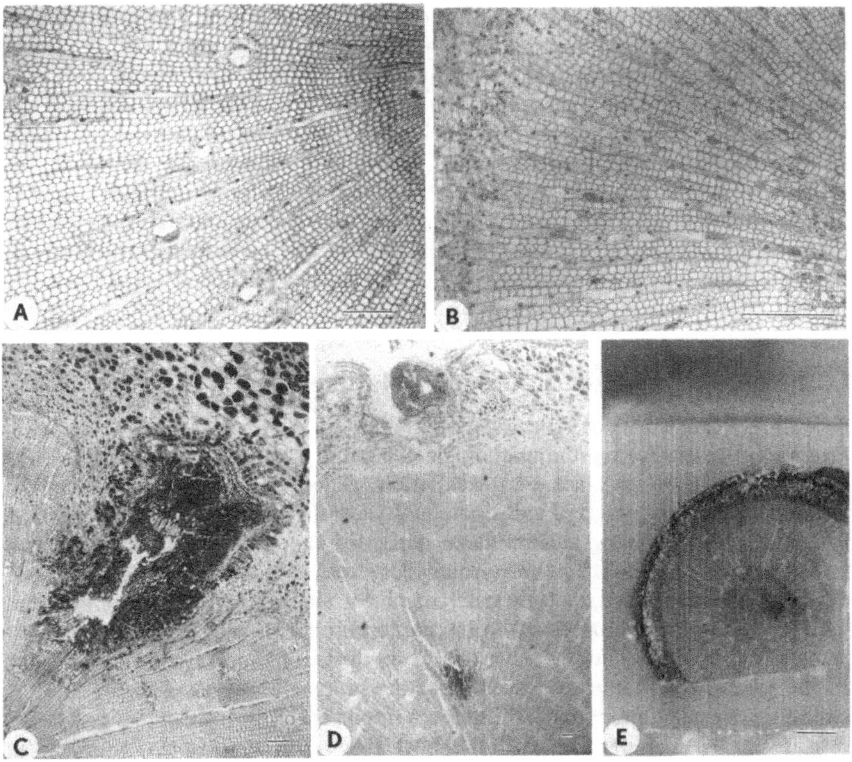

Fig. 12.1A–E. Transverse views of slash pine stems. **A** Normal (rust-free) stem. **B** Stem exhibiting large ray cells and an increased number of xylem rays typical of fusiform rust-gall tissue **C** A resistance zone interrupting the cambial cylinder and extending into the host xylem with the tissue of the resistance zone surrounded by normal appearing host tissues. **D** Resistance zone buried deep in xylem and remnants in the outer cortex, normal host tissue separating the two abnormal areas, and reestablishment of the cambial cylinder. **E** Macro view of buried resistance zone. *Bar* = 300 μm in **A** to **D**, 1 cm in **E**

Jewell et al. (1962) also introduced a new concept concerning the growth of a rust pathogen in the host gall tissue. Apical growth of hyphae in an inward radial direction ceased as the hyphal tips crossed the host cambium. The hyphae of the pathogen were observed to be continuous and intact across the cambium for a considerable distance into the xylem. This continuity was theorized to result from radial growth in hyphal length at or in the host cambium at the same or similar rate as newly differentiated xylem cells were produced by the invaded cambium. Thus, the pathogen is perpetuated by growing and extending itself with differentiating host tissues rather than ramifying by apical hyphal growth through mature secondary tissues, as indicated for *Cronartium ribicola* in *Pinus strobus* (Colley 1918, Struckmeyer and Riker 1951). This represented a type of intercalary growth of the pathogen in relation to the host cambial tissues.

Similar hyphal growth of *Cronartium guercuum* f.sp. *echinata*, the Eastern gall-rust pathogen, in the cambium of several species of *Pinus* confirmed the validity of the intercalary growth theory for the pathogen (Jewell 1969). This concept was substantiated by Peterson and Shurtleff (1965) and Krebill (1968). This growth pattern of the fusiform rust pathogen assisted in the determination of anatomical resistance reactions in Southern pine hosts observed in subsequent research.

12.3 Anatomical Resistance in 1- and 2-Year-Old Pines

In 1963, Snow et al. reported the survival of fusiform rust infected (galled) slash and loblolly pines after planting. The experimental seedlings were 1 year of age, and exhibited typical fusiform rust galls when planted. Certain seedlings, both slash and loblolly, exhibited macroscopic indications of recovery from the initial rust infection 2.5 years after planting. Microscopic examination of discolored tissue, evident at the center of surviving stems, revealed pathogen presence and tissue characteristics typical of fusiform rust galls, as described previously (Jackson and Parker 1958, Jewell et al. 1962). The host tissue between the discolored area and the bark of the survivors was considered normal and the pathogen was absent from these tissues. It was concluded that the survivors had, in some way, recovered from the initial infection and establishment of the fusiform rust pathogen. This paper was the first to document possible resistant reactions in Southern pines to fusiform rust. Siggers (1949) reported similar survival after planting fusiform rust-infected slash pine seedlings, but microscopic examinations of host tissue to determine the presence of the rust pathogen in the survivors were not made.

In 1972, Jewell and Snow published the first paper attempting to determine if an anatomical expression of a resistance mechanism by particular pine hosts was responsible for their recovery from fusiform rust infection. This research dealt with slash pine seedlings inoculated with the fusiform rust pathogen, and the gall-rust pathogen, *Cronartium guercuum* f.sp. *echinata* (Snow and Kais 1972). Certain seedlings were classed as galled at 6 months following inoculation, but at 1 year of age the same seedlings were classed as recovered. Microscopic examination revealed areas of abnormal tissue in or adjoining pith areas with the residual stem area comprised of normal host pine tissue. This occurred regardless of the pathogen involved. The abnormal areas (reaction areas) were theorized to originate from rust-infected procambial cells which, in turn, produced the vertical (reaction) parenchyma making up the reaction areas. This substantiated the theory that if rust infection occurred before vascular cambium development, the procambium, other meristems, and derivatives would be invaded by the rust pathogen (Jewell et al. 1962). It was concluded that development of the reaction areas was an expression of an anatomical resistance mechanism by the host pines.

These results were substantiated by papers dealing rather intensively with a study of resisted infections by fusiform rust and Eastern gall rust in slash pine seedlings (Jewell and Speirs 1976, Jewell et al. 1976). The seedlings examined were considered recovered, but exhibited certain macroscopic irregularities of the stem; crook, splitting of bark, discoloration, but no swelling of the stem cylinder. Internally, the seedlings had zones of abnormal living and nonliving host tissue, termed resistance zones (Fig. 12.1C). Reaction parenchyma (Jewell et al. 1962) separated the resistance zones from the normal host tissue. Hyphae of the pathogen were limited to the resistance zones where the pathogen was considered to be both functional, and nonfunctional depending on the location in the tissues of the zone. In 1-year-old seedlings, the resistance zones were more extensive radially toward the pith than in 2-year-old specimens. Also, the vascular cambium was interrupted by the resistance-zone tissue in 1-year-old seedlings, but in the 2-year-old ones the cambial cylinder had been reestablished. In the latter, normal host tissue consistently separated resistance zone tissue at or near the pith and at the cortex/phloem/cambium area (Fig. 12.1D,E). The pathogen involved appeared to be isolated in the resistance zones of the host regardless of age. This represented regression of the resistance zones from similar configuration in 1-year-old samples by the deposition of normal types of pine tissue in the stem cylinder. This normal appearing tissue originated from the reestablished rust-free vascular cambium. If the pathogen had been present in the cambium, abnormal or gall-type tissues would have been deposited on the xylem or phloem side of this meristem (Jewell et al. 1962), but this was not the case. It was concluded that seedlings with this tissue configuration would survive and, in time, produce a normal, mature pine.

Subsequently, Miller et al. (1976), although utilizing a different terminology, substantiated the previous research, describing the resistant reactions in slash pine and the relationship of the pathogen with the host cambium. Walkinshaw et al. (1984), working with 25 slash pine families, provided additional substantiation and descriptions of resistance zones in pine. In addition to the anatomical characteristics noted in the host pines, differences in growth of the pathogen within the cambium of resistant and susceptible pines were observed. Resistant specimens exhibited reduced numbers of haustoria, an observation reported previously by Speirs et al. (1976), and later by Walkinshaw and Jewell (1982).

12.4 Anatomical Resistance In Southern Pine Hybrid Progeny

In the early stages of Southern pine rust resistance research, shortleaf pine (*P. echinata* Mill.) was considered a significant source of resistance to fusiform rust, and initially this theory appeared to have significance (Henry and Bercaw 1956, Jewell 1959). However, inoculation studies indicated susceptibility of progeny from shortleaf × slash pine crosses (Jewell 1961a,b). Histopathological studies were conducted on these inoculated seedlings, which exhibited characteristic fusiform-shaped galls or a partial swelling of the stem. This latter feature reportedly is an indication of resistance (Jewell and Speirs 1976, Miller et al.

1976). Two distinct host reactions were described; typically susceptible, and pseudo-resistant (Jewell 1988). The typically susceptible hybrid specimens exhibited the characteristic anatomy of rust galls on susceptible slash or loblolly pine (Jackson and Parker 1958, Jewell et al. 1962). The pseudo-resistant (partial stem swelling) reactions varied from previous research on rust-infected pines. Abnormal and regressive tissue areas were observed, which were separated from the normal tissue by reaction parenchyma. The pathogen appeared, for the most part, to be isolated in the zone tissue. However, proliferation of certain parenchyma cells bordering the zones had occurred. The pathogen was observed commonly in these cells and appeared viable and functional. In this circumstance, the pathogen was considered to have escaped the resistance zone by invading these cells, which appeared to be meristematic and to act in place of the normal cambium tissue. It was concluded that the escapes would, with continued growth of the host, result in the development of typical gall anatomy around the entire stem circumference and the ultimate formation of a normal appearing fusiform rust gall. This potential had been earlier expressed for rust-resistant slash pine (Jewell 1978).

Jewell in 1988 published a histopathological study outlining three host/pathogen relationships observed among rust-inoculated progeny from back-crosses of (shortleaf × slash) × shortleaf pine parents. Two of these, susceptible and pseudo-resistant, had previously been described (Jewell 1984, Jewell and Walkinshaw 1985). The third type, described as resistant, exhibited host reactions very similar to those described in the original work of Snow et al. (1963) and reported by Jewell and Snow (1972), and later by Jewell and Speirs (1976). The specimens observed were 1 year from inoculation at 6 weeks of age. The resistant reactions were documented as abnormal cellular areas deep in the xylem cylinder and were associated with needle-trace bases at or near the pith. The needle-trace bases could be traced to corresponding foliar remnants at the stem surface. Normal or near-normal appearing host tissue separated the two tissue areas. Remains of the pathogen were present in the needle-base traces and the foliar remnants, indicating an initial but failed establishment of the pathogen following inoculation. Configuration of sclerenchyma in the outer cortex subtending the foliar remnants indicated potential sloughing off and loss of this tissue from the stem. Hutchinson (1935) recorded similar observations for affected tissue in Scots pine. Thus, the remaining indication of initial rust establishment became buried deep in the xylem cylinder by deposition of normal host tissue following the establishment of a rust-free vascular or cambium. No evidence of escape of the pathogen was observed in these resistant samples, and a normal growth process of such pines would be expected. Apparently, early in the host/pathogen association one of several possible processes occurred; the pathogen ceased to operate normally in stimulating the gall-developing process in the host, the host did not respond in a typically susceptible manner to the presence of the pathogen, a combination of these occurrences took place, or unknown factors caused the host/pathogen relationship to cease. Nevertheless, the resistance configuration was similar to that reported earlier for slash pine and, as such, is considered a viable process in pine survival from fusiform rust infection.

12.5 Anatomical Resistance in Juvenile Southern Pines

The results of the anatomical investigations of resistant reactions in 1 year and older slash and loblolly pines initiated similar research in juvenile pine hosts. These studies were to determine if macro- and microscopic evidence of resistance to fusiform rust was initiated in pines in their primary tissue stages.

Macroscopically, differences were found in resistant and susceptible slash pines 3 months following inoculation with the fusiform rust pathogen at 6 weeks of age (Jewell et al. 1980). Resistant pines had small, localized purple stem lesions, which originated at the base of primary needles. No stem swelling was macroscopically evident. Microscopically, resistance zones bordered by reaction parenchyma and pseudo-periderm were present at primary needle traces, which corresponded to the external purple stem lesions exhibited at primary needle bases. Dead or nonfunctional hyphae of the pathogen were observed among the resistance-zone cells. There was evidence of tissue regression in the resistance zones, indicating potential recovery of the host from effects of the pathogen. Susceptible pines had similar but larger stem lesions with swelling, and crooking of the stem occurred. Hyphae of the pathogen were well established in the susceptible host tissue. Similar studies with slash pine indicated earlier differences in macro- and microscopic expressions of rust resistance than had previously been reported (Speirs 1976, Speirs et al. 1976). Susceptible and resistant pines exhibited macroscopic symptoms of rust infection 15 and 13 days following inoculation, respectively. At this time, symptom expression occurred on the stems of resistant sources, while the primary needles of susceptible sources were involved, not the stems. Microscopic evidence of host reactions in both resistant and susceptible pines was discernible 8 to 9 days after inoculation. This was 5 to 8 days before the external stem symptoms appeared. By 15 days after inoculation, considerable differences in the resistant and susceptible host tissues were evident. Resistant samples exhibited localized cell abnormalities in cortex and pith with the pathogen appearing to be under stress, limited in extent, and often in stages of dissolution. Similar results with nearly complete hyphal degradation occurring 60 days following inoculation have been reported (Walkinshaw and Jewell 1982). In contrast, abundant hyphae and haustoria of the pathogen were present in the susceptible pine samples. Thirty-day infections revealed a normal gall-forming host/parasite relationship for slash pine and the fusiform rust pathogen. Similarly aged, resistant pines exhibited a localized area of degenerative tissue, in which the pathogen was isolated. A well-developed wound periderm surrounded these tissue zones. The origin of the periderm was considered to be the result of stimulation by the rust pathogen of ray parenchyma associated with resin ducts of the primary cortex. These parenchyma, apparently meristematic, were derived from base meristem cells (Speirs and Jewell 1976) of the primary plant, and possibly are involved in initiating the fusiform rust resistance process in juvenile slash pine.

Speirs (1976) proposed interesting conclusions concerning the development of fusiform rust resistance in juvenile slash pine. Resistance was associated with a rapid, localized reaction of the host to isolate the pathogen by metabolic and anatomical mechanisms in specific, degenerating tissue areas (resistance zones). Establishment of resistance may depend, in part, on the retention of the meristematic characteristics of particular parenchyma cells, such as resin duct parenchyma, until initiation of secondary growth in the host. An additional conclusion was that particular parenchyma cells, as well as the primary and secondary cambial tissues of susceptible and resistant pines, were functionally diverted to protecting the pathogen or to destroying or isolating the pathogen, respectively. Speirs (1976) also concluded that previous assumptions concerning the origin of resistance zones were probably correct (Jewell and Snow 1972, Jewell and Speirs 1976).

Pigmentation and number of stem lesions of juvenile slash and loblolly pine stems inoculated with fusiform rust differed between resistant and susceptible parent families (Lundquist and Luttrell 1982). Rates of pigmentation and lesion numbers were observed to increase with the resistant families, and differences between resistant and susceptible families were intensified by increases in the inoculum density of the pathogen. The assumption to be made here is that an increase in inoculum load will cause an increase in pigmentation and the number of stem lesions. This is a logical conclusion, but whether this is a measure of resistance is questionable, as many variables exist. Lundquist and Luttrell (1982) observed a periderm in both resistant and susceptible pine seedlings but did not pursue this tissue in relation to rust resistance, other than to report haustoria present where the pathogen apparently extended from an infected area into the uninvaded cortex beyond layers of phellum. These layers may be similar to reaction parenchyma and, if so, the escape of the fusiform rust pathogen from the confines of this tissue has been documented previously (Jewell 1978, 1988). Additional research on stem pigmentation and lesion development indicated that pigmentation was not the only resistance mechanism present in juvenile slash pine. Lundquist and Miller (1984) concluded that fusiform rust resistance in slash pine may be determined by the rate at which the pathogen spreads in the host, in contrast to the speed at which the host reacts to activate the inherent resistance mechanisms. Similar views were proposed by Speirs (1976).

12.6 Conclusions

The literature concerning resistance to fusiform rust in slash and loblolly pines, although diverse in nature and content, supports the initial premise of anatomical resistance mechanisms developed to isolate the rust pathogen in unique tissue areas in the pine host (Jewell and Snow 1972). This appears to be true of both juvenile pines with primary tissues and more mature hosts with fully developed secondary tissue systems. The research reported in the recent literature has been concerned with infections originating on pine seedlings that were, for the most part, inoculated with fusiform rust under controlled laboratory or other conditions at approximately 6 to 12 weeks of age. Seedlings at this age are in the primary stage of growth, and small amounts or no secondary tissues were accordingly present at the time of inoculation. Speirs (1976) has shown susceptible or resistant slash pine of this age to react rapidly to rust infection. For a short time, cellular abnormalities appeared in a similar manner in the infection/establishment process of the host and pathogen relationship. Due to some unknown "trigger" process, certain cells of particular pines established the beginnings of resistance zones, while susceptible host/pathogen relationships were developed in other pines. The process of resistance may result from a combination or complex of not only anatomical mechanisms but also biochemical and biophysical aspects of the host/pathogen association. The process may be so fragile that the division of a particular individual, affected host cell might determine resistance or susceptibility of the host pine, as indicated previously (Speirs 1976). The anatomy of the host reaction to the fusiform rust pathogen being quite similar, regardless of susceptibility or resistance, this occurrence may be an inherent feature of the defensive capabilities of the pine host. If so, this feature may be genetically controlled and capable of being manipulated either naturally or artificially.

Although types of host response or resistance reactions have been reported (Miller et al. 1976), these basically appear to be simply stages of resistance-zone formation. The one distinct characteristic of the resistance zones or host responses identified in anatomical fusiform rust resistance is that the pathogen is confined to the necrotic resistance-zone tissue and that, with continued normal growth of the host, recovery from infection by the pathogen can be expected. This has been shown with slash and other Southern pine hybrid progeny. This is not always true, as escapes from the zones have been documented (Jewell 1978, 1988, Lundquist and Luttrell 1982), but the opportunity for successful recovery is quite good among resistant families of slash and loblolly pine. The problem arises in the identification of pines with the potential for resistance from the initial symptoms expressed in progeny tests. This is an area of research requiring considerable attention.

References

Burdsall HH Jr, Snow GA 1977 Taxonomy of Cronartium quercuum and C. fusiforme. Mycologia 69:503–508

Colley RH 1918 Parasitism, morphology, and cytology of Cronartium ribicola. J Agr Res 15:619–660

Dodge BO, Adams JF 1918 Some observations on the development of Peridermium cerebrum. Mem Torr Bot Club 17:253–261

Hartig R 1874 Wichtige Krankheiten der Waldbäume. Springer, Berlin, 124 pp

Henry BW, Bercaw TE 1956 Shortleaf-loblolly hybrid pines free of fusiform rust after 5 years exposure. J For 54:779

Hutchinson WG 1935 Resistance of Pinus sylvestris to a gall-forming Peridermium. Phytopathology 25:819–843

Jackson LWR, Parker JN 1958 Anatomy of fusiform rust galls. Phytopathology 48:637–640

Jewell FF 1959 Disease resistance studies in tree improvement research. In: Proc 5th Southern forest tree improvement Conf, 11–12 June 1959, Raleigh, NC, 18–20

Jewell FF 1960 Inoculation of slash pine seedlings with Cronartium fusiforme. Phytopathology 50:48–51

Jewell FF 1961a Artificial testing of intra- and interspecies Southern pine hybrids for rust resistance. In: Proc 6th Southern forest tree improvement Conf, 7–8 June 1961, Gainesville, FL, 105–109

Jewell FF 1961b Infection of artificially inoculated shortleaf pine hybrids with fusiform rust. Plant Dis Rep 45:639–640

Jewell FF 1969 Evidence of hyphal growth of Cronartium quercuum in the pine-host cambium. Plant Dis Rep 53:793–794

Jewell FF 1978 Reactivation of the fusiform rust fungus in rust-resistant slash pine progeny. Proc Am Phytopathol Soc 5:168

Jewell FF 1984 Anatomical reaction to fusiform rust of interspecies progeny from crosses between shortleaf and slash pines. In: Proc Rusts of hard pines Working Party Conf S2.06.10 IUFRO; and USDA For Serv, Athens, GA, 331 pp

Jewell FF 1988 Histopathology of fusiform rust-inoculated progeny from (shortleaf × slash) × shortleaf pine crosses. Phytopathology 78:396–402

Jewell FF, Jewell DC, Walkinshaw CH 1980 Histopathology of the resistance-zones in juvenile slash pine to Cronartium quercuum f.sp. fusiforme. Phytopathol Mediterr 19:8–12

Jewell FF, Jewell DC, Walkinshaw CH 1982 Histopathology of anatomical mechanisms for resistance to fusiform rust in slash pine. Resistance to diseases and pests in forest trees. Pudoc, Wageningen, 503 pp

Jewell FF, Snow GA 1972 Anatomical resistance to gall-rust infection in slash pine. Plant Dis Rep 56:531–534

Jewell FF, Speirs DC 1976 Histopathology of one- and two-year old resisted infections by Cronartium fusiforme in slash pine. Phytopathology 66:741–748

Jewell FF, Speirs DC, Snow GA 1976 Resistant-reaction in slash pine to Cronartium fusiforme and C. quercuum. Proc Am Phytopathol Soc 3:214(Abstr)

Jewell FF, True RP, Mallett SL 1962 Histology of Cronartium fusiforme in slash pine seedlings. Phytopathology 52:850–858

Jewell FF, Walkinshaw CH 1985 Histopathology of fusiform rust-infected progeny of shortleaf × slash hybrids backcrossed to shortleaf pine. Phytopathology 75:1293(Abstr)

Krebill RG 1968 Histology of canker rusts in pines. Phytopathology 58:155–164

Lundquist JE, Luttrell ES 1982 Early symptomology of fusiform rust on pine seedlings. Phytopathology 72:54–57

Lundquist JE, Miller T 1984 Development of stem lesions on slash pine seedlings infected by Cronartium quercuum f.sp. fusiforme. Phytopathology 74:514–518

McKenzie MA 1942 Experimental autoecism and other biological studies of a gall-forming Peridermium on northern hard pines. Phytopathology 32:785–798

Miller T, Cowling EB, Powers HR Jr, Blalock TE 1976 Types of resistance and compatibility in slash pine seedlings infected by Cronartium fusiforme. Phytopathology 66:1229–1235

Peterson RS, Shurtleff RG Jr 1965 Mycelium of the limb rust fungi. Am J Bot 52:519–525

Phelps WR 1973 Fusiform rust incidence survey for 1971–1973. US Dep Agr For Serv SE Area State and Private For, 69 pp

Siggers PV 1949 Mortality in slash pine from planting seedlings with fusiform rust cankers. Miss Farm Res 12:1

Snow GS, Jewell FF, Eleuterius LN 1963 Apparent recovery of slash and loblolly pine seedlings from fusiform rust infection. Plant Dis Rep 47:318–319

Snow GA, Kais AG 1972 Technique for inoculating pine seedlings with Cronartium fusiforme. In: Biology of rust resistance in forest trees. Proc NATO-IUFRO Adv Stud Inst, US Dep Agr Misc Publ 1221:325–326

Speirs DC 1976 Anatomy of juvenile slash pine relating to resistance to Cronartium fusiforme. MS Thesis, Coll Life Sci, Louisiana Tech Univ, 54 pp

Speirs DC, Jewell FF 1976 Observations on the anatomy of the hypocotyl and epicotyl of slash pine seedlings. School For Louisiana Tech Univ Bull 9:29 pp

Speirs DC, Jewell FF, Walkinshaw CH 1976 Reaction to Cronartium fusiforme in resistant and susceptible juvenile slash pine. Phytopathology 66:213–214

Speirs DC, Walkinshaw CH, Jewell FF 1977 Rust resistant reaction in juvenile slash pine simulated by wounding. Proc Am Phytopathol Soc 4:141

Stewart A 1916 Notes on the anatomy of Peridermium galls. I. Am J Bot 3:12–22
Struckmeyer BE, Riker AJ 1951 Wound-periderm formation in white-pine trees resistant to blister rust. Phytopathology 41:276–281
True RP 1938 Gall development of Pinus sylvestris attacked by the Woodgate Peridermium and morphology of the parasite. Phytopathology 28:24–49
Walkinshaw CH 1978 Cell necrosis and fungus content in fusiform rust-infected loblolly, longleaf, and slash pine seedlings. Phytopathology 68:1705–1710
Walkinshaw CH, Ammon VD, Jewell FF 1984 Comparison of slash pine seedlings of varying fusiform rust resistance. In: Proc Rusts of hard pines Working Party Conf S2.06.10, IUFRO and US Dep Agr For Serv, Athens, GA, 331 pp
Walkinshaw CH, Jewell FF 1982 Atypical hyphae in resistant slash pines inoculated with Cronartium fusiforme. Phytopathology 72:930(Abstr)
Wolff R 1876 Aecidium pini Pers. und sein Zusammenhang mit Coleosporium senecionis Lev. Riga (Festschr Kaiserl Akad Wiss St. Petersburg zu ihrer hundertundfünfzigjährigen Gedenkfeier dargebracht von der Plenarconferenz der polytechnischen Schule zu Riga), 28 pp

Chapter 13

Anatomical and Physiological Aspects of Resistance to Dutch Elm Disease

G.B. Ouellette and D. Rioux

13.1 Introduction

Most elm species (*Ulmus* spp.), considered by many to be the most beautiful urban trees, are well adapted to the numerous stressful conditions in cities. Their major shortcoming, however, is their susceptibility to Dutch elm disease (DED). Although millions of elms have died since the first observation of DED about 70 years ago in Europe, they are not verging on extinction because they have a great capacity for regeneration, which explains why large elms affected by DED in woodland areas are gradually replaced by younger and smaller elm trees.

Despite the numerous studies conducted on DED, several aspects of this disease are still misunderstood. After presenting a review of elm anatomy pertaining to the ascent of sap and pathogenesis, this chapter describes our present knowledge of pathological and anatomical resistance mechanisms.

13.2 Wood Anatomy and the Ascent of Sap in Elm

Elm growth rings are usually composed of a band of earlywood containing one or two rows of large vessel elements, and of latewood comprising scattered small vessel elements and fibers. Apotracheal and paratracheal parenchyma cells are present within growth rings. Vessel length can vary from 1 cm at the branch apex to 5.5 m in the trunk and large branches (Liming 1934).

Xylem sap is transported almost equally in several growth rings of diffuse-porous species (e.g. *Acer* and *Betula*) but mainly in the last growth ring of ring-porous species (e.g. *Ulmus*, *Quercus*, *Fraxinus*). A study on *U. americana* demonstrated that 92% of the flow rate occurred in the outermost growth ring, and that the large vessel elements formed at the beginning of this growth ring were responsible for 96% of this conduction (Ellmore and Ewers 1985). Large vessels in older growth rings are rendered nonfunctional principally by embolism caused by winter freezing followed by thawing in the spring. Interested readers may refer to specialized works on this subject (Zimmermann 1983, Tyree and Sperry 1989).

In elm, the period of greatest susceptibility corresponds to the formation of these large vessels, a time when beetles also have great opportunity to reach and inoculate trees with propagules of *Ophiostoma ulmi*. The theory of the ascent of sap in ring-porous species also explains why they are more vulnerable to conditions that can break the continuity of the xylem sap column in large vessel elements (Zimmermann 1983).

13.3 Histopathology

As a prerequisite to describing resistance mechanisms in elms, the mode of development and action of the pathogen in incipient and later infection in trees with different degrees of susceptibility to DED must be established. The notion that DED pathogen development is limited to vessel elements until invaded xylem dies (as with other wilt diseases) has emerged from most reports of light microscope (LM) studies of host-pathogen relationships (Banfield 1968, Dimond 1970, Pomerleau 1970, Van Alfen and MacHardy 1978, Pegg 1985).

Factors that favored multiplication and spread of the pathogen in vessel elements were thus thought to be conducive to severe wilting (Tchernoff 1965, Elgersma 1969, 1973, 1982, Sinclair et al. 1975a,b, MacHardy 1978, Scheffer 1984, Bonsen et al. 1985, Pegg 1985, Beckman 1987). In this instance cytoplasmic alterations and death of parenchyma cells would result only as an indirect action of the pathogen; similarly, toxic compounds that the pathogen might produce would primarily contribute to pit chamber and vessel occlusion rather than affect directly living cells, and thus act as aggressins rather than as toxins (Van Alfen and Turner 1975, Russo et al. 1982, Scheffer 1984). Work at the ultrastructural level (MacDonald 1970, MacDonald and McNabb 1970, 1974, Krause and Wilson 1972, Ouellette 1978b,c,d, 1980, 1982, Scheffer 1984) and histochemical and immunocytochemical studies (Benhamou et al. 1985a,b, 1986, Ouellette and Benhamou 1987, Rioux and Ouellette 1989, 1991a,b) have added precision to earlier LM studies (e.g. Ouellette 1962a,b). Some of these more recent observations indicate that living cells of the xylem and even of the cambial area can be directly and primarily affected by the pathogen. Establishing the chronology of such events can, however, only be approximative and based on interpretation; any one section, for example, taken 10 days after inoculation may illustrate a range of variously aged alterations. To approximate conditions of natural infection, comparative observations must be made on samples taken relatively soon after inoculation and at various locations above the inoculation point to the tips of developing streaks. Pathogen development and action also differ with the type of tissue invaded and in recurring infections. The following descriptions take these factors into account.

Fig. 13.1. Light microscope observation (cross-section) showing pathogen development in xylem tissues. Recurrence of infection from one growth ring (*top part*) to the current growth ring (*bottom*). Fungal cells in large spring vessels and in the last formed xylem cells of the preceding year (*arrows*). *Bar* = 4 μm

Fig. 13.2a-e. Light microscope observations (**a–e** radial to tangential) showing pathogen development in xylem tissues. **a** Fungal hyphae and conidia (*arrows*) in vessel elements sampled

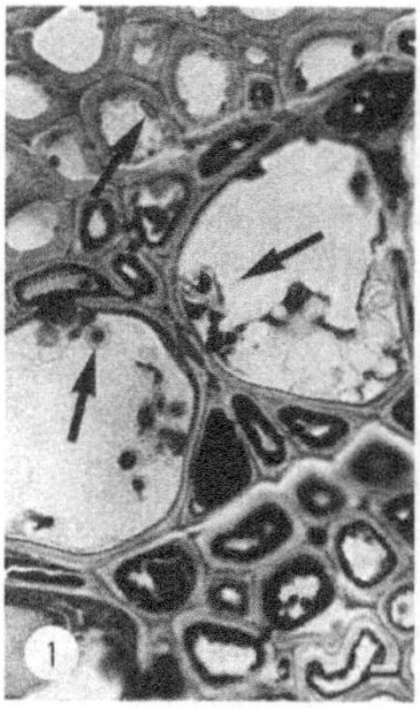

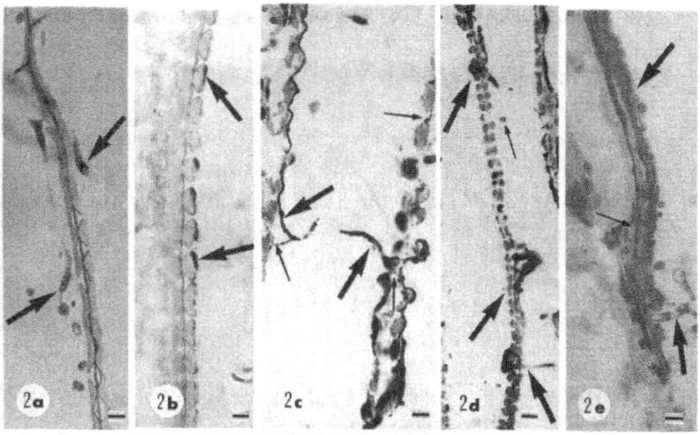

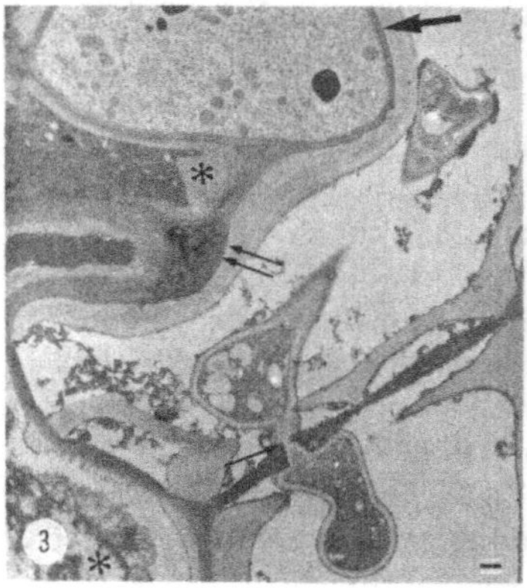

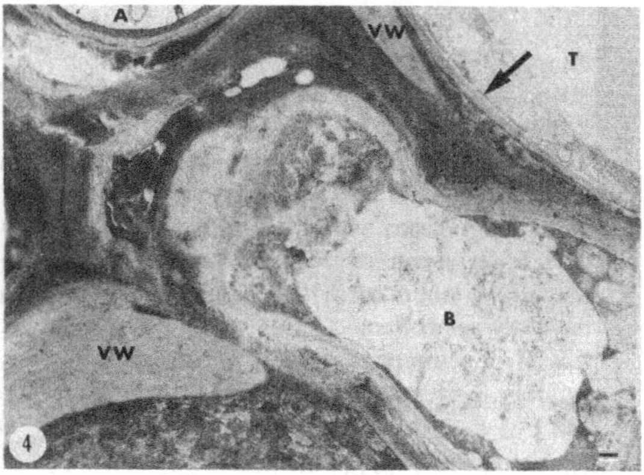

just above a developing streak in the xylem. The vessel lumina are free of other extraneous material. *Bar* = 4 μm. **b** Opaque material (coating) accumulating along vessel secondary wall thickenings (*arrows*) in young secondary xylem (occluding material is present in vessel element at the *left*). *Bar* = 4 μm. **c** Opaque material completely lining the walls of young metaxylem-like vessel elements. At times, this lining has the appearance of hyphae linked with spore-like structures (*arrows*). Tiny filaments bridge the coating between two adjacent vessel elements, extending often through the much altered vessel walls (*small arrows*). *Bar* = 4 μm. **d** Similar to **c**, but at a less advanced stage. Some opaque masses are linked with tiny filaments crossing the pits (*arrows*); free fungal cells (*small arrow*) in the vessel lumen. *Bar* = 3 μm. **e** Fungal cells associated with bands of opaque material lining a vessel wall (*arrows*). Similar bands are present in the wall itself (*small arrow*). *Bar* = 4 μm

Fig. 13.3. Fungal cells showing an outer opaque layer and inner more translucent one, in a distorted, recently differentiated vessel bordering the cambial area. Passage of hypha (*small arrow*) through the pit membrane of a bordered pit, without causing a zone of lysis or disruption. Host cells (*lower left*) are altered, some suspected to be vessel elements (*) having deposited thin portions of secondary walls. Other recently divided cells (*top part*) are hypertrophied and contain dense cytoplasm and opaque bodies. Their inner wall is also more opaque than normal (*arrow*). Opaque material in the thickened primary wall (*double arrow*). *Bar* = 3.5 μm

Fig. 13.4. Opaque material in altered walls of parenchyma cells situated between metaxylem vessel elements. Both parenchyma cells (*A*, *B*) have formed an irregular cell wall; cell *A* contains near-normal cytoplasm, while cell *B* contains altered cytoplasm. Tyloses were present in adjacent vessel elements; masses of opaque material and strands of fibrillar material are present between the tylosis (*arrow*) and vessel walls. *Bar* = 1 μm. *T* Tylosis; *VW* vessel wall

13.3.1 Pathogen Spread in Vessel Elements

Observations that *O. ulmi*, once introduced into a functional vessel element, moves rapidly upward with the sap stream to developing twigs (Banfield 1941, Ouellette 1960, Pomerleau 1970 and citations by these authors) have been generally accepted by researchers; some stress, however, that such knowledge is based on artificial inoculations (McNabb et al. 1970). Explanations for the downward spread of the pathogen are still mostly theoretical (Banfield 1968, Brener and Beckman 1968). If factors were present to hinder this downward movement, they would be at least as important as those related to its upward, tangential, or radial spread, particularly with regard to beetle inoculation in crotches of small branches. That only a few of the vessel elements are scored by feeding beetles, and that from such wounds the fungus can rapidly spread down trees of all sizes, indicate important aspects of tree susceptibility. Medium-sized American elm trees can be infected from one successful feeding wound by artificially contaminated *Scolytus multistriatus* adults (Ouellette 1960, 1962b). Although feeding wounds by *Hylurgopinus rufipes* have been recognized to be more effective (Banfield 1968), after the first year of infection, tree mortality is equally rapid following both types of beetle inoculation.

In both types of inoculation, damaged cells may favor multiplication of propagules (Ouellette 1962a, Banfield 1968, Pomerleau 1970) and thus produce a more pronounced mass inoculum effect. For the histopathological studies, we have always collected study material at a distance from such wounds or at the tips of developing streaks so as to obtain as real an insight as possible into the primary mode of action of the pathogen in invaded xylem tissues. Samples taken from trees with recurrent infections were also examined (Fig. 13.1).

Light microscope observations have often resulted in authors stating that fungal growth is very sparse in vessel lumina (for a review, see Banfield 1968). Although Ouellette (1962a) and Pomerleau (1970) reported prolific fungal growth in vessel elements, the fungus likely cannot grow across a functional water-saturated vessel; growth across or in the vessel lumen (Fig. 13.2a,c–e)

Fig. 13.5. Electron-opaque material (*arrows*) associated with the alteration of the primary wall of a parenchyma cell near a metaxylem vessel (*V*). A normal but irregular wall layer (*small arrow*) delimits the cell containing remnants of cytoplasm. *Bar* = 0.4 μm

Fig. 13.6. Bands of opaque material (*arrows*) traversing the membrane and border of a pit; one band is continuous with a small fungal cell which was probably continuous with the microhypha (*MH*) that extended for a long distance into the vessel lumen. Other small fungal cells are present in the adjoining vessel. *Bar* = 0.5 μm

Fig. 13.7. A cavity in a paratracheal parenchyma cell wall (*arrow*) in continuity with a fungal cell (*F*) containing altered cytoplasm; this cell is delimited by several irregular, distended, and broken wall layers, the innermost layer surrounds likely an endocell. *Bar* = 0.3 μm

Fig. 13.8. A very tiny hypha (*arrow*) appressed to a vessel wall. Another larger fungal cell with an outer opaque layer and inner more translucent one can be seen (*arrowhead*). An additional secondary wall-like layer has been deposited in the adjacent parenchyma cell. The pit membrane is intact and covered by a moderately opaque coating; both contain strands or masses of more opaque material. *Bar* = 0.4 μm. *ASL* Additional secondary layer; *PC* parenchyma cell

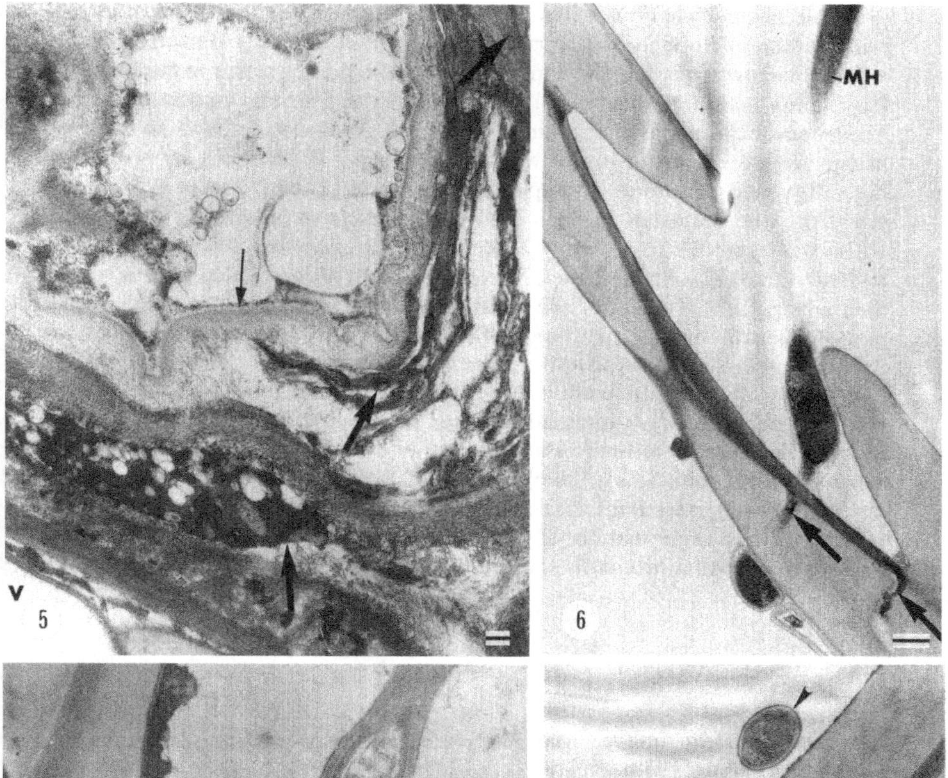
V
5
MH
6
F
7
ASL
PC
8

would occur only when the column tension is released from cessation of evapotranspiration or through disease action (Beckman et al. 1962). Theoretically, in vessel lumina with fast sap movement, rapidity should be greater in the center of the column than along the vessel wall interface; as a result, conidia and other extraneous material in the column would be gradually directed to or even become appressed to the vessel wall interface (Fig. 13.2b,e). Every portion of the wall would constitute a trapping site, with pit chambers as exits, favoring growth of the pathogen through pit membranes from one vessel element to the next (Figs. 13.2d, 13.3) (Ouellette 1962a, Banfield 1968, MacDonald and McNabb 1970, Pomerleau 1970, Krause and Wilson 1978, Elgersma and Steerenberg 1978, Ouellette 1978b, Scheffer and Elgersma 1982). Whether the pathogen readily invades paratracheal parenchyma cells or other cell types has been debated. In ultrastructural studies by Ouellette (1982), fungal cells were observed relatively infrequently in paratracheal parenchyma cells. However, host tissue damage was observed to occur in association with masses of cytoplasmic-like material in intercellular areas, or even between wall layers of the same cell (Figs. 13.4, 13.5) (Ouellette 1978d, 1982). This material was strongly labeled following injection of tritiated thymidine, whereas host nuclei were not labeled in these same mature tissues, although occasionally a few grains were present over host mitochondria (Ouellette 1978d).

13.3.2 Pathogen Morphology

Conditions favoring dimorphism and coremia formation have received much attention (Hubbes 1975, Kulkarni and Nickerson 1981, Dalpé 1983); the levuriform state has been considered a key factor in the invasion of vessel elements (Banfield 1968), whereas the coremial state is important for pathogen dissemination. Typical conidia are also produced in early infections in apparently nonoccluded vessel elements (Figs. 13.2a,e) (Ouellette 1962a,b, Pomerleau 1970). The reported formation of microendospores (Ouellette and Gagnon 1960) has been questioned (Pomerleau 1970, Sansome and Brasier 1973), but structures similar to tiny spores are nevertheless present in invaded elm and nonhost tissues (Fig. 13.6) (Ouellette 1978b,c, Rioux and Ouellette 1989). This requires further study. The production by *O. ulmi* of microhyphae (Ouellette 1962b, Casagrande and Ouellette 1971, Ouellette and Benhamou 1987), similar to those reported for other fungi (Liese and Schmid 1964, Peek et al. 1972, Leightley and Eaton 1977, Hale and Eaton 1985a,b), may be of great importance. Microhyphae are often less than 0.2 μm in diameter and can extend appreciable distances through or between cell wall layers, alongside vessel walls, or through vessel lumina (Figs. 13.6, 13.9). Numerous tiny bands or channels are often present in diseased elms linked with larger fungal cells (Figs. 13.10a–c), thus corresponding to microhyphae that extend across host walls, particularly in fibers. Transmission electron microscope (TEM) observations are necessary, however, to show that such channels are attributable to passage of microhyphae (Fig. 13.7 and unpublished observations). TEM obser-

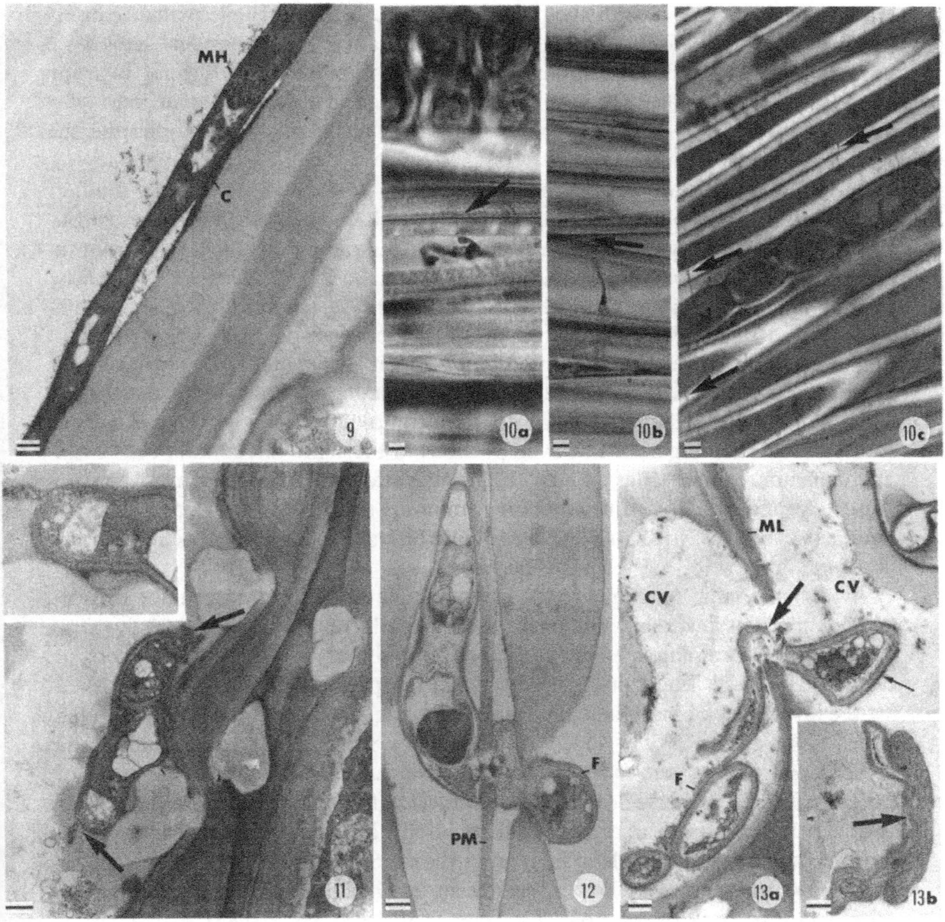

Fig. 13.9. A microhypha (*MH*) with very thin wall layers (particularly the inner, translucent one), extending without septa for a long distance, appressed to a vessel wall; a band of opaque material (corresponding to a thin coating, *C*) situated at times between both. *Bar* = 0.4 μm

Fig. 13.10 a–c Microhyphae crossing fiber walls, (*arrows*) some in continuation with larger fungal cells (**a** and **b**) **c** Examined under polarized light. Longitudinal sections observed with the light microscope. *Bars* = 4.4 μm

Fig. 13.11. Alteration and impregnation of fungal walls with opaque material (*arrows* and *inset*). Thickening of vessel wall and wall apposition in the parenchyma cell (*lower right*) are noticeable. *Bar* = 0.8 μm

Fig. 13.12. Penetration of fungal cell (*F*) via a very thin band of opaque material of a bordered pit membrane (*PM*) (the membrane on each side is translucent and slightly collapsed). Numerous vesicles and lamellar structures in fungal cell next to the point of penetration. *Bar* = 0.9 μm

Fig. 13.13a,b. Large cavities (*CV* in secondary wall layer) in fiber cell walls associated with microhyphae (*arrow*). **a** Discontinuities in fungal walls (*F*, fungal cell) through the middle lamella (*ML*, *arrow*) or elsewhere in portions of its outer opaque layer (*small arrow*). *Bar* = 0.4 μm. **b** Contorted fungal wall material, some of it apparently enclosing a portion of the host wall (*arrow*). *Bar* = 0.3 μm vations have also disclosed the regular presence of such small hyphae closely appressed to vessel walls (Figs. 13.8, 13.9), a feature that cannot be seen with certainty with LM. Such microhyphae could also permit an efficient invasion of vessel elements; their barely detectable presence may explain in part why the pathogen can be isolated from throughout the invaded xylem, despite the reported scarcity of fungal cells in vessels (Banfield 1968, Elgersma 1969).

Most fungal cells in cell lumina or traversing host cell walls usually contain large vacuoles and abundant membrane networks, some as lomasomes (Figs. 13.11, 13.12). Disorganization or distention of the walls of fungal cells present in invaded vessel lumina often occurs, even when these cells contain apparently normal cytoplasm, liberating therein some of their constituents (Fig. 13.11) (Ouellette 1978b). Possibly related to such alterations is the completely disorganized state of the protoplasm observed in many cells (Figs. 13.7, 13.13). Such cells usually contain masses of electron-opaque material, some associated with membranes, which is similar to that completely filling other cells (Fig. 13.6). As nearby fungal cells or even portions of the same cell contain cytoplasm that appears normal (Figs. 13.8, 13.12), the effect of fixation cannot explain these alterations. Also, the common formation of endocells or extra walls in such cells (Fig. 13.7) is considered evidence that such alterations occurred prior to sampling.

Walls in young pathogen cells normally consist of an external electron-opaque fibrillar layer and an inner, more translucent layer (Figs. 13.3, 13.8, 13.9, 13.13) (Benhamou et al. 1987), or sometimes two alternating layers of each (Fig. 13.14). The translucent layer is often partially or completely masked by the presence of perpendicular, electron-opaque strands extending through both layers (Figs. 13.11, 13.15) (Ouellette 1978b). Ouellette considered the presence of such strands and opaque material outside the cell to be mostly of fungal origin, as observed in other systems (Leightley and Eaton 1980, Foisner et al. 1985, Hale and Eaton 1985b, Heath 1990); it likely accounts for the dense outside staining of such cells in light microscopy. Such extracellular material was considered to be gummy host constituents in DED (Pomerleau 1970) and in other wilt diseases (for example, see Pegg et al. 1976, Pegg 1985).

Even though TEM revealed recognizable microhyphae in host walls close to invaded vessels soon after inoculation, it is doubtful whether at this stage they contribute much to pathogen spread except in vessels. Microhyphae are often associated with pronounced cell wall breakdown of fibers (Figs. 13.10, 13.13), predominantly in the nondiscolored latewood of the preceding year. Small chips from these nondiscolored zones consistently yield pure cultures of the pathogen (Ouellette, unpublished observations). This indicates the potential of microhyphae in spreading the fungus from one growth ring to the next, possibly in early spring or late fall, through gaps in the barrier zones (Sect. 13.6).

For this reason, the margins of tissue invasion as established by the extent of tissue discoloration (Banfield 1968) may not always be reliable. Spotty discoloration in invaded xylem may nevertheless be an expression of restricted xylem invasion (Tchernoff 1965, Sinclair et al. 1975a, Svihra 1980).

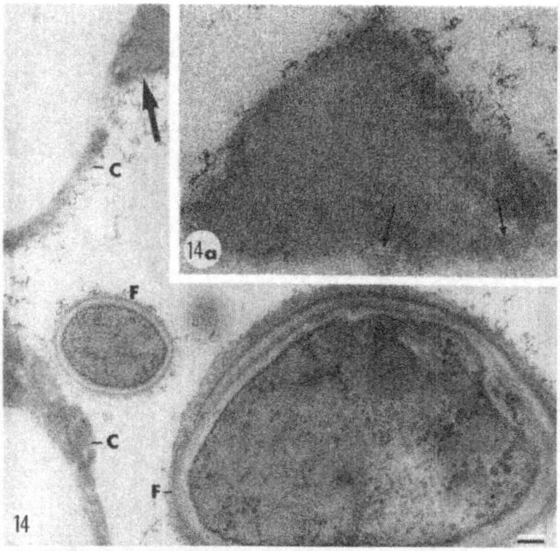

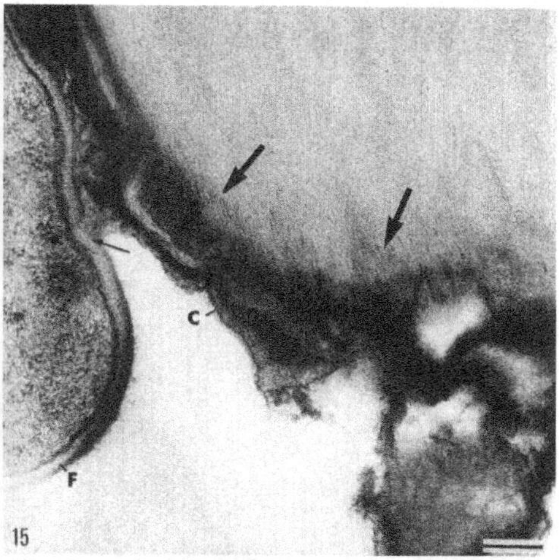

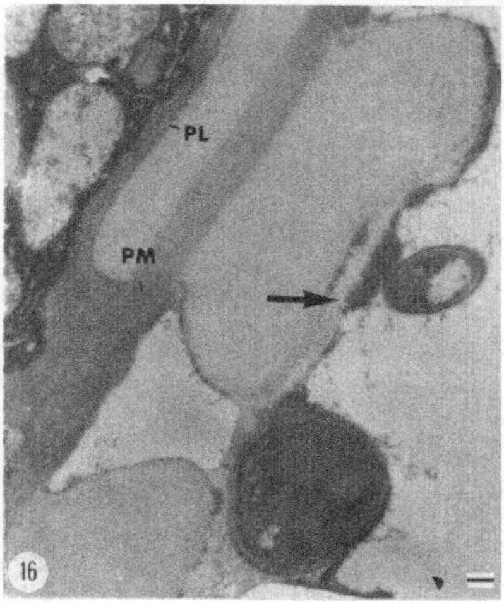

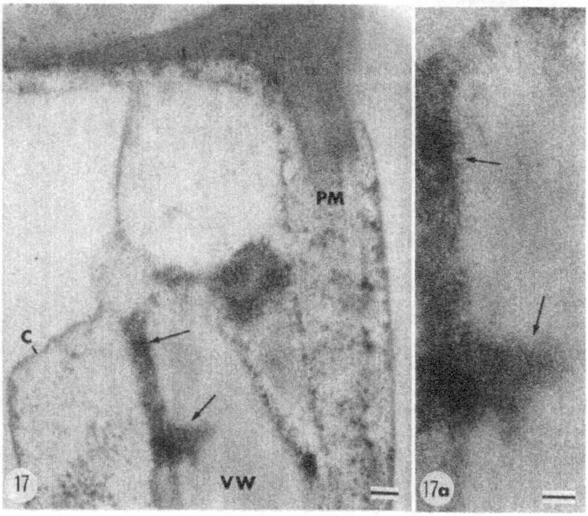

Fig. 13.14. Pathogen cells and coating material in a vessel shortly after inoculation. The coating (*C*) material is composed of bodies of opaque material and arrays of fibrillar strands, some of which extend directly into the host cell wall (*arrow*) (*small arrows* in **a**, enlarged portion; *bar* = 0.1 μm). Large fungal cell (*F*) surrounded by a sheath of fibrillar material and delimited at times by a four-layered cell wall, with alternating translucent and more opaque layers. *Bar* = 0.2 μm

Fig. 13.15. Next to a pit border, an opaque coating (*C*) from which numerous opaque strands (*arrows*) radiate into the host wall. The fungal cell wall (*F*) nearby is impregnated with similar strands (*small arrow*). More translucent bands, meandering through or surrounding the opaque material lining the vessel wall, also have the appearance of the fungal wall. *Bar* = 0.1 μm

Fig. 13.16. A similar association as in Fig. 13.15 showing the beginning of vessel wall alteration (*arrow*). Pit membranes (*PM*) are intact but apparently impregnated with opaque material. Adjoining parenchyma cells contain much electron-opaque cytoplasm, with a dilated reticulum and altered mitochondria. The protective layer (*PL*) is thickened and more opaque than normal. *Bar* = 0.3 μm

Fig. 13.17. Coating material (*C*) in a metaxylem vessel delimited toward the vessel lumen by an opaque border, and containing fibrillar and opaque bodies. The more opaque layer next to the wall (*VW*) may be composed of an aggregation of these bodies together with the fibrillar matter. The opaque material is also present in the wall as arrays of fibrillar strands (*small arrows* in **a**, enlarged portion; *bar* = 0.1 μm); the pit membrane (*PM*) is permeated with similar material. *Bar* = 0.2 μm

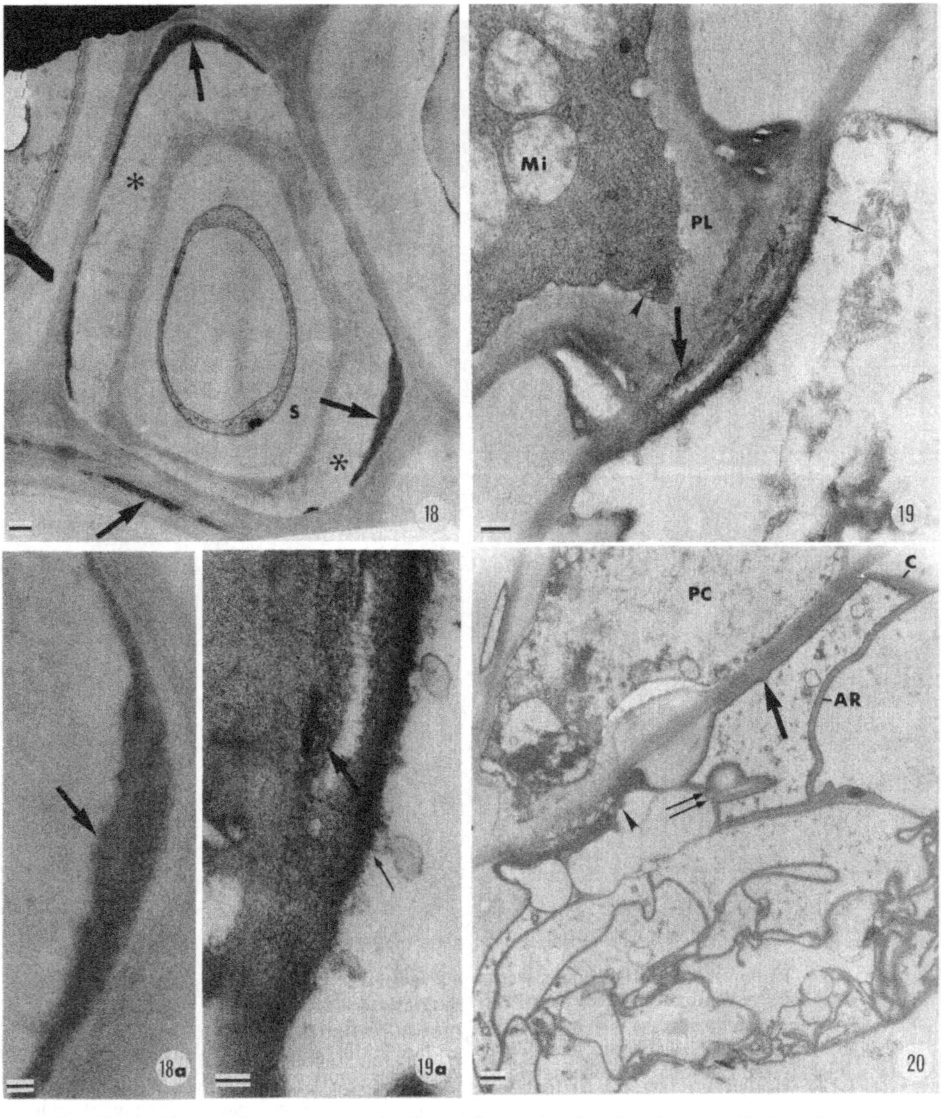

Fig. 13.18. Masses of opaque material (*arrows*) associated with primary wall alteration of a xylem cell of *U. pumila*; similar material is present also in the middle lamella of this and other surrounding cells. This material comprises small opaque particles and fine, fibrillar more translucent material (**a**, enlarged portion; *bar* = 0.1 μm). A pronounced apparently void space (*) is present between the altered wall and the opaque material. A normal secondary wall (*S*) also delimits the normal cytoplasm (cf. Figs. 13.43, 13.53–13.55). *Bar* = 0.3 μm

Fig. 13.19. Altered pit membrane of a half-bordered pit associated with opaque matter (*arrow*) (**a**, enlarged portion; *bar* = 0.1 μm). The membrane is also perpendicularly permeated by arrays of fibrillar strands (*small arrows*). Altered plasmalemma (*arrowhead*) of the parenchyma cell is shown near the pit membrane; the protective layer (*PL*) is thickened in this area. *Mi* mitochondria; *bar* = 0.2 μm.

Fig. 13.20. Alveolar reticulum (*AR*) continuous with coating material (*C*). This material is affixed to the almost intact pit membrane (*arrow*) but recurved toward the vessel lumen over the other pit (*double arrow*) showing an altered membrane (*arrowhead*). *PC* Parenchyma cell; *bar* = 0.4 μm

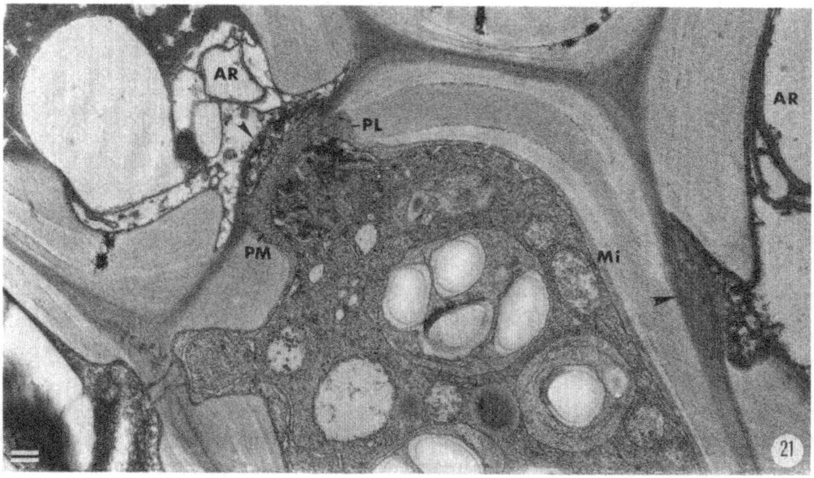

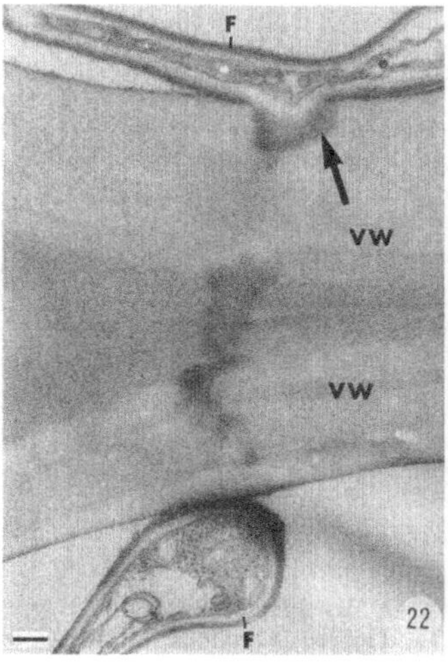

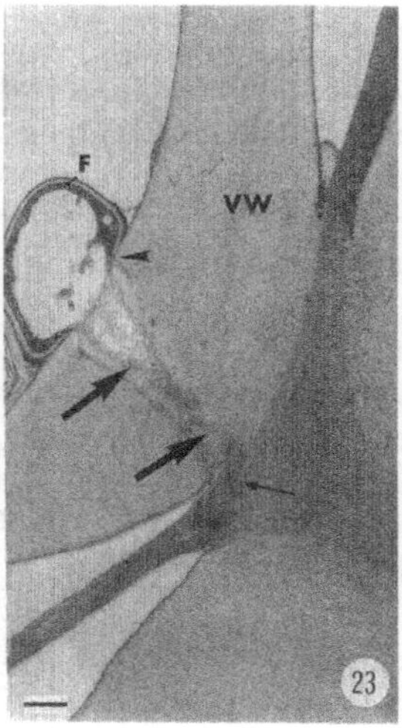

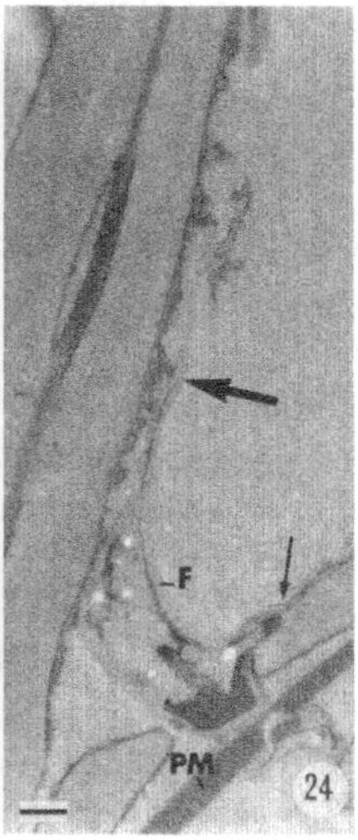

Fig. 13.21. Masses and bands of opaque material located in the cytoplasm of a parenchyma cell, just next to a pit membrane (*PM*); this membrane and underlying protective layer (*PL*) are permeated by bands of opaque material. The membranes (*arrowheads*) are altered as in Fig. 13.19. Note mitochondria (*Mi*) with altered cristae and alveolar networks (*AR*) in adjoining vessel elements. *Bar* = 0.4 μm

Fig. 13.22. A wavy band of opaque material continuous with the outside wall layer of a fungal cell (*arrow*) bridging two fungal cells through the walls (*VW*) of two vessel elements. *Bar* = 0.2 μm

Fig. 13.23. A median view of a penetrating small hypha continuous with a fungal cell (*F*) in the vessel lumen. Host wall material (*arrows*) seems to be situated between the penetration channel and the outer opaque bands, one of which is continuous with the outer wall of the fungus cell (*arrowhead*). The opaque bands also appear to traverse the pit membrane (*small arrow*). *VW* Vessel wall; *bar* = 0.4 μm

Fig. 13.24. A medium-sized fungal cell (*F*) continuous with tiny hyphae, extending at one end along the vessel wall (*arrow*), sideways as a small peg infringing on the wall of a pit border (*small arrow*), and at the other end through the pit membrane (*PM*) of that bordered pit. *Bar* = 0.3 μm

13.3.3 Host Wall Alterations

Until recently, alteration of host vessel and parenchyma cell walls by *O. ulmi* has received limited attention, despite the isolation of wall-degrading enzymes from the pathogen (Beckman 1956, Husain and Dimond 1958). As for other wilt pathogens (Pegg 1985), these enzymes are produced in greater quantities in the presence of host residues or certain saccharides than in regular cultures (Biehn and Dimond 1971, Bilodeau 1973, Colombani 1988). As yet, there is no consensus on the role of such enzymes as a direct cause of wilting (Pegg 1985).

Unable to isolate appreciable quantities of pectic or cellulolytic enzymes from infected material, Elgersma (1969) concluded that such enzymes are not as important in DED as in other vascular diseases. Yet, in agreement with Ouellette (1962a,b) and Gagnon (1967a), Scheffer and Elgersma (1982) observed the breakdown of vessel walls, not necessarily in immediate contact with cells of *O. ulmi*; breakdown was more pronounced with aggressive than with less aggressive isolates.

Alteration of host cell walls, traceable to lysis around normal hyphae in living tissues (Ouellette 1978b) and similar to growth in sterilized wood (Benhamou et al. 1986), is exceptional. Wall alterations in early infections are nevertheless quite noticeable, linked with bands or masses of osmiophilic matter (Ouellette 1978c,d, 1982). These alterations are particularly visible in primary walls (Figs. 13.4–13.5), in secondary walls of vessel elements and adjacent cells (Figs. 13.6, 13.14a, 13.18), and in pit membranes of half-bordered pits (Figs. 13.19–13.21). Significant chipping of secondary walls of vessel elements (Ouellette 1978d), their disruption and that of adjacent cell walls (Ouellette 1978a), associated with bands or strands of the opaque matter (Fig. 13.2c), constitute striking examples of alteration. Obviously, such disruptions account for the wall pieces that Gagnon (1967a) commonly observed in vessel lumina of invaded elm xylem. Unmodified gelatinous layers of fibers are also traversed easily and digested extensively by the pathogen (Fig. 13.13a,b).

Although the opaque material may be observed to be loosely linked with fungal cells (Figs. 13.22, 13.23) (Ouellette 1978d), it can often extend for long distances without an apparent association with them (Figs. 13.18, 13.34). Discernible elements of this material are small opaque particles in the range of 100 nm and stacks of small lamellar-like structures that can extend into the host wall in groups or individually (Figs. 13.14a, 13.15, 13.17a) (Ouellette 1978d). Similar structures also bridge fungal cell cytoplasm with extracellular opaque material (Sect. 13.3.2).

The margins of penetration channels of host walls by the pathogen are delimited by diffuse bands of opaque material that are continuous with the outside wall layer of the fungal cells (Figs. 13.22, 13.23) (Fig. 5 in McNabb 1971); such bands also contain membranous structures. A similar association was observed in other soft rot fungi (Leightley and Eaton 1980, Foisner et al. 1985, Hale and Eaton 1985a,b). These authors indicate that such membranes and opaque material are probably of fungal origin and constitute a transport system for wood-degrading enzymes, permitting wall penetration. Wall irregularities as described by them often occur around fungal cells in cavities in elm

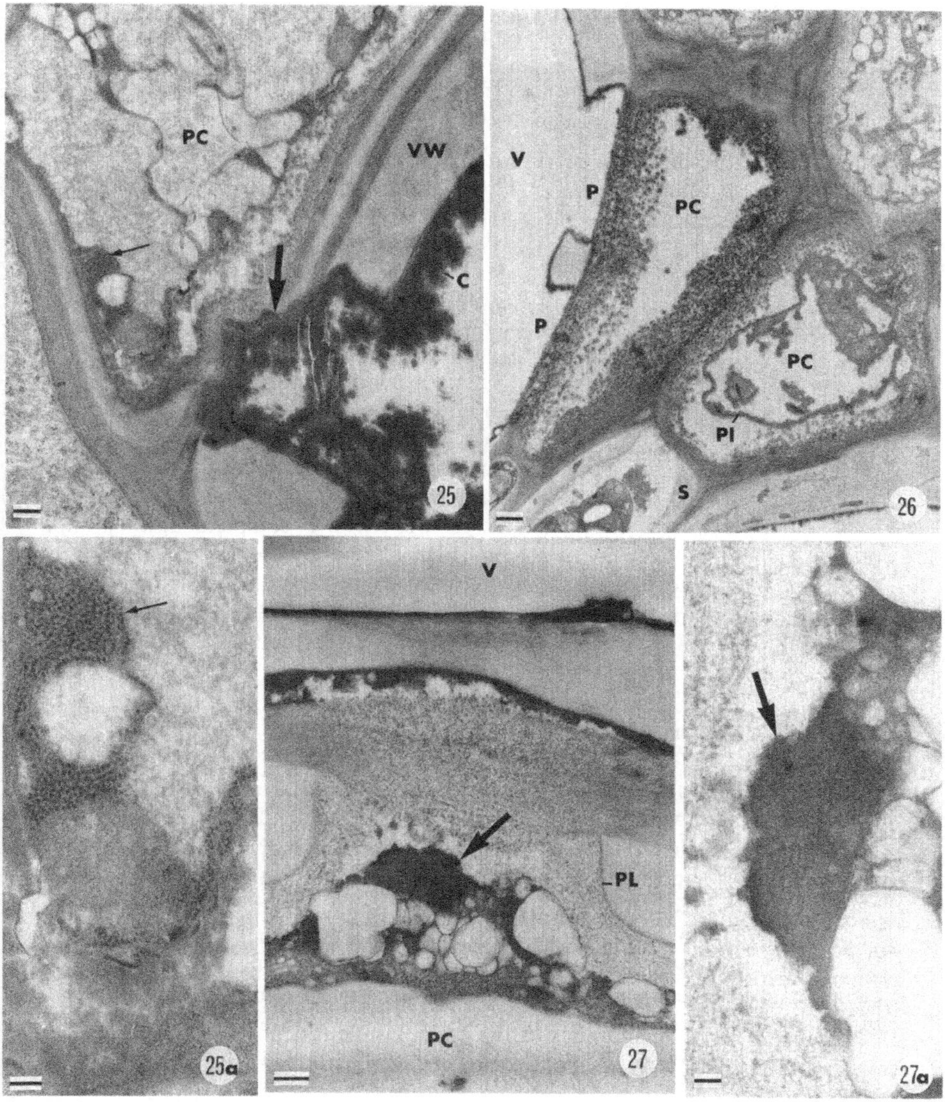

Fig. 13.25. Thick opaque coating (*C*) lining the walls (*VW*) of a secondary xylem vessel element and the membrane of the half-bordered pit; bulge of the membrane and protective layer (*arrow*) toward the inside of the parenchyma cell (*PC*). Masses of opaque small particles (*small arrows*; **a** enlarged portion; *bar* = 0.3 μm) around or throughout degraded portions of the cell cytoplasm. *Bar* = 0.9 μm)

Fig. 13.26. Altered parenchyma cells (*PC*) adjoining a metaxylem vessel (*V*) in an area that was covered by a layer of apparently normal secondary xylem; alterations were noted, however, in the cambial zone (as in Figs. 13.37–13.41). The cells underneath the vessel pits (*P*) have appreciably altered walls; other cells nearby have more normal primary walls, but lack or have only a thin secondary wall (*S*). Most of these cells contain altered cytoplasm but often with recognizable emplacement of the variously configured and altered plasmalemma (*Pl*). *Bar* = 0.2 μm

Fig. 13.27. Masses of opaque material located between the protective layer (*PL*) and the plasmalemma of a paratracheal parenchyma cell (*PC*); this material is composed of small opaque particles (*arrow*; **a** enlarged portion; *bar* = 0.1 μm) and is linked with similar material surrounding vacuolar areas of the cytoplasm. *Bar* = 0.3 μm

(Fig. 13.13a); they have been interpreted as indicating that typical wall layers around these cells have been laid down only after cavity formation.

Pit membranes of bordered pits of secondary xylem vessels are penetrated without noticeable digestion or rupture (MacDonald and McNabb 1970) even when these membranes (Figs. 13.3, 13.6, 13.12, 13.24) are impregnated or covered by electron-opaque protective substances (Robb et al. 1987). Pit membranes may also be penetrated by apparently unbound bands of opaque material (Figs. 13.6, 13.12). In metaxylem vessels, a more complete disintegration of pit membranes, between vessel elements, or between vessel elements and parenchyma cells, takes place (Ouellette 1978c).

Pit membranes of half-bordered pits in secondary xylem rarely disintegrate. Instead, they become distended to various degrees up to the point of disruption, toward the vessel lumen, in association with opaque and lamellar-like material (Figs. 13.19, 13.21). Inward displacement may also occur in association with this material, particularly toward the altered cytoplasm (Fig. 13.25) of parenchyma cells. Alteration of primary walls is of two types: a complete disintegration or impregnation by opaque material (Fig. 13.26), and an often pronounced distension similar to that of pit membranes (Figs. 13.4, 13.5). Similar opaque material also occurs in the middle lamella (Ouellette 1978d, 1982), at times in association with collapsed cells or other types of cell alterations (Sect. 13.3.4).

13.3.4 Effect on Parenchyma Cells

Various types of protoplasmic alterations occur in axial and ray parenchyma cells near invaded vessels soon after inoculation (Ouellette 1978a,b,c,d, 1982). The first noticeable effect is a localized retraction and/or alteration of the plasmalemma adjacent to the pit. Complete retraction and pronounced cytoplasm disorganization may rapidly follow in many cells, in greater proportions in younger than in older tissues (Figs. 13.26, 13.28) (Ouellette 1978c, 1982). Further alterations in other cells, including some at a distance from vessel elements, evolve less suddenly and less drastically. Examples of these changes (Ouellette 1978a,b,c, 1982) are: disruption of mitochondrial cristae (Figs. 13.19, 13.29); distension of membranes of the endoplasmic reticulum and the nuclear envelope and irregularly shaped nuclei containing large patches of heterochromatin (Figs. 13.16, 13.28, 13.29); stratified, dense, and noticeably more opaque cytoplasm (Fig. 13.30); considerable hypertrophy to pronounced degradation of plastids, dilation or increased opacity of their lamellae, and accumulation of globuli (Figs. 13.31, 13.32); a pronounced localized or general vacuolation of the cytoplasm (Fig. 13.4) which may become very electron-opaque (Figs. 13.16, 13.27, 13.30, 13.32); and numerous dictyosome profiles.

An astonishing peculiarity in affected parenchyma cells is the presence of zones of opaque material meandering through their less dense or relatively unaltered cytoplasm (Figs. 13.21, 13.33). This matter, containing opaque particles and arrays of lamellar structures, is at times continuous with other opaque material in the periplasm. Dense compact masses or bands of opaque material accumulate in periplasmic areas of a large number of cells, irrespective

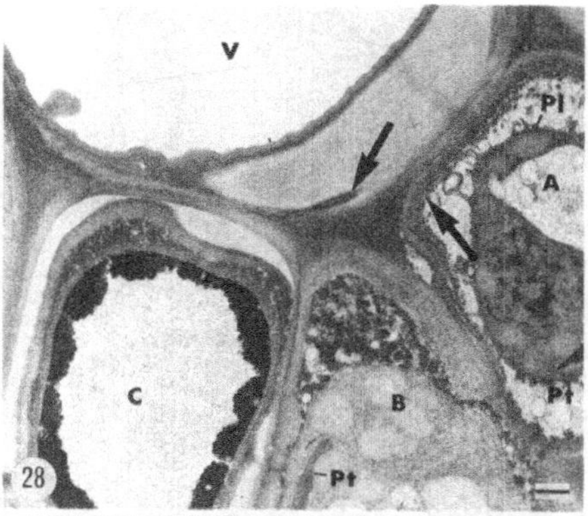

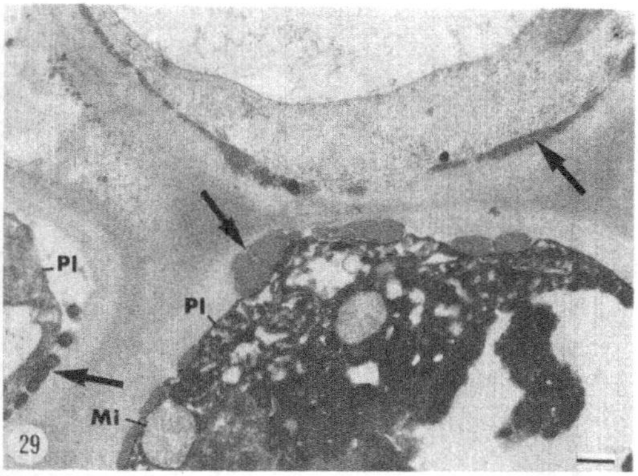

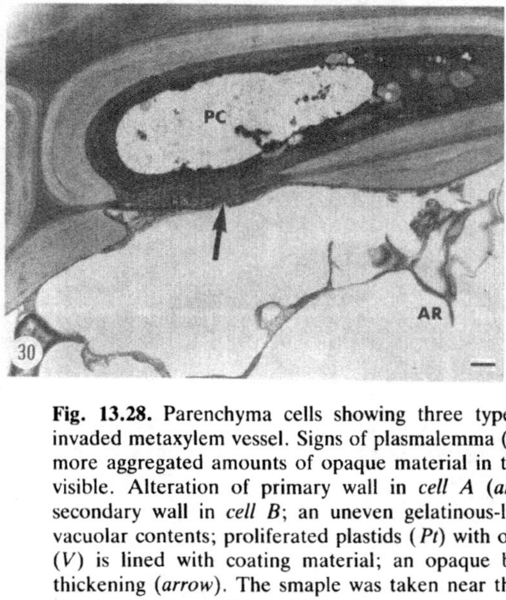

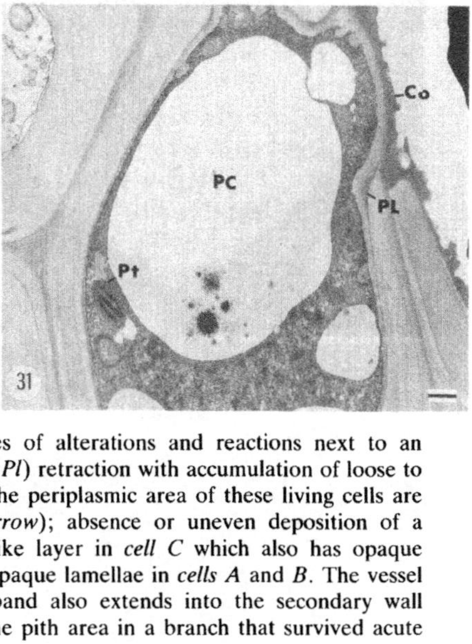

Fig. 13.28. Parenchyma cells showing three types of alterations and reactions next to an invaded metaxylem vessel. Signs of plasmalemma (*Pl*) retraction with accumulation of loose to more aggregated amounts of opaque material in the periplasmic area of these living cells are visible. Alteration of primary wall in *cell A* (*arrow*); absence or uneven deposition of a secondary wall in *cell B*; an uneven gelatinous-like layer in *cell C* which also has opaque vacuolar contents; proliferated plastids (*Pt*) with opaque lamellae in *cells A* and *B*. The vessel (*V*) is lined with coating material; an opaque band also extends into the secondary wall thickening (*arrow*). The smaple was taken near the pith area in a branch that survived acute disease symptoms. *Bar* = 0.6 μm

Fig. 13.29. Higher magnification of masses of opaque material (*arrows*) accumulating in periplasmic areas of cells showing either a fairly normal cytoplasm (*above*), opaque vacuolated cytoplasm and altered mitochondria (*Mi*) (*below*), or retracted, medium-dense cytoplasm (*left*). *Pl* Plasmalemma; *bar* = 0.3 μm

Fig. 13.30. Paratracheal parenchyma cell (*PC*) with opaque, dense cytoplasm in a twig that showed only mild symptoms. The pit membrane is altered on the vessel side (*arrow*) but impregnated by opaque material on the parenchyma cell side. Alveolar reticulum (*AR*) in vessel lumen. *Bar* = 0.6 μm

Fig. 13.31. A thickened, slightly impregnated protective layer (*PL*) is shown in this parenchyma cell (*PC*) with dense and vacuolated cytoplasm. The pit membrane on the vessel side is covered by a coating (*Co*) layer. Plastids (*Pt*) altered as in Fig. 13.28. *Bar* = 0.6 μm

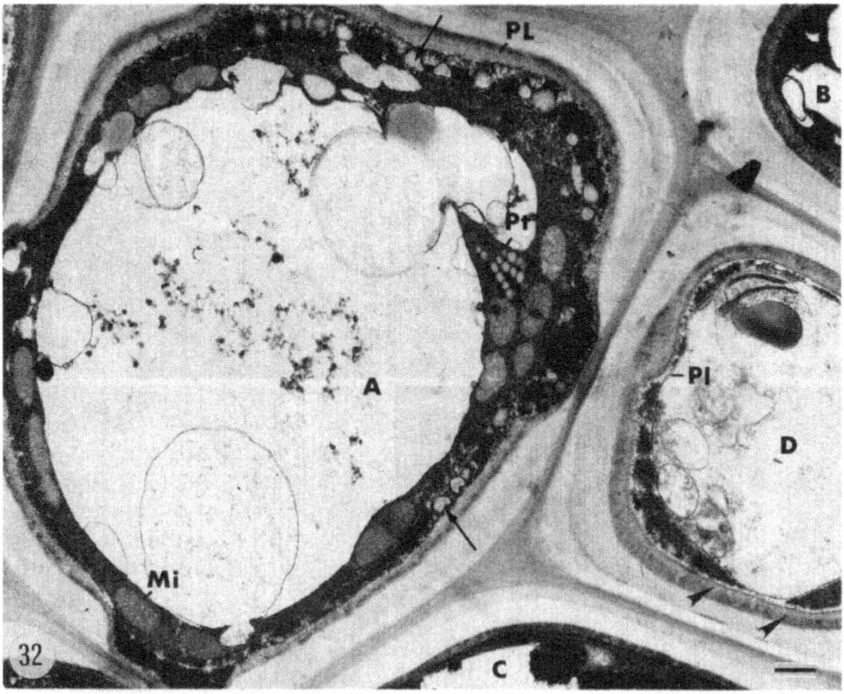
PL
B
Pr
Pl
A
D
Mi
C
32

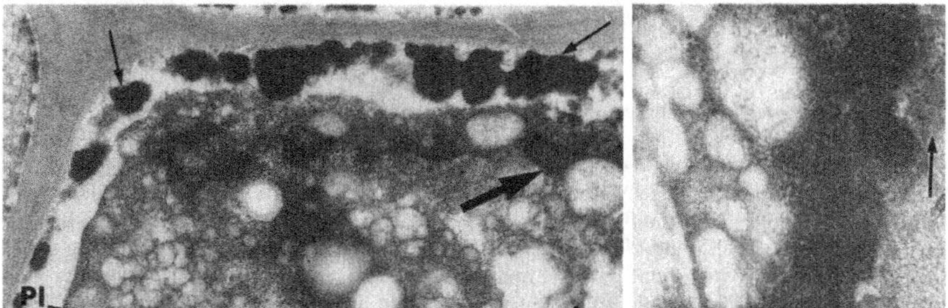
Pl
33
33a

of the condition of their cytoplasm (Figs. 13.25, 13.27, 13.29, 13.32). The high opacity of this material hampers perception of its detailed structures; it often discloses, however, arrays of lamellar structures and small opaque particles (Figs. 13.18a, 13.27a). Similar material is also present extending long distances through the middle lamella and secondary walls of bordering cells, and ending in the periplasm of such cells (Fig. 13.34); neither wall is ruptured (Fig. 13.34a), indicating that this material did not spread as a result of disruption of the cell walls.

13.3.5 Vessel Coating Material

Layers of coating material are deposited along vessel walls in infected elms, as in other plants affected by wilt diseases (MacDonald 1970, Pomerleau 1970, Czaninski et al. 1975, Ouellette 1978b,c, Hiratsuka and Takai 1982). Studies of verticillium-infected plants revealed that phenolic and lipidic compounds are components of this material (Mace et al. 1978, Robb et al. 1979, Pegg 1985, Newcombe and Robb 1989). Others (Gagnon 1967a, Beckman 1987) mention that it also comprises pectinic material originating from pit membrane breakdown and as secretions from parenchyma cells. The coating was also designated in elm as a gummy material, implying its origin from the host (MacDonald 1970, Pomerleau 1970, Czaninski et al. 1975). Ouellette (1978b,c) reports that much of the clogging material in vessel lumina can be traced to alterations or secretions from fungal cells; there is little indication that such material originates from pit membrane disintegration. Other workers (Van Alfen and Turner 1975, Scheffer 1984) also recognize that this material may contain compounds (vivoaggressins) from the pathogen.

The coating is progressively thicker and often more noticeably layered as infection progresses (Figs. 13.4, 13.20, 13.21, 13.25, 13.31) (Ouellette and Rioux, unpublished observations); the presence of lamellar structures and opaque particles (Figs. 13.14–13.17, 13.35), suggesting a cytoplasmic appearance, coincides with early vessel invasion. As mentioned in Section 13.3.3, this material is often associated with vessel wall breakdown and other types of

---

Fig. 13.32. Various reactions of parenchyma and fiber-like cells in an inoculated twig that showed mild symptoms. *Cell A* The cytoplasm is dense, opaque, vacuolated, with small vesicles close to the plasmalemma (*Pl*) or which seem to open on to the periplasm (*small arrows*) which contains masses of opaque material and similar vesicular bodies; protective layer (*PL*) is permeated by opaque material. *Cells B* and *C*: similar to *A*, but with a more opaque protective layer and or vacuolar contents. *Cell D* Presence of a gelatinous-like layer containing perpendicularly oriented arrays of opaque fibrillar material, some continuous with opaque material in the periplasm, and through an inner thin layer (*arrowheads*); remnants of plastids (*Pt*) and cytoplasm in the cell lumen. *Mi* Mitochondria; *bar* = 0.5 μm

Fig. 13.33. Masses of opaque material in the periplasm of several contiguous cells, in early infection of differentiating tissues. Bands of compact opaque material (*arrows*) are present in some of these cells, surrounded by highly vacuolated cytoplasm. Some of these bands were continuous with the opaque material (*small arrows*) in the periplasm. At higher magnification, tight arrays of tubular-like structures are discernible in this material (**a**, *small arrows*; *bar* = 0.1 μm). *Pl* Plasmalemma; *bar* = 0.2 μm

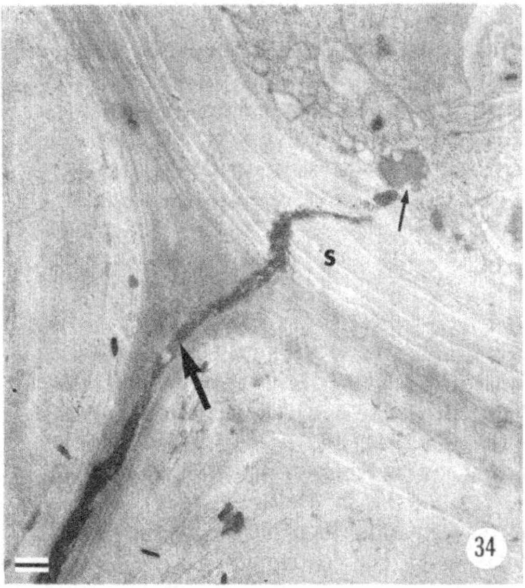
S
34

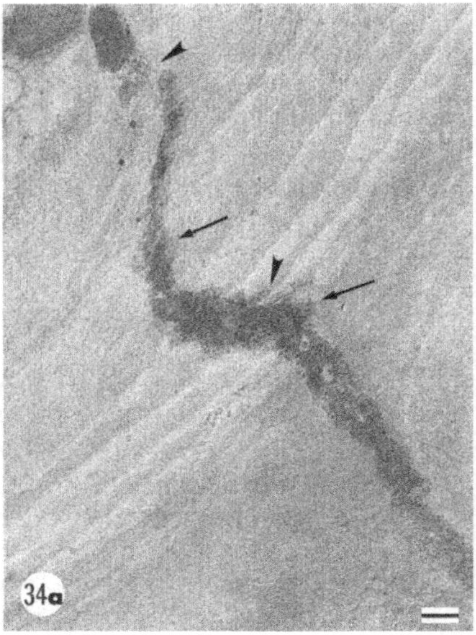
34a

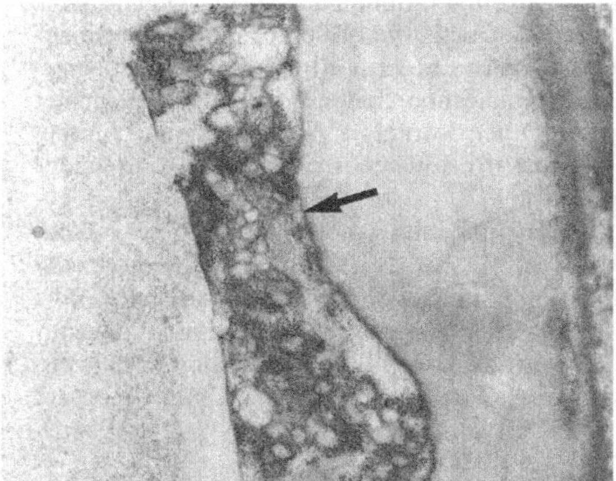
35

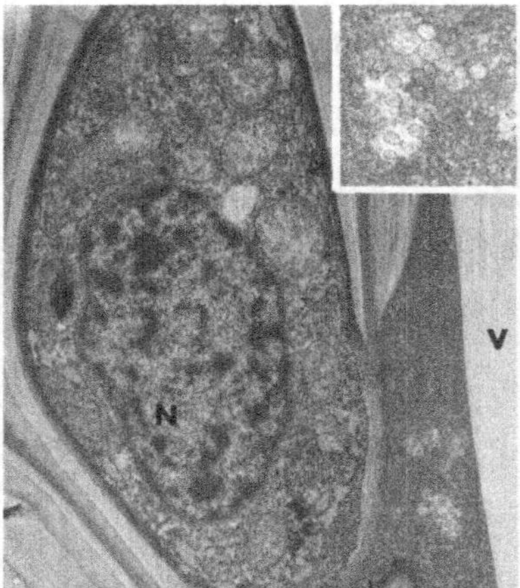
N
V
PL
P
36

clogging material. Apparent differences in the coating have been described (Takai and Hiratsuka, 1984); these could be an expression of different development stages of the same intervening entities, one of which could be of fungal origin (Ouellette 1978b,d, and unpublished observations). Other extraneous matter in vessel lumina could also constitute elements of variation. The validity of categorizing these differences may be questioned, particularly when revealed solely by scanning EM observations, as proposed by Takai and Hiratsuka.

The coating over bordered pit membranes is uniform and continuous with that along vessel walls; over half-bordered pit membranes, it may be thin, discontinuous, or, together with the distended pit membrane, recurved toward the vessel lumina, often in association with alveolar forms (Figs. 13.19–13.21, 13.27) (Sect. 13.3.6). Such a coating may be overlaid by masses of fibrillogranular material (Fig. 13.36) and vesicular structures (inset). Occasionally the coating over the pit membrane is less opaque, disclosing the presence of some fine fibrillar material; similar material is also present in the pit membrane.

13.3.6 Tyloses and Alveolar Networks

Prolific tylosis production was reported in early works on DED (see Ouellette 1962a, Banfield 1968). Ouellette identified many globose structures with tylosis configurations in vessel lumina that he denoted as swollen empty fungal cells. Others referred to these structures as gummy bubbles (Pomerleau 1970).

Observations with the TEM show that tyloses are characteristically delimited by a fibrillar layer similar to and continuous with the inner protective layer of the subtending parenchyma cell (MacDonald and McNabb 1974, Ouellette 1980). Secondary walls and inner suberin-like layers can also form in tyloses (Ouellette 1980). Other globose structures are also present in vessel lumina; they are demarcated by thin or thick layers of material that is quite different from that of the tylosis wall (Figs. 13.20, 13.21, 13.30). The term alveolar reticula is proposed to denote these variously configured structures instead of bubbles as described previously (Ouellette 1980). Some workers report abundant

Fig. 13.34. Band of opaque material (*arrow*) in the middle lamella of a few cells and the secondary wall (*S*) of a living cell; this band is also in line with opaque masses in the periplasm or outer portion of the cytoplasm (*small arrow*), associated with alteration of the plasmalemma. **a** At higher magnification (*bar* = 0.2 μm), opaque particles and fibrillar strands, some extending into the host wall (*small arrows*), can be seen in the opaque material. Normal host wall material is located between discontinuities in this band, or between its lateral extensions (*arrowheads*). *Bar* = 0.4 μm

Fig. 13.35. A band of material (*arrow*) composed of small opaque particles intermixed with vesicular-like structures and other bodies, which extended almost the whole length of a vessel wall, apparently through some fibrillar clogging material in the vessel lumen. Pit membranes of this vessel were altered (*right*). *Bar* = 0.3 μm

Fig. 13.36. An opaque protective layer (*PL*) of a paratracheal parenchyma cell containing a dense nucleus (*N*), with vacuolated cytoplasm in which altered plastids and mitochondria can be seen. The pit (*P*) chamber on the vessel (*V*) side is filled with opaque fibrillar matter intermixed with vesicular bodies (*inset*). *Bar* = 0.4 μm tylosis formation in infected elm trees independent of the periods after inoculation or of the species selected (Elgersma 1969, 1982, MacDonald and McNabb 1974, Bonsen et al. 1985). In the three elm species studied by Ouellette (1980 and unpublished observations), tylosis formation is rare during early infection in the period of highest tree susceptibility to the disease. In contrast, alveolar reticula form very quickly during this period (Ouellette 1980, Ouellette and Rioux, unpublished observations).

The reticulum structural matter (RSM), formed of apposed layers, the thinnest in the range of 100 nm, becomes thicker as infection progresses. Often continuous with the coating, as seen in more advanced infections, RSMs also are structurally similar to it. When tyloses appear in vessel lumina, RSMs are compressed as meandering entities between the tylosis and vessel wall (Fig. 13.4). Their expansion into balloon-like structures occurs in unconfined areas, likely under the effect of gaseous matter (Pegg 1985), as suggested by their association with distention of pit membranes and the overlaying coating toward the vessel lumen.

13.3.7 Effect on Cambium and Differentiating Cells

The difficulty in preserving the cambial area in preparing samples for microscope observations is a deterrent to such studies. However, demonstrating a direct effect by the pathogen on the cambial tissue would promote a better understanding of DED pathogenesis and defense mechanisms. Ouellette (1960, 1962a, 1982) observed the presence of the pathogen in differentiating cells several days after inoculation, in association with various types of alterations in that region. Zones of periderm-like tissue, or small to large pitted areas in the xylem that could have resulted from damage to the cambium, are commonly observed in trees that have been infected for more than 1 year (Ouellette, unpublished observations). Shigo (1982) also believes that the final stage of the disease is a massive attack on the cambium. Barrier zone formation (BZF) (Sect. 13.6) was observed as a reaction to the presence of the pathogen near the cambial area. These observations may justify questioning the point of view held by several workers that the effect on the cambial area results from wilting or other remote physiological disorders. The following observations show that the pathogen does have a direct effect on the cambial area, most appreciably in young branches and twigs.

A few days after inoculation, partial to more complete collapse of cambial cells, which can be traced back to invaded, deformed vessels nearby (Figs. 13.3, 13.37), often occurs at the tips of developing streaks. Some of the adjacent ray or axial parenchyma cells completely disintegrate, but the cytoplasm in others appears normal or in a reactive state. Such alterations may even extend through rows of cells toward the inner bark, interspersed with distorted cells that contain a virtually complete secondary wall or other type of inner layer (Figs. 13.38, 13.39). Some of these cells are identifiable as vessel elements, and some may even contain apparent tyloses (Fig. 13.37). These vessel elements may contain masses of altered cytoplasm and may be characterized by secondary walls and

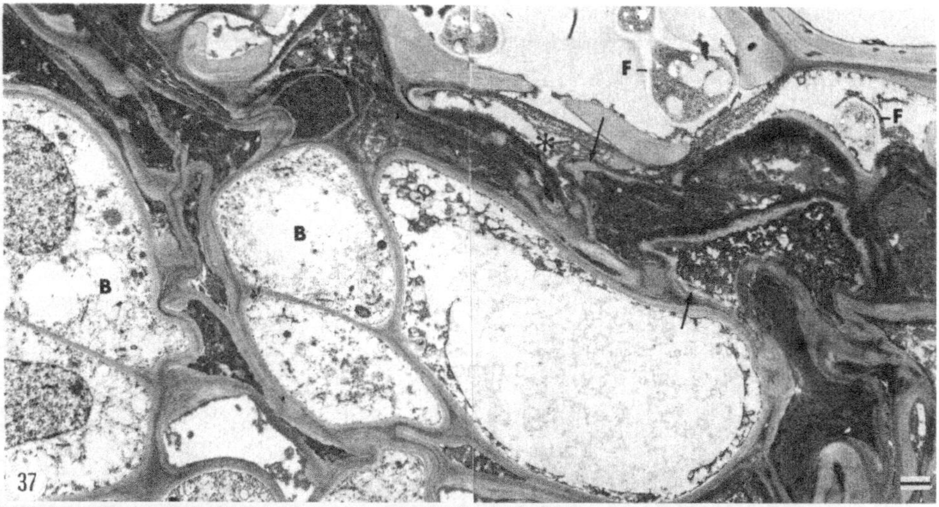

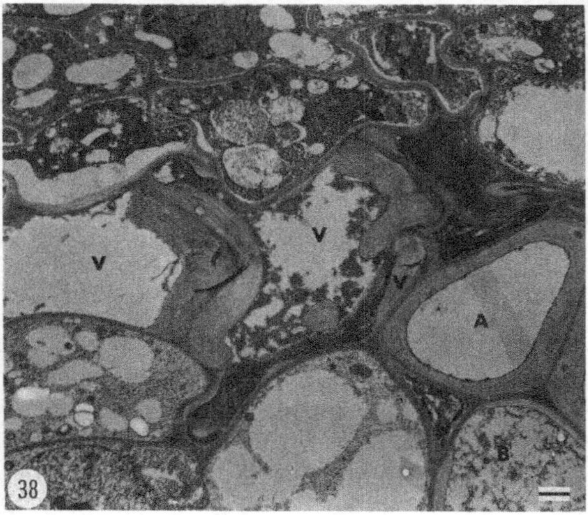

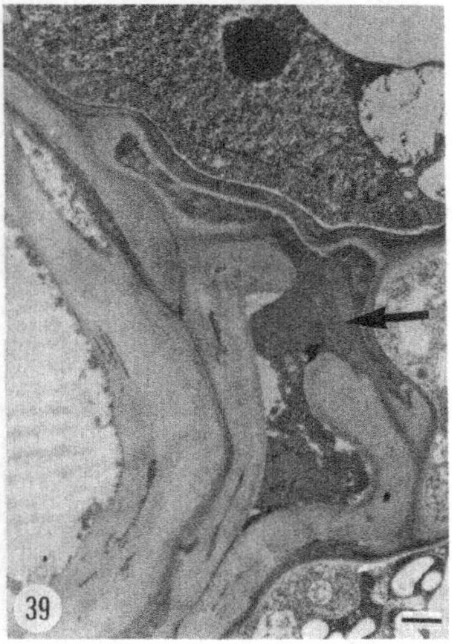

Fig. 13.37. Presence of pathogen cells in often distorted vessel elements next to the cambial area. Some of these host cells might have become vessel elements (*) containing tylosis-like structures (*small arrows*). Some living cells are present through and outside the layers of contorted host cells, with indications that they have recently divided (*B*). *F* Fungal cell; *bar* = 0.8 μm

Fig. 13.38. In an altered cambial area, distorted vessel (*V*) elements and a few cells with normal secondary walls (*A*, *B*) are surrounded by distorted cells with variously altered but not degraded contents. The secondary wall of the vessels was often thicker than usual. *Bar* = 0.9 μm

Fig. 13.39. Similar distorted vessels in differentiation separated from the inner xylem by layers of altered cells as in Fig. 13.38. A thick pit membrane (*arrow*) is shown between two such vessel elements which border hypertrophied ray cells (*above*). *Bar* = 0.7 μm

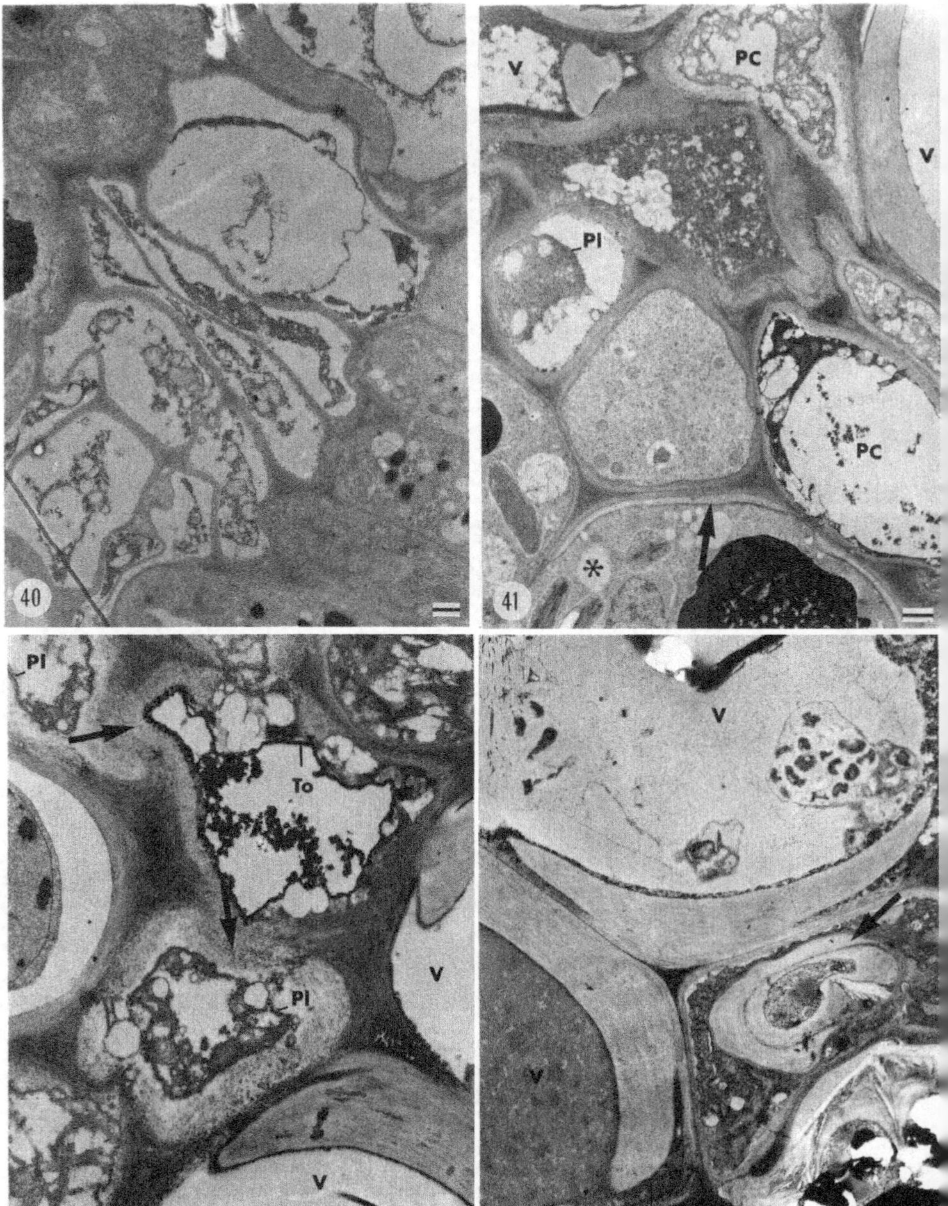
40
V
PC
V
Pl
PC
*
41
Pl
To
Pl
V
V
42
V
V
43

pit membranes that are appreciably thicker than normal (Figs. 13.38, 13.39). The early effect on tiers of cambial daughter and mother cells abutting on recently differentiated invaded vessel elements is localized, indicating that the irritating factor is easily propagated from one cell to the next (Fig. 13.40); cytoplasmic reactions seem to occur concomitantly in the adjacent cells. In these cases, cell deterioration obviously cannot be ascribed to mechanical rupture or improper fixation.

Similar tissue alterations have been observed not only at the cambium level in branches or twigs, but often at several locations throughout the xylem, including borders of the medullary sheath (Figs. 13.28, 13.41, 13.42). Other types of alterations noticeable in these include wall alterations and/or abnormal wall deposition in affected cells, and the absence or delayed formation of secondary walls in other apparently normal cells. Many of the cells may appear dead, but they seem to maintain some activity via cytoplasmic remnants and membranes, including the plasmalemma.

13.4 Wilt Mechanism: Overview

According to prominent workers in plant wilt diseases, various factors may operate simultaneously to bring about pathological wilting (Dimond 1970, Pegg 1985). For these diseases, determining the key factors of pathogen action and host reactions is essential. Various explanations for wilt mechanisms have been presented (Dimond 1970, Van Alfen and MacHardy 1978, Pegg 1985), but because such mechanisms may be dissimilar in different plant species affected by the same pathogen (Robb et al. 1975a,b), a theory applying to all cases would be misleading. In this respect, our knowledge of DED is only fragmentary. In most theories on wilt mechanisms in elm (Dimond 1970, Van Alfen and MacHardy 1978, Beckman 1987), little direct importance is attributed to alterations of parenchyma cells, such as those expressed by excretions of electrolytes

---

Fig. 13.40. Next to an invaded vessel element adjacent to the cambial zone, a row of dead cells (*center*), comprising even the phloem initials, is bordered by rows of living cells with dense cytoplasm and often abundant opaque material. *Bar* = 0.8 μm

Fig. 13.41. Parenchyma cells (*PC*) rich in cytoplasm and opaque material (*) next to rows of parenchyma cells bordering invaded vessel elements near the pith. A thin secondary wall (*arrow*) is present in some of these but not in others. There appears to be a gradation in the intensity of parenchyma cell alterations from the vessel element to these more normal cells. Retraction of the plasmalemma (*Pl*), as in Fig. 13.42, is indicative of a less damaged condition. *V* Vessel; *bar* = 0.7 μm

Fig. 13.42. Rows of altered cells near invaded metaxylem vessels (*V*), in which the locations of the altered plasmalemma (*Pl*) and or tonoplasts (*To*) are recognizable and apparently associated with wall deposition or apposition (*arrows*). *Bar* = 0.2 μm

Fig. 13.43. Formation of a type of endocell (*arrow*) in a parenchyma cell adjacent to completely or partially occluded metaxylem vessel elements (*V*). The endocell is limited by two disrupted secondary wall-like layers. A wall reaction is visible in the cell below, also containing abundant opaque matter. *Bar* = 0.6 μm or other substances, including phytotoxic compounds, into the sap stream, or to the action of toxins produced by the pathogen. Tyloses, gels, gums, coating material, and so on, have been considered as important vessel occluding agents.

Extraneous components in vessel lumina, initially favoring pathogen growth, would contribute to increased sap viscosity and resistance to sap flow. According to Newbanks et al. (1983), embolism from vapor produced by "enzymatic degradation of vessel walls" would precede fungal growth in vessel elements and their occlusion. These conclusions were based, however, on LM observations, and fungal entities may have been overlooked (Sect. 13.3.2). Whether impaired conduction due to vessel occlusion can be restored above the blockage point is also debated (Dimond 1970, Zimmermann and McDonough 1978, Beckman 1987). Gels or gums in vessel elements may eventually dissolve, presumably helping to restore vessel function (Ouellette 1978c). If wilting is related to dysfunction of vessel elements and if it is linked with occlusion, establishment of the origin and nature of the occluding components is essential.

These could be of host or pathogen origin or of both. For some authors studying the presence of compounds originating from the fungus (polysaccharides, phenols, enzymes, and so on), these would contribute mostly to clogging of pit membranes or to increased impermeability of vessel walls (Dimond 1970, Russo et al. 1982, Scheffer 1984, Beckman 1987, Newcombe and Robb 1989).

Of the various possible toxins isolated from the pathogen, cerato-ulmin, a peptide with a low-molecular weight (Takai et al. 1983), and a peptidorhamnomannan with a molecular weight over 100 kD (Salemink et al. 1965, Strobel et al. 1978, Scheffer 1982) have been the most studied. Scheffer's attempts (1984) to relate virulence of *O. ulmi* to cerato-ulmin production reveal exceptions, preventing the establishment of a close correlation between these two factors. Although the toxin was detected by immunological tests (Nordin et al. 1987), it was not localized at the proper time and place (Russo et al. 1982, Richards and Takai 1988), despite Pegg's (1985) statement to the contrary.

Scheffer and Elgersma (1981) used immunological techniques to detect the high-molecular weight toxin in diseased elms 2 weeks after inoculation. Positive reactions occurred with immunofluorescence only on fungal cells in colonized vessel elements (Scheffer et al. 1987). Monoclonal antibodies were obtained against this toxin. When they were labeled for LM and TEM observations, the toxin was revealed in vessel elements and adjacent cells soon after inoculation (Benhamou et al. 1985a,b). These observations show that the toxin soon accumulates in pit membranes and in the protective layer of parenchyma cells in association with several types of alterations similar to those described in Section 13.3.4. Later, this toxin occurs in intercellular spaces and other cells further from the vessel. Thus, significant progress has been made in demonstrating the production and detection of toxins in diseased tissues, but more work is required to confirm the dynamics of these toxins in disease development.

Whatever the situation, the following points require special attention for understanding symptom expression in DED. As external symptoms are expressed only when shoots have been invaded (Ouellette 1962b, Pomerleau 1968, Dimond 1970, MacHardy and Beckman 1973), these tree parts should be studied more thoroughly. Dimond indicates that, due to the fewer and smaller vessel elements in these tree parts, the plugging effect would be much more drastic than in larger tree parts (Fig. 13.43). As Figs. 13.38 and 13.39 suggest, some of this occlusion material might even originate from cytoplasm of young vessel elements. Similarly, the pathogen's effect on living cells could be greater in shoots than in stem parts because of a possible difference not only in the physiological functions of these cells but also due to the consequences of their being infected; indeed, the proportion of cells that need to be altered to cause a lasting, detrimental disturbance could be reached more quickly in shoots than in larger branches. Moreover, a difference in function may exist between such cells in each part, and disturbances of differentiating cells will likely result in a more profound, more lasting effect on tree physiology that those from alterations of mature cells. These considerations underline the necessity of recognizing that different mechanisms may be involved in acute and chronic disease symptoms, the effect of vessel occlusion probably being most prominent in the former. In other words, acute symptom expression of the disease could be associated with the absence of defense reactions; such reactions could occur in the chronic stage of the disease, but these would only be partial or of insufficient duration to assure recovery.

13.5 Defense Mechanisms

To obtain the ideal DED control, such as producing elm stocks with acceptable degrees of resistance, thorough knowledge of defense mechanisms against the disease is essential. Such resistance factors have been sought in studies of inter- and intraspecific differences in elm reactions to DED (McNabb et al. 1970, Sinclair et al. 1975a). Resistance to DED appears to be under polygenic control (Lester and Smalley 1972), indicating that many mechanisms are likely involved.

Concepts of DED pathogenesis based solely on knowledge of events occurring in vessel elements have obviously influenced notions on defense mechanisms. In this view, axial and ray parenchyma cells contribute to the production of inhibitory substances and to physical confinement of the pathogen in vessel lumina, some living cells being sacrificed to save others (Beckman 1987). The following scenario should also be considered: assuming direct action on these tissues by the pathogen, a search for cells with better resistance that form exclusion barriers, despite possibly extensive pathogen colonization of vessel elements in the initial stages of the disease, may be worthwhile. In this section, we present a summary of notions expressed in favor of a more active role of disease resistance mechanisms. From the CODIT model (Shigo and Marx 1977), wall 1 will be considered as a factor preventing or delaying vertical spread of the pathogen, walls 2 and 3 will be considered as impeding tangential and radial spread, respectively, and wall 4 as pertaining mainly to BZF. Factors delaying or preventing recurrence of the disease, despite extensive pathogen invasion of the preceding growth ring (Banfield 1968), will be considered as related to walls 3 or 4. Activities of parenchyma cells that could have a bearing on wall 1 have been presented in Section 13.3.4.

13.5.1 Anatomical Aspects

The lack or appreciable reduction of external symptom expression concomitant or not with localization of pathogen invasion of vessel elements, is considered as an expression of disease resistance in host elm or in nonhost species (McNabb et al. 1970). Remission of affected trees following the appearance of external symptoms has also been related to impaired pathogen invasion and to localization of xylem discoloration (Tchernoff 1965, Sinclair et al. 1975a, Sinclair and Larsen 1980).

Some authors (Elgersma 1969, Sinclair et al. 1975a) have recovered more propagules of the pathogen from inoculated, susceptible trees than from resistant ones. However, using inoculated, excised branches, Sinclair et al. (1975a) found pathogen growth equally abundant in all selections. This observation has favored hypotheses that differences in susceptibility are primarily attributable to differences in the tree's ability to impede rapid and extensive pathogen invasion of vessel elements and hence to localize infection. For a comprehensive review of the subject, refer to MacHardy (1978).

13.5.1.1 Anatomical Markers

Differences in vessel groupings within the growth rings were proposed to explain differences in susceptibility to DED, not only between trees (Pope 1943, McNabb et al. 1970), but also at definite periods in the growing season (Pomerleau 1965, Banfield 1968). The pathogen is believed to be able to invade new vessel elements only by passage through pits from an adjacent colonized vessel. The greater the number of vessel interconnections, the greater and more rapid the radial and tangential spread would be. Other workers have considered differences in both vessel length and diameter as the main factors in DED resistance (Elgersma 1970), with vessel diameter being likely the key factor in restricting vertical pathogen spread (McNabb et al. 1970, Sinclair et al. 1975b). By comparing the range of vessel diameters in rings 2 and 3 in branches from susceptible and resistant elm selections, Sinclair et al. (1975b) found a nearly linear relation between susceptibility and the proportions of vessels with diameters of 30 to 80 μm and more. Selecting 50 μm as an arbitrary lower value, these authors determined the percentage of vessels with diameters of 50 μm and greater. Trees with a vessel diameter index (VDI = percentage of vessels falling in that range) of less than 12 were noticeably resistant. A VDI of 12 or below was considered to be a good anatomical marker for selection of disease resistance. These authors mention that some resistant trees may nevertheless be rejected when this scale is used, indicating that resistance might also be linked to other fundamental types of defense mechanisms. In this respect, a low VDI does not necessarily exclude the possibility that such types of mechanisms are also involved.

13.5.1.2 Vessel Blockage

This resistance mechanism (Beckman 1987) is congruent to and reinforces the anatomical aspects. It is expressed as an accumulation of gums, gels, or phenols (Pegg 1985) and the formation of abundant tyloses that would restrict pathogen colonization. To be efficient in trees, this mechanism must be rapid, operate in advance of fungal invasion, and be persistent enough to last until the most susceptible growth period of the tree has elapsed. It is not evident that all these prerequisites are met for DED.

Vessel length does not seem to be significantly different between susceptible elms and more resistant ones, and no difference in initial spore movement has been recorded (McNabb et al. 1970). Differences in the propagules recovered from susceptible and more resistant selections are detectable only after day 4 following inoculation (Elgersma 1969). Ouellette (1980) (Sect. 13.1.6) found that tyloses are not formed to any appreciable extent in *U. pumila, U. americana, U. montana*, or X-ray treated material (Ouellet and Pomerleau 1965) during the period of highest tree susceptibility. Tyloses may form more abundantly later in the season, and their reactions to infection may be analogous to those of paratracheal parenchyma cells, including the formation of an inner suberin-like layer (Ouellette 1980). Differentiation between tyloses and alveolar reticula may be responsible for this apparent disagreement with other authors (Ouellette 1980, Ouellette and Rioux, unpublished observations). In a study on growth regulator effects on resistance, Brener and Beckman (1968) attributed an enhanced resistance of TCPA (2,3,6-trichlorophenylacetic acid)-treated trees to a more rapid and abundant tylosis formation. Magnifications of their illustrations of tyloses are too low to confirm whether these are actually tyloses or alveolar forms. In either case, the effect of this regulator could have been to make paratracheal parenchyma cells more resistant to other key detrimental factors, the formation of these structures in vessels being secondary manifestations. It is also doubtful that these observations provide sufficient evidence that susceptibility is induced by prevention of tylosis formation.

Gel formation in vessel lumina of plants infected with wilt diseases was ascribed to pectinic material (Sect. 13.3.5). In ultrastructural studies, we have sometimes observed material similar to that identified as gels, accumulating in chambers of half-bordered pits. Although some indications of continuity of this material with parenchyma cells were obtained (Rioux and Ouellette 1989, and unpublished observations), conclusive evidence that much of it comes from breakdown products of pit membranes is lacking. A positive reaction for pectinic material in vessels colonized by *O. ulmi* does not necessarily imply that it is of host origin, since the fungus also produces compounds that are positive for pectinic material (Benhamou 1989).

Gels, alveolar reticula, and associated coating material are among the first components detected in invaded vessels (Ouellette 1978b, and other unpublished observations), and are likely related to noticeable darkening of vessel walls (Elgersma 1969). Such material was considered to be comprised of oxidized and polymerized phenolic compounds (Dimond 1970). Phenols were detected in vessel lumina of diseased elms by Gagnon (1967b). Opaque vessel coating material thought to be similarly composed was also shown to contain lipid material in verticillium-infected plants (Robb et al. 1979, Newcombe and Robb 1989). Coating material in elm vessels may contain additional constituents (Sect. 13.3.5) different from the opaque material impregnating primary walls, middle lamella, and pit membranes (Street et al. 1986) that occur as normal constituents between collapsed elements of the protoxylem (Fig. 13.11). Nevertheless, swollen pit membranes do not often appear to be as easily penetrated as the more translucent ones. Like others, we found that some of the extraneous opaque compounds present in vessel elements are probably of host origin, but the pathogen can also produce phenolic compounds (Claydon et al. 1974); furthermore, many pathogen cells liberate their contents (e.g. enzymes and other proteins) into vessel lumina (Fig. 13.11) (Ouellette 1978a).

Considering that coatings and gels may be comprised of compounds from both the host and the pathogen, these compounds may be heterogeneous and may differ not only between vessel elements, but also within the same vessel element. Also, as coating material is associated with the alteration of vessel walls and adjacent cells at the tips of developing streaks, as in older infected parts, it is difficult to determine in infected elms whether blockage from gel and coating material results in a protective or a harmful effect. In any event, gels in vessel lumina do not seem to persist, and the vessel wall coating can be easily penetrated by the pathogen (Ouellette, unpublished observations).

13.5.2 Physiological Aspects

13.5.2.1 Composition of the Xylem Sap

Xylem sap is composed of low concentrations of sugars, nitrogenous compounds, and inorganic substances. Fungi that cause wilt diseases can grow well in a poor nutrient solution because their nutritional requirements are quite simple (Puhalla and Bell 1981, Pegg 1985).

In DED, the concentration of sugars in the xylem sap appears to influence resistance. Certain manipulations, such as keeping plants in the dark, induce a decrease in sugar content and resistance, while spraying foliage or injecting sugar solutions increase resistance (Zentmyer et al. 1946, Feldman et al. 1950, Caroselli and Feldman 1951). No differences in sugar concentration were found, however, between resistant and susceptible trees (Elgersma 1967, Singh and Smalley 1969a).

Other studies have focused on the nature and amount of nitrogenous compounds present in vessel elements in an attempt to link their presence to susceptibility or resistance. For instance, during the period of maximum susceptibility of the elm, the xylem sap contains a greater proportion of amino acids than later in the growing season when elms are more resistant to DED (Singh and Smalley 1969c). The inoculation of *U. americana* seedlings also induced a noticeable increase in the concentration of nitrogenous compounds (Singh and Smalley 1969b). This link between a high proportion of amino acids and susceptibility does not seem conclusive since resistance elms have, on average, more nitrogenous compounds than susceptible ones (Elgersma 1967, Singh and Smalley 1969a). These investigators reported that a high proline content appeared to be associated with resistance to DED.

In a review on the nutritional aspects of wilt diseases, Bell and Mace (1981) concluded that it is very difficult to establish a clear relation between these aspects and pathogenesis. Kessler (1966) demonstrated that four wilt fungi, including *O. ulmi*, were able to grow on the xylem sap of five tree species. No positive correlation was found between the proportion of growth and the susceptibility of the host.

13.5.2.2. Phytoalexins

The accumulation of mansonones as a possible resistance mechanism to DED was reported for the first time by Overeem and Elgersma (1970). Perhaps because of their extraction procedures (Bell and Mace 1981), these authors did not find any significant differences in the detection of mansonones E and F between a susceptible clone (Belgica) and a resistant one (390) of *U. hollandica*. Contrary to these first results, it has been demonstrated that phytoalexins accumulate rapidly and intensely in the resistant *U. pumila* (Duchesne et al. 1984). Phytoalexins have also been detected in appreciable quantities in American elm seedlings after inoculation with a nonaggressive isolate of *O. ulmi* (Hubbes and Jeng 1981, Jeng et al. 1983). These authors noted that this inoculation induced a greater resistance to aggressive isolates of the pathogen.

Aggressive and nonaggressive isolates of *O. ulmi* induced maximum mansonone accumulation in *U. pumila* (Duchesne et al. 1986) 2 weeks after inoculation; in *U. americana*, the maximum accumulation induced with a nonaggressive isolate (Duchesne et al. 1985) was at 4 weeks, but at 6 weeks with an aggressive isolate (Duchesne 1985, Duchesne et al. 1985). These results suggest that the production of phytoalexins is delayed by inoculation with aggressive isolates of the pathogen in susceptible trees, and that their accumulation is more pronounced and occurs earlier in resistant trees than under the susceptible conditions, paralleling results presented in review publications for other plant diseases (Keen 1981, Mansfield 1983, Rouxel 1989). After peak accumulation, the quantity of mansonones declined considerably in both elm species the following days (Duchesne et al. 1985, 1986).

Rapid phytoalexin accumulation may thus be very important in delaying invasion in the initial stages of infection and in remaining efficient until other mechanisms of resistance, such as BZF, take over. One must also consider that it may not be so much the quantity of any phytoalexin produced that counts, but mainly its presence at the right time and place. Developing histochemical tests to localize such compounds would be very useful to better understand their role in disease development.

13.5.2.3 Growth Substances

During DED development, as in other wilt diseases, an increase in auxin concentrations was noted (Beckman 1956). Pathogens may contribute to this phenomenon, as many can produce auxins in vitro (Pegg and Selman 1959, Fenn et al. 1977, 1978). Phenol accumulations in the xylem tissues, which can inhibit indoleacetic acid (IAA) oxidase, may allow auxin concentrations to build up (Sequeira 1964, Mace and Solit 1966).

Knowledge of the role of hormones in plant diseases is fragmentary. IAA would increase the plasticity of pit membranes, thus promoting tylosis formation (Beckman et al. 1953, Bornmann et al. 1967). In healthy plants, IAA influences the division and differentiation of cambial cells (Savidge and Wareing 1981). It is easy to speculate that variations in auxin concentration might have a role in barrier formation, an anatomical change that appears associated with DED resistance (Sect. 13.6).

The action of IAA may also be intercorrelated with that of other compounds. For example, IAA and some synthetic auxins induce the production of ethylene (Abeles 1973), which in turn may affect the production of abscisic acid, tylosis formation, and certain enzymes involved in resistance to various diseases (Gentile and Matta 1975, Pegg 1976). Very little is known about the possible functions of other types of hormones (e.g. cytokinins) in DED resistance.

Other compounds with hormone-like properties, which apparently increased resistance to DED, have also been tested in elm. Sodium 2,3,5,6-tetrachlorobenzoate (Beckman 1959) and aminotrichlorophenylacetate isomers (Edgington 1963) provide a degree of protection. The use of 1,2,5-trichlorophenylacetate also increased resistance to artificial inoculation with *O. ulmi* (Smalley 1962, Brener and Beckman 1968) and is thought to be related to enhanced tylosis formation and/or a delay in earlywood formation in treated trees (Sect. 13.5.1.2).

13.5.2.4 Possible Reactions Related to Phytotoxins

In other plant diseases, mechanisms that may be linked to resistance to phytotoxin action are: insensitivity to phytotoxins, inhibition of their synthesis, or their inactivation. The importance of these mechanisms has also been suggested in investigations of wilt diseases (Bell and Mace 1981). As the role of toxins in DED pathogenesis is still controversial (Sect. 13.4), it is understandable that such mechanisms have never been invoked to explain DED resistance. As more work on toxins is carried out, it is probable that eventually such a mechanism will be taken into consideration for screening resistance in certain elm species.

13.5.2.5 Influence of Environmental Factors

Most work on wilt diseases has had a direct bearing on the host or the pathogen, but very few studies have dealt with the influence of environmental conditions, which probably also have an effect on some of the physiological processes linked to DED resistance. Among other factors, intermediate temperatures (26 to 29 °C), short photoperiods, and high moisture levels may lead to reduced efficiency of resistance mechanisms to DED (Bell and Mace 1981). In the future, the role of environmental factors should be considered more attentively in studies concerning forest pathology, especially in relation to the additional stress caused by the numerous factors associated with pollution.

13.5.3 Cytological Aspects

In addition to tylosis formation, various types of reactions linked with the diverse cell alterations described in Section 13.3 (Ouellette 1978a,b,c,d, 1980, 1981a,b, 1982) may occur in twigs and branches of infected elms. When observed in branches surviving acute or chronic symptoms of DED in susceptible or more resistant trees, some of these reactions may represent resistance factors or recovery attempts. Categorization of these reactions is not easily done because the same reactions might occur in more successful infections as well as in less successful ones, the difference in efficacy in each being attributable to a difference in timing and in degree. Despite these shortcomings, enough supportive evidence exists to consider that a few of these reactions are possible indicators of defense reactions.

13.5.3.1 Mature Parenchyma Cells

These cells comprise axial and ray parenchyma cells that have completed or are completing formation of a secondary wall at the onset of infection. The first noticeable reaction to attack is an increased deposition of a compact inner layer, corresponding to the protective layer in cells adjoining vessels (Figs. 13.16, 13.19, 13.31). This increased wall deposition is first noted in the pit region, then on the whole side facing the vessel lumen, and finally around the entire cell. This layer, formed in cells containing little altered to even more diversely altered cytoplasm, also becomes impregnated with electron-opaque material (Figs. 13.30, 13.32, 13.36). Regardless of whether it is thin or thick, this opaque wall layer will not loosen, as is the case when it remains more translucent, apparently in conjunction with sudden or more progressive cell alteration (Figs. 13.5, 13.27).

A thick layer of opaque matter is commonly present in the periplasm. Some of the vacuolar areas in this layer appear to be linked with cytoplasmic vacuoles (Fig. 13.32). However, it is generally too dense to distinguish its ultrastructure, which has been observed to be heterogeneous and to contain lamellar structures and small opaque particles (Figs. 13.25a, 13.27a). Similar opaque material bordering gelatinous layers is also linked to the deposition of lamellar-like structures oriented perpendicularly to the wall (Fig. 13.32) (Ouellette 1981b). The formation or accumulation of such opaque material could be related to increased peroxidase activity (Gagnon 1968).

Formation of papilla-like structures, which may eventually occupy the greater part of the cell lumen (Ouellette 1982), and other wall modifications, also occur. One or two extra inner secondary-like layers, separated by an opaque protective layer of varying thickness, may be deposited within cells (Figs. 13.8, 13.44,

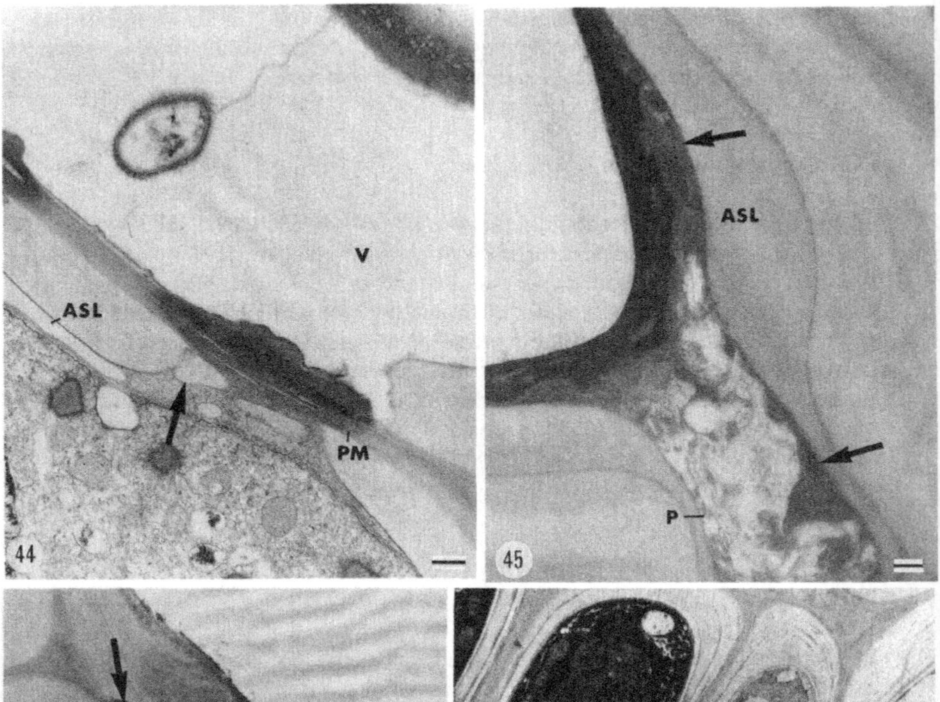
V
ASL
PM
44
ASL
P
45
VW
PC
46
PC
V
PI
47

13.45). A suberin-like layer may eventually be added or deposited on the first protective layer (Ouellette 1978a). Concomitantly, or subsequently, similar additional layers are deposited in contiguous cells, including ray parenchyma. Pit membranes of half-bordered pits of parenchyma cells containing these extra layers are also impregnated or thickened on the vessel side, sometimes as an indication of recovery from slight to more pronounced alteration (Fig. 13.44).

In samples from branches having withstood or recovered from initial infection, many of the cells contain very dense and opaque cytoplasm (Figs. 13.30, 13.32, 13.46). Their large vacuoles do not necessarily contain appreciable quantities of opaque material. The closer to invaded vessels these cells are, the more evident is the expression of resistance; in *U. pumila*, complete rows of such cells can border the paratracheal, sometimes ostensibly altered parenchyma cells (Fig. 13.47). These are usually less altered than in *U. americana* and can form large papillae, even in tyloses, in response to attack (Ouellette 1982). After an early attack, also in *U. pumila*, rows of cells with opaque contents may be deposited over invaded vessels; new series of ray parenchyma, rich in opaque material, may even be formed (Fig. 13.48).

The cytological aspects of parenchyma cells with opaque contents in the nonhosts (Sect. 13.6) studied are similar to those just described. They are also similar in parenchyma cells in tissue cultures challenged with culture filtrates of *O. ulmi* (Pijut et al. 1990).

13.5.3.2 Fibers

In fibers, various cell wall modifications also occur resembling those described for parenchyma cells (Ouellette 1981b). Considered to be an expression of resistance, these cells are particularly noticeable in barrier zones (Sect. 13.6). A reaction regularly seen in such cells, which may contain some altered cytoplasm, is the formation of a thickened gelatinous layer impregnated with amorphous, slightly to more opaque material, or arrays of lamellar-like opaque structures oriented parallel across the layer (Figs. 13.32, 13.49). Such a layer may be separated from the subtending layer, the secondary wall layer, usually, by a band of more opaque material. An opaque, thin to thick layer merging with a suberin-

←

Fig. 13.44. Formation of an additional but incomplete secondary wall (*ASL*) (*arrow*) in a paratracheal parenchyma cell. The slightly altered pit membrane (*PM*) on the vessel (*V*) side is thickened. *Bar* = 0.5 μm

Fig. 13.45. A complete additional secondary layer (*ASL*) in a ray cell. Masses of opaque matter (*arrows*) intermixed with more translucent matter in the pit (*P*) chamber, and in the periplasmic area elsewhere around the cell. *Bar* = 0.2 μm

Figl. 13.46. Much thickened and opaque secondary wall of a vessel member over pit chambers (*arrows*). Neighboring parenchyma cells (*PC*) contain opaque cytoplasm. From an *U. americana* branch that survived mild symptoms. *VW* Vessel wall; *bar* = 0.6 μm

Fig. 13.47. Rows of parenchyma cells (*PC*) with opaque contents in *U. pumila*, bordering a paratracheal parenchyma cell showing pronounced retraction of its plasmalemma (*Pl*) next to the pits and abundant fibrillar material in the periplasmic area. *V* Vessel; *bar* = 0.6 μm

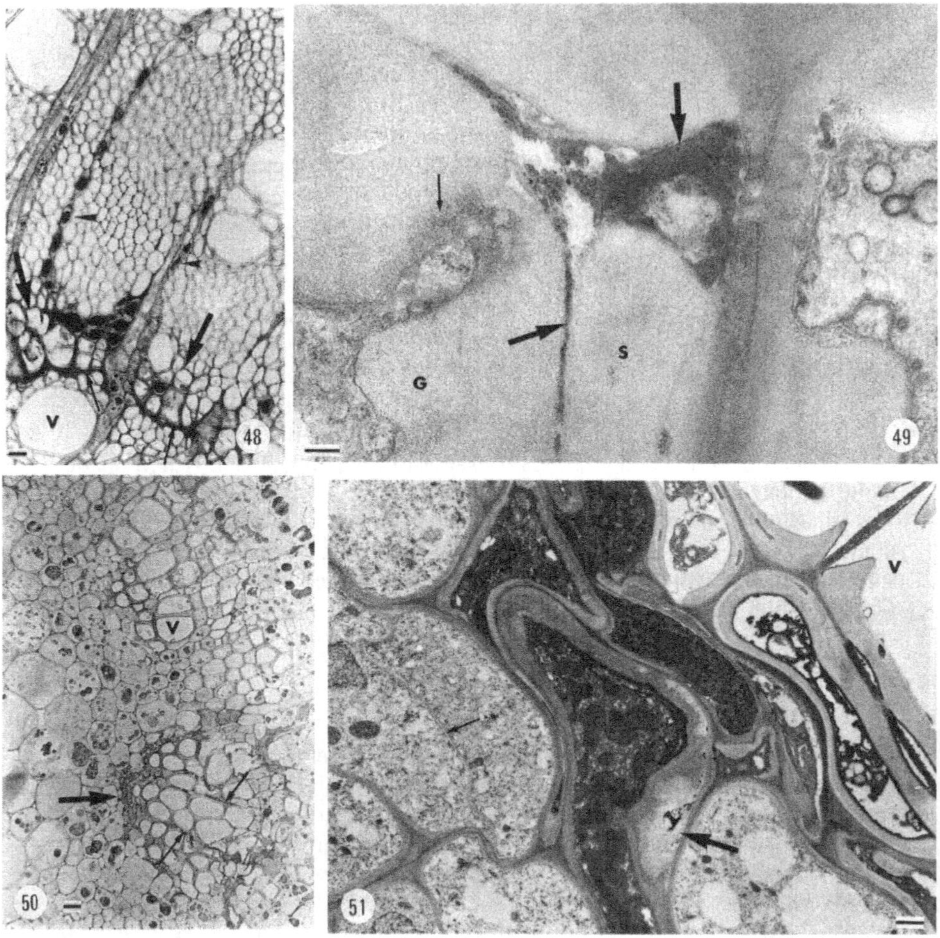

Fig. 13.48. Barrier zone formation in *U. pumila* next to a row of invaded vessels (*V*). The barrier zone is characterized by a band of cells larger than others (*arrows*) and delimited by densely stained walls above a band containing at times collapsed cells (*small arrows*); above these rows of cells, other cells almost filled with opaque contents are present. Rows of cells (*arrowheads*) with similar contents extend outwards into the xylem, suggesting new uniseriate rays. Light microscope observations; *bar* = 2.2 μm

Fig. 13.49. Presence of opaque matter (*arrows*) with small opaque particles in a pit chamber and between the secondary wall (*S*) and the gelatinous-like layer (*G*) in a fiber. Traces of opaque, small fibrillar material (*small arrows*) are visible in the latter. *Bar* = 0.2 μm

Fig. 13.50. Two vessel (*V*) bundles that were invaded early in the development of the twig, as indicated by rows of collapsed cells near the pith (*arrow*). Other collapsed cells lacking secondary walls, and bands of opaque material located intercellularly (*small arrows*) surround or extend through one of the bundles (*below*) to the cambial zone at the right. Some collapsed cells and a number of irregularly divided cells are present at this location. Although many fungal cells were observed in vessel elements of the other bundle, it seems to be much less affected. *Bar* = 2.2 μm

Fig. 13.51. Newly dividing cells (*small arrow*) next to or between a band of altered cells in the cambial area. Large cavities with traces of opaque material are present in the primary wall of like layer may then also form, on which an extra gelatinous layer may be added. Compartments may form in such cells, which may also be delimited by the same arrays of wall layers (Fig. 13.43) Histochemical tests (Sect. 13.6) have helped to confirm the nature of some of these layers.

13.5.3.3 Vessel Elements

Secondary vessel walls may also become thickened or contain an inner extra layer not traceable to tyloses (Ouellette 1981a). These depositions often seal off pit apertures (Fig. 13.46) and chambers of bordered and half-bordered pits. Pronounced thickening of pit membranes also occurs.

These reactions probably take place during or at the end of differentiation of vessel elements. When these cells retain their normal shape, their wall modifications may afford some resistance to pathogen invasion. It is doubtful that they serve such a purpose when vessel elements are entirely deformed (Fig. 13.39).

Portions of primary walls, situated between secondary thickenings in protoxylem vessels, may also become thickened, but apparently through secretion activity of bordering parenchyma cells (Fig. 13.11, and Ouellette unpublished observations).

13.5.3.4 Cambial and Differentiating Cells

In response to attack of the cambium and inner bark tissues, a range of cellular reactions, which are not always easy to decipher, can occur. In severe cases, one or more layers of altered cells, which may extend for some distance, demarcate the cambial area, overlaying the lastly differentiated, often contorted parenchyma cells and vessel elements (Figs. 13.3, 13.37, 13.50, 13.51). New wall material may have formed around some apparent cell compartments in the partially or completely collapsed cells present in those layers. Preserved cells, often in a state of division, are present above or in these layers, as well as vessel-like elements with secondary walls. Hypertrophy of ray cells (Fig. 13.39) results in many kinds of disorders. Less severe attacks are characterized by fewer collapsed or less distorted cells (Fig. 13.38). Many of the apparently reacting cells contain dense cytoplasm or opaque matter in their vacuoles (Figs. 13.40, 13.41).

Many cells do not have secondary walls or possess only a very thin one (Figs. 13.41, 13.42). In twigs, such cells are also present near the pith region either in initial or recurrent infections (Fig. 13.52). The first rows of cells bordering invaded vessels may deposit primary wall-like material that is fairly compact or poorly oriented and pervaded by strands or masses of opaque material (Figs. 13.4, 13.5, 13.41). Often the lumen of such cells is nearly empty and demarcated by an opaque thin band, the location of which corresponds to that of the plasmalemma (Fig. 13.42). In comparison, neighboring cells and those further from the vessel element retain a more normal appearance; little to more abundant material may be present in the periplasm (Fig. 13.29). Odd-shaped

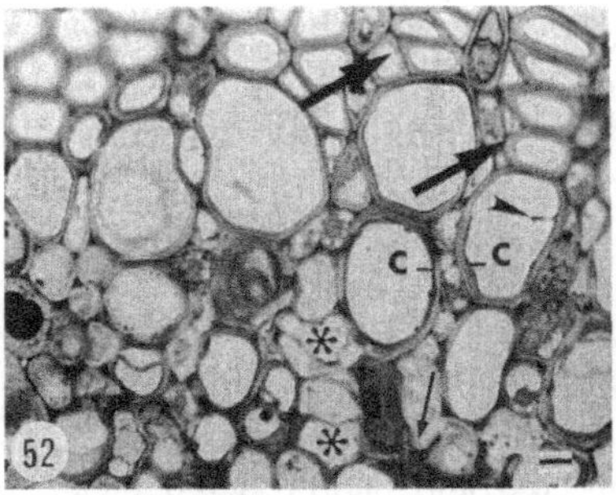

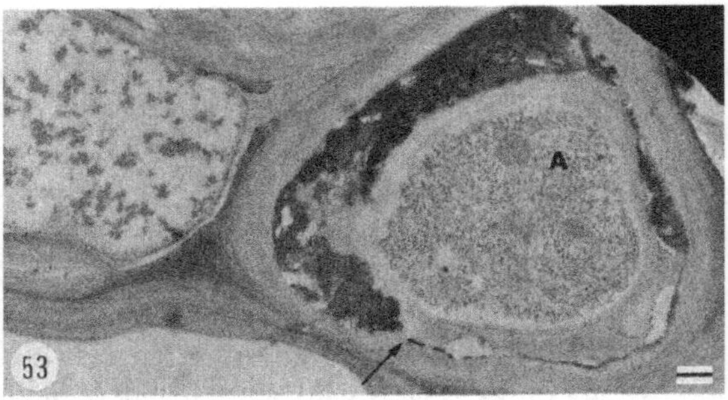

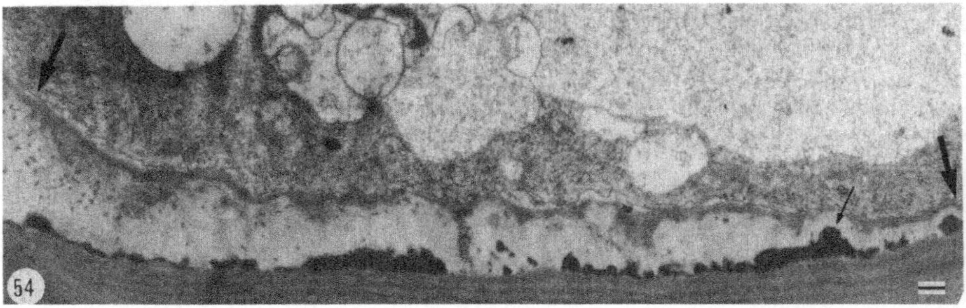

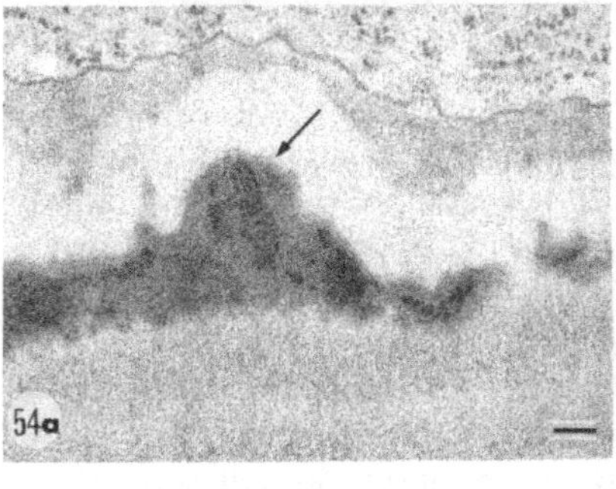

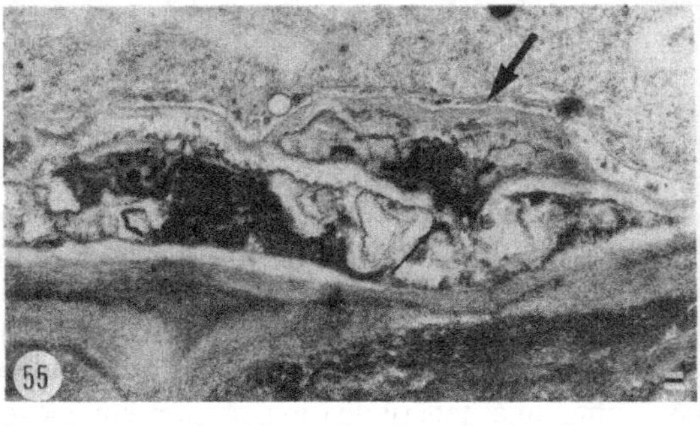

Fig. 13.52. Various types of xylem cell alterations near the pith area in a branch taken from a recurrent infection. Most prominent are collapsed cells or cells lacking a secondary wall (*). Some opaque bands extend some distance around such cells (*small arrow*). Cells typical of a barrier zone are (*arrows*) present outside the affected area; another band of apparently normal xylem cells separated these from another affected area (as in Fig. 13.51). Some of the vessels are coated (*C*) with opaque material, some of which extends into the vessel lumen, suggesting a small hypha (*arrowhead*). *Bar* = 3 μm

Fig. 13.53. A type of endocell (*A*) with an irregular wall but normal cytoplasm in a parenchyma cell near the cambial area. At some locations, the irregularities are continuous with the mother wall (*small arrow*), but most places are separated from it by bands or masses of opaque material. *Bar* = 0.4 μm

Fig. 13.54. In a living cell of an inoculated twig of *U. americana*, new wall-like material was deposited outside a retracted portion of its plasmalemma (*between arrows*). Opaque material in the periplasm is composed of a fibrillar matrix and opaque particles the size of ribosomes (*small arrow*; **a** enlarged portion; *bar* = 0.4 μm). Tangential section. *Bar* = 2 μm

Fig. 13.55. A situation similar to that shown in Fig. 13.54, but with a second zone (*arrow*) having formed over the first; the wall layer circumscribing the outer mass of opaque material is traversed by a band of similar material (*small arrow*). This cell is adjacent to altered cells. *Bar* = 0.5 μm vessels, some completely square, are present among ray cells and may correspond to abnormal differentiation.

Some types of endocells are often present in some cells in the invaded xylem area; they are bounded by primary wall-like material or even layers of secondary wall (Figs. 13.43, 13.53); it is suspected that these wall layers are formed around the retracted plasmalemma, circumventing portions of better protected cytoplasm (Fig. 13.18). Similarly, wall material may be formed in bordering shorter retracted portions of the plasmalemma, excluding opaque (at times cytoplasmic-like) material present in the periplasm or in the primary wall (Fig. 13.54); this even occurs in succession (Fig. 13.55).

These reactions in tissues bordering proto- and metaxylem are indicative of attempts at recovery as the altered cells were covered by cells containing a normal, dense cytoplasm with vacuole-filled opaque material (Fig. 13.41). These cells, some without secondary walls and probably related to necrophylactic tissue (Figs. 13.50, 13.51), may be the origin of the more normal xylem tissue that forms subsequently.

In *U. pumila*, invaded vessels are surrounded by layers of cells accumulating opaque material (Fig. 13.56). This situation does not differ significantly from circumscribed discoloration in American elm except that the cells in *U. pumila* are smaller. Opaque bands several μm thick and similar to those occurring in American elm (Fig. 13.50) are also noticeable over the invaded areas or in areas contiguous to them that may extend for long distances, not only tangentially but also radially. Close to invaded vessels or further out into the xylem, some bands may completely surround several contiguous, apparently still intact cells (Fig. 13.57). The opaque matter is intermingled with labyrinths of wall-like material, suggesting small collapsed cells. Some of this material is clearly present in the middle lamella and primary walls of those cells, which suggests that some of the meandering stretch of wall material might have originated by the disruption of bands from the primary wall or from some repeated exclusion of opaque material as indicated in Fig. 13.58 (Ouellette 1982), even in the apparently collapsed cells. Surprisingly, the bordering cells generally appear normal, having formed normal and even extra wall layers, either in fibers or in parenchyma cells (Fig. 13.59). These observations underline that if the opaque material is cytoplasmic-like, as we have often observed, then it is preserved from degradation, regardless of whether it originated from the host or from the pathogen. Regardless of origin, when present in the periplasm or in the primary wall, this material is not considered as normal by the cell, as it is excluded from it, however, with apparently less consequence in *U. pumila* than in *U. americana*. Significantly fewer parenchyma cells that seem to linger on before becoming disorganized have been observed in the former than in the latter.

13.5.4 Concluding Remarks

The numerous types of cell alterations and reactions in DED-infected elms cannot be accounted for solely by an indirect action of the pathogen. Dis-

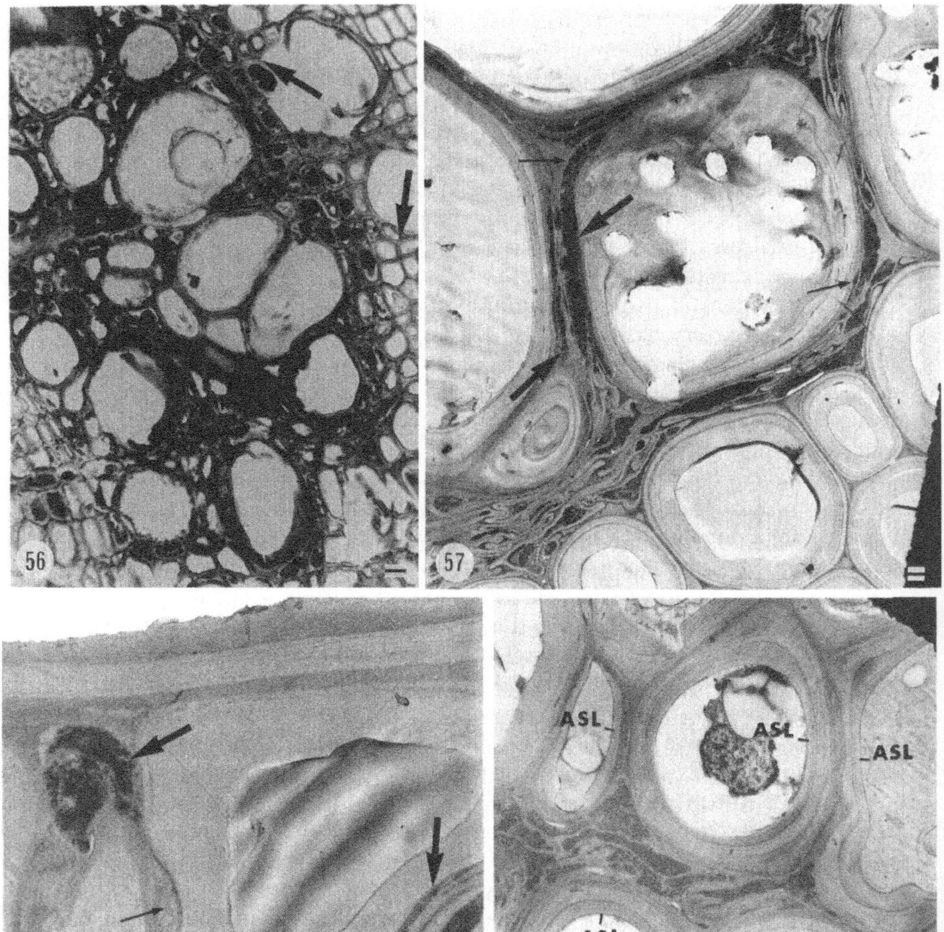
56
57
58
59
ASL
ASL
ASL
ASL
ASL

criminating between the various reactions in order to assign a pivotal function to any of them is not possible at this stage. They may all represent attempts to counteract or delay pathogen action. To select the reaction that appears the most common and related to resistance, the rapid increase in density and opacity of cytoplasmic contents in parenchyma cells would have to stand out; possibly related to this factor and equally as important would be modifications of wall layers in fibers and parenchyma cells (Sect. 13.5.2).

These observations and considerations also suggest the occurrence of hypersensitive reactions. If these are defined as a rapid cell death, triggering chemical or anatomical transformation of neighboring cells, the attacks and reactions at the cambial level would fit the definition and be analogous to similar observations in other plant wilt diseases (Pennypacker and Nelson 1972, Baayen 1988). Whether or not constituents of the pathogen are directly involved in the collapse of so many cells, it is obvious that in *U. pumila* such an action is better withstood or overcome than in *U. americana*. With regard to reactions of attacked cells present in the inner xylem and sometimes near the pith region, it is more problematical to apply criteria of hypersensitivity. Determining when such cells are actually dead, i.e., incapable of metabolic functions, is difficult. Many living cells with very dense and opaque cytoplasm might, in light microscopy observations, be classified as dead, but still be metabolically active or simply in a quiescent condition. Despite the uncertainty in classifying these types of cell reactions as hypersensitive or not, they nevertheless constitute a kind of barrier allowing maintenance or subsequent deposition of more healthy cells. The various reactions described would thus be beneficial as long as this subsequent deposition of functional tissue adequately compensates the deleterious effect of the impaired functions of the altered cells. In response to culture filtrates of *O. ulmi*, cytoplasm in cells of callus tissue culture obtained from a resistant elm selection also becomes dense and rich in opaque components (Pijut et al. 1990).

Figs. 13.56–13.59. Reactions in *U. pumila* to inoculation with *O. ulmi*

Fig. 13.56. Barrier zone formation (*arrows*) above a severely invaded zone of earlywood xylem. Cells subtending the barrier zone have abundant opaque contents and walls. Light microscope observations; *bar* = 3 μm

Figs. 13.57–13.59. Bands of opaque matter in intercellular areas or in walls (*arrows*) of parenchyma cells and fibers, sometimes completely encircling some of these. One to two additional secondary wall-like layers (*ASL*) have been deposited in surrounding cells (**Fig. 13.59**; *bar* = 0.8 μm). Whereas some of the bands of wall-like material intermixed with the opaque material might correspond to those of crushed cells, others probably correspond to sloughed off wall portions (*small arrows* in **Fig. 13.57**, *bar* = 0.9 μm; and **Fig. 13.58**, *bar* = 0.8 μm). They also may possibly be additional wall layers formed by the attacked cell (cf. Figs. 13.4, 13.15, 13.18, 13.37, 13.43, 13.54, 13.55).

13.6 Nonhost Trees

The use of nonhost plants in plant disease study has allowed us to improve our knowledge of these diseases (Heath 1980, 1981, Ride 1985). Few studies of DED pathogen development in inoculated nonhost species have been undertaken (Smucker 1941, Smalley and Riker 1962, Holmes 1966), and they have not provided details linking histological changes to resistance. This section proposes certain comparisons between microscopic changes in elm (described previously) and those induced in certain nonhost trees following artificial inoculation with *O. ulmi*. Three nonhost species (*Prunus penslyvanica*, *Populus balsamifera*, and *Sorbus americana*) were selected as representatives of the three groups that could be classified according to their increasing degree of resistance to *O. ulmi*. The results appeared in recent publications (Rioux 1989, Rioux and Ouellette 1989, 1991a,b). The aggressivity of the isolate used did not influence the microscopic changes observed.

As in *U. americana*, one of the first changes observed was the presence of alveolar reticular (AR) in vessel elements of *P. pensylvanica* by the 3rd day following inoculation. TEM examination revealed that the alveolar-delimiting matter (RSM) is often in continuity with thin coating material. By the 8th day after inoculation, ARs appeared intermixed with gels. Although the use of many histochemical reagents disclosed little information about the chemical nature of gels, at times they reacted positively to toluidine blue O (Rioux and Ouellette 1989), suggesting that phenols are one of their constituents. These phenols might confer a better stability of gels against the action of the pathogen. Gels showing this type of reaction were very rarely observed in inoculated American elm.

At about 5 days after inoculation, the cambial zone often appeared abnormal, and many cambial cells contiguous to the invaded xylem were stained brown and seemed slightly collapsed (Rioux and Ouellette 1989). Even when affected by the presence of *O. ulmi*, some of these cells survived, or new rows of healthy cells were generated and thereafter regularly formed a barrier zone (Sect. 13.6). When external symptoms were occasionally observed, histological examination revealed that all the xylem tissues around the branch were invaded, and that BZF was incomplete.

In the invaded xylem of *P. balsamifera*, tylosis formation was observed in vessel elements by the 8th day after inoculation, but gels and coating material were absent. A few days later, many parenchyma cells bordering such vessels, as well as some tyloses, appeared to have accumulated phenols (Rioux and Ouellette 1989). The tangential and longitudinal spread of *O. ulmi* was always restricted.

In *S. americana*, traces of gels and a few cells of *O. ulmi* were present in vessel elements; the most noticeable reaction was the rapid xylem accumulation throughout the invaded xylem (3 days after inoculation in vigorous branches) of what seemed to be phenols (Rioux and Ouellette 1989). These compounds might be responsible for the resistance observed with this nonhost. Phytoalexin production following inoculation appears to be a general response of nonhost plants (Heath 1981). In addition, considering that the normal xylem sap of different species was an equally good medium for the growth of four wilt fungi (Kessler 1966), including *O. ulmi*, it is probable that the initial differences in the pathogen action in these nonhosts are related to differences either in the nature of the phytoalexins formed in response to it, in the rapidity of production, and/or in the quantity produced. This first reaction could then retard colonization by *O. ulmi*, giving the tree enough time to form anatomical barriers in advance of the pathogen.

One of the most noticeable defense responses of the nonhosts was BZF, which regularly occurred in *P. pensylvanica* and *P. balsamifera* within 15 days after inoculation above the inoculation wound. Barrier zones were generally continuous and surrounded all the invaded xylem (Fig. 60). Parenchyma cells and gum canals were identified as the principal components of *P. pensylvanica* barrier zones, whereas fibers formed the main cell type in barrier zones of *P. balsamifera*.

The obvious BZF in these nonhosts incited us to reconsider it in *U. americana*. Other researchers report that special cells were formed in diseased elms (Buisman 1935, Banfield 1968), but these bands were not called barrier zones at that time. BZF was described as a defense mechanism in naturally infected American elms when the cambium was not killed by the infection (Shigo and Tippet 1981). BZF was studied in more detail by Bonsen et al. (1985) who found that they were much more frequent in larger parts of the tree than in small branches, in several *Ulmus* species. The poor availability of energy reserves in small branches might explain this difference in BZF.

The results of our LM and TEM observations and a series of histochemical tests follow. BZF occurred in American elm around 30 days after inoculation. BZF was usually discontinuous over the invaded xylem, even when the inoculated branch survived the infection (Fig. 13.61). It was then possible to observe many vessel elements containing various occluding structures and/or cells of *O. ulmi* that were not separated from the cambium by barrier zone cells (see Figs. 1, 10, 11 in Rioux and Ouellette 1991a). Elm branches inoculated with the non-aggressive isolate did not show external symptoms of wilting when samples were collected (by the 60th day after inoculation), except that leaves were smaller than normal. The number of invaded vessel elements, usually covered by barrier zone cells, was lower and more dispersed than in small branches inoculated with the aggressive isolate. Barrier zones contained parenchyma cells and fibers. Sometimes parenchyma cells predominated, but at other times fibers were more frequent. Intermediate situations also occurred where portions of each cell type appeared similar. Samples from naturally infected branches showed somewhat different histological changes, whereas all the xylem around the artificially inoculated branches was generally invaded; noncoalescing spots of invaded xylem were usually seen in samples from natural infection. These areas were often surrounded by barrier zone cells.

Among the histochemical tests used (Table 1 in Rioux and Ouellette 1991a), the phloroglucinol-HCL test for lignin and the various Sudan dyes for lipid compounds (e.g. suberin) gave good results. The method followed by Biggs (1984) using phloroglucinol-HCL and violet light excitation also permitted easy localization of the suberized cells.

60

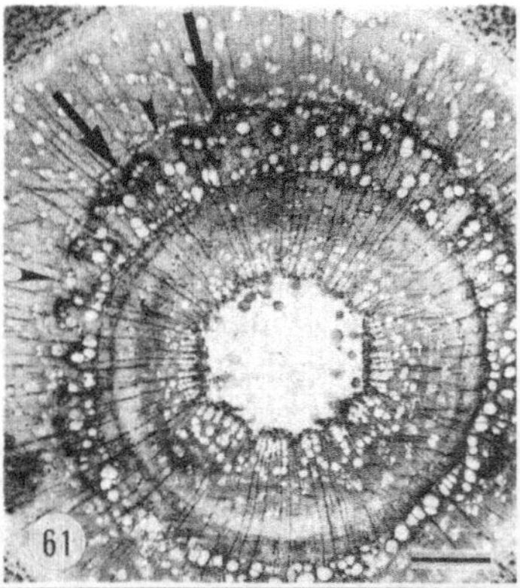
61

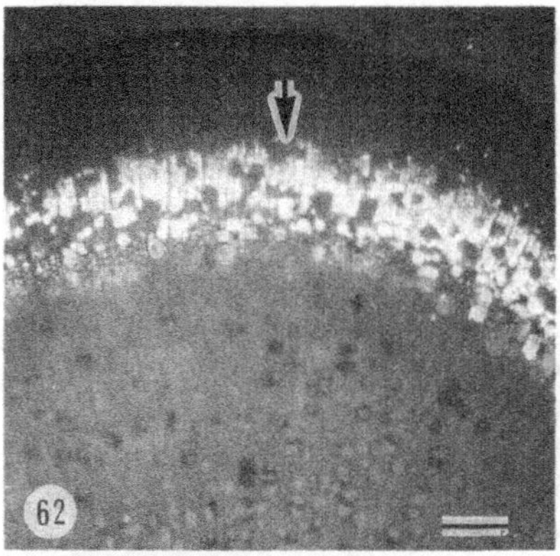
62

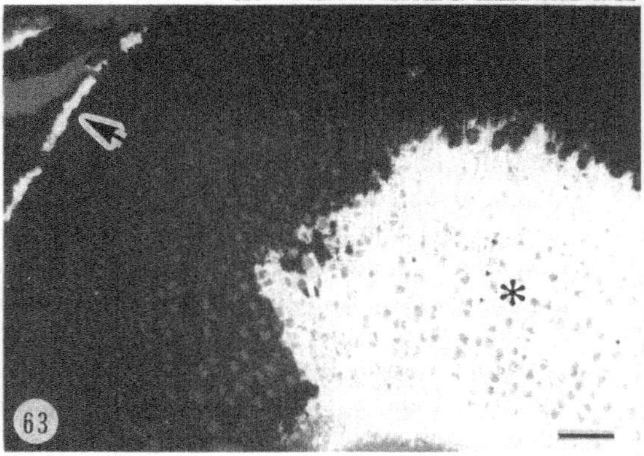
63

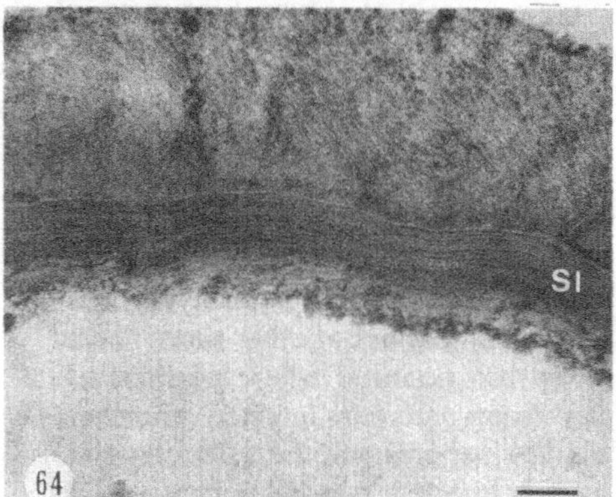
SI
64

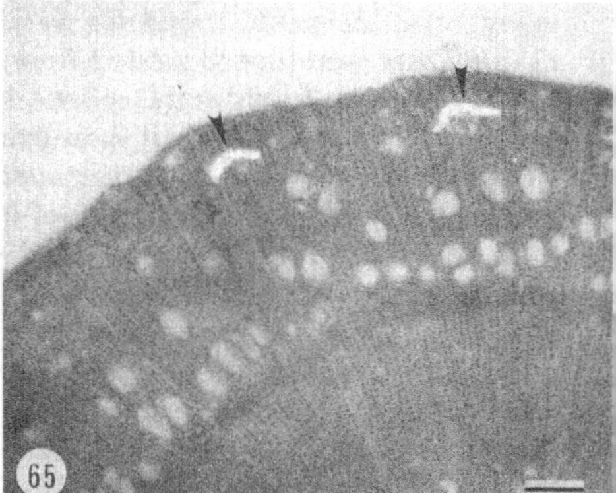
65

Barrier zone cells of *P. balsamifera* and *P. pensylvanica* reacted strongly with phloroglucinol-HCL (see Figs. 28 and 31 in Rioux and Ouellette 1991a). In *U. americana*, in addition to barrier zone cells, other tissue that encompassed the invaded vessel elements also showed a positive reaction for lignin (see Figs. 24 and 25 in Rioux and Ouellette 1991a).

Various types of observations due to different fluorescence techniques were obtained (Figs. 13.62, 13.63, 13.65). Another noticeable constituent detected in barrier zone cell walls was suberin. A large band of suberized cells was easily detected in barrier zones of *P. balsamifera* after treatment of sections with phloroglucinol-HCI (Fig. 13.62); in *P. pensylvanica*, however, such cells were usually sparse (Fig. 44 in Rioux and Ouellette 1991a), but, at times, contiguous cells showed strong positive reactions for suberin, as with Sudan black B (Fig. 43 in Rioux and Ouellette 1991a). TEM examination often permitted observation of the lamellar appearance of inner wall layers in barrier zone cells.

Other results relative to the CODIT model were obtained. In *P. balsamifera*, a reaction zone delimiting the tangential spread of *O. ulmi* (wall 3) was as obvious as BZF (Figs. 31, 34, 50 in Rioux and Ouellette 1991a). Wall 3 was apparently formed by more lignified cells, by suberized ray cells, and by vessel elements containing tyloses. A band of suberized cells was even observed adjacent to the pith that seemed to have originated from the dedifferentiation of some living cells of the medullary sheath (Figs. 35 and 49 in Rioux and Ouellette 1991a). Thus, the invaded xylem of *P. balsamifera* can become completely encompassed by lignified and suberized tissues. In *P. pensylvanica*, the formation of a wall 3 was less apparent, and there was no suberized band next to the pith. An aqueous solution of KMnO4, originally used because of its affinity for lignin (Hepler et al. 1970), allowed us to determine that the tissues found within the invaded xylem of these nonhost species were relatively impermeable to the solution. This was particularly evident in *P. balsamifera* when dry sections cut with the cryomicrotome were observed under blue or violet light illumi-

←

Fig. 13.60. A continuous barrier zone (*between arrows*) surrounding the invaded xylem of *P. balsamifera*. *Bar* = 300 μm

Fig. 13.61. Barrier zone cells (*arrows*) showing discontinuities (*arrowheads*) in the xylem of an *U. americana* branch that survived inoculation with an aggressive isolate of *O. ulmi*. *Bar* = 400 μm

Figs. 13.62–13.64. Samples collected from inoculated *P. balsamifera* branches.

Fig. 13.62. Suberized barrier zone cells (*arrow*) are easily detected under violet illumination after treatment of sections with phloroglucinol-HCl. *Bar* = 150 μm

Fig. 13.63. Sample treated with an aqueous solution of $KMnO_4$; the area circumscribed by the barrier zone fluoresced (*) under violet or blue light illumination, whereas the remaining parts are darkened, indicating that the former was impervious to the solution. Phloem fibers (*arrow*) were not penetrated by the solution. *Bar* = 150 μm

Fig. 13.64. The lamellar appearance of a suberized wall layer (*Sl*) is evident in this barrier zone fiber. *Bar* = 0.1 μm

Fig. 13.65. In *U. americana* xylem, two bands of suberized cells (*arrowheads*) of a discontinuous barrier zone are shown; this debarked section was treated with phloroglucinol-HCl and observed under violet light illumination. *Bar* = 120 μm. Sections of Figs. 13.62, 13.64, and 13.65 were obtained with a cryomicrotome from fresh samples nation (Fig. 13.63). Tissues permeated by the solution took on a blackish coloration, whereas others retained their usual appearance and showed strong autofluorescence.

Ultrastructural examination revealed that additional wall layers were generally formed in barrier zone cells. When histochemical tests for suberin were strongly positive, as with *P. balsamifera*, the lamellation typical of suberin was evident (Fig. 13.64). Other observations, such as the formation of additional secondary wall-like layers and the accumulation of electron-opaque material (suspected to be phenolic) within barrier zone cells, are described elsewhere (Rioux and Ouellette 1991b).

In *U. americana*, the use of phloroglucinol-HCI to quench lignin autofluorescence followed by observation under violet light also permitted the localization of a few suberized cells (Fig. 13.65) that very likely would not have been noted as such with routine microscopic techniques. These cells were dispersed with other cell types, mainly fibers, in such discontinuous barrier zones. Discontinuous BZF observed in the more susceptible situation probably allowed the pathogen to progress further and to succeed in invading the entire circumference of the branch xylem (as shown in Fig. 24 in Rioux and Ouellette 1991a). Some continuous barrier zones of *U. americana* were at times observed which were comprised mostly of fibers. These fibers had a slightly suberized internal wall layer (Fig. 26 in Rioux and Ouellette 1991a). In all species studied, the Prussian blue test (Pearce 1972) revealed that the interior of most barrier zone cells was filled with phenolic compounds (Figs. 29 and 42 in Rioux and Ouellette 1991a).

Concerning possible mechanisms of resistance against DED, the use of nonhost species allowed us to formulate the following hypothesis: a first reaction to limit pathogen colonization of xylem tissues would be the rapid production of phytoalexins. Such phytoalexins have been isolated and characterized from elm species. As mentioned in Section 13.5.2.2, however, these may not provide complete resistance to DED. This first response might be more rapid and/or equally intense in nonhosts than in elm. Differences in BZF may be just as important, however, to explain differences in susceptibility (Rioux and Ouellette 1989). If the response in elm were as efficient as in the nonhost in forming strongly lignified and/or suberized cells around the invaded xylem elements, elms might be less susceptible to DED than at present. In incipient infection the pathogen could spread and develop equally well in both the susceptible and more resistant trees. Although reactions could occur that subsequently restrain vertical spread of the pathogen (wall 1), these might not be effective without rapid and extensive lignification and seemingly interrelated suberization of cell walls in barrier zone formation (wall 4); differences in tree susceptibility could then be ultimately linked to differences in these last two reactions. Regarding the recurrence of the disease, it is obvious that the formation of barrier zones would be the most durable and effective reaction to pathogen attack, and may perhaps play an important role in assuring satisfactory recovery. Finding means that would enhance the formation of these barrier zones, which may be possible in the future, would be promising to combat this disease.

13.7 General Conclusions

This review shows that more than one mechanism may be implicated in DED resistance. Resistance appears more pronounced when these mechanisms occur soon after the onset of infection and with sufficient intensity. Therefore, it is often difficult to distinguish with certainty the role a particular mechanism plays in DED resistance.

The role of toxins and possible defense reactions triggered by toxins should be the object of future investigations. The availability of recent, more sophisticated histochemical and cytochemical techniques, concomitantly with the use of the various biochemical techniques that allow better insight into plant physiology, should permit clarification of the role that some molecules play in the pathogenesis of DED. In addition, genetic studies of the pathogen could undoubtedly improve our knowledge of the mechanisms by which the pathogenicity and virulence of this pathogen are expressed. Accordingly, such studies should shed some light on the defense reactions that a tree should possess in order to cope with DED. Although a miracle cure to DED has not yet been found, the ultimate goal of the investigations cited in this chapter, i.e., to improve elm resistance either by traditional or by molecular genetic techniques, may be attained in the future.

References

Abeles FB 1973 Ethylene in plant biology. Academic Press, New York, London, 302 pp

Baayen RP 1988 Fusarium wilt of carnation. Disease development, resistance mechanisms of the host and taxonomy of the pathogen. PhD Thesis, Univ Amsterdam, 165 pp

Banfield WM 1941 Distribution by the sap stream of spores of three fungi that induce wilt disease of elm. J Agr Res 62:637–681

Banfield WM 1968 Dutch elm disease recurrence and recovery in American elm. Phytopathol Z 62:21–60

Beckman CH 1956 Production of pectinase, cellulases, and growth-promoting substances by Ceratostomella ulmi. Phytopathology 46:605–609

Beckman CH 1959 Dutch elm disease control with polychlorobenzoic acid. Phytopathology 49:227 (Abstr)

Beckman CH 1987 The nature of wilt diseases of plants. Am Phytopathol Soc Press. St Paul, Minnesota, 175 pp

Beckman CH, Halmos S, Mace ME 1962 The interaction of host, pathogen, and soil temperature in relation to susceptibility to Fusarium wilt of bananas. Phytopathology 52:134–140

Beckman CH, Kuntz JE, Riker AJ, Berbee JG 1953 Host responses associated with the development of oak wilt. Phytopathology 43:448–454

Bell AA, Mace ME 1981 Biochemistry and physiology of resistance. In: Mace ME, Bell AA, Beckman CH (eds) Fungal wilt diseases of plants. Academic Press, New York, London, 431–486

Benhamou N 1989 Ultrastructural study of galacturonic acid distribution in some pathogenic fungi using gold-complexed Aplysia depilans gonad lectin. Can J Microbiol 35:349–358

Benhamou N, Chamberland H, Ouellette GB, Pauze FJ 1987 Ultrastructural localization of β-(1→4)-D-glucans in two pathogenic fungi and in their host tissues by means of an exoglucanase-gold complex. Can J Microbiol 33:405–417

Benhamou N, Lafontaine JG, Joly JR, Ouellette GB 1985b Ultrastructural localization in host tissues of a toxic glycopeptide produced by Ophiostoma ulmi, using monoclonal antibodies. Can J Bot 63:1185–1195

Benhamou N, Ouellette GB, Gardiner RB, Day AW 1986 Immunocytochemical localization of antigen-binding sites in the cell surface of two ascomycetes fungi using antibodies produced against fimbriae from Ustilago violacea and Rhodotorula rubra. Can J Microbiol 32:871–883

Benhamou N, Ouellette GB, Lafontaine JG, Joly JR 1985a Use of monoclonal antibodies to detect a phytotoxic glycopeptide produced by Ophiostoma ulmi, the Dutch elm disease pathogen. Can J Bot 63:1177–1184

Biehn WL, Dimond AE 1971 Effect of pectin source and sugars on polygalacturonase production by Ceratocystis ulmi. Phytopathology 61:745–746

Biggs AR 1984 Intracellular suberin: occurrence and detection in tree bark. IAWA Bull 5:243–248

Bilodeau M 1973 Étude de la carboxyméthylcellulase produite par Ceratocystis ulmi. MS Thesis, Univ Laval, Sainte-Foy, 83 pp

Bonsen KJM, Scheffer RJ, Elgersma DM 1985 Barrier zone formation as a resistance mechanism of elms to Dutch elm disease. IAWA Bull 6:71–77

Bornmann CH, Spurr AR, Addicott FT 1967 Abscission, auxin and gibberellin effects on the developmental aspects of abscission in cotton. Am J Bot 54:125–135

Brener WD, Beckman CH 1968 A mechanism of enhanced resistance to Ceratocystis ulmi in American elms treated with sodium trichlorophenylacetate. Phytopathology 58:555–561

Buisman C 1935 The anatomy of wood of elms infected with Graphium ulmi. Plantenzickten 41:104–120

Caroselli NE, Feldman AW 1951 Dutch elm disease in young elm seedlings. Phytopathology 41:46–51

Casagrande F, Ouellette GB 1971 A technique to study the development in wood of soft rot fungi and its application to Ceratocystis ulmi. Can J Bot 49:155–159

Claydon N, Grove JF, Hosken M 1974 Phenolic metabolites of Ceratocystis ulmi. Phytochemistry 13:2567–2571

Colombani F 1988 Étude de la β-glucosidase produite par Ceratocystis ulmi. MS Thesis, Univ Laval, Sainte-Foy, 277 pp

Czaninski Y, Pomerleau R, Moreau M 1975 Ultrastructure du xylème de l'orme atteint par le Ceratocystis ulmi (Buism.) C. Moreau. C R Acad Sci Paris Ser D 280:853–856

Dalpé Y 1983 L'influence de la carence en pyridoxine sur la morphologie et l'ultrastructure cellulaire de Ceratocystis ulmi. Can J Bot 61:2079–2084

Dimond AE 1970 Biophysics and biochemistry of the vascular wilt syndrome. Annu Rev Phytopathol 8:301–322

Duchesne LC 1985 Phytoalexin accumulation and resistance to Dutch elm disease. M Sc Thesis, Univ Toronto, 111 pp

Duchesne LC, Hubbes M, Jeng RS 1986 Mansonone E and F accumulation in Ulmus pumila resistant to Dutch elm disease. Can J For Res 16:410–412

Duchesne LC, Jeng RS, Hubbes M 1984 Effect of Ceratocystis ulmi strain aggressiveness on phytoalexin accumulation in Ulmus americana. Can J Plant Pathol 6:261 (Abstr)

Duchesne LC, Jeng RS, Hubbes M 1985 Accumulation of phytoalexins in Ulmus americana in response to infection by a nonaggressive and an aggressive strain of Ophiostoma ulmi. Can J Bot 63:678–680

Edgington LV 1963 A chemical that retards development of Dutch elm disease. Phytopathology 53:349 (Abstr)

Elgersma DM 1967 Factors determining resistance of elms to Ceratocystis ulmi. Phytopathology 57:641–642

Elgersma DM 1969 Resistance mechanisms of elms to Ceratocystis ulmi. Meded Phytopathol Lab Willie Commelin Scholten 77:1–84

Elgersma DM 1970 Length and diameter of xylem vessels as factors in resistance of elms to Ceratocystis ulmi. Neth J Plant Pathol 76:179–182

Elgersma DM 1973 Tylose formation in elms after inoculation with Ceratocystis ulmi, a possible resistance mechanism. Neth J Plant Pathol 79:218–220

Elgersma DM 1982 Susceptibility and possible mechanisms of resistance to Dutch elm disease. In: Kondo ES, Hiratsuka Y, Denyer WBG (eds) Proc Dutch elm disease, Symp and Worksh, Winnipeg, Manitoba, Oct 5–9, 1981, 169–177

Elgersma DM, Steerenberg PA 1978 A study of the development of Ophiostoma ulmi in elms with immunological techniques. Neth J Plant Pathol 84:127–133

Ellmore GS, Ewers FW 1985 Hydraulic conductivity in trunk xylem of elm, Ulmus americana. IAWA Bull 6:303–307

Feldman AW, Caroselli NE, Howard FL 1950 Chemicals for Dutch elm disease therapy. Phytopathology 40:8–9 (Abstr)

Fenn P, Durbin RD, Kuntz JE 1977 Synthesis of tryptophol and O-acvetyltryptophol from tryptophan by Ceratocystis fagacearum. Phytochemistry 16:899–901

Fenn P, Durbin RD, Kuntz JE 1978 Conversion of tryptophan to indole-3-acetic acid and other metabolites by Ceratocystis fagacearum. Physiol Plant Pathol 12:297–309

Foisner R, Messner K, Stachelberger H, Röhr M 1985 Wood decay by basidiomycetes: extracellular tripartite membranous structures. Trans Br Mycol Soc 85:257–266

Gagnon C 1967a Histochemical studies on the alteration of lignin and pectic substances in white elm infected by Ceratocystis ulmi. Can J Bot 45:1619–1623

Gagnon C 1967b Polyphenols and discoloration in the elm disease investigated by histochemical techniques. Can J Bot 45:2119–2124

Gagnon C 1968 Peroxidase in healthy and diseased elm trees investigated by the benzidine histochemical technique. Can J Bot 46:1491–1494

Gentile IA, Matta A 1975 Production of and some effects of ethylene in relation to Fusarium wilt of tomato. Physiol Plant Pathol 5:27–37

Hale MD, Eaton RA 1985a The ultrastructure of soft rot fungi. I. Fine hyphae in wood cell walls. Mycologia 77:447–463

Hale MD, Eaton RA 1985b The ultrastructure of soft rot fungi. II. Cavity-forming hyphae in wood cell walls. Mycologia 77:594–605

Heath MC 1980 Reactions of nonsuscepts to fungal pathogens. Annu Rev Phytopathol 18:211–236

Heath MC 1981 Nonhost resistance. In: Staples RC, Toenniessen GH (eds) Plant disease control. Wiley & Sons, New York, 201–217

Heath MC 1990 In vitro formation of haustoria of the cowpea rust fungus Uromyces vignae in the absence of a living plant cell. II. Electron microscopy. Can J Bot 68:278–287

Hepler PK, Fosket DE, Newcomb EH 1970 Lignification during secondary wall formation in Coleus: an electron microscopic study. Am J Bot 57:85–96

Hiratsuka Y, Takai S 1982 Scanning electronmicrographs of cerato-ulmin (CU) and Ceratocystis ulmi induced vascular symptoms in elm. In: Kondo ES, Hiratsuka Y, Denyer WBG (eds) Proc Dutch elm disease, Symp and Worksh, Winnipeg, Manitoba, Oct 5–9, 1981, 195–202

Holmes FW 1966 Resistance of Hemiptelea davidii to Ceratocystis ulmi. Phytopathology 56:881–882 (Abstr)

Hubbes M 1975 Terpenes and unsaturated fatty acids trigger coremia formation by Ceratocystis ulmi. Eur J For Pathol 4:129–137

Hubbes M, Jeng RS 1981 Aggressiveness of Ceratocystis ulmi strains and induction of resistance in Ulmus americana. Eur J For Pathol 11:257–264

Husain A, Dimond AE 1958 The function of extracellular enzymes of Dutch elm disease pathogen. Nat Acad Sci Proc 44:594–601

Jeng RS, Alfenas AC, Hubbes M, Dumas M 1983 Presence and accumulation of fungitoxic substances against Ceratocystis ulmi in Ulmus americana: possible relation to induced resistance. Eur J For Pathol 13:239–244

Keen NT 1981 Evaluation of the role of phytoalexins. In: Staples RC, Toenniessen GH (eds) Plant disease control: resistance and susceptibility. Wiley & Sons, New York, 155–177

Kessler KJ Jr 1966 Xylem sap as a growth medium for four tree wilt fungi. Phytopathology 56:1165–1169

Krause CR, Wilson CL 1972 Fine structure of Ceratocystis ulmi in elm wood. Phytopathology 62:1253–1256

Kulkarni RK, Nickerson KW 1981 Nutritional control of dimorphism in Ceratocystis ulmi. Exp Mycol 5:148–154

Leightley LE, Eaton RA 1977 Mechanisms of decay of timber by aquatic microorganisms. BWPA Annu Convent, 221–251

Leightley LE, Eaton RA 1980 Micromorphology of wood decay by marine microorganisms. In: Oxley TA, Allsopp D, Becker G (eds) Biodeterioration. Proc 4th Int Biodeterioration Symp. Pitman, London, 83–88

Lester DT, Smalley EB 1972 Response of Ulmus pumila and Ulmus pumila x rubra hybrids to inoculation with Ceratocystis ulmi. Phytopathology 62:848–852

Liese W, Schmid R 1964 Über das Wachstum von Bläuepilzen durch verholzte Zellwände. Phytopathol Z 51:385–393

Liming FG 1934 A preliminary study of the lengths of the open vessels in the branches of the American elm. Ohio J Sci 34:415–419

MacDonald, WL 1970 Electron microscopy of elm infected with Ceratocystis ulmi (Buism) C. Moreau. PhD Thesis, lowa State Univ, Ames, IA

MacDonald WL, McNabb HS Jr 1970 Fine-structural observations of the growth of Ceratocystis ulmi in elm xylem tissue. BioScience 20:1060–1061

MacDonald WL, McNabb HS Jr 1974 Electron microscope observations of Ceratocystis ulmi-induced tylosis development in Ulmus. Eur J For Pathol 4:2–10

Mace ME, Bell AA, Stipanovic RD 1978 Histochemistry and identification of flavanols in Verticillium wilt-resistant and -susceptible cottons. Physiol Plant Pathol 13:143–149

Mace ME, Solit E 1966 Interactions of 3-indoleacetic acid and 3-hydroxytyramine in Fusarium wilt of banana. Phytochemistry 56:245–247

MacHardy WE 1978 Mechanisms of resistance. In: Sinclair WA, Campana RJ (eds) Dutch elm disease, perspectives after 60 years. Search agriculture, vol 8 (5). Plant pathology. Cornell Univ Agric Exp Stn, New York; State Coll Agr; and Life Sci, Ithaca, NY, 25–26

MacHardy WE, Beckman CH 1973 Water relations in American elm infected with Ceratocystis ulmi. Phytopathology 63:98–103

Mansfield JW 1983 Antimicrobial compounds. In: Callow JA (ed) Biochemical plant pathology. Wiley & Sons, New York, 237–265

McNabb HS Jr 1971 A new look at Dutch elm disease control. Ames For 58:14–18

McNabb HS Jr, Heybroek HM, MacDonald WL 1970 Anatomical factors in resistance to Dutch elm disease. Neth J Plant Pathol 76:196–204

Newbanks D, Bosch A, Zimmermann MH 1983 Evidence for xylem dysfunction by embolization in Dutch elm disease. Phytopathology 73:1060–1063

Newcombe AG, Robb J 1989 The chronological development of a lipid-to-suberin response at Verticillium trapping sites in alfalfa. Physiol Mol Plant Pathol 34:47–58

Nordin JH, Mason TL, Smith LL, Willmann PA, Richards WC, Takai S 1987 Use of an enzyme-linked immunosorbent assay with murine ascitic antibodies to screen microorganisms for production of cerato-ulmin, a toxin of Ceratocystis ulmi. Phytopathology 77:96–100

Ouellet CE, Pomerleau R 1965 Recherches sur la résistance de l'orme d'Amérique au Ceratocystis ulmi. Can J Bot 43:85–96

Ouellette GB 1960 Studies of host and pathogen in relation to the infection process of the Dutch elm disease, caused by Ceratocystis ulmi (Buism.) C. Moreau. PhD Thesis, Cornell Univ, Ithaca, NY, 123 pp

Ouellette GB 1962a Morphological characteristics of Ceratocystis ulmi (Buism.) C. Moreau in American elm trees. Can J Bot 40:1463–1466

Ouellette GB 1962b Studies on the infection process of Ceratocystis ulmi (Buism.) C. Moreau in American elm trees. Can J Bot 40:1567–1575

Ouellette GB 1978a Unusual cell wall layers in elm parenchyma of secondary xylem. Can J Bot 56:2109–2113

Ouellette GB 1978b Fine structural observations on substances attributable to Ceratocystis ulmi in American elm and aspects of host cell disturbances. Can J Bot 56:2550–2566

Ouellette GB 1978c Ultrastructural observations on pit membrane alterations and associated effects in elm xylem tissues infected by Ceratocystis ulmi. Can J Bot 56:2567–2588

Ouellette GB 1978d Light and electron microscopic studies on cell wall breakdown in American elm xylem tissues infected with Dutch elm disease. Can J Bot 56:2666–2693

Ouellette GB 1980 Occurrence of tyloses and their ultrastructural differentiation from similarly configured structures in American elm infected by Ceratocystis ulmi. Can J Bot 58:1056–1073

Ouellette GB 1981a Ultrastructural cell wall modifications in secondary xylem of American elm surviving the acute stage of Dutch elm disease: vessel members. Can J Bot 59:2411–2424

Ouellette GB 1981b Ultrastructural cell wall modifications in secondary xylem of American elm surviving the acute stage of Dutch elm disease: fibres. Can J Bot 59:2425–2438

Ouellette GB 1982 Comparative ultrastructural observations of reactions of xylem parenchyma cells to Ceratocystis ulmi infection in Ulmus spp. In: Kondo ES, Hiratsuka Y, Denyer WBG (eds) Proc Dutch elm disease, Symp and Worksh, Winnipeg, Manitoba, Oct 5–9, 1981, 267–297

Ouellette GB, Benhamou N 1987 Use of monoclonal antibodies to detect molecules of fungal plant pathogens. Can J Plant Pathol 9:167–176

Ouellette GB, Gagnon C 1960 Formation of microendospores in Ceratocystis ulmi (Buism.) C. Moreau. Can J Bot 38:235–241

Overeem JC, Elgersma DM 1970 Accumulation of mansonones E and F in Ulmus hollandica infected with Ceratocystis ulmi. Phytochemistry 9:1949–1952

Pearce AGE 1972 Histochemistry theoretical and applied. Livingstone, London, 1130–1132

Peek RD, Liese W, Parameswaran N 1972 Infection and degradation of spruce root bark by Fomes annosus. Eur J For Pathol 2:104–115

Pegg GF 1976 The response of ethylene-treated tomato plants to infection by Verticillium albo-atrum. Physiol Plant Pathol 9:215–226

Pegg GF 1985 Life in a black hole. The micro-environment of the vascular pathogen. Trans Br Mycol Soc 85:1–20

Pegg GF, Gull K, Newsam RJ 1976 Transmission electron microscopy of Verticillium albo-atrum hyphae in xylem vessels of tomato plants. Physiol Plant Pathol 8:221–224

Pegg GF, Selman IW 1959 An analysis of the growth response of young tomato plants to infection by Verticillium albo-atrum. II. Production of growth substances. Ann Appl Biol 47:222–231

Pennypacker BW, Nelson PE 1972 Histopathology of carnation infected with Fusarium oxysporum f.sp. dianthi. Phytopathology 62:1318–1326

Pijut PM, Lineberger RD, Domir SC, Ichida JM, Krause CR 1990 Ultrastructure of Ulmus americana cells cultured in vitro and exposed to the culture filtrate of Ceratocystis ulmi. Phytopathology 80:764–767

Pomerleau R 1965 The period of susceptibility of Ulmus americana to Ceratocystis ulmi under conditions prevailing in Quebec. Can J Bot 43:787–792

Pomerleau R 1968 Progression et localisation de l'infection par le Ceratocystis ulmi dans l'orme d'Amérique. Phytopathol Z 63:301–327

Pomerleau R 1970 Pathological anatomy of the Dutch elm disease. Distribution and development of Ceratocystis ulmi in elm tissues. Can J Bot 48:2043–2057

Pope SA 1943 Some studies on the Dutch elm disease and the causal organism. PhD Thesis, Cornell Univ, Ithaca, NY 66 pp

Puhalla JE, Bell AA 1981 Genetics and biochemistry of wilt pathogens. In: Mace ME, Bell AA, Beckman CH (eds) Fungal wilt diseases of plants. Academic Press, New York London 145–192

Richards WC, Takai S 1988 Production of cerato-ulmin in white elm following artificial inoculation with Ceratocystis ulmi. Physiol Mol Plant Pathol 33:279–285

Ride JP 1985 Non-host resistance to fungi. In: Fraser RSS (ed) Mechanisms of resistance to plant diseases. Nijhoff, Boston, 29–61

Rioux D 1989 Changements structuraux et histochimiques associés à l'action d'Ophiostoma ulmi chez des plantes hôtes et non-hôtes. PhD Thesis, Univ Laval, Sainte-Foy, 157 pp

Rioux D, Ouellette GB 1989 Light microscope observations of histological changes induced by Ophiostoma ulmi in various nonhost trees and shrubs. Can J Bot 67:2335–2351

Rioux D, Ouellette GB 1991a Barrier zone formation in host and nonhost trees inoculated with Ophiostoma ulmi. I. Anatomy and histochemistry. Can J Bot 69:2055–2073

Rioux D, Ouellette GB 1991b Barrier zone formation in host and nonhost trees inoculated with Ophiostoma ulmi. II. Ultrastructure. Can J Bot 69:2074–2083

Robb J, Brisson JD, Busch L, Lu BC 1979 Ultrastructure of wilt syndrome caused by Verticillium dahliae. VII. Correlated light and transmission electron microscope identification of vessel coatings and tyloses. Can J Bot 57:822–834

Robb J, Busch L 1975a Ultrastructure of wilt syndrome caused by Verticillium dahliae. I. In chrysanthemum leaves. Can J Bot 53:901–913

Robb J, Busch L, Brisson JD, Lu BC 1975b Ultrastructure of wilt syndrome caused by Verticillium dahliae. II. In sunflowers leaves. Can J Bot 53:2725–2739

Robb J, Powell DA, Street PFS 1987 Time course of wall-coating secretion in Verticillium-infected tomatoes. Physiol Mol Plant Pathol 31:217–226

Rouxel T 1989 Les phytoalexines et leur intervention dans la résistance hypersensible aux champignons phytopathogènes. Agronomie 9:529–545

Russo PS, Blum FD, Ipsen JD, Abul-Hajj YJ, Miller WG 1982 A novel delivery mechanism for cerato-ulmin in elm trees. In: Kondo ES, Hiratsuka Y, Denyer WBG (eds) Proc Dutch elm disease. Symp Worksh, Winnipeg, Manitoba, Oct 5–9, 1981, 203–214

Salemink CA, Rebel H, Kerling LCP, Tchernoff V 1965 Phytotoxin isolated from liquid cultures of Ceratocystis ulmi. Science 149:202–203

Sansome E, Brasier CM 1973 Intracellular spore formation in Ceratocystis ulmi. Trans Br Mycol Soc 61:588–590

Savidge RA, Wareing PF 1981 Plant-growth regulators and the differentiation of vascular elements. In: Barnett JR (ed) Xylem cell development. Castle House, Tunbridge Wells, 192–236

Scheffer RJ 1982 A phytotoxic glycopeptide produced by Ophiostoma ulmi in several elm species and clones. In: Kondo ES, Hiratsuka Y, Denyer WBG (eds) Proc Dutch elm disease. Symp and Worksh, Winnipeg, Manitoba, Oct 5–9, 1981, 215–223

Scheffer RJ 1984 Dutch elm disease: aspects of pathogenesis and control. PhD Thesis, Univ Amsterdam, 89 pp

Scheffer RJ, Elgersma DM 1981 Detection of a phytotoxic glycopeptide produced by Ophiostoma ulmi in elm by enzyme-linked immunospecific assay (ELISA). Physiol Plant Pathol 18:27–32

Scheffer RJ, Elgersma DM 1982 A scanning electron microscope study of cell wall degradation in elm wood by aggressive and non-aggressive isolates of Ophiostoma ulmi. Eur J For Pathol 12:25–28

Scheffer RJ, Liem JI, Elgersma DM 1987 Production in vitro of phytotoxic compounds by non-aggressive and aggressive isolates of Ophiostoma ulmi, the Dutch elm disease pathogen. Physiol Mol Plant Pathol 30:321–335

Sequeira L 1964 Inhibition of indoleacetic acid oxidase in tobacco plants infected by Pseudomonas solanacearum. Phytopathology 54:1078–1083

Shigo AL 1982 Dutch elm disease: a CODIT perspective. In: Kondo ES, Hiratsuka Y, Denyer WBG (eds) Proc Dutch elm disease. Symp and Worksh, Winnipeg, Manitoba, Oct 5–9, 1981, 151–168

Shigo AL, Marx HG 1977 Compartmentalization of decay in trees. Agr Inf Bull US Dep Agr For Serv 405:73 pp

Shigo AL, Tippett JT 1981 Compartmentalization of American elm tissues infected by Cerastocystis ulmi. Phytopathology 65:715–718

Sinclair WA, Larsen AO 1980 Localization of Dutch elm disease in 10-yr-old white elm clones from resistant parents. Plant Dis 64:203–205

Sinclair WA, Zahand JP, Melching JB 1975a Localization of infection in American elms resistant to Ceratocystis ulmi. Phytopathology 65:129–133

Sinclair WA, Zahand JP, Melching JB 1975b Anatomical marker for resistance of Ulmus americana to Ceratocystis ulmi. Phytopathology 65:349–352

Singh D, Smalley EB 1969a Nitrogenous and carbohydrate compounds in the xylem sap of Ulmaceae species varying in resistance to Dutch elm disease. Can J Bot 47:335–339

Singh D, Smalley EB 1969b Nitrogenous compounds in the xylem sap of American elms with Dutch elm disease. Can J Bot 47:1061–1065

Singh D, Smalley EB 1969c Nitrogenous compounds in the xylem sap of Ulmus americana: seasonal variations in relation to Dutch elm disease susceptibility. For Sci 15:299–304

Smalley EB 1962 Prevention of Dutch elm disease by treatments with 2,3,6-triclorophenylacetate. Phytopathology 52:1090–1091

Smalley EB, Riker AJ 1962 Tropical members of the Ulmaceae resistant to Dutch elm disease. Univ Wisconsin, For Res Notes 77:4 pp

Smucker SJ 1941 Susceptibility of Planera and several elm species to Ceratostomella ulmi. Phytopathology 31:21 (Abstr)

Street PFS, Robb J, Ellis BE 1986 Secretion of vascular coating components by xylem parenchyma cells of tomatoes infected with Verticillium alboatrum. Protoplasma 132:1–11

Strobel GA, Van Alfen N, Hapner KD, McNeil M, Albersheim P 1978 Some glycopeptides from Ceratocystis ulmi, the Dutch elm disease pathogen. Biochim Biophys Acta 538:60–75

Svihra P 1980 Dutch elm disease in California. Univ Cal Agr Sci Leaflet 21189:7 pp

Takai S, Hiratsuka Y 1984 Scanning electron microscope observations of internal symptoms of white elm following Ceratocystis ulmi infection and cerato-ulmin treatment. Can J Bot 62:1365–1371

Takai S, Richards WC, Stevenson KJ 1983 Evidence for the involvement of cerato-ulmin, the Ceratocystis ulmi toxin, in the development of Dutch elm disease. Physiol Plant Pathol 23:275–280

Tchernoff V 1965 Methods for screening and for the rapid selection of elms for resistance to Dutch elm disease. Acta Bot Neerl 14:409–452

Tyree MT, Sperry JS 1989 Vulnerability of xylem to cavitation and embolism. Annu Rev Plant Physiol Mol Biol 40:19–38

Van Alfen NK, MacHardy WE 1978 Symptoms and host-pathogen interactions. In: Sinclair WA, Campana RJ (eds) Dutch elm disease, perspective after 60 years. Search agriculture, vol 8 (5). Plant pathology. Cornell Univ Agr Exp Stn, New York; State Coll Agric; and Life Sci, Ithaca, NY, 20–25

Van Alfen NK, Turner NC 1975 Influence of a Ceratocystis ulmi toxin on water relations of elm (Ulmus americana). Plant Physiol 55:312–316

Zentmyer GA, Horsfall JG, Wallace PP 1946 Dutch elm disease and its chemotherapy. Conn Agr Exp Stn Bull 498, 70 pp

Zimmermann MH 1983 Xylem structure and the ascent of sap. Springer, Berlin Heidelberg New York, 143 pp

Zimmermann MH, McDonough J 1978 Dysfunction in the flow of food. In: Horsfall JG and Cowling EB (eds) Plant disease – an advanced treatise. Vol III. How plants suffer from disease. Academic Press, New York London, 117–140

Chapter 14

Resistance in Aspen to Hypoxylon Canker

P.D. Manion and D.H. Griffin

14.1 Introduction

Hypoxylon canker of aspen has many of the features of an ideal system for studying woody plant defense mechanisms against fungi. The primary hosts, trembling and European aspens (*Populus tremuloides* and *P. tremula*) occur over broad geographic areas of the northern hemisphere, providing an immense array of genetic and environmental diversity for the development of this system. Less commonly affected hosts such as *P. grandidentata, P. trichocarpa, P. balsamifera, P. alba*, and others provide a foundation of plant materials with different levels of resistance and susceptibility. The capacity to generate hybrids among these and other nonhost poplars further expands the foundations of test plant materials.

The pathogen *Hypoxylon mammatum* provides an equally broad array of pathogenic capabilities between collections (Valentine et al. 1975, Griffin and Manion 1986) and within a single ascus (French and Manion 1975). The fungus occasionally is found as a pathogen on *Salix* (French and Juzwik 1984), *Alnus* (French et al. 1969), and *Acer* (Bier 1939) species and occurs widely as a saprobe on a broad range of species (Miller 1961). Although the fungus can be readily grown in culture, the ascospore stage can only be collected from the field. This lack of a perfect stage in culture unfortunately limits simple genetic manipulation and testing of the pathogen, but parasexualism and protoplast fusion might be effective substitute procedures. Lack of a perfect stage in culture also presents some difficulties with spore germination and infection tests, as will be discussed later.

At the outset, it is important to acknowledge that some would question the importance of aspen or of Hypoxylon canker as a disease. There is a deep-rooted impression that aspen is a "weed species." Forestry training generally emphasizes either quality sawlogs or intensive plantation systems. This approach neglects the abundant and important natural aspen resource of the Great Lakes and northern Rocky Mountains regions. It neglects the current and future potential assets of aspens in these and other regions where nursery-grown aspen seedlings are currently a small but significant part of the planting stock. It also neglects the future potential for plantations of selected and genetically improved aspens and poplar hybrids.

Many have contributed to our present understanding of Hypoxylon canker (Manion and Griffin 1986). This chapter demonstrates how this research establishes a foundation for a model system for current and future investigations of mechanisms of host defenses against fungi in a perennial woody species. We will discuss: (1) symptoms and factors associated with Hypoxylon canker; (2)

phenotypic expressions of resistance and susceptibility; (3) chemical mechanisms of resistance and susceptibility; (4) host and pathogen responses to drought stress; and (5) toxic metabolites from *Hypoxylon mammatum*.

14.2 Symptoms and Factors Associated with Hypoxylon Canker

The disease, Hypoxylon canker (Fig. 14.1) appears initially as a yellowed discoloration of normally green aspen bark. Pathogen invasion of the phloem, cambium, and outer xylem results in blackened and mottled tissues. Expansion of the cankered area, over time, eventually results in girdling and death of the stem or branch. *H. mammatum* also decays and structurally weakens the wood, thereby contributing to stem breakage in the cankered region.

Although this chapter will emphasize the host defenses against *H. mammatum*, it is important to recognize that insect wounds caused by *Saperda inornata, S. calcarata, Oberea schaumii, and Magicicada septendecim* have been associated with Hypoxylon cankers in some locations (Bier 1940, Nord and Knight 1972, Anderson et al. 1979, Anderson and Martin 1981, Ostry and Anderson 1983). Host defenses against insects may therefore play a role in host defenses against the pathogen. Absence of insect association or even wounds in other situations (Manion 1975) would suggest that host responses to insects may mimic similar host responses to other factors, such as moisture stress, a factor that will be addressed in more detail later.

14.3 Phenotypic Expressions of Resistance and Susceptibility

Differing amounts of canker incidence in natural clones of aspens have commonly been interpreted as being related to resistance and susceptibility. This interpretation is complicated by the clonal habit of aspen and the ability of the clone to survive and expand while individual ramets die and decompose. Very different patterns of disease development in two clones (Table 14.1) demonstrate this. Initially, clone 604 was rated as more resistant than clone 609 (French and Manion 1976), but when the same two clones were surveyed 12 year later (Falk et al. 1989), canker incidences were reversed. Canker incidence within a natural clone may either increase or decrease over a particular time period. Decrease in canker incidence over time occurs because of a natural thinning. It would appear that diseased individuals in one of the clones were thinned from the population faster than the occurrence of new infections, and therefore disease incidence decreased.

Vegetative propagation and establishment of the clones in three replicated blocks of nine trees each in a garden plantation confirmed this hypothesis (Manion, unpublished). After 16 years of natural infection in the garden plantation, the total canker incidence was slightly higher in clone 609 (Table 14.1).

Fig. 14.1A–J. Hypoxylon canker. **A, B** Blackened lower stem cankers; **C** stem breakage at canker; **D** young canker associated with insect gall; **E** yellow discoloration of a young canker assocaited with small branch axil; **F** conidial pillars of *Hypoxylon mammatum*; **G** yellow discoloration on margin of expanding canker; **H** black and light mottling of infected bark tissue as seen in a knife cut at the lower portion of the figure; **I** closeup of *H. mammatum* perithecia embedded in fungus stroma. (**A, C, F, I** from Manion 1981)

Table 14.1. Comparisons of percent canker incidence and canker-caused mortality in two clones of aspen with four different surveys

Clone No.	604	609
	(%)	(%)
1972[a] Sample survery	25	44
1984[b] Total survey	47	24
Hypoxylon killed	23	15

Sixteenth and eighteenth year surveys in a garden plantation[c]

year	16	18	year 16	18
Living trees with cankers on:				
main stem	27	46	8	11
outer branch	34	31	42	31
Hypoxylon killed	4	8	23	27
Total cankered trees	65	85	73	69

[a] Transect sample survey from French and Manion (1975).
[b] Total clone survey from Falk et al. (1989). Clones 604 and 609 are Falk designations B00 and B10, respectively.
[c] Unpublished data from Manion (1990) garden plantation survey based on 26 trees in each clone.

More than half of the cankered trees had infections only on the outer branches. Clone 604 had a much lower mortality than clone 609.

Two years later the total incidence of Hypoxylon canker had dropped slightly in clone 609 due to the shedding of infected branches, while total incidence of disease increased substantially in clone 604 due to expansion of previously undetected main stem infections and expansion of branch infections into the main stem (Table 14.1).

Although both clones were susceptible to infection, they clearly responded very differently. Clone 604 initially displayed some level of resistance to canker expression and mortality. This would suggest that clone 604 is the more resistant clone. Clone 609 appeared to be more susceptible because of the higher incidence of stem mortality, but mortality of infected branches may reduce canker expansion from infected branches to the main stem.

Examination of the canker development responses to artificial inoculation in a garden experiment (Griffin et al. 1984) provides a possible mechanism for understanding these differences. Clone 609 had a high incidence of branch death response and a low incidence of callus formation following artificial inoculation, suggesting high susceptibility to mortality by girdling. Clone 604 had a low incidence of branch death and a higher incidence of callus formation. To maintain the trees for other studies, this experiment was terminated 2 years after inoculation by cutting off the inoculated branch, but presumably the infections of clone 604 could have spread to the main stem, where they would be tolerated rather than cause rapid girdling and death.

Rapid branch death of clone 609 may therefore contribute to the long-term survival of the clone by preventing stem infections. In a natural population, infected stems would be thinned early to allow space for noninfected stems with greater long-term potential. On the other hand, clone 604, with its resistance to invasion, allows more branch infections to spread eventually to the main stem. The individual infected tree may survive long enough to produce abundant seed, thereby maintaining the genotype, but this clone, as a population with many stem cankers, is more liable to future catastrophic girdling, breakage, and death.

This comparison of the two clones shows that characterizing resistance or susceptibility cannot be based on natural infection at one point in time. It also shows some of the possibilities for selection and breeding. A very short rotation system could utilize resistant genotypes like 604, while a longer rotation system would benefit from hypersensitive genotypes like 609.

Artificial inoculation of aspens in the field with mycelium of *H. mammatum* grown on grain normally results in cankers (Shea 1963), particularly if the inoculations are done in the spring or early summer. Since most trees become infected, canker expansion over time has been used to demonstrate variation in the host and the pathogen (French and Manion 1975). Although we assume that retardation of canker expansion is a form of resistance, if one considers the two clones above, it is possible that rapid expansion may also lead to hypersensitive resistance. The time period over which the canker enlargement is measured also affects the interpretation. Significant differences in canker length at 4 and 12 months postinoculation disappeared at 16 months, suggesting that canker length is not a reliable measure of resistance to *H. mammatum* (Griffin et al. 1984).

Callus formation and branch death are two other measurable responses to mycelial inoculation (Valentine et al. 1975, Griffin et al. 1984). Since nonhosts of *H. mammatum* respond by formation of callus, it is assumed that callus is a resistance response. The incidence of callus formation in aspen is usually rather low, but with some less virulent isolates it can be very high.

In a test of five isolates of *H. mammatum* against nine aspen clones, there was a moderately negative correlation of callus frequency with canker length and a moderately positive correlation of branch death with canker length (Griffin et al. 1984). Highly variable clonal and isolate branch death response suggests that branch death might be a genetically controlled resistance or susceptibility factor, but the application of this information would require considerations similar to those applied to the two naturally infected clones discussed above.

More extensive analysis of canker length, branch death, and callus formation led to the development of a multivariate ordination technique that utilizes all of the variables in a model for ranking relative resistance or susceptibility (Griffin and Manion 1986). The ordinations of nine clones, including 604 and 609 discussed above, showed that both were moderately to highly susceptible relative to the nine clones. Clone 604 ranked more resistant than clone 609, which was one of the two most suscetpible clones in the group tested. While ordination does not provide a statistical test for significance, it does provide a foundation for generating hypotheses for comparisons and testing.

14.4 Chemical Mechanisms of Resistance and Susceptibility

Chemical mechanisms of resistance and susceptibility are still rather poorly understood, even though both preformed fungal inhibitors and phytoalexins have been reported for aspen. The absence of a clearly documented site of infection complicates the interpretation of the reports. There is no consensus of where and when to sample and what tissues to assay. With these complications in mind, we will review the literature on the chemical identification and activity of host-produced compounds.

Interest in bark extractives as sources of raw chemical materials stimulated extensive research on extractives of aspen bark (Pearl 1969). A parallel line of research dealt with the chemistry of fungal inhibitory compounds from aspen bark (Hubbes 1962, 1966, 1969). Pyrocatechol, another unidentified phenol, and two glycosides extracted from bark inhibited growth of *H. mammatum* (Hubbes 1966).

It was later shown that hydrolysis of one of the glycosides yielded benzoic and salicylic acids. These are suggested to be more active inhibitors than the original compound but are sometimes stimulatory, depending on the availability of glucose and amino acids in the medium (Hubbes 1969). At pH 6, a measured pH for poplar bark (Hubbes 1964), benzoic acid had no activity against *H. mammatum* (Hubbes 1966). Since *H. mammatum* can metabolize benzoic acid (Hubbes and Mehran 1968), the possibility of detoxification is also indicated.

These studies began to elucidate some of the fungus inhibitory properties of aspen bark but generated many additional questions. It is recognized that the compounds may be altered by the extraction and sterilization process. Hubbes (1966) incorporated autoclaved bark meal with malt agar for his growth inhibition tests. Bark collected in the fall was suggested to be completely inhibitory to *H. mammatum*, whereas bark collected in the spring was not inhibitory, but there were inconsistencies associated with location, year of collection, genotype, and tissue sampled. Bark from the base of the trees was more inhibitory than bark from the upper portions. In contrast, French and Manion (1976) demonstrated that propylene oxide sterilized bark added to water agar consistently stimulated growth of *H. mammatum*. Their samples, collected at various times, were small twigs representing 1- to 3-year-old tissues, because these were assumed to be the points of infection (Manion 1975).

Growth inhibition tests potentially evaluate host defenses to invasion. Inhibition of fungal growth by bark, but lack of inhibition by wood media, led Hubbes (1964) to conclude that the fungus first invades the wood. In contrast, ascospore germination tests demonstrated that the fungus germinated better on unwounded bark than wounded bark, but both were lower than the controls (Hubbes and D'Astous 1967). It has been generally assumed that wounds are necessary for infection, but no one has been able to consistently induce infection with wounds inoculated with ascospores (Shea 1963). The only consistent infection by ascospores has been with unwounded, water-stressed tissue culture plantlets (Belanger et al. 1989b).

Although ascospore germination tests have evaluated inhibitory and stimulatory properties of bark tissue (French and Oshima 1959, Hubbes and D'Astous 1967, Anderson and French 1972, French and Manion 1976, Mahoney 1983), the only compound to be specifically tested was pyrocatechol (Hubbes and D'Astous 1967), which inhibited germination on malt agar and in distilled water. Earlier tests demonstrated a reduced activity of pyrocatechol against mycelial growth as the temperature approached the optimum for the fungus (Hubbes 1966). The interaction of inhibitors with temperature on spore germination has not been evaluated.

The absence of detailed studies on the role of various spore germination inhibitors and stimulators reflects to some degree the difficulties of obtaining consistently high viability in ascospore collections. Spore germination tests, with controls germinating less than 50%, are difficult to interpret but not uncommon. Unfortunately, stroma have to be collected in the field and bacterial and fungal contamination is often a problem. We have found that spores expelled from moistened stroma at 10 °C to clean glass slides in Petri dishes have good viability and minimum contamination problems (French and Manion 1976). Ascospores collected on glass slides can be stored at 10 °C for months. The stroma can be dried at room temperature, stored at 10 °C, and induced to produce additional crops of spores upon wetting.

We should recognize that although inhibitory activity of bark against mycelia of *H. mammatum* has been demonstrated, some of the inhibitory activity may be artifacts of the testing systems. The activities of the compounds that have been identified, such as pyrocatechol, benzoic acid, and salicylic acid, are affected by pH, temperature, available nutrients, and possibly other factors (Hubbes 1966, 1969). The inhibitory activity varies within and among trees and with season. Although never properly tested, it has been suggested that the roles of climatic, soil nutrition, and genetic factors need to be clarified (Hubbes 1966). It should be apparent also that some of the unknown compounds of these early tests deserve more thorough testing.

Phytoalexins, or dynamic host response chemicals, have been reported for aspen. Inoculation of debarked stem pieces of aspen with mycelia of *H. mammatum* or *Alternaria* sp. resulted in exudates that were inhibitory to the germination of ascospores of *H. mammatum* (Flores and Hubbes 1979). No exudates were formed following ascospore inoculation of debarked stem pieces, debarked noninoculated controls, and nondebarked materials. Phenolic glycosides were identified from the exudates through thin layer chromatography (Flores and Hubbes 1980). Hydrolysis and chromatography identified two sugars, possibly rhamnose and galactose, and three phenols.

This research suggests the occurrence of phytoalexins in this pathosystem. It is difficult to characterize the tissues which produced the exudates. It is also difficult to interpret the induction of phytoalexins by mycelia but not ascospores in such an artificial system.

Attempting to evaluate a more natural system, Kuntz (1987) screened for antifungal compounds in extracts of xylem tissues surrounding a naturally occurring canker. After methanol extraction of xylem tissue of a Hypoxylon cankered stem, a number of antifungal compounds were identified, including p-coumaric acid, cinnamic acid, and benzoic acid. Other antimicrobial compounds, tentatively identified as phenolic glycosides, were also isolated with this procedure, which involved spraying thin layer chromatography plates with a spore suspension of *Cladosporium cladosporioides*. Mycelial inoculation of branch pieces in the laboratory demonstrated an increase in all three compounds over a 3-week time period. Inoculated samples had higher levels of antifungal compounds than the controls, indicating phytoalexin-like behavior for these phenolic acids.

14.5 Host and Pathogen Responses to Drought Stress

We have a current working hypothesis that water stress in aspen induces increased levels of amino acids, especially proline, which in turn leads to a condition more favorable for infection and invasion by *H. mammatum*. The fungus grows more rapidly in the presence of proline and other specific amino acids (Griffin et al. 1986). The foundations of the hypothesis evolved from field and greenhouse observations that suggested a general relationship of higher canker incidence on water-stressed aspens (Day and Strong 1959, Bier and Rowat 1962, Bagga and Smalley 1974, Bruck and Manion 1980a).

A common response of plants to water stress is an increase in the level of proline (Hanson and Hitz 1982). The amino acid content of water-stressed aspen plantlets in tissue culture (Belanger et al. 1990), and in a hydroponic system (Griffin et al. 1991), demonstrated that proline changes differed among clones and in relation to the conditions of water stress. Higher levels of proline were detected in the tissue culture plantlets than in the hydroponic system, but even higher levels were detected in potted plants (Griffin et al. 1986). Increases in proline, alanine, and glutamine were correlated ($r = 0.89$, 0.86, and 0.96, respectively) with disease ratings in response to ascospore inoculations in a water-stressed tissue culture system. All three of these amino acids stimulate growth of *H. mammatum* (Griffin et al. 1986) and therefore suggest that other amino acids may affect drought stress stimulation.

14.6 Toxic Metabolites from *Hypoxylon mammatum*

A recent review by Pinon and Manion (1990) reports that five active compounds have been separated from the culture filtrates of *H. mammatum*. Two toxic compounds have been identified as a diterpene sulfate (Bodo et al. 1987) and a tetralone. The diterpene sulfate, hymatoxin A, is responsible for most of the activity and induces necrosis of sensitive clones at less than $2\,\mu M$ concentration. Although extracts of cankered plant tissue have been shown to contain toxins (Schipper 1978), the toxic extract has not been shown to contain any of the known toxins from culture filtrates (Genetet et al. 1989).

At first, the metabolites were assumed to function as toxins contributing to pathogenesis (Schipper 1978, Stermer et al. 1984). Later studies suggested that the metabolites may function as elicitors (Griffin and Manion 1985). Toxic metabolite sensitivity has been reported variously as highly correlated with disease levels in the field (Bruck and Manion 1980b), not correlated to disease levels in the field (French 1976, Belanger et al. 1989a), negatively correlated with canker length, and positively correlated with callus formation in experimental inoculations (Griffin and Manion 1985). The problem with these comparisons arises from the selection of clones to study and the observation that disease incidence in the field changes over time and in relation to environmental variables.

Belanger et al. (1989a) attempted to accommodate some of these variables in regression models of environmental variables and laboratory toxin assays but still could not definitively identify the pathotoxin or elicitor roles for the fungal metabolites. Direct comparisons of toxin assays with inoculation assays such as mycelial inoculation of trees in the field (French 1976, Griffin and Manion 1985) or the ascospore inoculation assay in tissue culture (Kruger and Manion 1990) demonstrated a lack of correlation between assay procedures. To properly interpret these findings it is also necessary to consider that disease resistance, as observed in a clonal field population, may occur because of extreme susceptibility or because of tolerance to the pathogen.

Our understanding of toxic metabolites of *H. mammatum* is still rather tentative. Since the identification of the compounds is recent, most of the research on *H. mammatum* toxins has utilized partially purified culture filtrates rather than the pure compounds. Individual fungal isolates produce differing levels of toxic metabolites (Schipper 1978, Griffin and Manion 1985). An array of different bioassay procedures have been used, including placing petioles of detached leaves in culture filtrates (Schipper 1978, Stermer et al. 1984), applying a drop of filtrate to a needle wound on a detached leaf (French 1976, Griffin and Manion 1985, Belanger et al. 1989a) or on a tissue-cultured plantlet (Belanger et al. 1989a), or by placing internode segments on culture filtrate-ammended tissue culture media (Pinon 1986a, Baker 1986). Although the research on toxic metabolites to date has to be interpreted carefully, it does demonstrate a fruitful line of future research, involving the interaction of fungal metabolites with host defense compounds and fungal stimulators.

14.7 Summary and Conclusions

Hypoxylon canker of aspen has many ideal features for studying woody plant defense mechanisms against fungi. This lethal disease system has developed over a broad geographic and species range. The wealth of fieldwork with this disease was not described in this report but it is appropriate to mention that a number of insects and environmental variables have been associated with disease incidence (Manion and Griffin 1986). The specific mechanisms of interaction for most of the field observations have never been addressed.

Although much of the work has concentrated on North American aspens, the more recent recognition of Hypoxylon canker in Europe (Pinon 1986b) has resulted in significant progress, particularly with pathogenicity on a number of plantation-grown poplars and has provided a major contribution with regard to the identity and activity of toxic metabolites from *H. mammatum* (Pinon and Manion 1990).

We conclude that disease incidence in the field is not necessarily a good measure of resistance to natural infection. Repeated observations indicate that both tolerance and hypersensitivity types of responses contribute to the impact of disease. Environmental and time-dependent variables affect the development of disease within a field clone.

Controlled inoculation experiments in the field are also difficult to interpret. Three host response variables have been used: (1) canker length, a measure of the ability of the host to reduce the rate of invasion; (2) branch death, a hypersensitive response to rapid invasion that results in localization of the pathogen in branches or rapid girdling and death of individuals infected in the main stem; and (3) callus formation, a typical host defense response to protect and restore tissues exposed by wounding. Each of these generates a different interpretation or ranking of clones. By combining these in a multivariate ordination analysis, a relative ranking of clones or fungal isolates was possible. Hypotheses generated by this procedure can be tested and evaluated by other procedures.

Chemical mechanisms of resistance and susceptibility have been suggested by examining extracts and exudates of infected plant material. Although some of the test procedures may generate artifacts, this research provides an initial indication of the chemical materials that may be involved in host defense. The research on dynamic chemical responses to infection also may be criticized because of the procedures but these studies have led to the development of a tissue culture system for evaluating chemical responses to ascospore infection. The ability to activate or inhibit the response through selection of clones and water stress conditions provide an ideal model system for more rapid progress.

The mechanisms of drought stress response and the role of these in infection and invasion are clearly in the early stages of characterization. Our initial hypothesis on the role of proline has some experimental foundation, but the correlative evidence needs more extensive and definitive testing. Drought stress also may inhibit defense responses.

Toxic metabolites from *H. mammatum* have been suggested of function as pathotoxins or as elicitors of host defense. These differences need to be reconciled through more detailed examination of the relationship of metabolite response and host defenses or pathogen invasion. The recent identification of a number of different toxic metabolites from *H. mammatum* provides an opportunity for more controlled testing of pathotoxin or elicitor properties. The tissue culture system also provides a tool for more controlled testing.

The specific lines of research, such as environmental variables affecting disease, phenotypic expressions of resistance and susceptibility, chemical mechanisms of resistance and susceptibility, toxic fungal metabolites as pathotoxins or exicitors, and host and pathogen responses to drought stress, have generally examined isolated parts of an obviously complex host pathogen system. Future research will have to integrate these into a more holistic model.

References

Anderson DL, French DW 1972 Germination of ascospores of Hypoxylon mammatum in living aspen. Can J Bot 50:1973–1974

Anderson GW, Martin MP 1981 Factors related to incidence of hypoxylon canker in aspen and survival of cankered trees. For Sci 27:461–476

Anderson NA, Ostry ME, Anderson GW 1979 Insect wounds as infection sites for Hypoxylon mammatum on trembling aspen. Phytopathology 69:476–479

Bagga DK, Smalley EB 1974 The development of hypoxylon canker of Populus tremuloides: role of interacting environmental factors. Phytopathology 64:658–662

Baker SS 1986 Testing Populus tremuloides tissue culture response to Hypoxylon mammatum metabolites as an in vitro bioassay for Hypoxylon canker resistance. MS Thesis, State Univ NY, Coll Environ Sci For, Syracuse NY, 135 pp

Belanger RR, Falk SP, Manion PD, Griffin DH 1989a Tissue culture and leaf spot bioassays as variables in regression models explaining Hypoxylon mammatum incidence on Populus tremuloides clones in the field. Phytopathology 79:318–321

Belanger RR, Manion PD, Griffin DH 1989b Hypoxylon mammatum ascospore infection of Populus tremuloides clones: effects of moisture stress in tissue culture. Phytopathology 79:315–317

Belanger RR, Manion PD, Griffin DH 1990 Amino acid content of water-stressed plantlets of Populus tremuloides clones in relation to clonal susceptibility to Hypoxylon mammatum in vitro. Can J Bot 68:26–29

Bier JE 1939 Hypoxylon canker of maple. For Chron 15:122–125.

Bier JE 1940 Studies in forest pathology III Hypoxylon canker of poplar. Can Dep Agr Tech Bull 27:40 pp

Bier JE, Rowat MH 1962 The relation of bark moisture to the development of canker diseases caused by native, facultative parasites. VIII. Ascospore infection of Hypoxylon pruinatum (Klotzsche) Cke. Can J Bot 40:897–901

Bodo B, Davoust D, Lecommandeur D, Rebuffat S, Genetet I, Pinon J 1987 Hymatoxin A, a diterpene sulfate phytotoxin of Hypoxylon mammatum, parasite of aspen. Tetrahedron Lett 28:2355–2358

Bruck RI, Manion PD 1980a Interacting environmental factors associated with the incidence of hypoxylon canker on trembling aspen. Can J For Res 10:17–24

Bruck RI, Manion PD 1980b Mamatoxin assay for genetic and environmental predisposition of aspen to cankering by Hypoxylon mammatum. Plant Dis 64:306–308.

Day MW, Strong FC 1959 A study of hypoxylon canker on aspen. Mich Agr Exp Stn Q Bull 41:870–877

Falk SP, Griffin DH, Manion PD 1989 Hypoxylon canker incidence and mortality in naturally occurring aspen clones. Plant Dis 73:394–397

Flores G, Hubbes M 1979 Phytoalexin production by aspen (Populus tremuloides Michx.) in response to infection by Hypoxylon mammatum (Wahl.) Mill. and Alternaria sp. Eur J For Pathol 9:280–288

Flores G, Hubbes M 1980 The nature and role of phytoalexin produced by aspen Populus tremuloides. Eur J For Pathol 10:95–103

French DW, Hodges CS Jr, Froyd JD 1969 Pathogenicity and taxonomy of Hypoxylon mammatum. Can J Bot 47:223–226

French DW, Juzwik J 1984 Hypoxylon canker of Salix spp. Phytopathology 74:840

French DW Oshima N 1959 Host bark characteristics and infection by Hypoxylon pruinatum (Klot.) Cke. For Sci 5:255–258

French JR 1976 Screening aspen for resistance to hypoxylon canker. PhD Thesis, Mich State Univ, East Lansing, 66 pp

French JR, Manion PD 1975 Variability of host and pathogen in hypoxylon canker of aspen. Can J Bot 53:2740–2744

French JR, Manion PD 1976 Hypoxylon mammatum ascospore germination and mycelial growth on bark and wood media from young branches of trembling aspen. Can J Bot 54:1438–1442

Genetet I, Pinon J, Bodo B, Rebuffat S 1989 Hypoxylon mammatum toxins: possible involvement in canker development on aspen. Ann Sci For 46 (Suppl):868–870

Griffin DH, Manion PD 1985 Host-pathogen interactions as measured by bioassay of metabolites produced by Hypoxylon mammatum with its host Populus tremuloides. Phytopathology 75:674–678

Griffin DH, Manion PD 1986 Ordination, a multivariate method for estimating relative resistance of Populus tremuloides and virulence of Hypoxylon mammatum. Phytopathology 76:1289–1293

Griffin DH, Manion PD, Valentine FA, Gustavson L 1984 Canker elongation, branch death, and callus formation as resistance or susceptibility responses in Populus tremuloides and virulence or avirulence characteristics of Hypoxylon mammatum. Phytopathology 74:683–687

Griffin DH, Quinn K, McMillen B 1986 Regulation of hyphal growth rate of Hypoxylon mammatum by amino acids: stimulation by proline. Exp Mycol 10:307–314

Griffin DH, Schaedle M, Manion PD, DeVit M 1991 Clonal variation in amino acid contents of roots, stems, and leaves of aspen (Populus tremuloides Michx.) as influenced by diurnal drought stress. Tree Physiol 8:337–350

Hanson AD, Hitz WD 1982 Metabolic responses of mesophytes to plant water deficits. Annu Rev Plant Physiol 33:163–203

Hubbes M 1962 Inhibition of Hypoxylon pruinatum by pyrocatechol isolated from bark of aspen. Science 136:156

Hubbes M 1964 New facts on host-parasite relationships in the hypoxylon canker of aspen. Can J Bot 42:1489–1494

Hubbes M 1966 Inhibition of Hypoxylon pruinatum (Klotzsche) Cke. by aspen bark meal and the nature of active extractives. Can J Bot 44:365–386

Hubbes M 1969 Benzoic and salicylic acids isolated from a glycoside of aspen bark and their effect on Hypoxylon pruinatum. Can J Bot 47:1295–1301

Hubbes M, D'Astous R 1967 Factors affecting the germination of spores of Hypoxylon pruinatum. Can J Bot 45:1145–1153

Hubbes M, Mehran AR 1968 Decarboxylation of benzoic acid by Hypoxylon pruinatum. Phytopathol Z 63:232–238

Kruger BM, Manion PD 1990 Relationship between in vitro ascospore and toxic metabolite bioassays of Populus tremuloides tissue culture plantlets. Phytopathology 80:977

Kuntz BL 1987 Phytoalexins in Populus tremuloides Michx. MS Thesis, State Univ NY, Coll Environ Sci For, Syracuse, 63 pp

Mahoney EM 1983 Investigations on the infection court of Hypoxylon mammatum on aspen. MS Thesis, State Univ NY, Coll Environ Sci For, Syracuse, 94 pp

Manion PD 1975 Two infection sites of Hypoxylon mammatum in trembling aspen (Populus tremuloides). Can J Bot 53:2621–2624

Manion PD 1981 Tree disease concepts. Prentice-Hall, Inglewood Cliffs, NJ

Manion PD, Griffin DH 1986 Sixty-five years of research on hypoxylon canker of aspen. Plant Dis 70:803–808

Miller JH 1961 A monograph of the world species of Hypoxylon. Univ Georgia Press, Athens, 158 pp

Nord JC, Knight FB 1972 The importance of Saperda inornata and Oberea schaumii (Coleoptera: Cerambycidae) galleries as infection courts of Hypoxylon pruinatum in trembling aspen, Populus tremuloides. Great Lakes Entomol 5:87–92

Ostry ME, Anderson NA 1983 Infection of trembling aspen by Hypoxylon mammatum through cicada oviposition wounds. Phytopathology 73:1092–1096

Pearl IA 1969 Water extractives of American Populus pulpwood species barks – a review. TAPPI 52:428–431

Pinon JD 1986a Test d'inhibition de l'activité cambiale du Peuplier par Hypoxylon mammatum: mise au point et application. Eur J For Pathol 16:230–238

Pinon JD 1986b Situation d'Hypoxylon mammatum en Europe. EPPO Bull 16:543–546

Pinon JD, Manion PD 1991 Hypoxylon mammatum and its toxins – recent advances in their relationships with poplars. Eur J For Pathol 21:202–209

Schipper AL Jr 1978 A Hypoxylon mammatum pathotoxin responsible for canker formation in quaking aspen Phytopathology 68:866–872

Shea KR 1963 Induction of hypoxylon cankers in aspen by artificial inoculations. For Sci 9:2–7

Stermer BA, Scheffer RP, Hart JH 1984 Isolation of toxins of Hypoxylon mammatum and demonstration of some toxic effects on selected clones of Populus tremuloides. Phytopathology 74:654–658

Valentine FA, Manion PD, Moore KE 1975 Genetic control of resistance to Hypoxylon infection and canker development in Populus tremuloides. In: Proc 12th Lake States For Tree Improv Conf, US Dep Agr For Serv, 132–146

Chapter 15

Role of Mycorrhizal Fungi in Tree Defense Against Fungal Pathogens of Roots

N.E. STROBEL and W.A. SINCLAIR

15.1 Introduction

The roots of most flowering plants are capable of entering into mycorrhizal symbiosis, which enhances mineral nutrition or in some species is essential for adequate nutrition. Zak (1964) suggested that ectomycorrhizal fungi, besides enhancing plant growth by other mechanisms, might also protect their feeder roots from attack by pathogens. Subsequent research led to evidence for root protection by these fungi (reviewed by Marx 1972, 1975) and by vesicular-arbuscular (VA) mycorrhizae (reviewed by Schenck and Kellam 1978, Dehne 1982). The influence of mycorrhizal fungi on disease caused by root-pathogenic fungi (and nematodes) is now known to range from inhibitory to stimulatory (Schenck and Kellam 1978, Dehne 1982). The influence of mycorrhizal fungi on plant growth and disease development depends on the species and sometimes the strains of mycorrhizal and pathogenic fungi involved, as well as on the host cultivar, soil conditions, etc. Despite this complexity, interest in the practical application of mycorrhizal fungi for root protection and growth stimulation remains strong (Cordell et al. 1987, Caron 1989, Duchesne et al. 1989c, Marx et al. 1989. Marx and Ruehle 1989, Kropp and Langlois 1990). Although a high degree of host-symbiont specificity may be theoretically tolerable for the protection of roots of woody plants (which remain in place for many years), commercial application of mycorrhizal fungi for root protection may hinge on the development of fungal strains that can function broadly as root protectants. An understanding of the bases of specificity and the mechanisms of root protection, which are intimately related, is essential for the formulation of principles that can guide the selection and development of mycorrhizal fungi for root protection. Toward this understanding, we discuss the interactions of mycorrhizal fungi with roots, root pathogens, and rhizosphere microflora in light of contemporary interpretations of the bases of disease resistance in plants.

We have not attempted a broad survey of the literature on root protection; rather, references have been selected to illustrate principles and mechanisms, with emphasis on induced resistance. Because mechanistic research on the protection of roots of woody plants by mycorrhizal fungi has been limited, we draw on studies of nonwoody species where necessary. We also point out research opportunities and new techniques to address them.

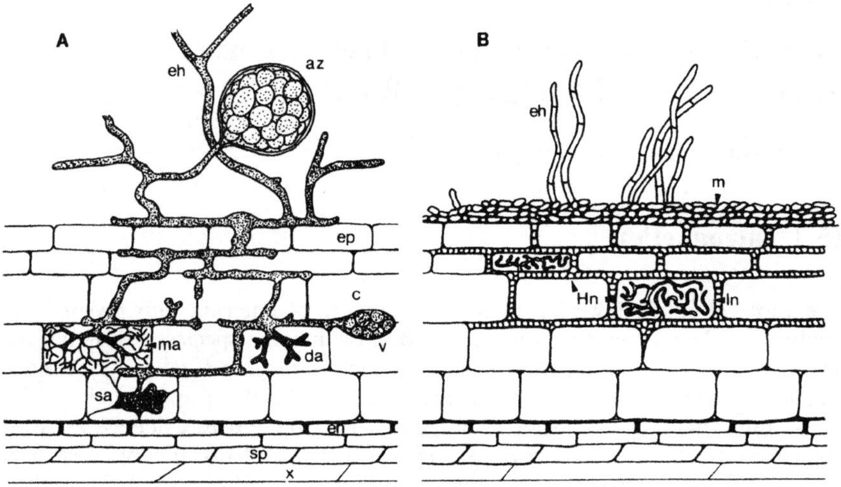

Fig. 15.1A,B. Schematic representations of longitudinal sections of **A** vesicular-arbuscular (VA) mycorrhiza and **B** ectomycorrhiza. Host structures represented include: *ep* epidermal; *en* endodermal; *c* cortical; *sp* stelar parenchyma; *x* xylem cells. **A** VA mycorrhizal structures external to the root surface include: *eh* extramatrical hyphae and *az* an azygospore. Structures within the root include intercellular and intracellular hyphae, a vesicle (*v*), and developing, mature, and senescent arbuscules (*da*, *ma*, and *sa*, respectively), which are within cortical cells. **B** Ectomycorrhizal structures external to the root surface include: *eh* extramatrical hyphae and *m* mantle (m). The Hartig net (*Hn*) consists of intercellular hyphae which form a labyrinthine network (*ln*) on the outer surfaces of cortical cell walls (Drawing prepared by M. Lindeberg, Department of Plant Pathology, Cornell University, whose contribution the authors gratefully acknowledge)

15.2 Overview of Mycorrhizal Symbiosis

Mycorrhizae are absorptive organs consisting of rootlets and filamentous fungi in symbiotic relationship. Plants that do not form mycorrhizae are unusual (Hirrel et al. 1978, Tester et al. 1987). Harley and Smith (1983) list seven kinds of mycorrhizae; this chapter will focus on the two principal types on woody plants, ectomycorrhizae and VA mycorrhizae.

The characteristic structures of an ectomycorrhiza are the Hartig net and the mantle (Fig. 15.1B). The Hartig net is a highly branched system of thin-walled fungal hyphae which develop among (but do not penetrate walls of) the cortical and epidermal cells of a root that has not undergone secondary growth. The Hartig net has a large surface area and is presumed to be the site of nutrient transfer between root and fungus. The mantle, or sheath, is comprised of densely packed hyphae, which are continuous with the Hartig net and enclose the rootlet. Extramatrical hyphae extend from the mantle into the soil and absorb water and nutrients.

The characteristic structures of a VA mycorrhiza are vesicles and arbuscules (Fig. 15.1A). Vesicles are round to ovoid swellings that contain lipids and are formed both inside and outside roots. Arbuscules are haustorium-like structures, formed by extensive dichotomous branching of intracellular hyphae, which invaginate the host plasmalemma. Like the Hartig net, arbuscules combine thin hyphal walls with a large surface area and are presumed to be the site of nutrient transfer between root and fungus. Absorption of water and nutrients from soil is again accomplished by extramatrical hyphae.

Ectomycorrhizae are largely restricted to woody plants. Harley and Smith (1983, Table 22) list 140 genera in 43 families containing species that form ectomycorrhizae. Most are forest dominants in the temperate and subarctic zones; a few grow in tropical and subtropical regions. VA mycorrhizae are formed by many woody species that do not form ectomycorrhizae and by the majority of herbaceous species. Woody plants in several genera (e.g., *Prunus*, *Populus*, *Salix*, *Acacia*, *Casuarina*, *Eucalyptus*) may have both VA and ectomycorrhizal associations. In *Eucalyptus*, both types have been found coexisting in individual root apices (Boudarga et al. 1990).

Fungi in many families of the Basidiomycotina and a few families of the Ascomycotina form ectomycorrhizae. These fungi can be grown in pure culture, but many grow slowly and do not form reproductive structures in the absence of host plants. In nature they grow only in association with living roots. VA mycorrhizal fungi are classified in three families of the order Glomales of the Zygomycotina (Morton and Benny 1990). These fungi have not yet been grown apart from host roots.

Mycorrhizal symbiosis is typically mutualistic, with both members deriving benefit from the association. Extramatrical hyphae of the mycorrhizal fungi exploit a greater volume of soil than do root hairs and thereby enhance the uptake by plants of phosphorus and other poorly mobile mineral nutrients (e.g., copper, zinc) from soil. Enhanced uptake of other nutrients, more readily mobile in soils, also has been reported (Tinker 1984). Mycorrhizae also function in the absorption of soil water by plants. In exchange for water and inorganic nutrients, the mycobionts receive organic compounds, especially simple carbohydrates, from their hosts.

Several factors may restrict the development of mycorrhizae in agricultural soils. High levels of nitrogen and phosphorus can suppress root colonization by mycorrhizal fungi. Soil sterilization by heat or biocides can destroy propagules of mycorrhizal fungi, and propagules may be absent from artificial rooting media. Cultivars of crop species vary widely in the degree to which they support mycorrhizal infections and in the degree of which their growth is enhanced by them.

15.3 Host-Symbiont Specificity and the Responses of Roots to Mycorrhizal Fungi

Host-fungus specificity, recognition, and compatibility in mycorrhizal symbioses have been discussed by Gianinazzi-Pearson (1984), Duddridge (1986a), and Anderson (1988).

15.3.1 Assessment of Specificity

The host range, or species specificity, of mycorrhizal fungi has been assessed through field observations and the study of plant-fungus interactions under controlled conditions. Field observations can establish the association of fungal reproductive structures (fruiting bodies of ectomycorrhizal fungi, spores of VA mycorrhizal fungi) with particular plant species. The presence of these structures signifies that the fungus is capable of completing its life cycle on a given host. Field observations have indicated the existence of both generalists and specialists among ectomycorrhizal fungi and their hosts (Molina and Trappe 1982a,b). Generalists can form associations with a wide range of symbionts, whereas specialists associate with a far narrower range of symbionts. VA mycorrhizal fungi and their hosts are primarily generalists.

In controlled studies, reproductive structures of VA mycorrhizal fungi typically form in a compatible plant-fungus combination, but ectomycorrhizal fungi often do not fruit in experimental pure culture systems. The evaluation of species specificity in these systems has been based primarily on the presence or absence of "normal" mycorrhizae. The normality of mycorrhizae can be assessed at the gross anatomical level (for ectomycorrhizae) and especially at the microscopic level, based on the presence of characteristic structures, i.e., the mantle and Hartig net for ectomycorrhizae, and vesicles and arbuscules for VA mycorrhizae. The viability of host and fungal cells is also assessed, generally on the basis of appearance. Full compatibility in the mycorrhizal symbiosis results in the maintenance of cellular viability. Recently, the vitality of experimental host/fungus combinations has been evaluated by quantifying the transfer of radiolabeled carbon and phosphorus between plant and fungus (Finlay 1989).

As has been the case with some plant pathogens, controlled inoculations with both VA mycorrhizal (Davis and Menge 1981) and ectomycorrhizal (Molina and Trappe 1982a,b) fungi have revealed broader host species ranges than were detected through field observations. Thus, biological specificity of mycorrhizal fungi is broader than their ecological specificity. Controlled inoculations also revealed that compatibility and incompatibility are not qualitative. Cytological evidence for "moderate incompatibility" has been found in some ectomycorrhizal associations (Molina and Trappe 1982b, Duddridge 1986a).

15.3.2 Responses of Roots to Incompatible Mycorrhizal Fungi

The interactions of nonhost roots with the ectomycorrhizal fungus *Suillus grevillei*, a mycosymbiont with a narrow host range, exemplify incompatibility. Duddridge (1986a) inoculated nonhost roots with this fungus and observed fungal development and host responses. On a few species, *S. grevillei* failed to colonize the rhizoplane. More often, incompatible interactions were characterized by the formation of a loose, fragmentary sheath and an irregular, weakly developed Hartig net. This partial colonization of roots by *S. grevillei* was accompanied by the accumulation of phenolic compounds in host cells and cell walls (and sometimes within the fungus), cell wall appositions, and degenerative changes in the cytoplasm and walls of the plant and fungus. Thus, plant responses to this ectomycorrhizal fungus resembled those to incompatible pathogens (Duddridge 1986a,b, Anderson 1988).

Moderate incompatibility appears to result when ectomycorrhizal fungi with narrow host ranges form morphologically normal mycorrhizae (compact sheath and complete Hartig net) in vitro on hosts with which the fungi are not normally found to associate in nature (Molina and Trappe 1982b, Duddridge 1986a). This moderate incompatibility is expressed as thickening and deposition of phenolics in host cell walls adjacent to fungal hyphae. Cytoplasmic degradation in plant and fungal cells are sometimes observed. Incompatibility of VA mycorrhizal fungi with roots is also indicated by abnormal fungal development and host responses. Many members of the families Brassicaceae and Chenopodiaceae do not form true VA mycorrhizae (Harley and Smith 1983). Hirrel et al. (1978) observed no infection of these plants unless they were grown in association with mycorrhizal plants (donors), in which case the incidence of infection was quite low (1–5%), and was generally restricted to moribund roots. Where infection was observed, the VA mycorrhizal fungus colonized both the cortex and stele, and although vesicles were sometimes formed, arbuscules were not. This pattern, atypical of VA mycorrhizae, indicated that the roots had little or no resistance to colonization, probably due to their moribund condition (Hirrel et al. 1978).

Incompatibility of VA mycorrhizal fungi with some members of these families may involve failure of the fungus to penetrate the root. Tommerup (1984) found that when the VA mycorrhizal fungus *Glomus caledonium* was inoculated as spores (no nurse plant) to *Brassica napus*, it formed appressoria that attached firmly to roots, but it then formed penetration pegs less frequently than on a compatible host, and about 25% of the pegs formed lacked cytoplasm. Incompatibility was not accompanied by hypersensitive-type plant responses (Tommerup 1984).

Tester et al. (1987) reasoned that the quantity and quality of root exudates (including potential preformed chemical inhibitors) probably play only a small role in failed penetration of nonhosts by VA mycorrhizal fungi, and proposed that control of penetration is exerted by interactions occurring at the level of the cell wall and/or the middle lamella. In contrast, Allen et al. (1989) found that incompatibility of the nonmycotrophic plant *Salsola kali* (Chenopodiaceae) with VA mycorrhizal fungi developed subsequent to fungal penetration and involved active host resistance. Although arbuscules were sometimes formed in roots, infected tissues turned brown and died several days later. Similar responses were obtained with VA mycorrhizal fungi of two genera, indicating that this may be a general response of *S. kali* to such fungi.

Nonmycorrhizal mutants of pea (*Pisum sativum*) and faba bean (*Vicia faba*), species that normally form VA mycorrhizae, were recently identified (Duc et al. 1989). In the nonmycorrhizal mutants, mycelial development was normal in the rhizosphere, and although appressoria were malformed, penetration occurred. However, colonization was limited to one or a few cells, and arbuscule formation was rarely detected. Underlying mechanisms for the incompatibility were not reported, but presumably involve elicitation of active host responses. Further study of these and other mutants (both plant and fungal) should provide important insights into the molecular bases of compatibility and incompatibility in the mycorrhizal symbiosis.

15.3.3 Responses of Roots to Compatible Mycorrhizal Fungi

Roots infected by compatible mycorrhizal fungi exhibit a variety of defense-related responses. Phenolic materials and terpenes accumulate in response to ectomycorrhizal fungi (for a review of this topic, see Duchesne et al. 1987). VA mycorrhizal fungi elicit transient increases in the activities of cell wall-bound peroxidase (Spanu and Bonfante-Fasolo 1988) and chitinase (Spanu et al. 1989) during early stages of mycorrhizal development and modest accumulation of phytoalexins (Morandi et al. 1984, Morandi 1989, Wyss et al. 1989), phenolic substances (Krishna and Bagyaraj 1984), and cyanoglucosides (Lieberei and Feldmann 1989) during later stages of the association. Responses to compatible fungi differ from those to incompatible fungi in their nature and intensity, however, and do not prevent root colonization and the formation of functional symbiotic infections. For example, although phenolic materials accumulate in host cortical cells in both compatible and incompatible ectomycorrhizal associations, the phenolics remain localized within host cells in the former, whereas in the latter these materials are deposited in host cell walls (Molina and Trappe 1982b, Duddridge 1986a), and may also migrate into the hyphae of the incompatible fungus (Malajczuk et al. 1982).

The expression of host defenses in compatible interactions with mycorrhizal fungi may play no role in the symbiosis (thus perhaps being vestigial in this regard) or may function in the regulation of the nature and degree of root colonization by mycobionts. Fungi that are typically intercellular in ectomycorrhizal hosts may colonize senescing root cells internally (Wilcox 1983), and these fungi routinely engage in intracellular colonization of roots of other host species, in which they form arbutoid mycorrhizae (Molina and Trappe 1982a). This intracellular colonization by ectomycorrhizal fungi presumably involves penetration of cell walls by means of fungal enzymes not produced in normal ectomycorrhizae, and indicates that the living host cells may play a role in regulating fungal colonization. A potential for pathogenesis by ectomycorrhizal fungi is indicated by observations of Duddridge (1986b). Amendment of axenic cocultures with soluble carbohydrate converted compatible ectomycorrhizal associations into incompatible associations involving host cell death. The potential for pathogenesis in certain VA mycorrhizal fungus-host combinations has also been established (Modjo and Hendrix 1986, Allen et al. 1989). The rapid turnover of arbuscules in VA mycorrhizal infections led Yawney and Schultz (1989) to suggest that this stage of the symbiosis is unstable. Thus, compatible mycorrhizal symbiosis may represent a tentative balance between host and fungus that is characterized, if not in part maintained, by persistent low-level expression of plant defensive mechanisms. The possible role of this chronically activated state of host defenses in root protection is discussed in Section 15.4.6.

Patterns of protein accumulation of both host and fungal origin in compatible ectomycorrhizal (Hilbert and Martin 1989) and VA mycorrhizal (Dumas et al.

1989) symbioses are now being studied. This work is allied with the study of pathogenesis-related proteins (reviewed by van Loon 1985). If the functions of specific proteins can be identified, this approach may contribute to our understanding of the molecular bases of compatibility, incompatibility, and moderate incompatibility between roots and mycorrhizal fungi. For example, some mycorrhizins (symbiosis-specific proteins elicited by VA mycorrhizal infection) of soybean were recently found to be immunologically cross-reactive and possibly identical with nodulins elicited by a nodulating bacterium (Wyss et al. 1990). The inability of certain mutants of pea to form either VA mycorrhizae or rhizobial nodules (Duc et al. 1989) may thus be due to the absence or altered expression of such symbiosis-related proteins. Furthermore, if protein patterns consistently associated with plant-mycobiont combinations that express enhanced resistance to root-pathogenic fungi can be identified, this information may find application in the rapid initial screening of naturally occurring and genetically engineered mycorrhizal fungi for their ability to protect roots from fungal pathogens. Such screening might be done, for example, on root cultures derived from plants transformed with T-DNA from *Agrobacterium rhizogenes* (Mugnier and Mosse 1987). The roles in induced resistance of specific proteins, whose expression is stimulated by mycorrhizal infection, may be evaluated by challenging, with fungal pathogens, transgenic plants that constitutively express these proteins at a high level. This approach has validated the supposition that chitinase can play a role in resistance to a fungal pathogen whose hyphal walls contain chitin (Broglie et al. 1990).

15.4 Mechanisms by Which Mycorrhizal Fungi Protect Roots

15.4.1 Synoptic View

Protection of roots from fungal pathogens by mycorrhizal fungi has generally been attributed to structural or chemical changes that occur prior to challenge with the pathogen. In each of the five mechanisms discussed by Marx (1972), for example, the relationship of the host to the pathogen is viewed as primarily passive or static, although changes induced by mycorrhizal fungi may confer resistance to pathogens. These mechanisms, conceived to explain root protection by ectomycorrhizal fungi, are: antibiotic production by the fungal symbiont, presentation of a mechanical barrier in the form of the mantle, production of chemical inhibitors by the host in response to the mycorrhizal fungus, alteration in chemical exudation of mycorrhizal as compared to nonmycorrhizal roots, and induction of protective microbial populations in the rhizosphere. We will briefly discuss examples of these mechanisms and will consider in detail the proposition that root protection may also involve induced changes in the recognition of and response to pathogens by roots. Plant tolerance to root pathogens due to enhanced uptake of nutrients, an important benefit of mycorrhizae under certain conditions (Smith 1988), will not be considered.

15.4.2 Antibiosis

Numerous cases of antibiotic production by ectomycorrhizal fungi in vitro have been reported. For examples, see the list compiled by Marx (1972). For antibiotics to play a role in root protection, they must accumulate in the rhizosphere and/or plant tissues in concentrations inhibitory to pathogens. Mycorrhizal fungi, because of their close association with roots, are well situated to obtain nutrients necessary for antibiotic production. Marx and Davey (1969a) found that diatretyne antibiotics produced by *Leucopaxillus cerealis* var. *piceina* were associated with the protection of mycorrhizal and adjacent nonmycorrhizal short roots of shortleaf pine (*Pinus echinata*) seedlings from *Phytophthora cinnamomi*. The diatretyne antibiotics were present in mycorrhizae, in the rhizosphere, and also in the adjacent nonmycorrhizal short roots, presumably due to translocation within the root system or uptake from the medium. Nonmycorrhizal short roots adjacent to mycorrhizae formed with fungi that did not produce antibiotics were not similarly protected. Protection from *P. cinnamomi* of nonmycorrhizal short roots adjacent to mycorrhizae was also observed on shortleaf and loblolly (*Pinus taeda*) pines inoculated with *Suillus luteus*, presumably due to antibiotics produced by the fungus (Marx 1969). Duchesne et al. (1988) observed that the ectomycorrhizal fungus *Paxillus involutus* can protect roots of red pine (*Pinus resinosa*) from *Fusarium oxysporum* f. sp. *pini* in vitro. This protection was closely linked to the production by *P. involutus* of antifungal substances external to the root (Duchesne et al. 1989d). These substances suppressed spore germination and induced structural and chemical alterations in the pathogen (Farquhar and Peterson 1990). Root protection was largely reproduced by incorporation of the principal antifungal substance, oxalic acid, into the culture system at a concentration (5.5 mM) comparable to that produced by *P. involutus* when grown in association with roots (Duchesne et al. 1989b). Production of antifungal chemicals and root protection from *F. oxysporum* f. sp. *pini* were not observed in associations of *P. involutus* with *Pinus banksiana* or *P. strobus*, and antifungal substances were not detected in association with the slight degree of protection of *P. sylvestris* afforded by this fungus (Duchesne et al. 1989a). This differential influence of *Pinus* species on root protection indicates yet another level of specificity that must be considered in the evaluation of mycorrhizal fungi that protect roots through the production of antibiotics. Soil chemistry and micobiota could also be expected to condition root protection mediated by oxalate, because oxalate concentrations may be decreased by precipitation with cations such as Ca^{2+} and Fe^{3+}, and by microbial degradation. However, Farquhar and Peterson (1990) report that root protection by *P. involutus* has also been obtained in a soil-culture system.

Antibiotics produced by ectomycorrhizal fungi might also protect roots indirectly, by inducing resistance to pathogens. Concentrations of oxalate, ranging from 0.06- to 3.5-fold greater than those found in the *P. resinosa* – *P. involutus* system of Duchesne et al. (1989b), elicit phytoalexin accumulation in soybean (*Glycine max*) (Favaron et al. 1988), and *P. involutus* is known to elicit extensive phenolic accumulation in a number of hosts (Molina and Trappe 1982b). Although *P. involutus* did not induce accumulations of phenolics

(Farquhar and Peterson 1990) or extractable antifungal materials (Duchesne et al. 1988) under the specific cultural conditions that have been reported, it is possible that these and other host resistance mechanisms elicited by oxalate and other metabolites of *P. involutus* may contribute to root protection under other conditions. Fungus-elicited, nontoxic phenolics do play a role in protection of primary roots of Douglas-fir (*Pseudotsuga menziesii*) from *F. oxysporum* by *Laccaria bicolor*, and this protection is dependent upon cultural conditions (Strobel and Sinclair 1991a,b). Polyene antibiotics, such as that found by Hillis and Ishikura (1969) among extractives of ectomycorrhizae of *P. menziesii*, interact with sterols of fungal membranes as a primary means of fungitoxicity (Fowlks et al. 1967). The saponin digitonin, which also interacts with sterols (Fowlks et al. 1967), has been found to induce the generation of superoxide anion and resistance to *Phytophthora infestans* in leaves of potato (*Solanum tuberosum*) (Doke and Chai 1985) and to induce phytoalexin accumulation, callose deposition, and resistance to *Phytophthora megasperma* f. sp. *glycinea* in roots of soybean (*Glycine max*) (Bonhoff and Grisebach 1988). Antibiotics with such dual effects are more likely to provide stable root protection than are those which act only as toxicants to fungi.

Some antibiotics produced by ectomycorrhizal fungi can be translocated to aerial portions of trees, where they may retain activity. An antibiotic produced by *Cenococcum geophilum* was detected in the foliage of two *Pinus* species but not in poplar leaves (Grand and Ward 1969). Translocatable antibiotics produced by *Suillus* sp. have been associated with lower incidence of needle diseases caused by fungi under field conditions (Garrdio et al. 1982). Controlled experiments would be necessary to learn whether the antibiotics actually suppressed the diseases.

The role of antibiotics produced by ectomycorrhizal fungi in protection against plant pathogens may be best evaluated by comparisons of wild-type antibiotic producers with mutants that differ only in their production of antibiotics. It may be possible, through gene transfer, to develop mycorrhizal fungi that produce copious amounts of translocatable antibiotics for systemic protection of trees from bacterial and fungal diseases. This approach might prove to be of limited utility, however, as pathogen populations resistant to the antibiotics would likely arise.

Several authors, after observing declines in pathogen populations in the rhizosphere of mycorrhizal plants, have suggested that extramatrical hyphae of VA mycorrhizal fungi may exert antibiosis against bacteria and fungi (Caron et al. 1986, Garcia-Garrido and Ocampo 1988, 1989). Selective changes in the bacterial populations of the rhizosphere of VA mycorrhizal plants have also been reported (see following section), but antibiotic production by VA mycorrhizal fungi has not been demonstrated.

15.4.3 Influence of Mycorrhizal Fungi on Root Exudation in Relation to Plant Disease

Nutrients exuding from roots can play an important role in the development of root disease by promoting or inhibiting germination, growth, and infection by root-invading fungi. A less direct effect of root exudates could result from their influence upon other organisms that might antagonize or stimulate pathogenic fungi in the rhizosphere. Decreased exudation would reduce the distance to which stimulatory concentrations of exudate components migrate through soil, and so reduce the potential for attack by pathogens. A change in the composition of exudates could also be important. A relatively higher ratio of sugar to organic N, for example, promotes mycelial growth of *F. solani* f. sp. *phaseoli* and delays or inhibits penetration of the pathogen into bean (*Phaseolus vulgaris*) roots, whereas a relatively lower ratio of these substances promotes rapid infection and disease development (Tousson et al. 1960).

Both VA and ectomycorrhizal fungi could cause alterations in the quantity and quality of nutrients exuding from roots. Little research has been done, however, on these alterations in relation to root diseases. Prior to infection, mycorrhizal fungi may be expected to consume nutritive exudates in the rhizosphere and rhizoplane, thereby competing with microbes for them. Upon infection, these fungi would have access to nutrients prior to their exudation and perhaps would contribute their own metabolites, further altering the composition of exudates. Ectomycorrhizal fungi that form thick mantles may have a great effect in this regard, as nutrients from mycorrhizal rootlets would pass through fungal tissue before becoming available to other organisms of the rhizosphere. Additionally, the composition of roots, and hence, of their exudates, could change in response to infection by the mycorrhizal fungi. Krupa's group, however, found little difference in the composition of exudates from nonmycorrhizal and ectomycorrhizal short roots of conifers, with the exception of volatiles (Krupa and Fries 1971), which will be discussed in Section 15.4.6. On the other hand, Marx and Davey (1969a,b) observed that encysted zoospores of *Phytophthora cinnamomi* germinated more slowly, and germ tubes elongated less vigorously on mycorrhizal than on nonmycorrhizal short roots, whether the roots were intact or cut. Nutrients may have played a role in this outcome, but it is possible that materials static or toxic to *P. cinnamomi* were produced by the mycorrhizal fungus or the infected host tissues.

Exudates from VA mycorrhizae, in comparison with uninfected roots, have not been shown to suppress pathogens. Hall and Finch (1974) found that infection of avocado (*Persea americana*) roots by VA mycorrhizal fungi did not alter their attractiveness to zoospores of *Phytophthora cinnamomi*. Menge et al. (1978) found that VA mycorrhizal roots of *Citrus* plants attracted 43% more zoospores of *P. parasitica* than did nonmycorrhizal roots. VA mycorrhizal infection did not protect against infection by the pathogen, and growth increases due to VA mycorrhizal infection were negated in the presence of the pathogen.

Phosphorus in VA mycorrhizal systems may affect root protection by influencing root exudation. Ratnayake et al. (1978) found that *Citrus* roots low in P contained less phospholipids and had more permeable cell membranes which exuded much greater amounts of amino acids and sugars than did P-sufficient roots. Early root protection experiments revealed that P-sufficient VA mycorrhizal plants sustained less frequent or severe root disease than did P-deficient nonmycorrhizal plants. More recently, both high- and low-P nonmycorrhizal plants have been included in such studies, with the outcome that improved P status, whether due to fertilization or VA mycorrhizal infection, often explains root protection (Davis and Menge 1980, Graham and Menge 1982, Smith 1988). A P-mediated change in root exudation could contribute to root protection in these systems, but this has not been established experimentally. The relation of P status to root protection by ectomycorrhizal fungi has received little or no attention.

15.4.4 Establishment or Induction by Mycorrhizae of a Protective Rhizosphere Microflora

The interactions of mycorrhizal fungi with rhizobacteria were recently reviewed by Linderman and Paulitz (1990). Both VA and ectomycorrhizal fungi can cause quantitiative and qualitative changes in the rhizosphere microflora (Rambelli 1973, Ames et al. 1984, Meyer and Linderman 1986b). Garcia-Garrido and Ocampo found that inoculation of tomato (*Lycopersicon esculentum*) roots with the VA mycorrhizal *Glomus mosseae* protected them against suppression of root and shoot growth by the plant-pathogenic bacteria *Erwinia carotovora* pv. *carotovora* (1988) and *Pseudomonas syringae* (1989). Growth suppression and its prevention were not due to microbial effects on the P content of plant tissues. Population density of the pathogenic bacteria in the rhizosphere declined during the experiment, and this decline was more rapid in the presence of mycorrhizal inoculum. The researchers therefore suggested that the extramatrical phase of *G. mosseae* may have promoted bacterial decline, possibly by production of antibiotics.

Mycorrhizal fungi possibly also induce the increase in other protective organisms in the rhizosphere, although little evidence for such selection has been reported. Meyer and Linderman (1986b) observed that leachates from rhizosphere soil of VA mycorrhizal plants supported the production of fewer sporangia and zoospores by *Phytophthora cinnamomi* than did leachates from rhizosphere soil of nonmycorrhizal plants. Malajczuk and McComb (1979) suggested that the differential susceptibility of *Eucalyptus calophylla* and *E. marginata* to *P. cinnamomi* in a disease-conducive soil might be a consequence of the differential association of these plant species with ectomycorrhizal fungi that select distinct rhizobacterial populations from the soil. But Harley and Smith (1983) commented, "We are no clearer about the importance of mycorrhizosphere organisms than about that of the rhizosphere populations of 'uninfected' roots."

Mycorrhizal fungi have been found to interact with selected strains of rhizobacteria that promote plant growth. Such bacteria are thought to enhance plant growth by several mechanisms. Growth-promoting strains may inhibit other rhizobacteria that are subclinical pathogens or that promote the development of fungal pathogens in the rhizosphere (Kloepper and Schroth 1981). Some growth-promoting strains inhibit fungal pathogens by competition for iron (Kloepper et al. 1980) and production of antibiotics (Kloepper and Schroth 1981, Thomashow et al. 1990).

Meyer and Linderman (1986a) studied the interactions of mixed species of VA mycorrhizal fungi with the plant growth-promoting bacterium *Pseudomonas putida* on roots of subterranean clover (*Trifolium subterraneum*) in a nonsterile soil. Whereas VA mycorrhizal infection had no influence on the rhizosphere populations of *P. putida*, the bacterium increased VA mycorrhizal development at early stages of plant growth. Plant growth and nodulation by indigenous *Rhizobium* inoculum were stimulated more by dual inoculation with VA mycorrhizal fungi and *P. putida* than by either of these organisms alone. Studies by Azcón (1989), Garbaye and Bowen (1989), and Paulitz and Linderman (1989) of interactions of mycorrhizal fungi with rhizobacteria also indicate the importance of evaluating interactive effects during the development of either rhizobacterial or mycorrhizal inocula.

15.4.5 Mantle as a Barrier

Ectomycorrhizal fungi ensheath roots with a mantle of mycelium that can be from 1–2 to as many as 30–40 hyphal diameters thick. Marx and Davey (1969a,b) and Marx (1970) concluded that the mantle can serve as a highly effective physical barrier to penetration of pine roots by *P. cinnamomi*. Mature ectomycorrhizae with complete mantles were fully protected from this pathogen, whereas nonmycorrhizal short roots were all infected. Mantles that were naturally incomplete or rendered incomplete by microsurgery did not protect root tips from the pathogen. Even superficial growth of ectomycorrhizal fungi in the absence of mycorrhizal infection could prevent penetration by *P. cinnamoni*, provided that the mycelium was densely packed. VA mycorrhizal fungi produce no mantle, and so cannot protect roots from pathogens by this mechanism.

15.4.6 Induced Resistance

Resistance to pathogens may be expressed by plant organs following exposure to microorganisms, their metabolites, or chemicals that elicit or stimulate defensive processes. Production of antimicrobial chemicals, formation of papillae and other wall appositions, and modification of cell walls with phenolic compounds are among the observed responses. In experiments where induction of resistance is separated in time from challenge by a pathogen, the defensive processes may be expressed fully prior to challenge, or they may be activated but not fully expressed until further stimulation by the pathogen. Plants are said to have been *sensitized* (Sinha and Hait 1982) if the activation results in more rapid and intense expression of resistance mechanisms after challenge than occurs in non-induced plants. Sensitized (otherwise susceptible) plants resemble resistant plants in the rapidity and intensity of their responses to pathogens. Indeed, Kuć (1990) has suggested that susceptible and resistant plants differ not in genes for resistance but in the activation of these genes. Although improved nutrition due to mycorrhizal infection may promote plant vigor and thus enhance defensive responses to pathogens relative to those of nutrient-deficient nonmycorrhizal plants (Wick and Moore 1982), we do not consider this effect of mycorrhizal fungi to constitute induced resistance.

Plant resistance to pathogens is characterized by the coordinated expression of multiple defensive mechanisms (Kuć and Caruso 1977, Kuć 1987), each of which taken alone would probably be insufficient to halt the progress of a pathogen. Most research on resistance induced by mycorrhizal fungi, however, has focused primarily on one mechanism, i.e., the production of antimicrobial substances. When Marx proposed in 1972 that this mechanism may be operating generally in root protection by ectomycorrhizal fungi, only a modest amount of evidence existed. It was known that orchid tubers produced antifungal substances in response to mycorrhizal infection by *Rhizoctonia* spp., and that these substances were associated with enhanced resistance to pathogens (for a review, see Arditti 1979). Some of these substances were recognized to be phytoalexins because of their absence from uninfected tubers. Foster and Marks (1966, 1967) demonstrated that cortical cells of roots accumulated phenolics in response to ectomycorrhizal infection, and they proposed that these phenolics play a regulatory role in root colonization, limiting the extent of the Hartig net and preventing pathogenic activity by the fungal symbiont. Hillis and Ishikura (1969) found that ectomycorrhizae of Douglas-fir contained larger amounts of flavanols (catechin, epicatechin, leucoanthocyanin) than did nonmycorrhizal roots and that mycorrhizae lacked several flavanones (poriolin, poriol, taxifolin, taxifolin-3′-β-glucoside) that were relatively abundant in long roots. Krupa and Fries (1971) found that ectomycorrhizal infection resulted in the elevated production of volatile terpenes and sesquiterpenes relative to uninfected roots. Melin and Krupa's (1971) finding that some of these materials were fungistatic to both mycorrhizal and pathogenic fungi led them to speculate that they might play roles in both the regulation of root colonization by mycorrhizal fungi and the induced resistance to mycorrhizal roots to pathogens. Subsequently, the accumulation of phenolics induced by ectomycorrhizal fungi was found to depend on the host-fungus combination. Piché et al. (1983) therefore suggested that induced phenolics do not play a universal role in the regulation of development of the Hartig net. Low concentrations of specific phenolics, perhaps not amenable to detection by histochemical means, and constitutive or inducible in the host, may influence mycorrhizal development not as toxicants but as signals, in the manner similar to the regulation of symbiotic genes of *Rhizobium* species by flavonoids (Rao 1990).

Little research on the role of terpenes has been reported in recent years. Sampangi and Perrin (1986) found an association of induced terpene accumulations with the protection of *Picea abies* roots from *Fusarium oxysporum* by an ectomycorrhizal fungus. Because these authors did not quantify the terpenes or demonstrate their toxicity to *F. oxysporum*, it is difficult to assess their role in root protection. Schisler and Linderman (1989) found that volatiles purged from soil around ectomycorrhizae formed by *Laccaria bicolor* or *Hebeloma* sp. on roots of Douglas-fir increased total bacterial populations but did not significantly alter populations of specific microbial groups that might affect plant growth, i.e., actinomycetes, chitinase producers, facultative anaerobes, phosphate solubilizers, or *Fusarium* species.

Duchesne et al. (1987) reviewed the evidence for a role of plant-produced antimicrobial compounds elicited by ectomycorrhizal fungi in three phenomena:

controlled ingress of compatible ectomycorrhizal fungi, inhibited colonization of roots by incompatible ectomycorrhizal fungi, and increased resistance of ectomycorrhizal and premycorrhizal roots to pathogens. The reviewers concluded that strong circumstantial evidence exists for a role of induced substances in all three phenomena. They also stressed the need for careful evaluation of such a role, including the use of an array of histochemical procedures of differing specificity for the detection of phenolics, and the need to demonstrate the occurrence of phenolics at the appropriate time and place in sufficient amounts to play their hypothesized roles. Our recent research has dealt with the nature, timing, location, and amount of induced phenolic accumulations associated with the protection of Douglas-fir roots from *Fusarium oxysporum* by *Laccaria bicolor* (Strobel and Sinclair 1991a,b).

15.4.6.1 Induction by *Laccaria bicolor* of Resistance to *Fusarium oxysporum* in Primary Roots of Douglas-Fir: The Role of Phenolics

In forest nurseries, *F. oxysporum* causes pre- and postemergence damping off, root rot, stunting, and death of first-year seedlings of Douglas-fir (Bloomberg 1971, Sinclair et al. 1975). Fatal infections occur in primary roots in the first month after sowing seed (Bloomberg 1973), before the emergence of lateral roots. The ectomycorrhizal fungus *L. bicolor* (formerly *L. laccata*; Vellinga and Mueller 1987) can protect Douglas-fir primary roots from *F. oxysporum* prior to mycorrhizal formation. In nursery beds (Sinclair et al. 1975) and some controlled environments (Stack and Sinclair 1975), this protection was expressed as reduced or delayed seedling mortality. Consistent with these findings, *L. bicolor* was found to rapidly colonize primary roots of Douglas-fir in soil (Brown and Sinclair 1981).

Sylvia and Sinclair (1983a,b), investigating the mechanism of root protection by *L. bicolor*, eliminated the possible contributions of the mantle as a physical barrier, protective rhizosphere microflora, and altered root exudates, and found evidence of only weak antibiosis of *L. bicolor* toward *F. oxysporum*. They did find a strong association of induced phenolic accumulations (condensed tannins) with enhanced resistance to *F. oxysporum* (Sylvia and Sinclair 1983b). Elongation of roots of plants grown with *L. bicolor* for 4 weeks was usually inhibited, and the inhibited roots developed extensive phenolic accumulations in their cortices. Such roots were seldom colonized internally by *F. oxysporum* or *Pythium ultimum* when subsequently challenged with these pathogens for 10 days, whereas non-inhibited roots and those not exposed to *L. bicolor* lacked phenolic accumulations and were extensively colonized by the pathogens. An inverse relationship was found between phenolic content (assessed microphotometrically as absorbance of light by osmiophilic granules within cortical cells) and colonization by *F. oxysporum*. Osmiophilic materials were essentially absent from root tissues extensively colonized by *F. oxysporum*. Induction of phenolic accumulation and resistance was also observed when *L. bicolor* was kept separate from roots by a dialysis membrane, indicating that extracellular metabolites of the fungus could elicit resistance. Materials that elicit the expression of plant defense mechanisms have been isolated from cell walls of mycorrhizal fungi (Anderson and Wood unpublished, cited by Coleman and Anderson 1985, Anderson 1988, Sauter and Hager 1989).

We tested the interpretations of Sylvia and Sinclair (1983b). Our initial hypothesis was that phenolics within the cortical cells were of primary importance in the induced resistance of Douglas-fir roots to *F. oxysporum*, as these intracellular phenolics contributed most of the light absorption observed by Sylvia and Sinclair (1983b) in osmium-stained roots free of hyphae of the pathogen (phenolic-infused walls, it was later learned, were only moderately osmiophilic). In keeping with the phytoalexin model, we hypothesized that these substances exerted their inhibitory effect on *F. oxysporum* via fungitoxicity or fungistasis, and that resistance to *F. oxysporum* should be quantitatively related to the amount of extractable phenolics in roots. Because *F. oxysporum* was not thought to induce phenolics (due to the absence of osmiophilic granules from tissues extensively colonized by the pathogen), we assumed that these materials were induced by *L. bicolor* and had to accumulate prior to challenge with the pathogen in order to play a role in disease resistance. It seemed, however, that the necessary exposure to *L. bicolor* prior to challenge with *F. oxysporum* should be briefer than the 2- to 4-week periods employed by Sylvia and Sinclair, if the results were to be relevant to field observations of root protection.

In our experiments, Douglas-fir seedlings were cultured axenically in Petri-dish systems. During the first phase (growth phase) of an experiment, primary roots of intact germlings were trained to elongate through a nutrient-agar slurry, amended or not with living hyphase of *L. bicolor*, for 1 week. For the second, or challenge, phase, plants were transferred to a second apparatus and inoculated with *F. oxysporum* by placing a mycelial plug of the pathogen 5 mm behind and to one side of the root apex. Controls were inoculated with an agar plug without the fungus, and challenge was carried out for 13 days.

It was necessary at the outset to establish the conditions under which *L. bicolor* could induce resistance of roots to *F. oxysporum* in the Petri-dish system. We learned that effective resistance (prevention of lesion formation) was expressed only against a relatively nonaggressive isolate of the pathogen, and only at a temperature (17 °C) below the optimum for disease development. Root protection was not observed at 22 °C or when roots were challenged with more aggressive isolates of the pathogen (Strobel and Sinclair 1991a).

In order to elicit a range of phenolic concentrations in roots, we allowed them to grow in the presence of varied amounts of mycelium of *L. bicolor* (Strobel 1989). Phenolic materials were then extracted from some roots, and the amounts of ethanol-soluble, ortho-dihydroxy phenolics and condensed proanthocyanidins were quantified spectrophotometrically. Other roots were challenged with *F. oxysporum* to evaluate the influence of increasing concentrations of *L. bicolor* and of phenolics on disease resistance. After challenge, roots were fixed in sodium cacodylate buffer containing glutaraldehyde and caffeine to prevent the leaching of phenolics (Mueller and Greenwood 1978), postfixed with 2% (w/v) osmium tetroxide, and prepared for microscopy.

We found that increases in extractable phenolics were associated with resistance, both not in a simple quantitative relationship. Root protection by

L. bicolor was most strongly expressed (lowest incidence of root necrosis) after exposure to doses of mycelium that induced less than maximal accumulations of extractable phenolics. Higher doses of *L. bicolor*, which severely inhibited root elongation, resulted in a susceptible condition despite elevated phenolic levels. Within diseased roots, osmiophilic granules, shown by histochemical studies of Sylvia and Sinclair (1983b) and ourselves to contain phenolics, were absent from cells internally colonized by *F. oxysporum* but were present in uninfected cells.

Catechin was found to be the predominant phenolic in ethanolic extracts of Douglas-fir roots, as had been reported by others (Hillis and Ishikura 1969, Stafford and Cheng 1980). It proved to be noninhibitory to growth of *F. oxysporum* in vitro, even at a concentration of 10^{-2} M, which exceeded the range we calculated to occur in roots. Therefore, the role of phenolics in root protection was not primarily that of strong fungitoxic or fungistatic agents.

Roots not developing lesions during a 13-day incubation with *F. oxysporum* were found to contain hyphae of the pathogen in their intercellular spaces (Strobel 1989, Strobel and Sinclair 1991a). The walls of cortical cells adjacent to these hyphae contained a tan-to-brown pigment. The apparent restriction of *F. oxysporum* to intercellular spaces was more frequent, and cortical wall pigmentation was more intense, in roots that had been exposed to *L. bicolor* for 1 week prior to challenge with the pathogen than in roots exposed only to *F. oxysporum*. Further study revealed that both *F. oxysporum* (Fig. 15.2A) and *L. bicolor* induce wall infusions. Those elicited by *L. bicolor* were present throughout the cortex, rather than exclusively adjacent to the few hyphae of this fungus that we observed in intercellular spaces. Histochemical tests showed that the wall infusions contained flavanols but not lignin. Granular structures in the cytoplasm also contained flavanols. In tissues degraded by the pathogen, histochemically detectable wall infusions were present at only modest levels, whereas extractable phenolics were largely absent.

If flavanolic infusion of cortical cell walls plays a role in induced resistance to *F. oxysporum*, this infusion would be analogous to pathogen-induced lignification, which in various plants inhibits cell wall degradation and penetration by fungal pathogens (Ride 1978, 1980, Vance et al. 1980, Hammerschmidt and Kuć 1982, Hammerschmidt et al. 1985). The process of lignification involves the action of peroxidase on phenolics (Harkin and Obst 1973). Therefore, we investigated the participation of peroxidase in flavanolic wall infusion. Two additional reasons for interest in peroxidase were as follows. First, some phenolic compounds, including flavanols, are not toxic to microorganisms or inhibitory to their enzymes (including those involved in degradation of plant cell walls) without prior enzymatic oxidation (Byrde et al. 1959, Loomis and Battaile 1966, Reddy and Mahadevan 1967, Hunter 1974, Urs and Dunleavy 1975, Leatham et al. 1980). Second, deposition of phenolic materials in epidermal and cortical cell walls had been observed in incompatible interactions between ectomycorrhizal fungi and nonhost roots (Molina 1981, Malajczuk et al. 1982, 1984, Molina and Trappe 1982b, Duddridge 1986a,b).

The wall infusions in Douglas-fir seedlings did appear to be formed by the action of peroxidase on flavanols. Histochemical tests showed that both peroxidase activity (Fig. 15.2B) and flavanols were present in cortical cell walls

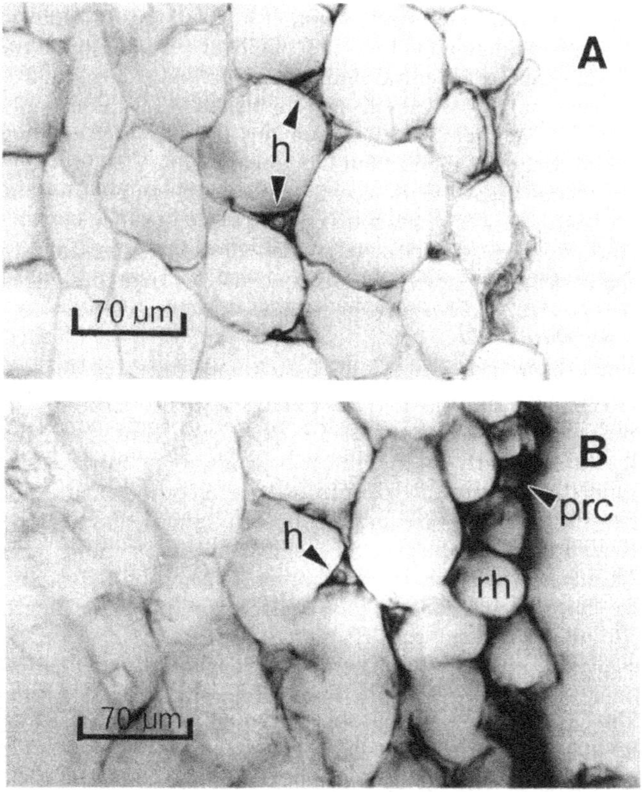

Fig. 15.2A,B. Wall infusions and peroxidase activity induced by *Fusarium oxysporum*. The figures represent transverse sections from fresh tissues of apparently healthy roots that had been challenged with *F. oxysporum* for 9 (**A**) or 10 (**B**) days, **A** Unstained section. Cortical cell walls adjacent to intercellular spaces containing hyphae (*h*) of the pathogen are pigmented. **B** Section incubated with 3,3′,5,5′-tetramethylbenzidine (TMB) for the detection of peroxidase activity. A positive reaction occurred in cortical cell walls adjacent to intercellular spaces containing hyphae (*h*) of *F. oxysporum*. A reaction with TMB also occurred in walls and cytoplasm of persistent root cap (*prc*) and rhizodermal (*rh*) cells (Light photomicrographs from Strobel 1989)

undergoing pigmentation in response to the presence of hyphae of *F. oxysporum* in intercellular spaces. These responses were observed in cortical cell walls adjacent to the pathogen and, less frequently, in walls one to two cells distant from detectable intercellular hyphae, but not in cortical cell walls more distant from the pathogen, or in uninoculated roots. Constitutive peroxidase activity was detected at the surface of uninoculated roots and incubation of roots with the flavanol, catechin, resulted in the formation of wall infusions at the root surface with staining properties identical with those elicited in the cortex by *F. oxysporum* and *L. bicolor*.

For flavanolic infusions to contribute to restriction of *F. oxysporum* to intercellular spaces, they must accumulate before the pathogen would normally penetrate cortical cell walls and begin intracellular colonization. Accordingly, we examined the chemical nature, location, and inducibility of these wall infusions by *L. bicolor* and *F. oxysporum* and the timing of infusion in relation to the normal timing of intracellular colonization by the pathogen. We observed that *L. bicolor* elicited flavanolic accumulations within the cytoplasm but not the walls of cortical cells during the 1-week induction period prior to challenge with *F. oxysporum*. Therefore, *L. bicolor* may have conditioned the timing and/or intensity of the subsequent flavanolic wall infusion response to *F. oxysporum*. In one experiment in which *L. bicolor* did not enhance wall infusion, it also did not protect roots from *F. oxysporum*.

Wall infusions adjacent to intercellular hyphae of *F. oxysporum* were initiated within 6 days after inoculation with this fungus (13 days after first exposure to *L. bicolor*). This timing was 3 to 4 days before *F. oxysporum* could normally be detected growing intracellularly in unprotected roots. The intracellular accumulation of flavanols in response to *L. bicolor* was likely a precursor of wall infusion, whether the infusion was induced by this fungus alone or by *L. bicolor* and *F. oxysporum* in sequence. The association of stunting with phenolic accumulation and resistance to *F. oxysporum*, as seen by Sylvia and Sinclair (1983a,b), may have been a consequence of decreased plasticity (hence reduced elongation) of infused cortical walls (Fry 1986).

Stafford (1988) suggested that the occurrence of flavanols or condensed tannins in cell walls results passively from the loss of integrity associated with cell aging or death. Our observations indicate that flavanolic wall infusion may also be an active, coordinated response of living cells to stress in the cortex of primary roots of Douglas-fir. First, flavanolic infusion was often highly localized, occurring in portions of cell walls adjacent to fungal hyphae, but not in opposite walls of the same cells. Second, the reaction with the peroxidase reagent, tetramethylbenzidine, at infusion sites, obtained without exogenous hydrogen peroxide, signaled endogenous, presumably metabolic, generation of hydrogen peroxide. Third, intact nuclei were observed in cells with flavanolic wall infusions. It seems likely that the flavanols were synthesized in cytoplasmic inclusions (Hradzina and Wagner 1985, Snyder and Nicholson 1990) where they could be detected by reaction with osmium tetroxide or phenolic reagents, and were then actively secreted to cell wall regions adjacent to hyphae of *F. oxysporum*. Localized deposition of phenolic materials (resembling lignin and suberin) into cortical cell walls was observed by Brammall and Higgins (1988a,b) in roots of a resistant cultivar of tomato (*Lycopersicon esculentum*) responding to hyphae of *Fusarium oxysporum* f. sp. *radicis-lycopersici* within intercellular

---

Fig. 15.3A–F. Influence of flavanolic infusions on the resistance of cortical walls to degradation by a commerical cellulase and by *F. oxysporum*. Figures represent transverse sections cut from tissues 2–5 mm proximal to the root cap. Control roots (**A, B, C**) received no pretreatment, whereas induced roots (**D, E, F**) were exposed intact to an autoclaved cellulase preparation for 30 h, prior to assays of resistance to degradation. Prior to sectioning, roots were killed by

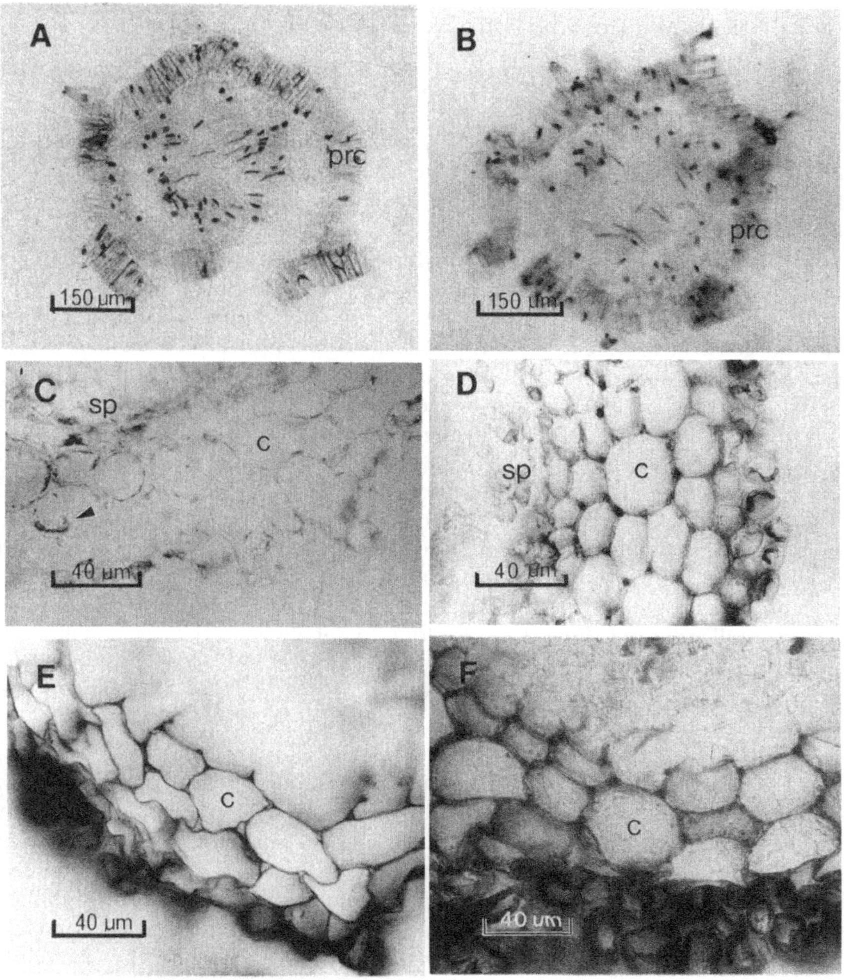

immersion in 70% ethanol at 80 °C for 10 min to preclude further wall infusion. Some sections were immediately stained with 2,4-dimethoxybenzaldehyde-HCl (DMB) for the detection of flavanols (**C, D**), whereas other sections were incubated with active cellulase for 24 h (**B, E**) or with *F. oxysporum* for 5 days (**A, F**). **A, B** Control sections incubated with *F. oxysporum* (**A**) or cellulase (**B**) and then stained with 0.01% (w/v) safranin O. A positive reaction (*dark gray* in these figures) indicates the presence of phenolics. Noninfused walls of stelar parenchyma and cortex have been completely degraded by *F. oxysporum* (**A**) and cellulase (**B**). Phenolic-containing walls of persistent root cap (*prc*) cells remain intact but have collapsed inward. **C, D** Sections stained with DMB. **C** A section of an untreated control root. A positive reaction with DMB (*dark gray* in these figures) is seen in the cytoplasm of a few cells (*arrowhead*). The walls of cortical (*c*) and stelar parenchyma (*sp*) cells lack flavanolic infusions, are unreactive with DMB, and are therefore faint. **D** A section of an induced root. Flavanolic infusions are present in all cortical (*c*) walls, but not in walls of stelar parenchyma (*sp*). **E, F** Sections from induced roots incubated with cellulase (**E**) or *F. oxysporum* (**F**) and then stained with 0.01% (w/v) safranin O. A positive reaction (*dark gray* in these figures) indicates the presence of phenolics. Noninfused stelar parenchyma walls have been completely degraded by cellulase (**E**) and *F. oxysporum* (**F**), but infused cortical walls (*c*) remain largely intact. *Bars* = 20 μm (Light photomicrographs from Strobel 1989) spaces, and this directed deposition of phenolics was associated with restriction of the pathogen to those spaces.

We hypothesized that flavanolic infusions elicited by *L. bicolor* might increase the resistance of cortical cell walls to degradation by *F. oxysporum*, but were unable to test this hypothesis directly, because elicitation of wall infusions by *L. bicolor* required a 7–10 day exposure period, during which time cortical cell walls underwent maturational changes that rendered them incompletely degradable (though still penetrable) by the pathogen, or by a commercial cellulase preparation. However, we noted that walls of immature cortical cells, in a region 2–5 mm proximal to the root cap, were highly susceptible to degradation by *F. oxysporum* (Fig. 15.3A) as well as the cellulase preparation (Fig. 15.3B). Further, we found that a 30-h exposure of intact roots to autoclaved cellulase elicited, in immature cortical cell walls of some roots, infusions that stained as intensely with dimethoxybenzaldehyde-HCl (a reagent for the detection of flavanols) as those elicited by *L. bicolor* (Fig. 15.3D). Therefore, we chose to examine the influence of flavanolic wall infusions on enzymatic digestibility in a model system comprised of transverse sections cut from the immature region of roots with intensely staining wall infusions elicited by autoclaved cellulase (Fig. 15.3D), or from roots not exposed to cellulase and lacking wall infusions (Fig. 15.3C). Prior to sectioning, roots were killed by immersion in 70% ethanol at 80°C for 10 min to preclude further wall modifications. Sections, 30 μm thick, were incubated with nonautoclaved ("live") cellulase (for 1 day) or with *F. oxysporum* (for 5 days). As hypothesized, flavanolic infusions protected cortical cell walls from degradation by cellulase (Fig. 15.3E) and *F. oxysporum* (Fig. 15.3F). In contrast, cortical cell walls lacking flavanolic infusions (controls) were completely degraded by cellulase (Fig. 15.3B) and by *F. oxysporum* (Fig. 15.3A). Walls of stelar parenchyma cells, which were never observed to develop flavanolic walls infusions (or maturational changes rendering them resistant to degradation), were completely degraded by both cellulase and *F. oxysporum*, whether or not roots had been exposed to elicitor (Fig. 15.3A,B,E,F). Moreover, flavanolic wall infusions elicited by the bacterium *Pseudomonas cepacia* also rendered roots resistant to degradation by a commercial pectinase preparation (Strobel and Sinclair, unpublished). Digestibility of intact roots by the commercial cellulase was enhanced, however, by inclusion of ascorbate (pH 5.5) (an inhibitor of the oxidation of flavanols by peroxidase) in the enzyme solution. Ascorbate also accelerated disease onset and increased disease severity when incorporated into the medium in which roots were challenged with *F. oxysporum*. Therefore, we conclude that flavanolic wall infusions confer resistance by rendering cortical cell walls relatively indigestible by exoenzymes of *F. oxysporum*, thus inhibiting their penetration by the pathogen.

Protection by *L. bicolor* of primary roots of Douglas-fir from *F. oxysporum* in nursery beds (Sinclair et al. 1975) was slight in comparison with the protection obtained in pot tests and in vitro (Sinclair et al. 1982, Sylvia and Sinclair 1983b). Our work (Strobel and Sinclair 1991a,b) indicated that the degree of root protection depends on the amount and duration of exposure of roots to *L. bicolor* prior to challenge by *F. oxysporum*. Therefore, protection from *F. oxysporum* and other fungal pathogens that cause damping-off and root rot of young seedlings in nursery beds may be enhanced by means that ensure early exposure of the roots to effective inoculum of mycorrhizal fungi that elicit resistance. Mortier et al. (1988) found that mycelial fragments of *Laccaria bicolor* entrapped in alginate beads gave greater mycorrhizal development, plant growth, and protection from diseases affecting older seedlings (a single benomyl drench was employed to control diseases in the early stages of plant growth) than did peat-vermiculite inoculum such as has been commonly used for inoculation with ectomycorrhizal fungi. Seed treatments might best ensure early, uniform exposure to optimal doses of resistance-inducing agents for protection against seedling diseases. Perhaps seeds could be encapsulated together with mycobiont spores or mycelium using a clay adhesive (Marx et al. 1984) or alginate gel. Preplant treatment of seeds with elicitors has been found to induce systemic resistance in the foliage of rice (*Oryza sativa*) (Trivedi and Sinha 1976), barley, and wheat (*Triticum aestivum*) (Yamada et al. 1990) to fungal pathogens. Exposure of woody plant seeds to elicitors derived from mycorrhizal or other fungi via seed soaks or gel entrapment might similarly result in a systemic resistance that extends into the young root system. Such possibilities seem worth investigating.

15.4.6.2 Relation of Incompatibility/Inefficiency to Induced Resistance

The possibility that incompatible ectomycorrhizal fungi might be more effective inducers of disease resistance than compatible fungi was suggested by Duchesne et al. (1987). Phenolic wall infusions are not characteristic of fully compatible interactions leading to normal ectomycorrhizal development, but are frequently observed in moderately to highly incompatible plant-fungus combinations (Molina 1981, Malajczuk et al. 1982, 1984, Molina and Trappe 1982b, Duddridge 1986a,b). Perhaps the primary roots of Douglas-fir are somewhat incompatible with *L. bicolor*. If such incompatibility occurs also in long roots, it could account for their nonconversion into mycorrhizae and could provide a mechanism by which ectomycorrhizal fungi could induce resistance in them.

Inoculation of the roots of American elm (*Ulmus americana*), red maple (*Acer rubrum*) and green ash (*Fraxinus pennsylvanica*) with the incompatible pathogen *Fusarium oxysporum* f. sp. *lycopersici* induced resistance to the vascular wilt pathogen *Verticillium dahliae* (R. Green, Purdue Univ., personal communication). Induction of resistance by an incompatible, to a compatible, *formae specialis* of *Fusarium oxysporum* has also been demonstrated in tomato, and this induced resistance is associated with elicitation of chitinase and β-1,3-glucanase (Matta et al. 1988), plant enzymes that degrade fungal cell walls. Thus, resistance can be induced in roots of woody plants by means known to be effective in herbaceous plants. It would be of interest to learn whether inducible chitinase and β-1,3-glucanase activities are also associated with induced resistance in these hardwoods.

Moderate incompatibility may have contributed to the protection, observed by Cardoso and Echandi (1987), of bean (*Phaseolus vulgaris*) seedling roots from *Rhizoctonia solani* by a *Rhizoctonia*-like fungus. The fungus did not penetrate beyond epidermal cells and elicited no hypersensitive-like response, but it did elicit the accumulation of ethanol-extractable substances which inhibited hyphal growth and sclerotial germination of *R. solani* in vitro. It is not known whether the substances were present in inhibitory concentrations in roots prior to attempted invasion by *R. solani*, or increased to inhibitory concentrations in response to the pathogen.

A high degree of host-symbiont incompatibility is not normally observed in species that commonly support VA mycorrhizae. Davis and Menge (1981) observed, however, that species of VA mycorrhizal fungi not normally associated with *Citrus* or not as nutritionally efficient as other species on *Citrus* provided greater protection (apparently greater resistance) of *Citrus* roots from *P. parasitica* than did the commonly associated and nutritionally efficient *G. fasciculatum*. If lack of association in nature and low nutritional efficiency are taken to indicate some degree of incompatibility between VA mycorrhizal fungi and their hosts, then this may be an example of root protection by moderately incompatible mycorrhizal fungi.

Root protection by nutritionally inefficient VA mycorrhizal fungi may be related to their effects on secondary metabolites and other resistance factors and processes in roots. Lieberei and Feldmann (1989) observed that an isolate of *Glomus etunicatum*, found to be relatively inefficient in the enhancement of growth of rubber (*Hevea brasiliensis*) tree seedlings, stimulated the accumulation of cyanogenic glucosides and had no effect on the content of the phenolic antimicrobial compound scopoletin in infected as compared with nonmycorrhizal roots, whereas a relatively efficient strain of *G. etunicatum* did not affect the content of cyanogenic glucosides and reduced the content of scopoletin to 25% that of nonmycorrhizal roots. The role of incompatible and/or inefficient mycorrhizal fungi in root protection deserves further attention.

15.4.6.3 Role of Sensitization in Induced Resistance

In that form of induced resistance known as sensitization, induced plants express defensive mechanisms at levels only moderately greater than those of noninduced plants prior to challenge with a pathogen, but expression increases more rapidly, upon challenge with a pathogen, in induced than in noninduced tissues (Sinha and Hait 1982, Chai and Doke 1987, Hait and Sinha 1987, Tuzun et al. 1989). Therefore, the expression of a plant defensive mechanism at only moderate levels in mycorrhizal tissues does not preclude a role for that mechanism in resistance induced by mycorrhizal fungi.

The role of host sensitization in root protection by mycorrhizal fungi has been little examined, but the necessary elicitation of plant defenses by several of these fungi has been demonstrated. VA mycorrhizal fungi can elicit the accumulation of classical antimicrobial chemicals such as cyanogenic glucosides (Lieberei and Feldmann 1989), phenolics (Krishna and Bagyaraj 1984), and phytoalexins (Morandi et al. 1984, Morandi 1989, Wyss et al. 1989). These materials, however, do not accumulate in amounts strongly inhibitory to colonization of roots by the mycorrhizal fungi or by pathogens. Phenolics elicited by *Glomus fasciculatum* in peanut (*Arachis hypogea*) roots did not inhibit mycorrhizal development, despite their apparent accumulation within hyphae and arbuscules of the mycorrhizal fungus (Krishna and Bagyaraj 1984). A concentration of catechol equivalent to the maximum amount of unidentified extractable ortho-dihydroxy phenolics induced by *G. fasciculatum* inhibited growth of *Sclerotium rolfsii* in vitro by only 25%. This modest inhibition suggests that phenolics induced by mycorrhizal infection prior to challenge did not play a major role in the protection of peanut plants from *S. rolfsii* observed by Krishna and Bagyaraj (1983). The amounts of phytoalexins elicited in soybean roots by VA mycorrhizal infection were quite low relative to those elicited by cell damage with copper sulfate (Morandi et al. 1984) or incompatible nematodes (Kaplan et al. 1980). Similarly, the amounts of cyanogenic glucosides and scopoletin elicited by VA mycorrhizal infection of rubber roots were less than those elicited by the pathogen *Thanatephorus cucumeris* (Lieberei and Feldmann 1989).

VA mycorrhizal infection induced an increase in lignin content of the stele of cucumber (*Cucumis sativus*) roots, and this increase was suggested to play a role in the enhanced resistance of VA mycorrhizal plants to vascular wilt fungi (Dehne and Schönbeck 1979). However, VA mycorrhizal infection does not stimulate lignification of the cortex (Codignola et al. 1989), although this tissue in VA mycorrhizal roots is sometimes strongly protected from colonization by fungal pathogens (Schönbeck and Dehne 1979, Dehn and Dehne 1986). That VA mycorrhizal fungi do not elicit toxic levels of defensive metabolites makes biological sense: these fungi do not damage host cells and have too broad a host range to carry the enzymatic potential to detoxify all the various types of phytoalexins, etc. that they might encounter. Similarly, elicitation of lignification in cortical cell walls, which would presumably restrict VA mycorrhizal development, would be maladaptive for these fungi.

VA mycorrhizal infection is also reported to increase activity of chitinase (Dehne and Schönbeck 1978, Spanu et al. 1989), an enzyme which can degrade hyphal walls and inhibit growth of fungi (Schlumbaum et al. 1986, Broekaert et al. 1988). A cell-wall fraction of the ectomycorrhizal fungus *Amanita muscaria* elicited increased chitinase activity in roots of spruce (*Picea abies*), a compatible host plant (Sauter and Hager 1989). The accumulation of chitinase in response to fungal pathogens is associated with cultivar resistance in tomato (*Lycopersicon esculentum*) (Benhamou et al. 1990) and induced resistance in melon (*Cucumis melo*) (Roby et al. 1988).

Close association of chitinase with hyphal walls is required for their degradation (Benhamou et al. 1990), yet Spanu et al. (1989) found that chitinase elicited in leek (*Allium porrum*) by the VA mycorrhizal fungus, *Glomus versiforme*, was not closely associated with fungal structures. However, chitinase bound readily to hyphae of the mycobiont that had been boiled in water for 1 h, indicating that materials extractable by this treatment ordinarily coat the hyphae and prevent attack by chitinase (Spanu et al. 1989). In other fungi, chitin is masked by glucans, and can be made accessible to chitinase by exposure to β-1,3-glucanase (Wessels and Marchant 1974, Benhamou et al. 1989, 1990). Although β-1,3-glucanase and chitinase are typically under coordinate regulation in plants (Vögeli et al. 1988), elicitation of chitinase without substantial attendant elicitation of β-1,3-glucanase was recently reported (Lotan and Fluhr 1990). The nature of the materials that mask chitin in walls of mycorrhizal fungi, and the influence of mycorrhizal infection on the activity of β-1,3-glucanase deserve attention.

The influence of mycorrhizal fungi on the timing and extent of expression of active host defensive responses to pathogens has been little investigated. Our research indicated that stimulation of flavanol synthesis and deposition in cortical cell walls played a role in the protection by *L. bicolor* of Douglas-fir roots from *F. oxysporum*. Possible evidence for host sensitization was also reported by Dehne and Schönbeck (1979). They found that although infection of tomato roots with *G. mosseae* or *F. oxysporum* f. sp. *lycopersici* resulted in only modest increases in ortho-dihydroxy phenols, infection by both fungi resulted in a 50% increase in these phenolics as compared with control roots exposed to no fungus. The role of enhanced host response to pathogens in root protection by mycorrhizal fungi deserves further investigation.

The physiological bases of sensitization would logically include accelerated and/or intensified recognition of and response to pathogens. We will consider some of the possibilities amenable to experimental evaluation.

Increases in the activity of chitinase and β-1,3-glucanase, elicited in root tissues by mycorrhizal fungi, could play a prominent role in induced resistance. These hydrolases may be directly inhibitory to fungal pathogens (Schlumbaum et al. 1986, Broekaert et al. 1988), and may be more rapidly mobilized to the vicinity of invading pathogenic fungal hyphae (Benhamou et al. 1990) in induced than in noninduced roots. These hydrolases might also enhance the recognition of and response to fungal pathogens by releasing wall fragments with the ability to elicit plant defenses (Dixon and Lamb 1990) at an earlier time and/or in greater abundance than would occur in root tissues not previously stimulated. Although stimulation of chitinase activity is limited, in some cases, to the first 3 weeks of VA mycorrhizal development (Spanu et al. 1989), this transient activation could provide ongoing protection/sensitization of young tissues if these were colonized soon after differentiation, perhaps by runner hyphae or internal spread of earlier infections.

Tissues sensitized by mycorrhizal fungi or other agents might also exhibit a lower threshold for response to elicitors of pathogen origin, increased expression of the suite of defensive mechanisms per unit of elicitor, and selectively enhanced expression of particular defense mechanisms (e.g., generation of superoxide anion, Doke 1983). Enhanced expression of defensive mechanisms in response to pathogens or their elicitors may be a consequence of the greater abundance of the relevant mRNAs, proteins, and organelles involved in their production and transport, in induced as opposed to noninduced tissues, or may involve more rapid de novo synthesis of these materials and structures.

It seems unlikely that sensitization of host root tissues is a consistent or permanent consequence of mycorrhizal infection, but the phenomenon may yet be exploitable for root protection once it is better understood. The degree of sensitization by mycorrhizal fungi, and its efficacy against pathogens, are likely to depend on those factors observed to condition root protection generally, i.e., host species and cultivar, species or strain of the mycorrhizal fungus, and species (and perhaps strain) of the pathogen. A degree of incompatibility between host and mycobiont could enhance sensitization.

The greatest amount of useful information about mechanisms involved in resistance induced by mycorrhizal fungi may be obtainable through studies comparing resistant cultivars with otherwise susceptible cultivars in which resistance has been induced. This comparative research should involve co-ordinated biochemical, histochemical, immunocytochemical, and molecular biological studies of the responses (with emphasis on kinetics) of susceptible, genetically resistant, and induced resistant plants to fungal pathogens and elicitors, perhaps with initial focus on β-1,3-glucanase, chitinase, and elicitors released by them from fungal hyphal walls. It may be possible to synthesize, or to obtain from natural sources (Koga et al. 1987), chemicals that selectively inhibit chitinases of plant (but not fungal) origin. Selective inhibitors of plant chitinases could help in the elucidation of the role of plant chitinases in mycorrhizal development and in genetic and induced resistance. The relationship of efficiency/compatibility of mycorrhizal fungi to induced resistance should also be studied.

The tomato – *F. oxysporum* f. sp. *radicis – lycopersici* (FORL) pathosystem may prove useful for these investigations for several reasons. Protection of tomato roots from FORL by a VA mycorrhizal fungus has been documented (Caron et al. 1985, 1986), and induced resistance may be operating in this system. The roles of phenolic wall infusions (Brammall and Higgins 1988a,b), β-1,3-glucanase (Benhamou et al. 1989), and chitinase (Benhamou et al. 1990) in resistance of tomato to FORL have been documented, and methods for their study are well developed. Systemic induced resistance to *Phytophthora infestans* has been demonstrated in tomato (Heller and Gessler 1986), and sensitization is likely to play a role in this induced resistance, as has been demonstrated for induced resistance to this pathogen in potato (Chai and Doke 1987). Evidence exists that systemic accumulation of chitinase (Toppan and Roby 1982) and resistance (Gessler and Kuć 1982) extends into roots of plants exhibiting systemic induced resistance.

Poplar (*Populus* sp.) may also be a useful host for such studies for the following reasons. Poplar forms both VA and ectomycorrhizae (Harley and Smith 1983), allowing comparative studies of root protection by VA and ectomycorrhizal fungi. Poplar can be clonally propagated, assuring genetic uniformity of test plants. Systemic accumulation of mRNAs encoding defense-related proteins, in response to localized wounding, has been demonstrated in poplar (Parsons et al. 1989); thus poplar may be amenable to the induction of systemic resistance by means employed for herbaceous plants. Suitable root pathogens for study on poplar must be identified, however, and clones with resistance to these pathogens may not have been identified.

15.5 Conclusion: An Expanded View of Root Protection by Mycorrhizal Fungi

Mycorrhizal fungi may protect roots by means that are either passive or active with regard to host response. Passive means discussed in this chapter include the production of antibiotics by the mycorrhizal fungus, and functioning of the mantle as a physical barrier to invading organisms. Systems involving host responses include induction of antimicrobial metabolites prior to challenge by pathogens, induction of lignified or flavanolic infusions in cortical walls, and sensitization of host defense systems such that the speed or magnitude of response to pathogens is enhanced. Alteration of root exudates and the development or induction of a protective rhizosphere microflora may occur passively or by active host responses. Although not a focus of this chapter, mycorrhizal enhancement of host vigor, with consequent increases in tolerance and/or resistance to pathogens, probably also plays a major role in root protection under nutrient-limited conditions.

Root protection by mycorrhizal fungi in nature probably involves a combination of mechanisms. In those systems involving induced resistance, a suite of defensive responses, coordinated and complementary in their expression, is probably involved. Modern genetic approaches may provide fundamental information leading to development of mycorrhizal fungi with enhanced ability to protect roots, particularly via induced resistance. For the immediate future, naturally occurring variants of mycorrhizal fungi with potential value in root protection should be further investigated to obtain baseline data for comparisons and for possible interim use in crop systems.

References

Allen MF, Allen EB, Friese CF 1989 Responses of the non-mycotrophic plant Salsola kali to invasion by vesicular-arbuscular mycorrhizal fungi. New Phytol 111:45–49

Ames RN, Reid CPP, Ingham ER 1984 Rhizosphere bacterial population responses to root colonization by a vesicular-arbuscular mycorrhizal fungus. New Phytol 96:555–563

Anderson AJ 1988 Mycorrhizae-host specificity and recognition. Phytopathology 78:375–378

Arditti J 1979 Aspects of the physiology of orchids. Adv Bot Res 7:421–655

Azcón R 1989 Selective interaction between free-living rhizosphere bacteria and vesicular-arbuscular mycorrhizal fungi. Soil Biol Biochem 21:693–644

Benhamou N, Grenier J, Asselin A, Legrand M 1989 Immunogold localization of β-1,3-glucanases in two plants infected by vascular wilt fungi. Plant Cell 1:1209–1221

Benhamou N, Joostein MHAJ, De Wit PJGM 1990 Subcellular localization of chitinase and of its potential substrate in tomato root tissues infected by Fusarium oxysporum f. sp. radicis-lycopersici. Plant Physiol 92:1108–1120

Bloomberg WJ 1971 Diseases of Douglas-fir seedlings caused by Fusarium oxysporum. Phytopathology 61:467–470

Bloomberg WJ 1973 Fusarium root rot of Douglas-fir seedlings. Phytopathology 63:337–341

Bonhoff A, Grisebach H 1988 Elicitor-induced accumulation of glyceollin and callose in soybean roots and localized resistance against Phytophthora megasperma f. sp. glycinea. Plant Sci 54:203–209

Boudarga K, Lapeyrie F, Dexheimeir J 1990 A technique for dual vesicular-arbuscular endomycorrhizal/ectomycorrhizal infection of Eucalyptus in vitro. New Phytol 114:73–76

Brammall RA, Higgins, VJ 1988a A histological comparison of fungal colonization in tomato seedlings susceptible or resistant to Fusarium crown and root rot disease. Can J Bot 66:915–925

Brammall RA, Higgins VJ 1988b The effect of glyphosate on resistance of tomato to Fusarium crown and root rot disease and on the formation of host structural defense barriers. Can J Bot 66:1547–1555

Broekaert WF, Van Parijs J, Allen AK, Peumans WJ 1988 Comparison of some molecular, enzymatic and antifungal properties of chitinases from thorn-apple, tobacco and wheat. Physiol Mol Plant Pathol 33:319–331

Broglie K, Broglie R, Holliday M, Chet I 1990 Transgenic tobacco plants with induced resistance to fungal attack. Phytopathology 80:1003

Brown AC, Sinclair WA 1981 Colonization and infection of primary roots of Douglas-fir seedlings by the ectomycorrhizal fungus Laccaria laccata. For Sci 27:111–124

Byrde RJW, Fielding AH, Williams AH 1959 The role of oxidized polyphenols in the varietal resistance of apples to brown rot. In: Pridham JB (ed) Phenolics in plants in health and disease. Pergamon, New York 95–99

Cardoso JE, Echandi E 1987 Nature of protection of bean seedlings from Rhizoctonia root rot by a binucleate Rhizoctonia-like fungus. Phytopathology 77:1548–1551

Caron M 1989 Potential use of mycorrhizae in control of soil-borne diseases. Can J Plant Pathol 11:177–179

Caron M, Fortin AJ, Richard C 1985 Influence of substrate on the interaction of Glomus intraradices and Fusarium oxysporum f. sp. radicis-lycopersici on tomatoes. Plant Soil 87:233–239

Caron M, Fortin JA, Richard C 1986 Effect of phosphorus concentration and Glomus intraradices on Fusarium crown and root rot of tomatoes. Phytopathology 76:942–946

Chai HB, Doke N 1987 Systemic activation of O_2^- generating reaction, superoxide dismutase, and peroxidase in potato plants in relation to induction of systemic resistance to Phytophthora infestans. Ann Phytopathol Soc Jpn 53:585–590

Codignola A, Verotta L, Spanu P, Maffei M, Scannerini S, Bonfante-Fasolo P 1989 Cell wall-bound phenols in roots of vesicular-arbuscular mycorrhizal plants. New Phytol 112:221–228

Coleman ME, Anderson AJ 1985 The role of elicitors in ectomycorrhizal formation. In: Molina R (ed) Proc 6th North Am Conf Mycorrhizae, Québec, 361

Cordell CE, Owen JH, Marx DH 1987 Mycorrhizae nursery management for improved seedling quality and field performance. In: Proc Intermountain For Nursery Assoc, Aug 10–14, 1987, 105–115

Davis RA, Menge JA 1980 Influence of Glomus fasciculatus and soil phosphorus on Phytophthora root rot of citrus. Phytopathology 70:447–452

Davis RM, Menge JA 1981 Phytophthora parasitica inoculation and intensity of vesicular-arbuscular mycorrhizae in citrus. New Phytol 87:705–715

Dehn B, Dehne HW 1986 Development of VA mycorrhizal fungi and interactions with Cochliobolus sativus in roots of gramineae. In: Gianinazzi-Pearson V, Gianinazzi S (eds) Physiological and genetical aspects of mycorrhizae. Inst Nat Rech Agron, Paris, 773–779

Dehne HW 1982 Interactions between vesicular-arbuscular mycorrhizal fungi and plant pathogens. Phytopathology 72:1115–1119

Dehne HW, Schönbeck F 1978 The influence of endotrophic mycorrhiza on plant diseases. III. Chitinase activity and ornithine cycle. J Plant Dis Prot 85:666–678

Dehne HW, Schönbeck F 1979 The influence of endotrophic mycorrhizae on plant diseases. II. Phenol metabolism and lignification. Phytopathol Z 95:210–216

Dixon RA, Lamb CJ 1990 Molecular communication in interactions between plants and microbial pathogens. Annu Rev Plant Physiol Plant Mol Biol 41:339–367

Doke N 1983 Involvement of superoxide anion generation in the hypersensitive response of potato tuber tissues to infection with an incompatible race of Phytophthora infestans and to the hyphal wall components. Physiol Plant Pathol 23:345–357

Doke N, Chai HB 1985 Activation of superoxide generation and enhancement of resistance against compatible races of Phytophthora infestans in potato plants treated with digitonin. Physiol Plant Pathol 27:323–334

Duc G, Trouvelot A, Gianinazzi-Pearson V, Gianinazzi S 1989 First report of non-mycorrhizal plant mutants (myc-) obtained in pea (Pisum sativum L.) and fababean (Vicia faba L.). Plant Sci 60:215–222

Duchesne LC, Campbell SE, Koehler H, Peterson RL 1989a Pine species influence suppression of Fusarium root rot by the ectomycorrhizal fungus Paxillus involutus. Symbiosis 7:139–148

Duchesne LC, Ellis BE, Peterson RL 1989b Disease suppression by the ectomycorrhizal fungus Paxillus involutus: contribution of oxalic acid. Can J Bot 67:2726–2730

Duchesne LC, Peterson RL, Ellis BE 1987 The accumulation of plant-produced antimicrobial compounds in response to ectomycorrhizal fungi: a review. Phytoprotection 68:17–27

Duchesne LC, Peterson RL, Ellis BE 1988 Interaction between the ectomycorrhizal fungus Paxillus involutus and Pinus resinosa induces resistance to Fusarium oxysporum. Can J Bot 66:558–562

Duchesne LC, Peterson RL, Ellis BE 1989c The future of ectomycorrhizal fungi as biological control agents. Phytoprotection 70:51–57

Duchesne LC, Peterson RL, Ellis BE 1989d The time-course of disease suppression and antibiosis by the ectomycorrhizal fungus Paxillus involutus. New Phytol 111:693–698

Duddridge JA 1986a The development and ultrastructure of ectomycorrhizas. III. Compatible and incompatible interactions between Suillus grevillei (Klotzsch) Sing. and 11 species of ectomycorrhizal hosts in vitro in the absence of exogenous carbohydrate. New Phytol 103:457–464

Duddridge JA 1986b The development and ultrastructure of ectomycorrhizas. IV. Compatible and incompatible interactions between Suillus grevillei (Klotzsch) Sing. and a number of ectomycorrhizal hosts in vitro in the presence of exogenous carbohydrate. New Phytol 103:465–471

Dumas E, Gianinazzi-Pearson V, Gianinazzi S 1989 Production of new soluble proteins during VA endomycorrhiza formation. Agr Ecosyst Environ 29:111–114

Farquhar ML, Peterson RL 1990 Early effects of the ectomycorrhizal fungus Paxillus involutus on the root rot organism Fusarium associated with Pinus resinosa. Can J Bot 68:1589–1596

Favaron F, Alghisi P, Marciano P, Magro P 1988 Polygalacturonase isoenzymes and oxalic acid produced by Sclerotinia sclerotiorum in soybean hypocotyls as elicitors of glyceollin. Physiol Mol Plant Pathol 33:385–395

Finlay RD 1989 Functional aspects of incompatible ectomycorrhizal associations. Agr Ecosyst Environ 28:127–131

Foster RC, Marks GC 1966 The fine structure of the mycorrhizas of Pinus radiata D. Don. Aust J Biol Sci 19:1027–1038

Foster RC, Marks GC 1967 Observations on the mycorrhizas of forest trees. II. The rhizosphere of Pinus radiata D. Don. Aust J Biol Sci 20:915–926

Fowlks E, Leben C, Snell JF 1967 Sterols in relation to the influence of nystatin on Pythium aphanidermatum and Colletotrichum lagenarium. Phytopathology 57:246–249

Fry SC 1986 Cross-linking of matrix polymers in the growing cell walls of angiosperms. Annu Rev Plant Physiol 37:165–186

Garbaye J, Bowen GD 1989 Stimulation of ectomycorrhizal infection of Pinus radiata by some organisms associated with the mantle of ectomycorrhizas. New Phytol 112:383–388

Garcia-Garrido JM, Ocampo JA 1988 Interaction between Glomus mosseae and Erwinia carotovora and its effects on the growth of tomato plants. New Phytol 110:551–555

Garcia-Garrido JM, Ocampo JA 1989 Effect of VA mycorrhizal infection of tomato on damage caused by Pseudomonas syringae. Soil Biol Biochem 21:165–167

Garrido N, Becerra J, Marticorena C, Oehrens E, Silva M, Horak E 1982 Antibiotic properties of ectomycorrhizae and saprophytic fungi growing on Pinus radiata roots. Mycopathologia 77:93–98

Gessler C, Kuć J 1982 Induction of resistance to Fusarium wilt in cucumber by root and foliar pathogens. Phytopathology 72:1439–1441

Gianinazzi-Pearson V 1984 Host-fungus specificity, recognition and compatibility in mycorrhizae. In: Verma DPS, Hohn TH (eds) Genes involved in plant microbe interactions. Springer, Berlin Heidelberg New York, 225–253

Graham JH, Menge JA 1982 Influence of vesicular-arbuscular mycorrhizae and soil phosphorus on take-all disease of wheat. Phytopathology 72:95–98

Grand LF, Ward WW 1969 An antibiotic detected in conifer foliage and its relation to Cenococcum graniforme mycorrhizae. For Sci 15:286–288

Hait GN, Sinha AK 1987 Biochemical changes associated with induction of resistance in rice seedlings to Helminthosporium oryzae with chemicals. J Plant Dis Protect 94:360–368

Hall JB, Finch HC 1974 Mycorrhiza in roots of avocado: effect upon chemotaxis of Phytophthora cinnamomi zoospores. Proc Am Phytopathol Soc 1:86 (Abstr)

Hammerschmidt R, Bonnen AM, Bergstrom GC, Baker KK 1985 Association of epidermal lignification with nonhost resistance of cucurbits to fungi. Can J Bot 63:2393–2398

Hammerschmidt R, Kuć J 1982 Lignification as a mechanism for induced systemic resistance in cucumber. Physiol Plant Pathol 20:62–71

Harkin JM, Obst JR 1973 Lignification in trees: indication of exclusive peroxidase participation. Science 180:296–298

Harley JL, Smith SE 1983 Mycorrhizal symbiosis. Academic Press, New York London, 483 pp

Heller WE, Gessler C 1986 Induced systemic resistance in tomato plants against Phytophthora infestans. J Phytopathol 116:323–328

Hilbert JL, Martin F 1989 Protein changes and the presence of ectomycorrhiza-specific polypeptides in the Pisolithus-Eucalyptus symbiosis. Agr Ecosyst Environ 28:181–184

Hillis WE, Ishikura N 1969 The extractives of the mycorrhizas and roots of Pinus radiata and Pseudotsuga menziesii. Aust J Biol Sci 22:1425–1436

Hirrel MC, Mehravaran H, Gerdemann JW 1978 Vesicular-arbuscular mycorrhizae in the Chenopodiaceae and Cruciferae: do they occur? Can J Bot 56:2813–2817

Hradzina G, Wagner GC 1985 Metabolic pathways as enzyme complexes: evidence for the synthesis of phenylpropanoids and flavonoids on membrane associated enzyme complexes. Arch Biochem Biophys 237:88

Hunter RE 1974 Inactivation of pectic enzymes by polyphenols in cotton seedlings of different ages infected with Rhizoctonia solani. Physiol Plant Pathol 4:151–159

Kaplan DR, Keen NT, Thomason IJ 1980 Association of glyceollin with the incompatible response of soybean roots to Meloidogyne incognita. Physiol Plant Pathol 16:309–318

Kloepper JW, Leong J, Teintze M, Schroth MN 1980 Enhanced plant growth by siderophores produced by plant growth-promoting rhizobacteria. Nature London 286:885–886

Kloepper JW, Schroth MN 1981 Relationship of in vitro antibiosis of plant growth-promoting rhizobacteria to plant growth and displacement of root microflora. Phytopathology 71:1020–1024

Koga D, Isogai A, Sakuda S, Matsumoto S, Suzuki A, Kimura S, Ide A 1987 Specific inhibition of Bombyx mori chitinase by allosamidin. Agr Biol Chem 51:471–476

Krishna KR, Bagyaraj DJ 1983 Interactions between Glomus fasciculatum and Sclerotium rolfsii in peanut. Can J Bot 61:2349–2351

Krishna KR, Bagyaraj DJ 1984 Phenols in mycorrhizal roots of Arachis hypogea. Experientia 40:85–86

Kropp BR, Langlois CG 1990 Ectomycorrhizae in reforestation. Can J For Res 20:438–451

Krupa S, Fries N 1971 Studies on ectomycorrhizae of pine. I. Production of volatile organic compounds. Can J Bot 49:1425–1431

Kuć J 1987 Plant immunization and its applicability for disease control. In: Chet I (ed) Innovative approaches to plant disease control. Wiley & Sons, New York, 255–274

Kuć J 1990 A case for self defense in plants against disease. Phytoparasitica 18:3–8

Kuć J, Caruso F 1977 Activated coordinated chemical defense against disease in plants. In: Hedin P (ed) Host plant resistance to pests. Am Chem Soc Symp Ser 62. Am Chem Soc Press, Washington, DC, 78–89

Leatham GF, King V, Stahmann MA 1980 In vitro protein polymerization by quinones or free radicals generated by plant or fungal oxidative enzymes. Phytopathology 70:1134–1140

Lieberei R, Feldmann F 1989 Physiological changes in roots colonized by vesicular arbuscular mycorrhizal fungi – reactions in mutualistic and parasitic interactions. Agr Ecosyst Environ 29:251–255

Linderman RG, Paulitz TC 1990 Mycorrhizal-rhizobacterial interactions: In: Hornby D (ed) Biological control of soil-borne plant pathogens. CAB, Wallingford UK, 261–283

Loomis WD, Battaile J 1966 Plant phenolic compounds and the isolation of plant enzymes. Phytochemistry 5:423–428

Lotan T, Fluhr R 1990 Xylanase, a novel elicitor of pathogenesis-related proteins in tobacco, uses a non-ethylene pathway for induction. Plant Physiol 93:811–817

Malajczuk N, McComb AJ 1979 The microflora of unsuberized roots of Eucalyptus calophylla R. Br. and Eucalyptus marginata Donn ex Sm. seedlings grown in soils suppressive and conducive to Phytophthora cinnamomi Rands. I. Rhizoplane bacteria, actinomycetes, and fungi. II. Mycorrhizal roots and associated microflora. Aust J Bot 27:235–272

Malajczuk N, Molina R, Trappe JM 1982 Ectomycorrhiza formation in Eucalyptus. I. Pure culture synthesis, host specificity and mycorrhizal compatibility with Pinus radiata. New Phytol 91:467–482

Malajczuk N, Molina R, Trappe JM 1984 Ectomycorrhizal formation in Eucalyptus. II. The ultrastructure of compatible and incompatible mycorrhizal fungi and associated roots. New Phytol 96:43–53
Marx DH 1969 The influence of ectotrophic mycorrhizal fungi on the resistance of pine roots to pathogenic infections. II. Production, identification, and biological activity of antibiotics produced by Leucopaxillus cerealis var. piceina. Phytopathology 59:411–417
Marx DH 1970 The influence of ectotrophic mycorrhizal fungi on the resistance of pine roots to pathogenic infections. V. Resistance of mycorrhizae to infection by vegetative mycelium of Phytophthora cinnamomi. Phytopathology 60:1472–1473
Marx DH 1972 Ectomycorrhizae as biological deterrents to pathogenic root infections. Annu Rev Phytopathol 10:429–454
Marx DH 1975 Role of ectomycorrhizae in the protection of pine from root infection by Phytophthora cinnamomi. In: Bruehl GW (ed) Biology and control of soil-borne plant pathogens. Am Phytopathol Soc, St Paul MN, 112–115
Marx DH, Cordell CE, Maul SB, Ruehle JL 1989 Ectomycorrhizal development on pine by Pisolithus tinctorius in bare-root and container seedling nurseries; pt 1: efficacy of various vegetative inoculum formulations; pt 2: efficacy of various vegetative and spore inocula. New For 3:45–66
Marx DH, Davey CB 1969a The influence of ectotrophic mycorrhizal fungi on the resistance of pine roots to pathogenic infections. III. Resistance of aseptically formed mycorrhizae to infection by Phytopthora cinnamomi. Phytopathology 59:549–558
Marx DH, Davey CB 1969b The influence of ectotrophic mycorrhizal fungi on the resistance of pine roots to pathogenic infections. IV. Resistance of naturally occurring mycorrhizae to infections by Phytophthora cinnamomi. Phytopathology 59:559–565
Marx DH, Jarl K, Ruehle JL, Bell W 1984 Development of Pisolithus tinctorius ectomycorrhizae on pine seedlings using basidiospore-encapsulated seeds. For Sci 30:897–907
Marx DH, Ruehle JL 1989 Ectomycorrhizae as biological tools in reclamation and revegetation of waste lands. In: Mahadevan A, Raman N, Natarajan K (eds) Mycorrhizae for green Asia. Proc 1st Asian Conf Mycorrhizae, 29–31 Jan 1988, Cent Adv Stud Bot, Univ Madras, Guindy Campus, India, 336–344
Matta A, Abbatista Gentile I, and Ferraris L 1988 Stimulation of 1,3-β-glucanase and chitinase by stresses that induce resistance to Fusarium wilt in tomato. Phytopathol Mediterr 27:45–50
Melin E, Krupa S 1971 Studies on ectomycorrhizae of pine. II. Growth inhibition of mycorrhizal fungi by volatile organic constituents of Pinus sylvestris (Scots pine) roots. Physiol Plant 25:337–340
Menge JA, Nemec S, Davis RM, Minassian V 1978 Mycorrhizal fungi associated with citrus and their possible interactions with pathogens. Proc Int Soc Citricult 1977, vol 3, 872–876
Meyer JR, Linderman RG 1986a Response of subterranean clover to dual inoculation with vesicular-arbuscular mycorrhizal fungi and a plant growth-promoting bacterium, Pseudomonas putida. Soil Biol Biochem 18:185–190
Meyer JR, Linderman RG 1986b Selective influence on populations of rhizosphere or rhizoplane bacteria and actinomycetes by mycorrhizas formed by Glomus fasciculatum. Soil Biol Biochem 18:191–196
Modjo HS, Hendrix JW 1986 The mycorrhizal fungus Glomus macrocarpum as a cause of tobacco stunt disease. Phytopathology 76:688–691
Molina R 1981 Ectomycorrhizal specificity in the genus Alnus. Can J Bot 59:325–334
Molina R, Trappe JM 1982a Lack of mycorrhizal specificity by the ericaceous hosts Arbutus menziesii and Arctostaphylos uva-ursi. New Phytol 90:495–509
Molina R, Trappe JM 1982b Patterns of ectomycorrhizal host specificity and potential among Pacific Northwest conifers and fungi. For Sci 28:423–458
Morandi D 1989 Effect of endomycorrhizal infection and biocides on phytoalexin accumulation in soybean roots. Agr Ecosyst Environ 29:303–305
Morandi D, Bailey JA, Gianinazzi-Pearson V 1984 Isoflavonoid accumulation in soybean roots infected with vesicular-arbuscular mycorrhizal fungi. Physiol Plant Pathol 24:357–364

Mortier F, LeTacon F, Garbaye J 1988 Effect of inoculum type and inoculation dose on ectomycorrhizal development, root necrosis and growth of Douglas-fir seedlings inoculated with Laccaria laccata in a nursery. Ann Sci For 45:301–310

Morton JB, Benny GL 1990 Revised classification of arbuscular mycorrhizal fungi (Zygomycetes) – a new order, Glomales, 2 new suborders, Glomineae and Gigasporineae, and 2 new families, Acaulosporaceae and Gigasporaceae, with an emendation of Glomaceae. Mycotaxon 37:471–491

Mueller WC, Greenwood AD 1978 The ultrastructure of phenolic-storing cells fixed with caffeine. J Exp Bot 29:757–764

Mugnier J, Mosse B 1987 Vesicular-arbuscular mycorrhizal infection in transformed root-inducing T-DNA roots grown axenically. Phytopathology 77:1045–1050

Parsons TJ, Bradshaw HD Jr, Gordon MP 1989 Systemic accumulation of specific mRNAs in response to wounding in poplar trees. Proc Natl Acad Sci USA 86:7895–7899

Paulitz TC, Linderman RG 1989 Interactions between fluorescent pseudomonads and VA mycorrhizal fungi. New Phytol 113:37–45

Piché Y, Peterson RL, Howarth MJ, Fortin JA 1983 A structural study of the interaction between the ectomycorrhizal fungus Pisolithus tinctorius and Pinus strobus roots. Can J Bot 61:1185–1193

Rambelli A 1973 The rhizosphere of mycorrhizae. In: Marks GC, Kozlowski TT (eds) Ectomycorrhizae. Academic Press, New York London, 299–349

Rao AS 1990 Root flavonoids. Bot Rev 56:1–84

Ratnayake M, Leonard RT, Menge JA 1978 Root exudation in relation to supply of phosphorus and its possible relevance to mycorrhizal formation. New Phytol 81:543–552

Reddy MK, Mahadevan A 1967 Effect of phenolic compounds on cellulase. Indian Phytopathol 20:265–267

Ride JP 1978 The role of cell wall alterations in resistance to fungi. Ann Appl Biol 89:302–306

Ride JP 1980 The effect of induced lignification on the resistance of wheat cell walls to fungal degradation. Physiol Plant Pathol 16:187–196

Roby D, Toppan A, Esquerré-Tugayé MT 1988 Systemic induction of chitinase activity and resistance in melon plants upon fungal infection or elicitor treatment. Physiol Mol Plant Pathol 33:409–417

Sampangi R, Perrin R 1986 Attempts to elucidate the mechanisms involved in the protective effect of Laccaria laccata against Fusarium oxysporum. In: Gianinazzi-Pearson V, Gianinazzi S (eds) Physiological and genetical aspects of mycorrhizae. Inst Nat Rech Agron, Paris, 807–810

Sauter M, Hager A 1989 The mycorrhizal fungus Amanita muscaria induces chitinase activity in roots and in suspension-cultured cells of its host Picea abies. Planta 179:61–66

Schenck NC, Kellam MK 1978 The influence of vesicular arbuscular mycorrhizae on disease development. Fla Agr Exp Stn Bull 798:1–16

Schisler DA, Linderman RG 1989 Response of nursery soil microbial populations to volatiles purged from soil around Douglas-fir ectomycorrhizae. Soil Biol Biochem 21:397–401

Schlumbaum A, Mauch F, Vögeli U, Boller T 1986 Plant chitinases are potent inhibitors of fungal growth. Nature (London) 324:365–367

Schönbeck F, Dehne HW 1979 Untersuchungen zum Einfluß der endotrophen Mycorrhiza auf Pflanzenkrankheiten. Pilzliche Sproßparasiten, Olpidium brassicae, TMV. Z Pflanzenkr Pflanzenschutz 86:103–112

Sinclair WA, Cowles DP, Hee SM 1975 Fusarium root rot of Douglas-fir seedlings: suppression by soil fumigation, fertility management, and inoculation with spores of the fungal symbiont Laccaria laccata. For Sci 21:390–399

Sinclair WA, Sylvia DM, Larsen AO 1982 Disease suppression and growth promotion in Douglas-fir seedlings by the ectomycorrhizal fungus Laccaria laccata. For Sci 28:191–201

Sinha AK, Hait GN 1982 Host sensitization as a factor in induction of resistance in rice against Drechslera by seed treatment with phytoalexin inducers. Trans Br Mycol Soc 79:213–219

Smith GS 1988 The role of phosphorus nutrition in interactions of vesicular-arbuscular mycorrhizal fungi with soilborne nematodes and fungi. Phytopathology 78:371–374

Snyder BA, Nicholson RL 1990 Synthesis of phytoalexins in sorghum as a site-specific response to fungal ingress. Science 248:1637–1639

Spanu P, Bonfante-Fasolo P 1988 Cell-wall-bound peroxidase activity in roots of mycorrhizal Allium porrum. New Phytol 109:119–124

Spanu P, Boller T, Ludwig A, Wiemkin A, Faccio A, Bonfante-Fasolo P 1989 Chitinase in roots of mycorrhizal Allium porrum: regulation and localization. Planta 177:447–455

Stack RW, Sinclair WA 1975 Protection of Douglas-fir seedlings against Fusarium root rot by a mycorrhizal fungus in the absence of mycorrhiza formation. Phytopathology 65:468–472

Stafford HA 1988 Proanthocyanidins and lignin connection. Phytochemistry 27:1–6

Stafford HA, Cheng TY 1980 The procyanidins of Douglas-fir seedlings, callus, and cell suspension cultures derived from cotyledons. Phytochemistry 19:131–135

Strobel NE 1989 Resistance induced by Laccaria bicolor to Fusarium oxysporum in roots of Douglas-fir seedlings: the influence of temperature and pathogen aggressiveness and the role of phenolic wall infusions. PhD Thesis, Cornell Univ, Ithaca NY, 152 pp

Strobel NE, Sinclair WA 1991a Influence of temperature and pathogen aggressiveness on biological control by Laccaria bicolor of Fusarium root rot of Douglas-fir. Phytopathology 81 (in press)

Strobel NE, Sinclair WA 1991b Role of flavanolic wall infusions in the resistance induced by Laccaria bicolor to Fusarium oxysporum in primary roots of Douglas-fir. Phytopathology 81 (in press)

Sylvia DM, Sinclair WA 1983a Suppressive influence of Laccaria laccata on Fusarium oxysporum and on Douglas-fir seedlings. Phytopathology 73:384–389

Sylvia DM, Sinclair WA 1983b Phenolic compounds and resistance to fungal pathogens induced in primary roots of Douglas-fir seedlings by the ectomycorrhizal fungus Laccaria laccata. Phytopathology 73:390–397

Tester M, Smith SE, Smith FA 1987 The phenomenon of "nonmycorrhizal" plants. Can J Bot 65:419–431

Thomashow LS, Weller DM, Bonsall RF, Pierson LS III 1990 Production of the antibiotic phenazine-1-carboxylic acid by fluorescent Pseudomonas species in the rhizosphere of wheat. Appl Environ Microbiol 56:908–912

Tinker PB 1984 The role of microorganisms in mediating and facilitating the uptake of plant nutrients from soil. Plant Soil 76:77–91

Tommerup IC 1984 Development of infection by a vesicular-arbuscular mycorrhizal fungus in Brassica napus L. and Trifolium subterraneum L. New Phytol 98:487–495

Toppan A, Roby D 1982 Chitinase activity in melon seedlings inoculated with Colletotrichum lagenarium (Pass.) Ell. et Halst. or treated with ethylene. Agronomie 2:829–834

Tousson TA, Nash SM, Snyder WC 1960 The effect of nitrogen sources and glucose on the pathogenesis of Fusarium solani f, sp. phaseoli. Phytopathology 50:137–140

Trivedi N, Sinha AK 1976 Resistance induced in rice plants against Helminthosporium infection by treatment with various fungal fluids. Phytopathol Z 86:335–344

Tuzun S, Rao MN, Vögeli U, Schardl CL, Kuć J 1989 Induced systemic resistance to blue mold: early induction and accumulation of β-1,3-glucanases, chitinases, and other pathogenesis-related proteins (b-proteins) in immunized tobacco. Phytopathology 79:979–983

Urs NVRR, Dunleavy JM 1975 Enhancement of the bactericidal activity of a peroxidase system by phenolic compounds. Phytopathology 65:686–690

Vance CP, Kirk TK, Sherwood RT 1980 Lignification as a mechanism of disease resistance. Annu Rev Phytopathol 18:259–288

Van Loon LC 1985 Pathogenesis-related proteins. Plant Mol Biol 4:111–116

Vellinga EC, Mueller, GM 1987 Taxonomic and nomenclatural notes on Laccaria B. & Br.-II. Laccaria bicolor, L. fraterna, and L. laccata var. pallidifolia. Persoonia 13:27–43

Vögeli U, Meins F, Boller T 1988 Coordinated regulation of chitinase and β-1,3-glucanase in bean leaves. Planta 174:364–372

Wessels JGH, Marchant R 1974 Enzymatic degradation of septa in hyphal wall preparations from a monokaryon and a dikaryon of Schizophyllum commune. J Gen Microbiol 83:359–368

Wick RL, Moore LD 1982 Interaction of endomycorrhizae and Thielaviopsis basicola black root disease of Japanese holly. Phytopathology 72:712 (Abstr)

Wilcox HE 1983 Fungal parasitism of woody plant roots from mycorrhizal relationships to plant disease. Annu Rev Phytopathol 21:221–242

Wyss P, Boller T, Wiemken A 1989 Glyceollin production in soybean during the process of infection by Glomus mosseae and Rhizoctonia solani. Agr Ecosyst Environ 29:451–456

Wyss P, Mellor PB, Wiemen A 1990 Vesicular-arbuscular mycorrhizas of wild-type soybean and non-nodulating mutants with Glomus mosseae contain symbiosis-specific polypeptides (mycorrhizins), immunologically cross-reactive with nodulins. Planta 182:22–26

Yamada T, Hiramoto T, Tobimatsu R, Shiraishi T, Oku H 1990 Elicitor-like substances present in barley and wheat seeds. J Phytopathol 128:89–98

Yawney WJ, Schultz RC 1989 Anatomy of a vesicular-arbuscular endomycorrhizal symbiosis between sugar maple (Acer saccharum Marsh.) and Glomus etunicatum Becker & Gerdemann. New Phytol 114:47–57

Zak B 1964 Role of mycorrhizae in root disease. Annu Rev Phytopathol 2:377–392

Chapter 16

Tree Defense Mechanisms Against Fungi Associated with Insects

K.F. Raffa and K.D. Klepzig

16.1 Introduction

Insect-vectored fungi cause some of the most serious problems facing tree protection. For example, bark beetles are typically listed as the foremost tree-killing agents in government surveys, but in fact these insects are intimately associated with fungal symbionts. From both an ecological and a physiological perspective, they are a complex. Likewise, root-feeding beetles and their associated black-staining fungi are becoming increasingly important as the conversion to plantation forestry intensifies. Insect-vectored fungi also cause severe losses to high-value urban trees, such as with Dutch elm disease.

Successful defense by the host against both insects and fungi is the rule rather than the exception. Symbiotic relationships between insects and fungi, however, can result in what Cherrett et al. (1989) have termed an "unholy alliance." That is, each agent can reduce the difficulties encountered by the other. For example, the low likelihood of a fungal propagule landing on a suitable infection court of a compatible host plant can be greatly increased by its association with an insect vector. Conversely, the toxicological barriers encountered by insects feeding on living plants can be reduced by fungal biochemistry.

Closely linked insect-fungal relationships also provide targets for pest management through enhanced host resistance, however. With such high interdependency, effective resistance against, or direct control of, one agent could often render the other ineffective.

16.2 Insect-Fungal Symbioses: Degrees of Association and Roles in Exploiting Host Trees

Insects and fungi that exploit woody plants have evolved many types of interactions that range from purely exploitative to strongly mutualistic. In some relationships, the plant plays a relatively secondary or passive role in the insect-fungal association. In other systems, increased ability to gain access to, overcome the defenses of, or digest the host plant appears to be a major advantage conferred to one or more members of the complex. The evolutionary origin of such associations are difficult to discern, but in most cases it appears that insects and fungi with some primitive capability of exploiting a portion of the host population gained a competitive advantage from a more consistent degree of association. In the more specialized relationships, derived morphological, chemical, and behavioral traits may facilitate the interaction, and a high level of interdependency may be involved.

Some of the principal benefits derived by insects and/or fungi from their associations are summarized in Table 16.1. This table is not intended to be all inclusive, as excellent reviews by Batra (1979), Wheeler and Blackwell (1984), and Wilding et al. (1989) are available. Rather, it provides some context for those relatively fewer cases where fungi are introduced by insects into living trees, and therefore encounter host defense mechanisms. The improved metabolism, survival, competitive status, and development of insects and fungi that may follow successful colonization will not be considered here, because they occur after the critical stages of host defense. Likewise, the attraction of insects to trees previously colonized by fungi will also not be considered, because it has no immediate bearing on tree defensive responses against pathogens.

In most cases, vector relationships are rather casual, in that there is no apparent benefit to the insect, and fungi benefit simply from the creation of and/or inoculation into infection courts. In other cases, insects and fungi may derive benefits that are marginal or delayed. For example, the oak bark beetles (Coleoptera: Scolytidae) *Pseudopityophthorus minutissimus* and *P. pruinosus* breed in killed trees. If brood trees are infected with *Ceratocystis fagacearum*, emergent adults can transmit the pathogen to healthy hosts during maturation feeding. Infection can kill the host, and thereby increase the food base for the beetles. *C. fagacearum* can also be transported by sap beetles (Coleoptera: Nitidulidae) that feed on sap and fungal mats. Once contaminated with conidia and ascospores, beetles can inoculate healthy trees during subsequent feeding on wound sites. Although both insect groups may benefit from *C. fagacearum*, these relationships are facultative: The oak bark beetles can colonize species not particularly susceptible to *C. fagacearum*, and sap beetles are attracted to almost any source of plant injury or microbial fermentation. Likewise, *C. fagacearum* may rely on insects for dispersal to distant hosts, but it can then spread to neighboring trees through root grafts without insect vectors.

Systems involving delayed benefits can become increasingly interdependent. For example, smaller European elm bark beetles, *Scolytus multistriatus*, and native elm bark beetles, *Hylurgopinus rufipes*, introduce *Ophiostoma ulmi* during adult feeding on healthy trees, and then breed in trees killed previously by *O. ulmi*. As with *C. fagacearum*, *O. ulmi* can subsequently progress through root grafts. However, both long-distance dispersal by *O. ulmi* and reproductive success by elm bark beetles are greatly facilitated by their high degree of host overlap. Likewise, *Dendroctonus valens* (Scolytidae) and *Hylobius radicis* (Curculionidae) introduce *Leptographium terebrantis* and *Leptographium procerum*, respectively, into *Pinus resinosa* roots (Klepzig et al. 1991). These fungi then spread through root grafts into healthy trees, thereby increasing their susceptibility to the stem-colonizing *Ips pini* (Coleoptera: Scolytidae) – *Ophiostoma ips* complex. The trees killed by *I. pini* in turn provide an increased food base for the root-invading beetle – *Leptographium* complex.

Among some highly intimate relationships, both species enter the tree simultaneously, and reproductive success is mutually dependent. For example, most of the "aggressive" bark beetles that mass attack trees are closely asso-

Table 16.1 Principal advantages to tree-inhabiting insects and fungi conferred from their association

	Example	Reference[a]
A. *Advantages to insects*		
1. Nutrition		
a. Mycophagy/fungivory	Ambrosia beetles	1
Mycophagy/fungivory	Anthomyid flies	2
Mycophagy/fungivory	Leaf-cutting ants	3
b. Enhanced wood digestion	Termites	4
Enhanced wood digestion	Bark beetles	5
2. Orientation to hosts		
a. Attraction/arrestment	Bark beetles	6, 7
by infected trees	Wood wasps	8
by infected trees	Root weevils	7
b. Pheromone production	Bark beetles	9
c. Host location by	Ichneumonid wasps	10, 11
entomoparasites	Cynipid wasps	11
3. Overcoming host resistance		
a. By nonvectored fungi	Bark beetles	12
b. By vectored fungi	Bark beetles	13, 14
By vectored fungi	Ants	15
By vectored fungi	Wood wasps	16
4. Protection	Scale insects	17
5. Microenvironment condition	Wood wasps	18
B. *Advantages to fungi*		
1. Transport	*Amylostereum*	8
Transport	*Ophiostoma*	5
Transport	*Leptographium*	19
Transport	*Ceratocystis*	20, 21
Transport	*Phytophthora*	22
Transport	*Cryphonectria*	23
2. Nutrition: pathogenesis	*Entomophthorales*	24
3. Fertilization	*Chrysomypa*	25
Fertilization	*Ceratocystis*	14
4. Cultivation	*Monilia*	26
Cultivation	*Attamyces*	27, 3
5. Overcoming host resistance	*Ophiostoma*	29
Overcoming host resistance	*Attamyces*	15

[a] 1, Norris (1979); 2, Hackman and Meinander (1979); 3, Cherrett et al. (1989); 4, Wood and Thomas (1989); 5, Whitney (1982); 6, Geizler et al. (1980); 7, Witcosky (1985); 8, Madden and Coutts (1979); 9, Brand et al. (1976); 10, Madden (1968); 11, Spradbery (1970 a,b); 12, Rudinsky (1962); 13, Hemingway et al. (1977); 14, Webber and Gibbs (1989); 15, Cherrett et al. (1989); 16, Coutts and Dolezal (1969); 17, Couch (1938); 18, Rawlings (1953); 19, Witcosky (1985); 20, Gibbs and French (1980); 21, Juzwik and French (1983); 22, Evans (1971); 23, Anagnostakis (1987); 24, Evans (1989); 25, Savile (1976); 26, Kok (1979); 27, Weber (1979); 28, Raffa and Berryman (1983a).

ciated with *Ophiostoma*. The insect-microbial complex must kill the plant to develop, or else be lethally confined. There are exceptions, as in occasional "strip" or "top" kills, but such trees are invariably killed by subsequent generations. The association between *Sirex noctilio* and *Amylostereum areolatum* on *Pinus radiata* also appears closely linked to the need to overcome host defenses. Like bark beetles, these fungus-carrying insects kill trees through chemically mediated mass attack and risk host responses that can confine the entire complex.

Some closely associated insect-fungal complexes do not require tree death for reproduction. Examples include the solitary bark beetles, *D. valens* and *D. terebrans* (Barras and Perry 1971, Owen et al. 1987, Klepzig et al. 1991), and the pine root weevils *H. radicis* and *H. rhizophagus* (Raffa and Hall 1988), which vector *Leptographium*.

16.3 Properties of Plant-Insect-Fungal Systems: A Cross Between Two Similar But Distinct Host-Parasite Relationships

Insects and fungi share many similarities in their relationships with host plants. For example, both utilize stored photosynthate for development, exploit host structures for shelter and optimal microclimate, and require special adaptations for compatibility with host conditions and availability. From an evolutionary perspective, both select for host ability to resist these actions, because continuous exploitation can cause direct mortality, increased susceptibility to other stress agents, and reduced competitive ability with other plants. Likewise, the resistance mechanisms used by plants to defend against both insects and fungi show many similarities. Presumably, this represents a sort of biochemical frugality, so that separate costs to plant growth and reproduction need not be incurred for each potential threat.

There are also substantial differences between insect-plant and fungal-plant interactions, however. These need to be understood, because insect-fungal complexes assume attributes of both species. These biological factors might also explain some of the differences between how entomologists and plant pathologists generally perceive host-parasite relationships.

One of the most important differences between insects and fungi is in the level of active searching during host selection. Fungi possess a variety of chemical, phenological, and morphological traits that may increase their likelihood of encountering compatible tissue, but these are only weakly analogous to the repertoire of insect behaviors involving directed locomotion and neurological control of perception and reaction. Perhaps most important, insects have the flexibility to land on a potential host, diagnose its suitability, and if it is suboptimal, leave to resume searching.

A fungal spore, on the other hand, is more commonly faced with a develop or perish situation. This distinction can have ecological and evolutionary consequences. For example, directed locomotion and behavioral responses allow some insects to more easily specialize on stressed hosts, just as many vertebrate predators do. The cues eliciting attack may be associated with a physiological condition, rather than a certain genotype within a host species or strict order of preference among several host species. There are some well-documented cases of stress predisposing plants to disease agents (Schoeneweiss 1975), of which *Armillaria* is a prime example in trees. However, the plant pathology literature appears to be more strongly dominated by examples of strict genetic control and specific recognition phenomena (e.g., Daly 1984, Gabriel and Rolfe 1990), as compared to entomological reviews that more commonly emphasize stress factors (e.g., Kozlowski 1969, Mattson and Haack 1987, Wallner 1987). Assuming these differences are biological and not just institutional, perhaps they arise partly from how selective pressures on individuals are mediated by varying degrees of behavioral options. If so, then associating with insects can increase the importance of environmental stress to fungi in their interactions with host trees. Likewise, the likelihood of evolving a strict linkage with insects may be greater among pathogens that are relatively stress-adapted. Many of the most damaging insect-fungus complexes involve vectors that show strong preferences for stressed trees, such as bark beetles, wood wasps, and root weevils, and corresponding fungi that are relatively weak pathogens, perhaps owing to a more recent saprophytic existence, such as *Ophiostoma* and *Leptographium* (Raffa and Berryman 1987, Smalley et al. 1991).

Insect behavior may also affect the relative importance of constitutive tree defenses. Plants commonly contain chemicals that deter insects (Miller and Strickler 1984). These preformed defenses can be very effective, both against insects not adapted to the host species, and against adapted insects that cannot tolerate the very high allelochemical concentrations present in some individual hosts. Although there are many examples of constitutive defenses inhibiting fungi, most refer to nonadapted pathogens, and none conclusively define a host-pathogen relationship (Stoessl 1983). Reviews over the past 25 years have been quite consistent in ascribing to preformed resistance a "general" protection from a wide range of potentially pathogenic species, but little role in determining which members of a host population will be colonized by which members of an adapted parasite population (e.g., Kuc 1966, Misaghi 1982, Stoessl 1983). Induced defenses are generally viewed as being needed to resist adapted species. Again assuming these differences are more biologically than institutionally based, behavioral repertoires may be partially responsible. That is, when viewed from the insect's perspective, repellency can be adaptive: Perception and avoidance of token stimuli associated with an inhospitable environment can confer a selective advantage. This capability is generally lacking in fungi. Another consequence of insect behavior to vectored fungi, then, may be that constitutive responses increase in their relative importance in overall defense.

A second major difference between insects and fungi concerns the role of morphological defenses, such as thick bark or a tough epidermis. These physical traits can pose formidable barriers to fungi (Campbell et al. 1980), which usually must either circumvent them by producing specialized structures or locating small openings such as stomata, or arrive at infection courts on a low probability-high numerical basis. Insects are equipped with sclerotized body parts and muscles that greatly facilitate direct penetration. The use of these tools, however, is hardly subtle from a cellular host-pathogen perspective. Mechanical shredding of live tissue necessitates that traumatic wounding accompanies every invasion, and so the highly sophisticated mechanisms evolved by pathogens to avoid and/or suppress host recognition (e.g., Albersheim and Valent 1974, Sequeira 1980) have little bearing. Moreover, free-living insects are typically contaminated with a broad and unpredictable array of randomly acquired microbes in addition to their normal symbionts. So, induced defenses are likely to be triggered by every encounter. Thus, the chances of an insect-microbial complex evolving into a flora unapparent to the host, particularly in a gene-for-gene fashion, seem very low.

Although host penetration by insects may facilitate fungal entry, it may also degrade the microhabitat. For example, when insects rip apart leaf cells or create wounds in bark, they also cause the tissue to desiccate. These infection courts may also become available to secondary fungi which are often strong competitors with primary pathogens. This may partially explain why the most intimate insect-fungal complexes are subcortical and rhizophagous despite the preponderance of insect folivores. Perhaps damage to leaf tissue caused by free-living folivores allows little opportunity for fungi that would be desiccated, consumed, and/or dropped to the ground before advancing ahead of the insect. Endoparasitic herbivores, by contrast, rely more heavily on and better maintain the host microenvironment during development.

A third difference is in the degree of injury to the host. Insects primarily reduce tree vigor through direct removal of photosynthate. Thus, a fairly high degree of successful insect feeding and breeding can usually be tolerated before severe symptoms are incurred. Exceptions include insects that interfere with translocation, such as subcortical or root-feeding species, and species that egest phytotoxic secretions. Although many fungi cause only local, relatively benign reactions, many contribute to systemic "disease." Thus, tolerance seems to be a more important component of insect- than fungal-plant interactions. A consequence to insects of associating with fungi then, is that they accrue both the benefits and disadvantages of disease agents.

Insects and fungi also differ in propagule longevity. Insect attacks are initiated by adults that oviposit onto new hosts, or by larvae that usually crawl from within the immediate vicinity (although some are windborne). Although these stages may diapause to avoid physical extremes, once they are in a host-seeking mode they must find plants very quickly to survive or reproduce. Adults of most phytophagous species are short-lived, and active larvae cannot survive long periods without food. Thus, each generation could undergo a sharp population decline due to unpredictable episodes of host unavailability. Although fungi also incur such risks, they often have the added dimension of being able to remain in a dormant state until conditions improve. These conditions may be physical, such as moisture-stimulated germination after prolonged quiescence, or biological as in responses to decreased host vigor among *Armillaria*. Such differences in propagule survival may affect plant-insect-fungal interactions. For example, insects and fungi have evolved effective responses to temporal changes (Agrios 1980) in host suitability, but their optimal timing may differ. This could be a major factor preventing the coupling and/or more intense association of many insect-fungal systems (Webber and Gibbs 1989). Conversely, some of the delayed-feedback benefits to insect vectors described earlier arise from the ability of the introduced fungi to remain quiescent until conditions become more suitable.

Other innate biological differences include higher propagule production, shorter generation times, and probable higher mutation rates of fungi than insects (Harris and Frederiksen 1984). All of these are likely to affect the interactions and genetic interplay among host and parasites.

16.4 Selected Examples of Tree Defenses Against Insect-Associated Fungi

16.4.1 Bark Beetle – *Ophiostoma* Interactions

Bark beetle (Coleoptera: Scolytidae) – *Ophiostoma* complexes are perhaps the most damaging of all forest pests. Although the most thoroughly studied host defenses are those of *Ulmus* against Dutch elm disease, primary bark beetles are primarily associated with conifers, including all of the species that kill large numbers of trees within their native range (Rudinsky 1962). Because defenses against *O. ulmi* are described in Chapter 13 (this Vol.), they are omitted here.

Only the small minority of scolytids that colonize living trees are considered here. These may be further separated based on whether they engage in pheromone-mediated mass attacks or are solitary. The former species primarily colonize the main stems of trees and are associated with blue-staining fungi. The latter species, which primarily colonize the lower stems and roots, are associated with black-staining fungi, and are considered with other root-infesting beetles in Section 16.4.2.

Bark beetle adults emerge from brood trees, disperse, select suitable hosts, and construct ovipositional galleries in the phloem. Successful colonization causes and is usually contingent on host death. The larvae emerge, tunnel through the cambium as they feed, and pupate in the outer bark (Fig. 16.1). Because a tree can only support one beetle generation, emerging adults must locate new suitable trees to complete the cycle.

Bark beetles are closely associated with a variety of yeasts, fungi, and bacteria, some of which assist in establishment and digestion (Whitney 1982). Some fungi are transported in specialized invaginated structures, called mycangia, while others are carried on the exoskeleton or within the gut. These mycangia can be quite elaborate and contain secretory cells that favor specific associates (Whitney 1982). Specialized beetle behaviors and associations with phoretic mites that transport spores from the phloem to outer bark (Moser 1985) may facilitate recontamination.

Some of the pathogens vectored by stem-infesting bark beetles include *Ophiostoma minus*, *O. ips*, *O. montium*, *O. clavigerum*, and *O. polonicum*. These fungi have demonstrated moderate pathogenicity in seedling inoculations and simulated mass attack, and when accompanied by severe wounding to the cambium (Mathre 1964, Horntvedt et al. 1983, Strobel and Sugawara 1986).

Fig. 16.1. *Pinus resinosa* responses to *Ips pini-Ophiostoma ips* complexes. *Upper left* Breeding galleries and fungal growth in successfully colonized tree; *upper right Ophiostoma ips*; *lower left* failed attack. Note necrotic reaction zone which contains *Ips* gallery, and healthy condition of cambium outside lesion; *lower right* response to controlled inoculation with *O. ips*. Note necrotic lesion and resin flooding as in response to natural attack.

Whenever healthy mature trees have been singly inoculated through small wounds equivalent to beetle entry sites, however, host reactions always confined the fungi. Tens of thousands of such simulated attacks have been reported in the literature without fungal establishment.

Shortly after the fungus reaches living tissue beneath the bark, a series of host cellular reactions is initiated. A necrotic lesion of killed cells forms in advance of the invading complex (Fig. 16.1). Lesion size around naturally resisted attacks is directly related to the length of the beetle gallery, indicating that lesions expand for as long as the invading complex continues to progress (Raffa and Berryman 1983a). After the beetle and fungus are confined, callus forms around the dead tissue, thereby containing the wound.

Concentrations of monoterpenes, resin acids, sesquiterpenes, and phenolics increase rapidly within the reaction zone (Raffa 1991). Monoterpene increases

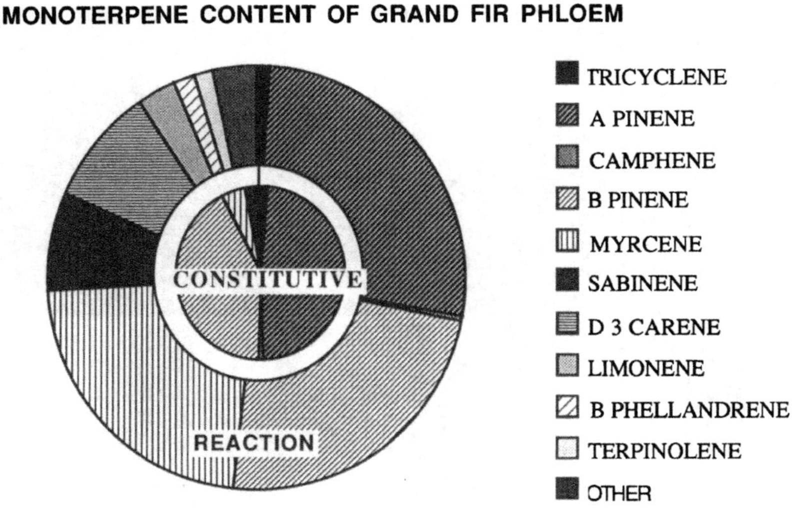

Fig. 16.2. Chemical changes in monoterpene fraction of *Abies grandis* following inoculation with *Trichosporium symbioticum*. The size of the area within each zone represents the absolute value of the terpene. The relative proportion within a circle denotes its percent abundance within the fraction. Defensive responses (full diameter) entail both absolute and relative changes. Data from Raffa and Berryman (1982b)

of over 500-fold within a few weeks have been reported (Hain et al. 1983, Cook and Hain 1985, 1986). Proportionate changes in the relative abundance of each allelochemical, and the appearance of compounds not present in constitutive phloem, accompany these quantitative changes (Fig. 16.2). It has also been proposed that cellular deterioration within the lesion can deprive the advancing fungus of essential nutrients (Berryman 1972).

Some of the chemicals that accumulate during this response are fungistatic. For example, extractives from *Pinus contorta* reaction, but not constitutive, phloem inhibit *C. montia* (= *O. ips*) and *O. clavigerum* growth (Shrimpton 1973a). Raffa et al. (1985) found that monoterpene concentrations within *Abies grandis* phloem 3–7 days after inoculation with *Trichosporium symbioticum* are inhibitory. These are not particularly toxic phytoalexins, however, at least when compared to other systems. Viable fungi can be recovered from within, but not beyond, the reaction zone a full year after inoculation (Wong and Berryman 1977, Raffa and Smalley 1988a).

Although adult beetles can usually tolerate resins from their normal host species (Smith 1963, 1965), monoterpene concentrations within reaction zones exceed the beetles' physiological tolerance within a few days (Raffa et al. 1985). The altered chemical environment is also toxic to the beetles' eggs and larvae (Raffa and Berryman 1983a,b). Bark beetles are often repelled once monoterpene concentrations rise, and may abandon the attack (Bordasch and Berryman 1977, Raffa and Berryman 1982a). Thereafter, any fungi in the phloem are without the benefit of further beetle activity, and have no more chance of success than the experimental single inoculations described above. In general, there is a disproportionately high increase in those chemicals most deleterious to the insects and fungi during induced responses (Raffa 1991).

Tree defensive responses can be elicited by the beetle-fungus complex, the fungus alone, mechanical wounding, and by purified compounds (Table 16.2). Each system has its unique properties, but certain generalizations can be made: First, host reactions to beetle-vectored fungi are not mere wound reactions. Responses to mechanical wounds yield less allelochemical accumulation, less necrosis, and no proportionate chemical changes or phytoalexin appearance. Secondly, responses to dead fungi are less extensive than responses to live fungi, but greater than responses to mechanical wounds. This suggests that fungal cell wall fragments are involved in, but not totally responsible for elicitation of host resistance. Work with defined elicitors supports this view. Chitosan, which is present in both fungal cell walls and insect exoskeletons, elicits chemical changes equivalent to fungal inoculation, but only moderate lesion formation in *Pinus contorta* (Miller et al. 1986a). Conversely, the wound-induced pectic fragment, proteinase inhibitor inducing factor (PIIF), elicits lesion formation equivalent to fungal inoculation, but only moderate chemical changes. Third, there is some specificity at the species level. *Pinus taeda*, for example, responds differently to *O. minus* than to *C. minor barrassii* (= *O. minus*) or to a mycangial basidiomycete (Paine and Stephen 1987a). Likewise, both *P. resinosa* and *P. banksiana* respond more strongly to *O. ips* than to *O. nigrocarpum*. Conversely, the reactions elicited by each fungus vary among tree species. In general, fungi elicit the most extensive responses when inoculated into their normal host species (Lieutier and Berryman 1988).

Berryman (1988) proposed that elicitation occurs in two phases. Initially, physical wounds caused by beetles release plant cell-wall fragments that induce increased metabolism in neighboring cells. For example, neighboring cells show an increase in the number of ribosomes, dictyosomes, mitochondria, endoplastic reticula, and plastids (Cheniclet et al. 1988). They also show increased starch and decreased lipid levels. Shortly thereafter, leucoplasts enlarge and secondary resin ducts begin to form. The second phase is elicited by microbial structural chemicals, such as chitosan, and/or perhaps metabolites. These elicitors stimulate mevalonic acid and phenylalanine pathway biosyntheses in parenchyma cells (Berryman 1988, Cheniclet et al. 1988, Lieutier and Berryman 1988). Biosynthesis is accompanied by activation of water-soluble enzymes such as phenyl transferase, and cell wall rupture. Products diffusing from cell walls activate cells ahead of the pathogen. Plant wound hormones, such as ethylene, may also contribute to autolysis (Raffa and Berryman 1982b, Popp et al. 1990). Lesion growth following controlled fungal inoculation occurs in a fixed sigmoid process (Raffa and Berryman 1982b, Raffa and Smalley 1988b,c).

Possessed with such defenses, trees might seem immune to bark beetle-vectored fungi. However, scolytid chemical communication systems enable them to engage in mass attacks, and thereby exhaust host physiology. Assisted by microbial symbionts, bark beetles enzymatically oxidize host monoterpenes via cyctochrome P450 pathways to terpene alcohols that attract other beetles

Table 16.2 Elicitation of defensive responses against fungi vectored by bark beetles. Defensive reactions are rated on a scale from + (low) to +++ (high). A (o) means no reaction, and a blank means data are not available. Lower stem feeding bark beetles are included

	Eliciting agent		Response		
Host	*Fungus*	*Treatment*	*Lesion*	*Chem. changes*	*References*[a]
P. contorta	*Ophiostoma clavigerum*	Live	+++	+++	1, 2, 3, 4
	–	Mechan. wound	+	+	4
	–	*Dendroctonus ponderosae* attack	+++	+++	1, 4, 5
	–	Cellulolytic enzymes	+++		6
	–	Chitosan	++	+++	4, 7
	–	PIIF	+++	++	4, 7
P. taeda	*O. minus*	Live	+++	+++	8, 9, 10, 11, 12, 13
	–	*D. frontalis* attack	+++		9
	–	Mechan. wound	+	+	8, 9, 12, 13
	O. minus barrassii	Live	+		14
	Basidiomycete	Live	+		14
Pinus ponderosa	*Ophiostoma minus*	Live	+++		15
	–	Mechan. wound	+		15
	O. clavigerum	Live	+++		7
	–	Chitosan	+++		7
	–	PIIF	+		7
P. monticola	*O. clavigerum*	Live	+++		7
	–	Chitosan	++		7
	–	PIIF	+		7
P. resinosa	*O. ips*	Live	+++	+++	16, 17
	O. ips	Autoclaved	++	++	16, 17
	O. nigrocarpum	Live	+	+	16, 17
	O. nigrocarpum	Autoclaved	+	+	16, 17
	–	*I. pini* attack	+++		16, 17
	–	*D. valens* attack	+++		16, 17
	Leptographium terebrantis	Live	+++	+++	17, 18
	L. terebrantis	Autoclaved	++	++	17, 18
	–	Mech. Wound	+	+	16, 17, 18

P. banksiana	*O. ips*	Live	+++	+++	16, 17
	O. ips	Autoclaved	++	++	16, 17
	O. nigrocarpum	Live	++	++	16, 17
	O. nigrocarpum	Autoclaved	++	++	16, 17
	–	*I. pini* attack	+++		16, 17
	L. terebrantis	Live	+++	+++	16, 17
	L. terebrantis	Autoclaved	++	++	16, 17
	–	Mech. wound	+	+	16, 17, 18
P. palustris	*O. minus*	Live		+++	10, 11
P. pinaster	*V.* spp.	Live		+++	19
	–	Mech. wound		+	19
Abies grandis	*Trichosporium symbioticum*	Live	+++	+++	20, 21, 22, 23
	–	*S. ventralis* attack	+++	+++	22
	–	Mech. wound	+	+	22
A. concolor	*T. symbioticum*	Live	+++	+++	24
	O. clavigerum	Live	0		7
	–	Chitosan	0		7
	–	PIIF	0		7
Larix occidentalis	*O. clavigerum*	Live	0		7
	–	Chitosan	0		7
	–	PIIF	0		7
Pseudotsuga menziesii	*O. clavigerum*	Live	0		7
	–	Chitosan	0		7
	–	PIIF	0		7
Thuja plicata	*O. clavigerum*	Live	0		7
	–	Chitosan	0		7
	–	PIIF	0		7
Picea abies	*O. polonicum*	Live	+++	+++	25, 26
	–	*I. typographus* attack	+++	+++	25

[a] 1, Reid et al. (1967); 2, Raffa and Berryman (1982b); 3, Shrimpton and Watson (1971); 4, Miller et al. (1986a); 5, Raffa and Berryman (1983a); 6, Shrimpton (1978); 7, Lieutier and Berryman (1988); 8, Gambliel et al. (1985); 9, Hain et al. (1983); 10, Cook and Hain (1985); 11, Cook and Hain (1986); 12, Paine et al. (1985); 13, Paine and Stephen (1987a); 14, Paine and Stephen (1987b); 15, Paine (1984); 16, Raffa and Smalley (1988 a,b); 17, Raffa and Smalley (unpubl. data); 18, Raffa and Smalley (1988c); 19, Cheniclet et al. (1988); 20, Russell and Berryman (1976); 21, Wright et al. (1979); 22, Raffa and Berryman (1982b); 23, Wong and Berryman (1977); 24, Ferrell (1988); 25, Christiansen and Horntvedt (1983); 26, Christiansen and Ericsson (1986).

Table 16.3 Evidence for role of mass attack in exhausting host defenses against bark beetle-fungal complexes

Measurement	Result	References
1. Examination of natural attacked trees	Surviving and killed trees can be separated based on whether attack density is above or below an apparent threshold	1, 2, 3, 4, 5
2. Control of attack density by artificially terminating natural attacks at varying intervals	Only colonization densities above a critical threshold are successful. Brood production increases in sigmoid fashion with increasing attack density	6
3. Comparison of tree defenses prior to and after natural attack	Resin flow and rate of lesion formation are reduced within 3 days after aggregation	6
4. Controlled inoculation with varying densities of fungi	Resin accumulation declines in sigmoid fashion with increasing inoculum density;	7, 8, 9
	tree death	8
5. Controlled inoculation with varying fungal load per inoculation	No effect	10

[a] 1, Hodges et al. (1979); 2, Keen (1938); 3, Berryman (1972); 4, Thalenhorst (1958); 5, Waring and Pitman (1983); 6, Raffa and Berryman (1983a); 7, Raffa and Berryman (1983b); 8, Horntvedt et al. (1983); 9, Christiansen (1985); 10, Paine and Stephen (1987b).

(Renwick and Vite 1970, Brand et al. 1975, 1976, Borden 1984). Each arriving beetle causes additional tissue damage, severs resin ducts, carries resin outside the tree using modified legs and elytra, and disseminates fungi throughout the cambium. Fungi further deplete tree defenses by interfering with host resin translocation (Whitney 1982) and secreting toxic isocoumarins that alter transpiration, disrupt membranes, and reduce water and resin flow (Hemingway et al. 1977). The inoculation density, rather than fungal load per inoculation, appears most important.

Defensive responses against fungi vectored by bark beetles are metabolically costly. Carbohydrates appear to provide the substrate for terpene biosynthesis, as Miller and Berryman (1985) measured a 30 and 15% decrease in soluble and reducing sugars, respectively, during induction. Croteau et al. (1972) determined that the synthesis of each monoterpene molecule requires 90 ATP molecules. Reduced sugar content can be observed 24–36 h after inoculation, whereas monoterpene increases require around 3 days (Miller and Berryman 1985).

The contribution of each beetle-fungus recruit generates a classical dosage-response curve. With increasing attack density, allelochemical levels fall below the concentrations required to resist the invading complex. This density defines a discrete "threshold of resistance" (Thalenhorst 1958, Berryman 1972, Raffa and Berryman 1983a) that is required for successful brood establishment, and is unique to each tree (Table 16.3). Beetles continue to arrive until the host's ability to resist is completely eliminated. Once a sufficiently high attack density has been reached, the tree's prior defensive potential is not manifested in fungal

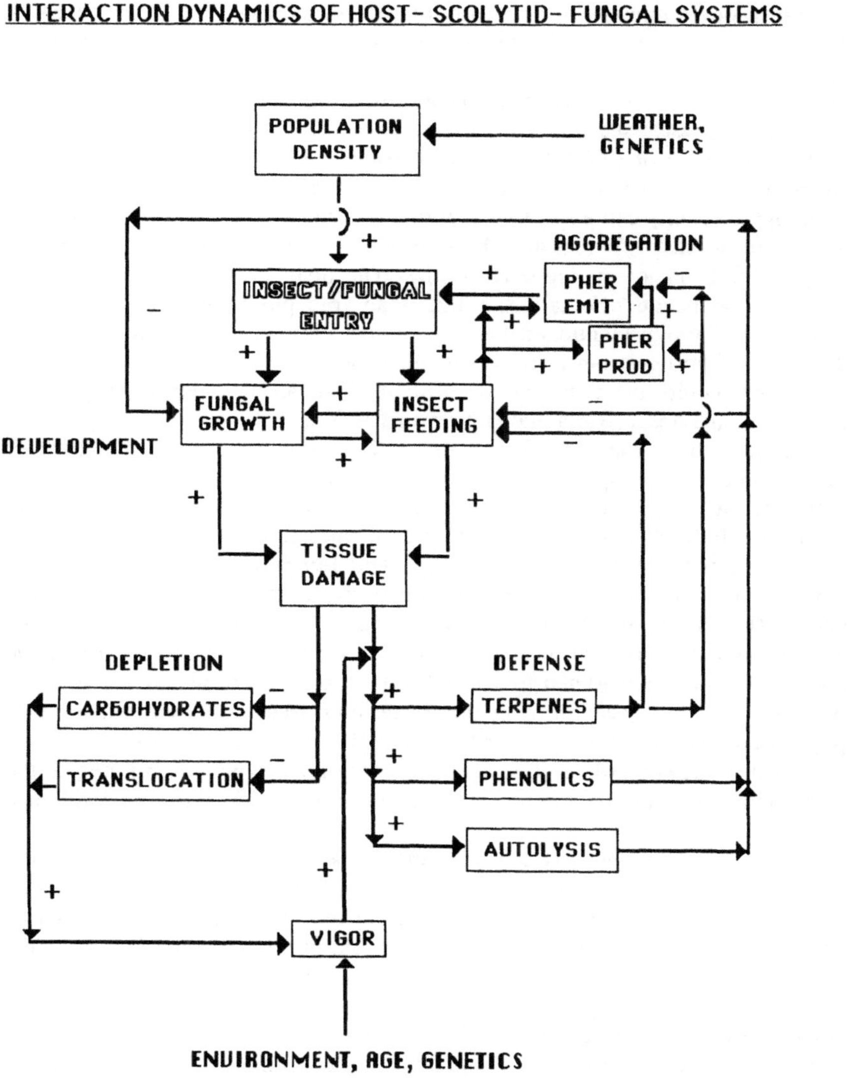

Fig. 16.3. Dynamics of host tree – bark beetle-fungal interactions. The outcomes of either tree survival or parasite reproduction are dictated by the interactions among conflicting rate processes. A (+) between two parameters indicates that an increase in the source box (independent variable) causes an increase in the recipient box (dependent variable). A (−) indicates a decrease in the same direction. See text for description

or insect development. In the case of *D. ponderosae – O. clavigerum – C. montia* (= *O. ips*) in *P. contorta* in northeastern Oregon, for example, around 40 attacks/m^2 are required to kill a tree, and 80 attacks/m^2 allow full larval survival. These attack densities are usually reached within 2–4 days for the more aggressive species.

The capability of scolytid – microbial complexes to convert host defensive chemicals into attractants may seem to render every tree vulnerable to attack. Although mortality can exceed 99% during outbreaks, such outbreaks are rare. There are at least two reasons why tree defense systems can function despite cooperative behavior among bark beetles. The first reason relates to selective pressures on beetle orientation, rather than tree physiology per se, and so will be only briefly mentioned here: Although each beetle contributes to the depletion of host resistance, it also reduces the available substrate per beetle. Trees that require more beetles to kill them than can be produced as progeny are suboptimal hosts (Raffa and Berryman 1987).

Secondly, trees may interfere with beetle communication. Although bark beetles produce pheromones upon exposure to monoterpenes in the laboratory, entry into a living tree does not guarantee that aggregation will occur. This cannot be fully explained by an absence of pheromone precursors or an insufficient number of flying beetles (Raffa and Berryman 1983a,b). Instead, it appears that a rapid flow of resin arising from the tree's constitutive defense system (in pines), and increasing during the induced response, may impede and/or mask the diffusion of beetle-produced volatiles from the gallery. If the invading beetle is not joined by conspecifics, it must either abandon the attack or be killed. This selects for avoidance of resinous trees, especially when beetle populations are low. Rapid landing by many beetles, such as during switching from nearby trees or when populations are very high, however, can negate the host's ability to interfere with beetle communication, and lead to successful mass attack.

Thus, the interaction between tree defenses and the number of bark beetle/fungal inocula is determined by a series of conflicting rate reactions depicted in Fig. 16.3. Parasite entry causes wounds in the host, and also initiates tree defenses. Some of these defenses can kill, inhibit, or repel the invading complex, but beetles modify allelochemicals to recruit additional arrivals. To a certain extent tree defenses may interfere with this attraction, but this capability is also depleted by high parasite densities. This interaction is basically unstable in that once a certain loop begins to predominate, it biomagnifies to the point of either tree death or parasite failure. It should be emphasized that the relationships in this scheme are not hypothetical, but have all been empirically demonstrated in at least one system. Using this general model, the relative rate, magnitude, and importance of each feedback process can be computed for each tree-insect-fungal interaction.

The balance among these conflicting rate processes is determined by the number of beetle recruits locally available, the genetic structure of the fungal and insect populations, climatic factors such as temperature that directly affect the beetles and fungi, and environmental, phenological, and genetic factors affecting host predisposition. When a longer time frame than that depicted in

Table 16.4 Stress factors associated with reduced rate and/or extent of tree defensive reactions to bark beetle-fungal complexes

Stress	Effect[a] Reduced defensive responses under controlled conditions	Increased susceptibility under natural conditions
Abiotic		
Water deficit	1, 2	3, 4, 5
Poor site condition	6, 7	3, 4, 8, 9
Artificial damage	10, 11	10, 11
Pollution	12 14	13
Lightning		15
Biotic	11, 16	
Crowding	21	17, 18, 19, 20
Defoliation	11	22, 23
Disease	25, 11	3, 4, 9, 24
Age		3, 4, 9, 20, 26

[a] References: 1, Vite and Wood (1961); 2, Vite (1961); 3, Keen (1938); 4, Rudinsky (1962); 5, Kalkstein (1976); 6, Lorio and Hodges (1968); 7, Paine and Stephen (1987a); 8, Waring and Pitman (1983); 9, Thalenhorst (1958); 10, Miller et al. (1986b); 11, Raffa and Berryman (1982b); 12, Johnson (unpubl. data); 13, Stark et al. (1968); 14, Blanche et al. (1985); 15, Coulson et al. (1983); 16, Amman et al. (1977); 17, Mason (1966); 18, Mitchell et al. (1983); 19, Nebeker and Hodges (1983); 20, Amman and Cole (1983); 21, Wright et al. (1979); 22, Wright et al. (1984); 23, Struble (1957); 24, Geizler et al. (1980); 25, Shrimpton (1973b); 26, Safranyik et al. (1975).

Fig. 16.3 is considered, the population densities and genetics of beetles, fungi, and trees become recipients of these feedback processes, rather than mere inputs.

A major factor affecting the extent and rate of host defense is physiological stress. A wide variety of abiotic and biotic stresses have been shown to be both correlated with tree losses to bark beetle – *Ophiostoma* complexes, and directly contributing to decreased measures of host resistance (Table 16.4). This relationship appears to arise from the direct physiological linkage between subcortical defenses and overall vigor (Berryman 1972, Safranyik et al. 1975, Mahoney 1978, Hodges et al. 1979, Waring and Pitman 1983, Miller et al. 1986b). For example, Shrimpton (1973a) compared the reaction tissue of successful and resisted attacks on *P. contorta*. All trees underwent an increase in monoterpenes and phenolics, with corresponding decreases in sugars. However, the living trees appeared to maintain higher sugar levels, even though their allelochemical increases were larger. Carbohydrate reserves in killed trees appeared to have been exhausted.

Both constitutive (Lorio 1986) and induced (Shrimpton 1973b, Paine 1984; Stephen and Paine 1985, Raffa and Smalley 1988a) defenses vary with season. Such seasonal changes may reflect differing allotments to defense at the expense of growth and other processes (Lorio 1986). Beetle phenological and diurnal rhythms, however, may likewise correspond to periods of minimal tree resistance (Raffa and Berryman 1987). Variation in defensive responses has also been reported in relation to height on the stem and prior inoculation, but the consistency and significance of these trends are not well understood (Reid and Shrimpton 1971, Paine et al. 1985, Miller et al. 1986a, Cook and Hain 1988).

16.4.2 Root Beetle – *Leptographium* Interactions

Black-stain root disease occurs on several species of pine, two species of hemlock, two species of fir and Douglas-fir throughout the western United States and British Columbia (Wagener and Mielke 1961, Goheen et al. 1978, Cobb 1988). The causal agent, *L. wageneri*, has been subdivided into three varieties (Harrington 1983, Harrington and Cobb 1986, 1987). Variety *wageneri* is primarily associated with pinyon pines, var. *ponderosum* with hard pines, and var. *pseudotsugae* with Douglas-fir. Local transmission of *L. wageneri* may occur through contact of diseased and healthy roots (Wagener and Mielke 1961, Smith 1967, Landis and Helburg 1976), through continuous xylem in root grafts, and through short-distance growth through the soil (Cobb 1988).

Infection by *L. wageneri* is characterized by a purplish-brown to black stain in the roots and lower trunk of affected trees (Wagener and Mielke 1961). Symptoms include reduced height and radial growth, needle chlorosis, and tree death (Cobb 1988). *Leptographium wageneri* is limited to the mature xylem tracheids, and is unable to colonize parenchyma cells (Hessburg and Hansen 1987). This unusual behavior is of diagnostic value (Cobb 1988), as other conifer-invading *Ophiostoma* and *Leptographium* spp. progress beyond the tracheids and thus produce a wedge-shaped pattern of staining rather than following the annual rings. During growth within the host, *L. wageneri* exhibits many of the characteristics of a wilt pathogen (Cobb 1988). It colonizes only active xylem tracheids, grows intercellularly through bordered pit pairs, induces tyloses, and can systemically colonize needle xylem in small trees. This fungus also produces metabolites such as orcinol and monomethyl ether, which both inhibit bacteria and reduce tree growth (Ayer et al. 1983). *Leptographium wageneri* is also incapable of breaking down cellulose (Smith 1969).

Related fungi and symptoms are involved in red pine decline and procerum root disease. Red pine decline affects plantation-grown, 20- to 40-year-old *P. resinosa* in the Great Lakes region of North America (Raffa and Hall 1988). Affected stands contain a circular area of dead trees, ringed by trees with reduced radial height, and foliar growth. Further from the epicenter, trees appear relatively healthy and vigorous. Mortality continues for several years as the edge trees die, and trees along the new perimeter begin to show stress symptoms. *Leptographium terebrantis* and/or *L. procerum*, and sometimes *O. huntii*, are found in the roots and lower stems of affected trees.

The pathogenicity of *L. terebrantis* is variable. Infection can cause lesions, staining (Wagener and Mielke 1961, Harrington and Cobb 1983, Raffa and Smalley 1988c), thin and chlorotic crowns (Highley and Tattar 1985, Rane and Tattar 1987), and tree death (Wingfield 1983). However, most mature *P. resinosa* and *P. banksiana* confine the fungus before any adverse effects on the tree can be detected (Raffa and Smalley 1988c).

Leptographium procerum is a relatively weak pathogen on some pine species, such as *P. resinosa* and *P. banksiana*. However, *L. procerum* can kill *P. strobus* and *P. taeda* that are under stress (Leaphart and Gill 1955, Lackner and Alexander 1982, 1983, Harrington and Cobb 1983, Wingfield 1983, Rane and Tattar 1987). Affected trees exhibit diffuse black streaking, resinosis, large basal cankers, and root lesions (Anderson 1980; Sinclair and Hudler 1980, Lackner and Alexander 1982, 1983, Wingfield 1983). In the crown, new shoots may be reduced or crooked, and needles uniformly wilt and die (Towers 1977, Lackner and Alexander 1982). Both *L. terebrantis* and *L. procerum* can colonize tracheids and rays, and therefore produce a wedge-shaped staining pattern in the wood.

Long-distance, interstand spread of *Leptographium* is via insect vectors that oviposit in roots and root collars of stumps (Barras and Perry 1971, Cobb 1988). For example, *H. nigrinus*, *H. macer*, *Pissodes fasciatus*, and *Steremnius carinatus* introduce *L. wageneri* into *P. menziesii* (Goheen and Cobb 1978, Witcosky et al. 1986a). The teleomorph, *O. wageneri* has only been observed within *H. macer* galleries, suggesting that insects bring together compatible mating types (Cobb 1988). The dependence of these vectors on stressed hosts is reflected in their increased activity in and attraction to site disturbances, such as road construction and precommercial thinning (Harrington 1983, Harrington et al. 1985, Witcosky et al. 1986b).

The root weevils *H. radicis*, *H. pales*, and *Pachylobius picivorus*, the lower stem-infesting bark beetle *D. valens*, and the root-infesting bark beetle *H. porculus*, are associated with red pine decline, and transport *L. terebrantis* and *L. procerum*. *Ophiostoma huntii* has been recovered from the pupal chambers and larval tunnels of *H. porculus* (Fig. 16.4). Within-stand spread by *Leptographium* appears facilitated by, but not dependent on, insect vectors. The introduction of *Leptographium* spp. into the extensively grafted root system of plantation *P. resinosa* by a vector apparently leads to progressive colonization of the roots (Fig. 16.5). Root infection then leads to tree decline and reduces host resistance to lethal attack by the stem bark beetle *I. pini* and its associate *O. ips* (Klepzig et al. 1991).

Resistance to *Leptographium* by root and root collar tissue closely resembles mechanisms against bark beetle-vectored fungi, and so will be only briefly described. Initial weevil invasion can be deterred by constitutive monoterpenes, particularly limonene. Invasion by the fungus elicits rapid resin accumulation and necrotic lesion formation in advance of its progress (Fig. 16.6). These resinous lesions invariably confine *Leptographium* that are either artificially inoculated or transported by caged beetles into healthy trees. Phenols, such as pinosylvin and its monomethyl ether, accumulate and inhibit fungal growth (Cobb 1988, Klepzig et al. 1990). Monoterpene concentrations rise to levels present in the reaction tissue against stem bark beetles. Trees moderately infected with *L. wageneri* may react more strongly to subsequent introduction of bark beetle-associated fungi than healthy trees (Cobb 1988).

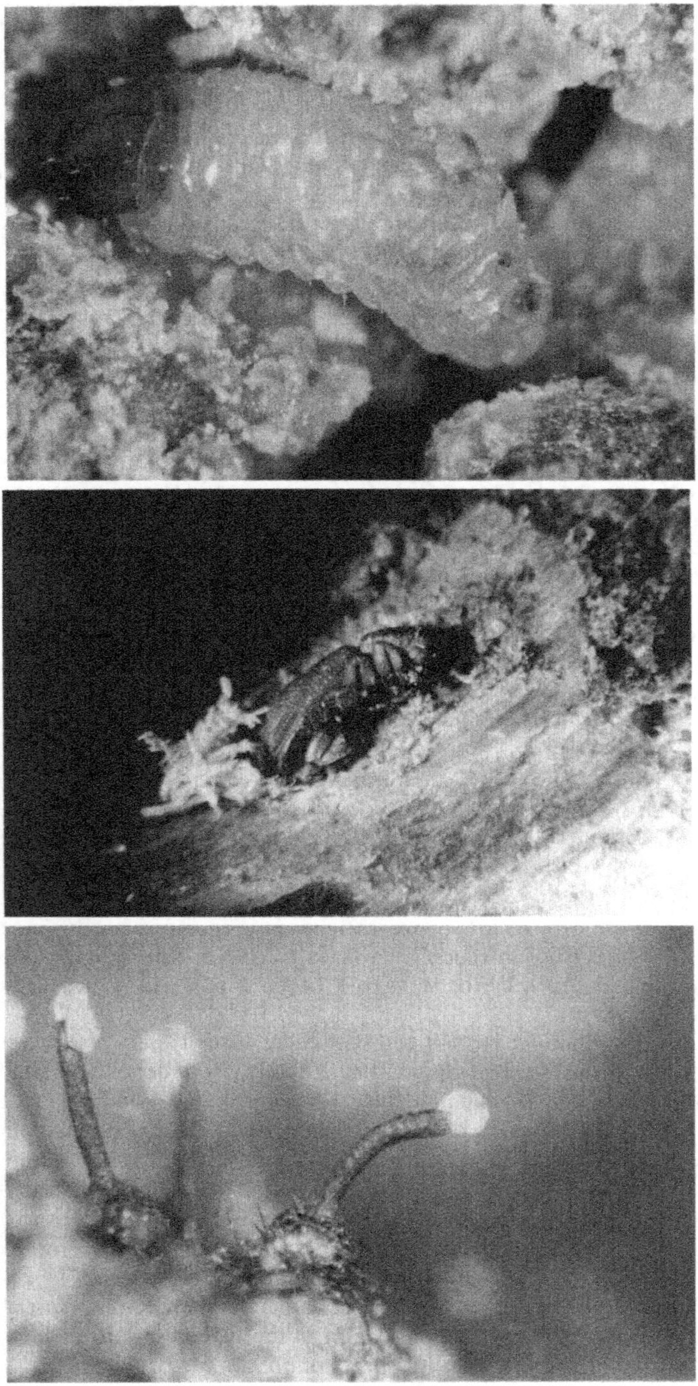

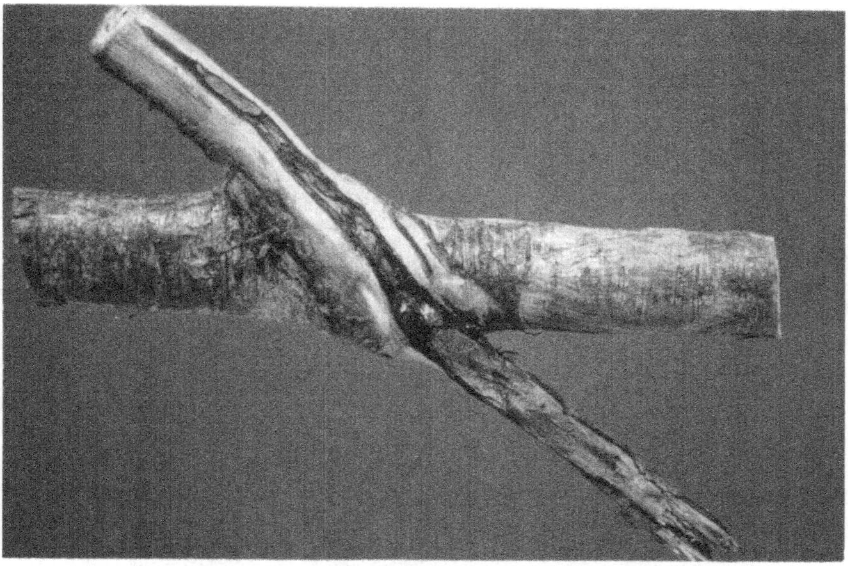

Fig. 16.5. Growth of *Leptographium* through root grafts of adjacent *Pinus resinosa* trees

Defensive reactions against root beetle – *Leptographium* complexes are not always successful, however (Mielke 1981). Several factors appear to contribute to fungal establishment, all of which resemble bark beetle – *Ophiostoma* interactions. First, the more virulent fungi, such as *L. wageneri* in *P. ponderosa* or *L. procerum* in *P. strobus*, appear capable of growing around the necrotic lesions of some trees (Hessburg and Hansen 1987). When colonization is successful, the initial increase in oleoresin pressure and yield is followed by a steady decline. The relative rates of tree fungistatic reactions versus fungal advancement appear critical in determining the outcome. Second, physiological stress appears to reduce the ability of host trees to react. Third, the beetle vectors contribute to the rate of fungal spread within hosts by excavating extensive galleries within the cambium. This appears to greatly increase the chances of *Leptographium* establishment when compared to simple inoculations, especially for the less virulent species. In addition, the beetles show a strong preference for stressed trees, and are attracted to volatiles associated with plant stress metabolism (Fatzinger 1985). Thus, the feedback structure among insect, fungus, and host is similar to that shown in Fig. 16.3, except that aggregation pheromones are not involved. This leads to a more stable interaction that can allow for simultaneous beetle

←

Fig. 16.4. Root-colonizing insects and fungi associated with red pine decline. *Above Hylastes porculus* larva in lateral roots; *center H. porculus* pupa in host; *below* fungal perithecia present alongside *H. porculus*

Fig. 16.6. *Pinus resinosa* responses to *Dendroctonus valens* – *Leptographium terebrantis* complexes; note similarities to reactions against stem attack in Fig. 16.1. *Upper left* Breeding galleries and fungal growth in successfully colonized trees; *upper right* *L. terebrantis*; *lower left* failed attack; *lower right* response to controlled inoculation with *L. terebrantis*

reproduction, fungal development, and host survival. Long-term infection, however, is likely to predispose trees to other, lethal agents.

16.4.3 Wood Wasp – *Amylostereum* Interactions

The woodwasp *Sirex noctilio* (Hymenoptera: Siricidae) vectors the wood-rotting fungus *Amylostereum areolatum* (Basidiomycetes). They were accidentally introduced into New Zealand around 1900, probably in timber imported from Europe, and thereafter into Australia around 1961 (Talbot 1977). They have caused extensive mortality of *P. radiata*, including up to 50% of the trees in Tasmania from 1950–1951 after a severe drought. The complex set of interactions present in this system illustrate the advantages as well as some of the costs involved in such intimate relationships.

The genus *Sirex* is limited to conifers, largely *Pinus*. Females possess specialized organs, called mycetangia, for transporting their associated fungi. These mycetangia consist of pear-shaped sacs surrounded by musculature, which empty anteriorly into the oviposition canal. Each sac contains large numbers of *A. areolatum* arthrospores within a mucus-like substance secreted from glandular organs near the base of the ovipositor (Francke-Grosmann 1967). When a female locates a suitable host tree, she drills one hole through the bark, and from there, several tunnels into the xylem, using her ovipositor. If only one tunnel is drilled, it usually contains fungus and mucus without any eggs. When multiple tunnels are drilled, several tunnels may contain one or more egg (and very few arthrospores), but the last tunnel always contains only fungus and/or mucus (Coutts 1969a,b, Coutts and Dolezal 1969). Upon hatching, *S. noctilio* larvae feed through the fungus-infested wood.

Larval nutrition is facilitated by fungal metabolism, and there is a direct relationship between final larval size and fungal performance (Madden and Coutts 1979). Larvae of a related woodwasp have been shown to benefit from cellulases and xylanases produced by their associated *Amylostereum* spp. (Kukor and Martin 1983). Female larvae also possess specialized mycetangia for transmitting *A. areolatum* within the host. These intersegmental cuticular invaginations in the hypopleural region contain deep crytae and large glandular cells, which secrete a fatty matrix around the spores (Francke-Grosmann 1967). Upon pupation this matrix hardens and entraps the masses of spores in waxy plates. These plates are believed to be taken up into the mycetangia of the emerging female by retractive movements of the ovipositor as in *Sirex juvencus* (Francke-Grosmann 1967, Talbot 1977). Within the mycetangia, the wax is solubilized and fungal growth is stimulated by cellular secretions including the mucus (Talbot 1977). Thus, females emerging from attacked trees have a full complement of fungi with which to begin the oviposition process.

Both insects and fungi benefit from this association. The fungus obtains dispersal and access to a food source without having to penetrate protective tissues (Talbot 1977). In addition to digestive degradation, woodwasps benefit from fungal conditioning of the tree, which promotes optimal larval development. However, there are detriments to being consistently associated with the same fungus. Volatiles associated with fungal establishment attract egg- and early-instar insect parasitoids, and volatiles from mature fungal colonies attract additional parasitoids to late-instar larvae that would otherwise be well hidden deep within wood (Madden 1968, Spradbery 1970a,b). Also, the nematode *Deladenus siricidicola* has a dimorphic life cycle based on the *S. noctilio* – *A. areolatum* association (Bedding 1972). When fungus-infested wood is abundant, juveniles become free-living mycetophagous adults. However, when an active *S. noctilio* larva is present, the juvenile nematodes develop into parasitic adults which infest the female ovaries and male testes of pupating wasps. Adult males may reproduce normally but parasitized females oviposit sterile egg cases filled with juvenile nematodes into their egg tunnels.

Constitutive components of *P. radiata* resistance tend to be associated with general tree vigor (Talbot 1977). Factors such as high resin flow (Madden 1974), high resin content, and high moisture have been associated with reduced woodwasp attack, reduced fungal growth, and reduced larval survival, respectively (Talbot 1977). Conversely, volatiles associated with injured and/or stressed hosts, such as certain monoterpenes, alcohols, aldehydes, ketones, acetates, phenols, and ethers are attractive to females (Simpson and McQuilkin 1976, Talbot 1977). Induced defensive reactions occur in response to both the mucus secreted by *Sirex* and the fungus. *Sirex* mucus elicits a hypersensitive reaction, involving the premature senescence of some foliage and consequent removal of some mucus toxin (Fong and Crowden 1973). Monoterpenes and the polyphenols pinosylvin and its monomethyl ether accumulate at the oviposition sites. These chemicals can intoxicate *Sirex* eggs and larvae (Madden and Coutts 1979) and inhibit fungal growth (Hillis and Inoue 1968). They also disrupt ray cells in advance of the fungus, and thereby prevent further spread (King 1966, Hillis and Inoue 1968). Polyphenol production does not accompany physical wounding, and is dependent on current photosynthate rather than stored starch as the carbon source (Madden 1988).

Although *Sirex* and *Amylostereum* affect the tree differently, they can sometimes combine to overcome host resistance. Injection of *Sirex* mucus into the tree causes increased stem respiration, reduced bark starch, reduced photosynthate translocation and, eventually, increased susceptibility to *A. areolatum* (Coutts 1969a, Fong and Crowden 1973). The fungus then invades the sapwood and thereby restricts the flow of sap to the tree crown. This causes a leaf water deficit and reduced local wood moisture (Coutts 1969b). The resulting wood moisture, around 60–70% dry weight, is optimal for larval tunneling (Madden and Coutts 1979). If wood moisture falls below 20% however, larval survival is reduced (Talbot 1977).

Madden and Coutts (1979) proposed a model whereby tree stress or damage is followed by stomatal water loss, reduced water tension, inhibited translocation, respiratory stress in the phloem, and reduced osmotic and turgor status. These conditions attract female *S. noctilio* and elicit oviposition. Changes caused by mucus and fungus lead to ruptured xylem vessels and aeration of the xylem. This creates optimal conditions for *A. areolatum*, which in turns favors *S. noctilio* larval development. The low levels of arthrospores deposited with the eggs, combined with the inoculation of the fungus into separate tunnels, can prevent a heavy resin flow into the egg tunnel. The fungus stimulates resin flow initially, but it substantially impairs host reactions to the woodwasp/fungal attack before reaching the egg tunnels (Coutts and Dolezal 1969).

Sirex noctilio and *A. areolatum* interact with host physiological properties in a dynamic fashion (Madden and Coutts 1979). Females respond to trees with high osmotic potentials in the phloem sap by forming only single tunnels initially. Introduction of the mucus and spores along with the initial stress of the attack, progressively reduces osmotic pressure. Wasps respond to this decline in resistance by creating drills with higher proportions of multiple tunnels that contain an increasing number of eggs. Female *S. noctilio* thus increase the chances of progeny survival by incorporating host physiology into their oviposition behavior.

16.4.4 Scale Insect–*Nectria* Interactions

The organisms responsible for beech bark disease are capable of little damage to their host trees individually, but together they cause extensive damage in Europe and North America. In both regions the disease results from the creation of infection courts by a scale insect, *Cryptococcus fagisuga* (Homoptera: Eriococcidae), and subsequent colonization by a fungus in the Ascomycotina, *Nectria* spp. In Europe, *Nectria coccinea* infects *Fagus sylvatica*, and in North America, *Nectria coccinea* var. *faginata* infects *F. grandifolia*. The girdling action of cankers produced by these fungi ultimately causes tree death (Houston et al. 1979). The disease has been known in Europe since before 1849, with the first outbreak occurring in 1920 (Shigo 1964). Since their introduction to North America in 1890, *C. fagisuga* and *N. coccinea* var. *faginata* have advanced throughout the northeastern United States (Houston et al. 1979).

Within stands, beech bark disease has been divided into "fronts" (Shigo 1964). The "advancing front" contains scattered low levels of scale insects followed by the "killing front", which consists of high levels of *C. fagisuga* infestation, severe *Nectria* cankering, and high tree mortality, and the "aftermath zone", containing young and old defective trees and both *C. fagisuga* and *Nectria* at endemic levels (Houston et al. 1979).

Within-tree colonization patterns by *C. fagisuga* vary from small groups in protected areas, such as patches of lichens and mosses on bark fissures, to massive infestations covering the entire stem (Shigo 1964, Houston et al. 1979). Initial feeding by *C. fagisuga* induces a series of subcortical changes that facilitate subsequent feeding, growth, and development by the scales (Wainhouse and Howell 1983a, Wainhouse et al. 1988). A zone of rapid cellular proliferation, which apparently enhances nutrition for the scales, develops within the parenchyma near the stylet tips (Lonsdale 1983, Wainhouse and Gate 1988). This reaction appears to be stimulated by growth hormones egested during penetration (Fink and Braun 1980). Subsequent crawlers preferentially feed at these attack sites (Wainhouse et al. 1988). *Cryptococcus fagisuga* may also feed at the sites of previous attacks by other scale insects, and thereby exploit the resultant cavities and callus (Shigo 1964). Nonspecific defenses induced by *C. fagisuga* feeding include the development of secondary necrophylactic periderm, and death of the isolated tissues (Braun 1976, 1977). The resultant dead outer bark may pose a barrier that prevents stylet penetration to the living parenchyma (Wainhouse et al. 1988). However, this generalized reaction provides only partial defense against the scales, is not correlated with resistance, and may actually favor *Nectria* establishment (Ostrofsky and Blanchard 1983, Wainhouse et al. 1988). That is, bark fissures may form at the juncture of wound periderm with phloem rays, thus providing access to additional *C. fagisura* and *Nectria*. Severe scale infestations are colonized shortly thereafter by airborne *Nectria* spp. Fungal infection closely follows patterns of scale infestation (Houston et al. 1979).

There is some evidence that scale infestations may also lower host resistance to *Nectria* infection, rather than merely providing infection courts (Lonsdale 1980). These infections are often followed by the appearance of "slime flux", a cambial wound response containing bacteria and saprophytic bark fungi (Shigo 1964, Houston et al. 1979). Secondary organisms such as scale predators, some of which may vector the fungus to other *C. fagisuga* populations, and mycoparasitic fungi subsequently arrive at the cankers. Eventually the bark is killed, and decay fungi gain entrance to the tree (Shigo 1964). Once the *Nectria* cankers coalesce, the tree dies.

Trees relatively untouched in a killing front, and those surviving the aftermath zone, have been examined for resistance to *Nectria*. Shigo (1964) hypothesized that differences in tree genetics and host vigor, as evidenced by the rate of branch stub healing, may be important in resistance. Evidence for a genetic component to resistance was provided by Perrin (1984), who demonstrated that longer *Nectria* lesions occurred in progeny from diseased mother trees than in progeny derived from healthy trees. Wainhouse and Deeble (1980) observed clonal variation in susceptibility among beeches in a seed orchard. Further evidence for heritable resistance was provided by Wainhouse and Howell (1983a,b), who found that putatively resistant trees were immune to artificial infestation. Although the mechanism of resistance is not well understood, both histological and chemical responses appear important. Longsdale (1983) found that resistant and susceptible trees differed in their patterns of stone cell formation and lignification in the bark parenchyma. Ostrofsky et al. (1984) found that constitutive bark phenolic levels did not differ between susceptible and resistant trees. However, wounded bark from susceptible trees contained lower phenolic levels than wounded bark in resistant trees, suggesting an important role of phenolics in resistance. No qualitative differences in phenolics were found between wounded and unwounded tissue. Although *N. coccinea* var. *faginata* seems to be tolerant of beech phenolics, Ostrofsky et al. (1984) suggest that these compounds may inhibit fungal growth within the bark tissue. They further hypothesize that increased phenolic levels in the bark may act as a dynamic barrier, retreating as the fungus grows but still slowing the rate of advance.

Genetic variation within the scale population (Wainhouse et al. 1988) and environmental factors can greatly affect patterns of tree mortality. In particular, various lichen species can either favor or impede scale establishment (Houston et al. 1979, Houston 1983).

16.5 Commonalities Among Tree Defense Mechanisms to Insect-Associated Fungi

Although the specific attributes of tree-insect-fungal interactions vary among systems, there also appear to be some commonalities.

Constitutive chemical defenses function largely to deter feeding or reduce attraction by insects. This can prevent fungal colonization by simply orienting the vector elsewhere. Constitutive fungitoxic chemicals have been isolated in some trees (Dix 1974), but their direct importance against insect-vectored fungi is not well established.

Induced defenses have been observed against all insect-vectored, tree-colonizing fungi examined. They typically involve both histological and chemical alterations. Trees can respond to invasion by forming lesions in advance of fungi, leaving the tissue within these lesions unsuitable for further fungal or insect growth. Chemical changes include a rapid increase in the local concentrations of preformed chemicals by both transport and de novo synthesis, alterations in the relative ratios of these compounds, and the appearance of phytoalexins not present in constitutive tissue. Degradation of potential nutrients in advance of the fungi may also be an important component of defense (Wong and Berryman 1977).

Induced defenses may be elicited by either or, more commonly, both members of the complex. Unlike many plant-fungal systems, induced defenses are initiated in almost all encounters that involve vectors. The extent and rate of the response, rather than its occurrence, often separate successful from failed defense. These parameters are often reduced by plant stress. The invading insects and fungi usually possess mechanisms for depleting host responses, sometimes by acting together.

Constitutive and induced responses can function in a cohesive fashion. For example, preformed morphological and chemical defenses may delay penetration by the invading complex long enough for induced responses to become effective. In addition, ecological factors such as association with certain lichens (by the beech scale) (Houston et al. 1979), epiphytes (Carroll 1988), ectomycorrhizae (Marx 1972), and plant community structure (Raffa and Berryman 1987) can greatly influence the behavior of the vector, pathogen, or complex.

16.6 Use of Tree Resistance Mechanisms in Managing Insect-Associated Fungi

The broad array of tree defenses against insect-vectored fungi provide several approaches that can be exploited for environmentally sound pest management. These opportunities include prediction, prevention, and control.

Devising reliable relationships between tree physiological traits and susceptibility to pests offers the potential for improving our predictive capabilities, and thereby taking corrective actions more quickly. In some cases, tree susceptibility can be estimated indirectly, by exploiting correlations between environmental conditions and pest outbreaks that are physiologically based. This is especially important in systems that have rapid self-amplifying properties, such as bark beetle-blue stain and root beetle-black stain interactions. For example, drought and precommercial thinning are clearly predisposing agents to bark beetle and root beetle outbreaks, respectively (Rudinsky 1962, Witcosky et al. 1986b). Such episodes should be followed by intense, focused monitoring. Likewise, the potential resistance of individual trees can be assayed by measuring such parameters as root carbohydrates for wood borers, reactions to salivary secretions against wood wasps, and induced chemical alterations following inoculation by bark beetle symbionts. Unfortunately, these methods are extremely labor-intensive, poorly suited for field assays performed by stand managers, and dependent on large sample sizes because of high variability. Some simpler assays have been devised, such as the use of periodic growth and sapwood-to-crown ratios against bark beetles (Mahoney 1978, Waring and Pitman 1983), but these are further removed from the resistance mechanism itself, and thus subject to even higher variation. The development of easy, reliable, and rapid measures of resistance remains one of the major gaps in our ability to predict losses by insect-associated fungi. Such assays should be vigorously pursued, particularly through use of the emerging molecular technologies.

Our knowledge of tree defense systems suggests that several preventive tactics may be feasible, with the optimal method varying with the specific features of each system. For example, the direct linkage between stress and susceptibility to bark beetle – *Ophiostoma* complexes implies that any method that increases overall vigor could be beneficial. This is a tremendous advantage that is not common to many systems. With many forest and agricultural pests, cultural tactics impose tradeoffs between improved protection from the target pest, and operational ease, optimal plant productivity, and escape from other pests. Prescribed preventive practices against bark beetle-fungal complexes, such as thinning, judicious site selection, avoidance of logging injury, watering of high-value trees, and early rotation, are advantageous from almost any perspective.

Even in systems with a high potential for silvicultural management, however, some caution is required. For example, thinning directly enhances resistance against bark beetle – *Ophiostoma* complexes, but it may also favor establishment of root beetle – *Leptographium* complexes in plantations (Harrington et al. 1983), or dissemination of mistletoe in unmanaged stands (Amman and Cole 1983). Infection by either group could subsequently reduce resistance to the intended bark beetle – *Ophiostoma* target. Therefore, a thorough understanding of multiple feedback components regulating forest ecosystems is required. An integrated approach to bark beetle-*Ophiostoma* and root beetle-*Leptographium* management in red pine, for example, might involve planting mixed species in a staggered block fashion to disfavor extensive root grafting, together with removing stumps during thinning and/or sanitation, to prevent *Leptographium* spread.

Enhanced heritable resistance may be the best strategy where pests attack nonadapted hosts. A thorough understanding of both the physiology and genetics of tree resistance mechanisms should accompany all breeding programs. Without a knowledge of desired traits, breeding programs are costly, slow, and filled with innumerable false leads. An ability to conduct meaningful resistance assays that directly measure tree defensive capability, and identification of good juvenile-mature correlations are especially important in systems where older trees are typically attacked. To increase the chances of success, mechanisms that repel the vector should also be considered, given the preponderance of this mechanism in nature.

Resistance can also be enhanced through genetic engineering. However, this poses the hazard of immune pest biotypes, which could not only reduce efficacy, but also destabilize existing tree-parasite relationships (Raffa 1989). Therefore, a thorough knowledge of existing tree defense systems should be a prerequisite to deploying transgenic resistance. Understanding tree defense mechanisms provides both a means of avoiding pest cross-resistance to transgenic and existing traits, and a powerful tool for confronting pests with variable mosaics of defensive properties.

Our ability to manipulate tree defensive reactions is currently insufficient for its use as a curative tool against insect-associated fungi. However, there are at least two possibilities, the use of elicitors to better activate host defenses, and the use of insects as vectors of biological control agents such as hypovirulent strains. Although a number of chemicals that seem to enhance resistance levels have been identified, their modes of action and possible application methods remain unclear. This approach may eventually be practical for high-value trees.

Control strategies involving cytoplasmic hypovirulent isolates of pathogens rely on dispersal and transmission through the pathogen population (Russin et al. 1984, Brasier 1986). The potential for insect vectors to introduce these hypovirulent isolates into the target pathogen population is particularly attractive. Insect behavioral and morphological adaptations responsible for transmission efficieny may also be valuable for using vectors as biological control agents.

Russin et al. (1984) found that 69 species of insects, mostly Coleoptera, were capable of carrying *Cryphonectria parasitica*. Some carried naturally occurring cytoplasmic hypovirulent isolates. Since wounding the host tree is not necessary for transmitting hypovirulence to this pathogen, these insects may function as effective vectors, and may already contribute to dispersal. Isolates of *O. ulmi* possessing hypovirulent characteristics, including slow, abnormal growth, reduced reproductive fitness, and cytoplasmically transmitted "d-factors" associated with double-stranded RNA, have also been described (Rogers et al. 1986). The best opportunities for employing these hypovirulent isolates are during phases of the disease cycle directly involving the beetle (Brasier 1986). Unfortunately, the factors that reduce virulence in *O. ulmi* may also limit the use of vectors. The d-factors would likely reduce the ability of *O. ulmi* to colonize, ramify, and persist in the bark until the next generation of beetles emerge, remain viable on beetles searching for and feeding on new hosts, and infect healthy xylem via beetle feeding (Brasier 1986). Moreover, the isolate carried may not be vegetatively compatible with the predominant strain in the field. Successful application would be most likely as part of small-scale programs on a local level involving sanitation, beetle trapping, and the planting of resistant elms (Brasier 1986).

Vector-fungal associations that are not highly evolved may provide the best opportunities for transmitting hypovirulent strains. In an intimate association, there is likely to be strong selection pressure for access to insect vectors and for insects carrying a sufficient load of virulent fungi. The interdependencies that arise in these associations, e.g., among bark beetles and *Ophiostoma*, may ultimately prove detrimental to the success of hypovirulent strains. In casual associations, however, the vector would be more likely to flourish and achieve transmission and dispersal. Further research is needed to evaluate the feasibility of insects as vectors of hypovirulent strains.

Although tree resistance mechanisms offer powerful tools for forest protection, their use should be integrated with other tactics. Much research is required to fully realize their potential, but greater knowledge of tree defense mechanisms could provide the foundation for new environmentally safe, economical, and stable approaches to protection from insect-vectored fungi.

Acknowledgements. This work was supported by the University of Wisconsin-Madison College of Agricultural and Life Sciences. The critical reviews of Dale Norris, Department of Entomology, and Eugene Smalley, Department of Plant Pathology, UW-Madison, and John Gibbs, Forestry Commission, Alice Holt Lodge, UK, are greatly appreciated.

References

Agrios GN 1980 Escape from disease. In: Horsfall JG, Cowling EB (eds) Plant disease. Academic Press, New York London, 18–38

Albersheim P, Valent BS 1974 Host-pathogen interactions. VII. Plant pathogens secrete proteins which inhibit enzymes of the host capable of attacking the pathogen. Plant Physiol 53:684–687

Amman GD, Cole WF 1983 Mountain pine beetle dynamics in lodgepole pine forests, pt 2. Population dynamics. US Dep Agr For Serv Gen Tech Rep INT-145:59

Amman GD, McGregor MD, Cahill DB, Klein WH 1977 Guidelines for reducing losses of lodgepole pine to the mountain pine beetle in unmanaged stands in the Rocky Mountains. US Dep Agr For Serv Gen Tech Rep INT-36 19 pp

Anagnostakis SL 1987 Chestnut blight: the classical problem of an introduced pathogen. Mycologia 79:23–37

Anderson RL 1980 Method for recovering Verticicladiella procera from basal cankers of white pine. Plant Dis 64:680

Ayer WA, Browne LM, Lovell SH 1983 Biologically active metabolites of a Verticicladiella species. Phytochemistry 22:2267–2271

Barras SJ, Perry T 1971 Leptographium terebrantis sp. nov. associated with Dendroctonus terebrans in loblolly pine. Mycopathol Mycol Appl 43:1–10

Batra LR 1979 Insect-fungus symbiosis: nutrition, mutualism, and commensalism. Halsted, Chichester

Bedding RA 1972 Biology of Deladenus siricidicola (Neotylinchidae) an emtomophagous-mycetophagous nematode parasitic in Siricid woodwasps. Nematologica 482–493

Berryman AA 1972 Resistance of conifers to invasion by bark beetle-fungal associations. BioScience 22:598–602

Berryman AA 1988 Towards a unified theory of plant defense. In: Mattson WJ, Levieux J, Bernard-Dagan C (eds) Mechanisms of woody plant defenses against insects: Search for pattern. Springer, Berlin Heidelberg New York, 39–56

Blanche CA, Hodges JD, Nebeker TE 1985 Changes in bark beetle susceptibility indicators in a lightning-struck loblolly pine. Can J For Res 15:397–399

Bordasch RP, Berryman AA 1977 Host resistance to the fir engraver beetle, Scolytus ventralis (Coleoptera: Scolytidae). 2. Repellency of Abies grandis resins and some monoterpenes. Can Entomol 109:95–100

Borden JH 1984 Semiochemical-mediated aggregation and dispersion in the Coleoptera. In: Lewis T (eds) Insect communication. Academic Press, New York London, 123–149

Brand JM, Bracke JW, Markovetz AJ, Wood DL, Browne LE 1975 Production of verbenol pheromone by a bacterium isolated from bark beetles. Nature (London) 254:137–137

Brand JM, Bracke JW, Britton LN, Markovetz AJ, Barras JS 1976 Bark beetle pheromones: production of verbenone by a mycangial fungus of Dendroctonus frontalis. J Chem Ecol 2:195–199

Brasier CM 1986 The d-factor in Ceratocystis ulmi: its biological characteristics and implications for Dutch elm disease. In: Buck KW (ed) Fungal virology. CRC, Boca Raton, 305 pp

Braun HJ 1976 Das Rindensterben der Buche, Fagus sylvatica L., verursacht durch die Buchenwollschildlaus Cryptococcus fagi Bar. I. Die Anatomie der Buchenrinde als Basis-Ursache. Eur J For Pathol 6:136–146

Braun HJ 1977 Das Rindensterben der Buche, Fagus sylvatica L., verursacht durch die Buchenwollschildlaus Cryptococcus fagi Bar. II. Ablauf der Krankeit. Eur J For Pathol 7:76–93

Cambell LL, Huang JS, Payne GA 1980 Defense at the perimeter: the outer walls and the gates. In: Horsfal JG, Cowling EB (eds) Plant disease. Academic Press, New York London, 103–120

Carroll G 1988 Fungal endophytes in stems and leaves: from latent pathogen to mutualistic symbiont. Ecology 69:2–9

Cheniclet C, Bernard-Dagan C, Pauly G 1988 Terpene biosynthesis under pathological conditions. In: Mattson WJ, Levieux J, Bernard-Dagan C (eds) Mechanisms of woody plant defenses against insects: Search for pattern. Springer, Berlin Heidelberg New York, 117–130

Cherrett JM, Powell RJ, Stradling DJ 1989 The mutualism between leaf-cutting ants and their fungus. In: Wilding N, Collins NM, Hammond PM, Webber JF (eds) Insect-fungus interactions. Proc 14th Symp R Entomol Soc London in collaboration with Br Mycol Soc. Academic Press, New York London, 93–120

Christiansen E 1985 Ips/Ceratocystis–infection of Norway spruce: what is a deadly dosage? Z Angew Entomol 99:6–11

Christiansen E, Ericsson A 1986 Starch reserves in Picea abies in relation to defense reaction against a bark beetle transmitted blue-stain fungus, Ceratocystis polonica. Can J For Res 16:78–83

Christiansen E, Horntvedt R 1983 Combined Ips Ceratocystic attack on Norway spruce and defensive mechanisms of the trees. Z Angew Entomol 96:110–118

Cobb FW Jr 1988 Leptographium wageneri cause of black-stain root disease: a review of its discovery, occurrence and biology with emphasis on pinyon and ponderosa pine. In: Harrington TC, Cobb FW Jr (eds) Leptographium root diseases on conifers. APS, St Paul 144 pp

Cook SP, Hain FP 1985 Qualitative examination of the hypersensitive response of loblolly pine, *Pinus taeda* L, inoculated with two fungal associates of the southern pine beetle, Dendroctonus frontalis Zimmermann (Coleoptera: Scolytidae). Environ Entomol 14:396–400

Cook SP, Hain FP 1986 Defensive mechanisms of lobolly and shortleaf pine against attack by southern pine beetle, Dendroctonus frontalis Zimmermann, and its fungal associate, Ceratocystis minor (Hedgecock) Hunt J Chem Ecol 12:1397–1406

Cook SP, Hain FP 1988 Wound response of loblolly and shortleaf pine attacked or reattacked by Dendroctonus frontalis Zimmermann (Coleoptera: Scolytidae) or its fungal associate, Ceratocystis minor (Hedgecock) Hunt. Can J For Res 18:33–37

Couch JN 1938 The genus "Septobasidium." Univ North Carolina Press, Durham

Coulson RN, Hennier PB, Flamm RO, Rykiel EJ, Hu LC, Payne TL 1983 The role of lightning in the epidemiology of the southern pine beetle. A Angew Entomol 96:182–193

Coutts MP 1969a The mechanism of pathogenicity of Sirex noctilio on Pinus radiata II. effects of S. noctilio mucus. Aust J Biol Sci 22:1153–1161

Coutts MP 1969b The mechanism of pathogenicity of Sirex noctilio on Pinus radiata I. effects of the symbiotic fungus Amylostereum sp. (Theolophthoraceae). Aust J Biol Sci 22:915–524

Coutts MP, Dolezal JE 1969 Emplacement of fungal spores by the woodwasp, Sirex noctilio, during oviposition. For Sci 15:412–416

Croteau R, Burbott AJ, Loomis DW 1972 Apparent energy deficiency in mono- and sesquiterpene biosynthesis in peppermint. Phytochemistry 11:2937–2948

Daly JM 1984 The role of recognition in plant disease. Annu Rev Phytopathol 22:372–307

Dix NJ 1974 Identification of a water-soluble fungal inhibitor in the leaves of Acer platanoides L. Ann Bot (London) 38:505–514
Evans HC 1971 Transmission of Phytophthora pod rot by invertebrates. Nature (London) 232:346–347
Evans HC 1989 Mycopathogens of insects of epigeal and aerial habitats. In: Wilding N, Collins NM, Hammond PM, Webber JF (eds) Insect-fungus interactions. Proc 14th Symp R Entomol Soc London in collaboration with Br Mycol Soc. Academic Press, New York London, 205–288
Fatzinger CW 1985 Attraction of the black turpentine beetle (Coleoptera: Scolytidae) and other forest coleoptera to turpentine-baited traps. Environ Entomol 14:768–775
Ferrell GT 1988 Wound-induced oleoresins of Abies concolor: is it part of host resistance to the fir-engraver, Scolytus ventralis? In: Mattson WJ, Levieux J, Bernard-Dagan C (eds) Mechanisms of woody plant defenses against insects. Springer, Berlin Heidelberg
Finks S, Braun HJ 1980 Die Bedeutung anatomischer Veranderungen bei der Buchenrindennekrose. Ann Sci For 37:333–343
Fong LK and Crowden RK 1973 A comparative study of the phytotoxic effects of siricid woodwasps on conifers. Ann Appl Biol 75:309–320
Francke-Grosmann H 1967 Ectosymbiosis in wood inhabiting insects. In: Henry SM (ed) Symbiosis, vol 2 Academic Press, New York London, 141–203
Gabriel DW, Rolfe BG 1990 Working models of specific recognition in plant-microbe interactions. Annu Rev Phytopathol 28:365–391
Gambliel HA, Cates RD, Caffey-Moquin M, Paine TD 1985 Variation in the chemistry of loblolly pine in relation to infection by the blue-stain fungus. In: Branham SJ, Thatcher RC (eds) Integrated pest management. Res Symp Proc, Asheville, NC, 177–185
Geizler DR, Gara RI, Driver CH, Gallucci VF, Martin RE 1980 Fire, fungi, and beetle influences on a lodgepole pine ecosystem of south-central Oregon. Oecologia 46:239–243
Gibbs JN, French DW 1980 The transimission of oak wilt. US Dep Agr For Serv Pap NC-185, 17 pp
Goheen DJ, Cobb FW Jr 1978 Occurrence of Verticicladiella wageneri and its perfect state, Ceratocystis wageneri sp. nov., in insect galleries. Phytopathology 68:1192–1195
Goheen DJ, Cobb FW Jr, McKibbin GN 1978 Influence of soil moisture on infectivity of ponderosa pine by Verticicladiella wageneri. Phytopathology 68:913–916
Hackman W, Meinander M 1979 Diptera feeding as larvae on macro-fungi in Finland Ann Zool Fenn 16:50–83
Hain FP, Mawby WD, Cook SP, Arthur FH 1983 Host conifer reaction to stem invasion. Z Angew Entomol 96:247–256
Harrington TC 1983 Verticicladiella wageneri: taxonomy and vector relations. PhD Thesis, Univ Cal, Berkeley, 113 pp
Harrington, TC, Cobb FW Jr 1983 Pathogenicity of Leptographium and Verticicladiella spp. isolated from roots of western North American conifers. Phytopathology 73:596–599
Harrington TC, Cobb FW Jr 1986 Varieties of Verticicladiella wageneri. Mycologia 78:562–567
Harrington TC, Cobb FW Jr 1987 Leptographium wageneri var. pseudotsugae, var. nov. cause of black-stain root disease on Douglas-fir. Mycotaxon 30:501–507
Harrington TC, Cobb FW Jr, Lownsberry JW 1985 Activity of Hylastes nigrinus, a vector of Verticicladiella wageneri, in thinned stands of Douglas-fir. Can J For Res 15:519–523
Harrington TC, Reinhart C, Thornburgh DA, Cobb FW Jr 1983 Association of black-stain root disease with precommercial thinning of Douglas-fir. For Sci 29:12–14
Harris MK, Fredericksen RA 1984 Concepts and methods regarding host plant resistance to arthropods and pathogens. Annu Rev Phytopathol 22:247–272
Hemingway RW, GW McGraw, SJ Barras 1977 Polyphenols in Ceratocystis minor-infected Pinus taeda: fungal metabolites, phloem and xylem phenols. J Agr Food Chem 25:717–720
Hessburg PF Sr, Hansen EM 1987 Pathological anatomy of black stain root disease of Douglas-fir. Can J Bot 65:962–971
Highley L, Tattar TA 1985 Leptographium terebrantis and black turpentine beetles associated with blue stain and mortality of black and scots pines on Cape Cod, Massachusetts. Plant Dis 69:528–530

Hillis WE, Inoue T 1968 The formation of polyphenols in trees. IV. The polyphenols formed in Pinus radiata after Sirex attack. Phytochemistry 7:13–22

Hodges JD, Elam WW, Watson WF, Nebeker TE 1979 Oleoresin characteristics and susceptibility for four southern pines to southern pine beetle attacks. Can Entomol 11:889–896

Horntvedt R, Christiansen E, Solheim H, Wang S 1983 Artificial inoculation with Ips typographus-associated bluestain fungi can kill healthy Norway spruce trees. Medd Norsk Inst Skogforsk 38:1–20

Houston DR 1983 Influence of lichen species on colonization of Fagus grandifolia by Cryptococcus fagisuga: preliminary observations from certain Nova Scotia forests. In: Houston D, Wainhouse D (eds) Proc IUFRO Beech bark disease working party conference, Hamden, Connecticut. USDA FS Gen Tech Rep WO-37, 105–108

Houston DR, Parker EJ, Lonsdale 1979 Beech bark disease: patterns of spread and development of the initiating agent Cryptococcus fagisuga. Can J For Res 9:336–344

Houston DR, Parker EJ, Perrin R, Lang KJ 1979 Beech bark disease: a comparison of the disease in North America, Great Britain, France and Germany. Eur J For Pathol 9:199–211

Juzwik J, French DW 1983 Ceratocystic fagacearum and C. piceae on the surface of free-flying and fungus-mat-inhibiting nitidulids. Phytopathology 73:1146–1168

Kalkstein LS 1976 Effect of climatic stress upon outbreaks of the southern pine beetle. Environ Entomol 5:653–658

Keen FP 1938 Insect enemies of western forests. US Dep Agr Misc Publ 273 280 pp

King JM 1966 Some aspects of the biology of the fungal symbiont of Sirex noctilio. Aust J Bot 14:25–30

Klepzig KD, Raffa KF, Smalley EB 1991 Association of insect-fungal complexes with red pine decline in the Lake States. For Sci 37:1119–1139

Klepzig KD, Smalley EB, Raffa KF 1990 Fungistatic activity of extracts from red pine towards Leptographium terebrantis (Abstract). Mycol Soc Am Newslett 41:2

Kok LT 1979 Lipids of ambrosia fungi and the life of mutualistic beetles. In: Batra LR 1979: Insect-fungus symbiosis: nutrition, mutualism, and commensalism, Halsted, Chichester, 33–52

Kozlowski TT 1969 Tree physiology and pest insects. J For 67:118–123

Kuc J 1966 Resistance of plants to infectious agents. Annu Rev Microbiol 20:337–370

Kukor JJ, Martin MM 1983 Acquisition of digestive enzymes by siricid woodwasps from their fungal symbiont. Science 220:1161–1163

Lackner AL, Alexander SA 1982 Occurrence and pathogenicity of Verticicladiella procera in Christmas tree plantations in Virginia. Plant Dis 66:211–212

Lackner AL, Alexander SA 1983 Root disease and insect infestations on air-pollution sensitive Pinus strobus and studies of pathogenicity of Verticicladiella procera. Plant Dis 67:679–681

Landis TD, Helburg LB 1976 Black stain root disease of pinyon pine in Colorado. Plant Dis Rep 60:713–717

Leaphart CD, Gill LS 1955 Lesions associated with pole blight of western white pine. For Sci 1:232–239

Lieutier F, Berryman AA 1988 Elicitation and defensive reactions in conifers. In: Mattson WJ, Levieux J, Bernard-Dagan C (eds) Mechanisms of woody plant defenses against insects. Springer, Berlin Heidelberg New York, 313–320

Lonsdale D 1980 Nectria infection of beech bark in relation to infestation by Cryptococcus fagisuga Lindiger. Eur J For Pathol 10:161–168

Lonsdale 1983 Wood and bark anatomy of young beech in relation to Cryptococcus attack. In: Houston D, Wainhouse D (eds) Proc IUFRO Beech bark disease working party conference, Hamden, Connecticut. USDA FS Gen Tech Rep. WO-37, 43–49

Lorio PL Jr 1986 Growth-differentiation balance: a basis for understanding southern pine beetle-tree interactions. For Ecol Manag 14:259–273

Lorio PL Jr, Hodges JD 1968 Microsite effects on oleoresin exudation pressure of large loblolly pines. Ecology 49:1207–1210

Madden JL 1968 Behavioral responses of parasites to the symbiotic fungus associated with Sirex noctilio F. Nature (London) 218:189–190

Madden JL 1974 Oviposition behavior of the woodwasp, Sirex noctilio F. Aust J Zool 22:341–351
Madden JL 1988 Sirex in Australia. In: Berryman AA (ed) Dynamics of forest insect populations: patterns, causes, implications. Plenum, New York, 407–429
Madden JL, Coutts MP 1979 The role of fungi in the biology and ecology of woodwasps (Hymenoptera: Siricidae). In: Batra LR (ed) Insect-fungus symbiosis: nutrition, mutualism, and commensalism. Halstad, New York, 165–174
Mahoney RL 1978 Lodgepole pine/mountain pine beetle risk classification methods and their application. In: Berryman AA, Amman GD, Stark RW, Kibbee DL (eds) Theory and practice of moutain pine beetle management in lodgepole pine forests. Univ Idaho Press, Moscow, 106–113
Marx DH 1972 Ectomycorrhizae as biological deterrents to pathogenic root infections. Annu Rev Phytopathol 10:429–454
Mason RR 1966 Dynamics of Ips populations after summer thinning in a loblolly pine plantation: with special reference to host tree resistance. PhD Thesis Univ Michigan Annu Anbor, 152 pp
Mathre DE 1964 Pathenogenicity of Ceratocystis ips and Ceratocystis minor to Pinus ponderosa. Contrib Boyce Thompson Inst 22:363–388
Mattson WJ, Haack RA 1987 The role of drought in outbreaks of plant-eating insects. BioScience 37:110–118
Mielke ME 1981 Pathogenicity of Verticicladiella ponicillata (Grosm.) Kendrick to northern Idaho conifers. For Sci 27:103–110
Miller JR, Strickler KL 1984 Finding and accepting host plants. In: WJ Bell, RT Carde (eds) Chemical ecology of insects. Sinauer, Sunderlund, Massachusetts, 127–157
Miller RH and Berryman AA 1985 Energetics of conifer defense against bark beetles and associated fungi. In: Safranyik L (ed) The role of the host in the population dynamics of forest insects. Proc IUFRO Conf, Barff, Alberta, 13–23
Miller RH, Berryman AA, Ryan CA 1986a Biotic elicitors of defense reactions in lodgepole pine. Phytochemistry 25:611–612
Miller RH, Whitney HS, Berryman AA 1986b Effects of induced translocation stress and bark beetle attack (Dendroctonus ponderosae) on heat pulse velocity and the dynamic wound response of lodgepole pine (Pinus contorta var. latifolia). Can J Bot 64:2669–2674
Misaghi IJ 1982 Physiology and biochemistry of plant-pathogen interactions. Plenum, New York 287 pp
Mitchell RG, Waring RH, Pitman GB 1983 Thinning lodgepole pine increases tree vigor and resistance to mountain pine beetle. For Sci 209:204–211
Moser JC 1985 Use of sporothecae by phoretic Tarsonemus mites to transport ascospores of coniferous blue stain fungi. Trans Br Mycol Soc 84:750–753
Nebeker TE, Hodges JD 1983 Influence of forestry practices on host susceptibility to bark beetles. Z Angew Entomol 96:194–208
Norris DM 1979 The mutualistic fungi of xyleborine beetles. In: Batra LR (ed) Insect-fungus symbiosis, nutrition, mutualism, and commensualism. Halsted, Chichester, 53–63
Ostrofsky WD, Blanchard RO 1983 Characteristics and development of necrophylactic periderms in mature bark of American beech. In: Houston D, Wainhouse D (eds) Proc IUFRO Beech bark disease working party conference, Hamden, Connecticut. USDA FS Gen Tech Rep WO-37, 69–79
Ostrofsky WD, Shortle WC, Blanchard RO 1984 Bark phenolics of American beech (Fagus grandifolia) in relation to beech bark disease. Eur J For Pathol 14:52–29
Owen DR, Lindahl KQ, Wood DL, Parmeter JR 1987 Pathogenicity of fungi isolated from Dendroctonus valens, D. bravicomis, and D. ponderosae to ponderosa pine seedlings. Phytopathology 77:631–636
Paine TD 1984 Seasonal response of ponderosa pine to inoculation of the mycangial fungi from the western pine beetle. Can J Bot 62:551–555
Paine TD, Stephen FM 1987a Influence of tree stress and site quality on the induced defensive system of loblolly pine. Can J For Res 17:569–571
Paine TD, Stephen FM 1987b Fungi associated with the southern pine beetle: avoidance of induced defense response in loblolly pine. Oecologia 74:377–379

Paine TD, Stephen FM, Cates RG 1985 Induced defenses against Dendroctonus frontalis and associated fungi: variation in loblolly pine resistance. In: Branham SJ, Thatcher RC (eds) Integrated pest management. Res Symp Proc, Asheville, NC, 169–176

Perrin PR 1984 La variabilité de la sensibilité du Hetre (Fagus sylvatica L.) à Nectria coccinea, un des agents de la maladie de L'écorce. Eur J For Pathol 14:321–325

Popp MP, Johnson JD, Wilkinson RC, Massey TL 1990 Stimulation of ethylene and monoterpene production by bark beetle vectored fungi in slash pine (Pinus elliottii var. elliotii). Ini 11th N Am For Biol Worksh, Athens Ga

Raffa KF 1989 Genetic engineering of trees to enhance resistance to insects: evaluating the risks of biotype evolution and secondary pest outbreak. BioScience 39:524–534

Raffa KF 1991 Induced defenses in conifer-bark beetle systems. In: Tallamy DW, Raupp MJ (eds) Phytochemical induction by herbivores. Academic Press, New York London, 245–276

Raffa KF, Berryman AA 1982a Gustatory cues in the orientation of Dendroctonus ponderosae (Coleoptera: Scolytidae) to host trees. Can Entomol 114:97–104

Raffa KF, Berryman AA 1982b Accumulation of monoterpenes and associated volatiles following fungal inoculation of grand fir with a fungus transmitted by the fir engraver Scolytus ventralis (Coleoptera: Scolytidae). Can Entomol 114:797–810

Raffa KF, Berryman AA 1982c Physiological differences between lodgepole pines resistant and susceptible to the mountain pine beetle and associated microorganisms. Environ Entomol 11:486–492

Raffa KF, Berryman AA 1983a The role of host plant resistance in the colonization behavior and ecology of bark beetles (Coleoptera: Scolytidae). Ecol Monogr 53:27–49

Raffa KF, Berryman AA 1983b Physiological aspects of lodgepole pine wound responses to a fungal symbiont of the mountain pine beetle. Can Entomol 115:723–734

Raffa KF, Berryman AA 1987 Interacting selective pressures in conifer-bark beetle systems: a basis for reciprocal adaptations? Am Nat 129:234–262

Raffa KF, Berryman AA, Simasko J, Teal W, Wong BL 1985 Effects of grand fir monoterpenes on the fir engraver beetle (Coleoptera: Scolytidae) and its symbiotic fungi. Environ Entomol 4:552–556

Raffa KF, Hall DJ 1988 Seasonal occurrence of pine root collar weevil, Hylobius radicis Buchanan (Coleoptera: Curculionidae), adults in red pine stands undergoing decline. Great Lakes Entomol 21:69–74

Raffa KF, Smalley EB 1988a Seasonal and long-term responses of host trees to microbial associates of the pine engraver, Ips pini. Can J For Res 18:1624–1634

Raffa KF, Smalley EB 1988b Response of red and jack pines to inoculation with microbial associates of the pine engraver, Ips pini. Can J For Res 18:581–586

Raffa KF, Smalley EB 1988c Host resistance to invasion by lower stem root infesting insects of pine: response to controlled inoculations with the fungal associate Leptographium terebrantis. Can J For Res 18:675–681

Rane KK, Tattar TA 1987 Pathogenicity of blue-stain fungi associated with Dendroctonus terebrans. Plant Dis 71:879–883

Rawlings GS 1953 Rearing of Sirex noctilio and its parasite Ibalia leucospoides. For Res Notes NZ 1:20–34

Reid RW, Shrimpton DM 1971 Resistance response of lodgepole pine to inoculation with Europhium clavigerum in different months and at different heights on stem. Can J Bot 49:349–351

Reid RW, Whitney HS, Watson JA 1967 Reactions of lodgepole pine to attack by Dendroctonus ponderosae Hopkins and blue stain fungi. Can J Bot 45:1115–1126

Renwick JAA, Vite JP 1970 Systems of chemical communication in Dendroctonus. Contrib Boyce Thompson Inst 24:283–292

Rogers HJ, Buck KW, Brasier CM 1986 Transmission of double-stranded RNA and a disease factor in Ophiostoma ulmi. Plant Pathol 35:277–287

Rudinsky JA 1962 Ecology of Scolytidae. Annu Rev Entomol 7:327–348

Russell CE, Berryman AA 1976 Host resistance to the fir engraver beetle. I. Monoterpene composition of Abies grandis pitch blisters and fungus-infected wounds. Can J Bot 54:14–18

Russin JS, Shain L, Nordin GL 1984 Insects as carriers of virulent and cytoplasmic hypovirulent isolates of the chestnut blight fungus. J Econ Entomol 77:838–846

Safranyik L, Shrimpton DM, Whitney HS 1975 An interpretation of the interaction between lodgepole pine, the mountain pine beetle and its associated blue stain fungi in western Canada. In: Baumgartner DM (ed) Management of lodgepole pine ecosystems. Washington State Univ Co-op Ext Serv, Pullman, 406–428

Savile DBO 1976 Evolution of the rust fungi (Uredinales) as reflected by their ecological problems. Evol Biol 9:137–207

Schoeneweiss DF 1975 Predisposition, stress, and plant disease. Annu Rev Phytopathol 13:193–211

Sequeira L 1980 Defenses triggered by the invader: recognition and compatibility phenomena. In: Horsfall JG, Cowling EB (eds) Plant disease: an advanced treatise. Academic Press, New York London, 179–200

Shigo AL 1964 Organism interactions in the beech bark disease. Phytopathology 54:263–269

Shrimpton DM 1973a Extractives associated with the wound response of lodgepole pine attacked by the mountain pine beetle and associated microorganisms. Can J Bot 51:527–534

Shrimpton DM 1973b Age- and size-related response of lodgepole pine to inoculation with Europhium clavigerum. Can J Bot 51:1155–1160

Shrimpton DM 1978 Resistance of lodgepole pine to mountain pine beetle infestation. In: Berryman AA, Amman GD, Stark RW Kibbee DL (eds) Theory and practice of mountain pine beetle management in lodgepole pine forests. Univ Idaho Press, Moscow, 64–76

Shrimpton DM, Watson JA 1971 Response of lodgepole pine seedlings to inoculation with Europhium clavigerum, a blue stain fungus. Can J Bot 49:373–375

Simpson RF, McQuilkin RM 1976 Terpenes of the bark oil of Pinus radiata. Phytochemistry 15:328–329

Sinclair WA, Hudler GW 1980 Tree and shrub pathogens new or noteworthy in New York State. Plant Dis 64:590–592

Smalley EB, Raffa KF, Proctor RH, Klepzig 1991 Tree responses to infection by species of Ophiostoma and Ceratocystis. Chpt. 20. In: Wingfield MJ, Seifert KA, Webber SF (eds) Ceratocystis and Ophiostoma: taxonomy, ecology, and pathology. APS, St Paul

Smith RH 1963 Toxicity of pine resin vapors to three species of Dendroctonus bark beetles. J Econ Entomol 56:837–831

Smith RH 1965 Effects of monoterpene vapors on the western pine beetle. J Econ Entomol 58:509–510

Smith RS Jr 1967 Verticicladiella root disease of pines. Phytopathology 57:935–938

Smith RS Jr 1969 The inability of Verticicladiella wageneri to break down cellulose (Abstr). Phytopathology 59:1050

Spradbery JP 1970a The biology of Ibalia drewseni Borries (Hymenoptera: Ibalidae), a parasite of siricid woodwasps. Proc R Entomol Soc London 45:104–113

Spradbery JP 1970b Host finding by Rhyssa persuasoria (L.) an Ichneumonid parasite of Siricid woodwasps. Anim Behav 18:103–114

Stark RW, Miller PR, Cobb FW, Wood DL, Parmeter JR, Zavarin E 1968 Photochemical oxidant injury and bark beetle (Coleoptera:Scolytidae) infestation of Ponderosa Pine. Cal Agr Exp Stn 39:121–152

Stephen FM, Paine TD 1985 Seasonal patterns of host tree resistance to fungal associates of the southern pine beetle. Z Angew Entomol 99:113–122

Stoessl A 1983 Secondary plant metabolites in preinfectional and postinfectional resistance. In: Bailey JA, Deverall BJ (eds) The dynamics of host defense. Academic Press, New York London, 71–122

Strobel GA, Sugawara 1986 The pathenogenicity of Ceratocystis montia to lodgepole pine. Can J Bot 64:113–116

Struble GR 1957 The fir engraver, a serious enemy of western true firs. US Dep Agr Prod Res Rep 11, 18 pp

Talbot PHB 1977 The Sirex-Amylostereum-Pinus association. Annu Rev Phytopathol 15:41–54

Thalenhorst W 1958 Grundzüge der Populationsdynamik des groβen Fichtenborkenkäfers Ips typographus L. Schriftren Forstl Fak Univ Göttingen 21:126 pp

Towers B 1977 The occurrence of Verticicladiella procera in Pennsylvania: 1976. Plant Dis Rep 61:477

Vite JP 1961 The influence of water supply on oleoresin exudation pressure and resistance to bark beetle attack in Pinus ponderosa. Contrib Boyce Thompson Inst 21:37–66

Vite JP, Wood DL 1961 A study on the applicability of the measurement of oleoresin exudation pressure in determining susceptibility of second-growth ponderosa pine to bark beetle infestation. Contrib Boyce Thompson Inst 21:67–78

Wagener WW, Mielke JL 1961 A staining-fungus disease of ponderosa, jeffrey, and pinyon pines. Plant Dis Rep 45:831–835

Wainhouse D, Deeble R 1980 Variation in susceptibility of beech (Fagus spp.) to beech scale (Cryptococcus fagisuga). Ann Sci For 37:279–289

Wainhouse D, Gate IM 1988 The beech scale. In: Berryman AA (ed) Dynamics of forest insect populations: patterns, causes, implications. Plenum, New York, 67–85

Wainhouse D, Gate IM, Lonsdale D 1988 Beech resistance to the beech scale: a variety of defenses. In: Mattson WJ, Levieux J, Bernard-Dagan C (eds) Mechanisms of woody plant defenses against insects: search for pattern. Springer, Berlin Heidelberg New York, 277–293

Wainhouse D, Howell RS 1983a Intraspecific variation in beech scale populations and in susceptibility of their host Fagus sylvatica. Ecol Entomol 8:351–359

Wainhouse D, Howell RS 1983b Distribution of attack by beech scale Cryptococcus fagisuga in beech progeny trials. In: Houston D, Wainhouse D (eds) Proc IUFRO Beech bark disease working party conference, Hamden, Connecticut. USDA FS Gen Tech Rep WO-37, 33–37

Wallner WE 1987 Factors affecting insect population dynamics: differences between outbreak and nonoutbreak species. Annu Rev Entomol 32:317–340

Waring RH, Pitman GB 1983 Physiological stress in lodgepole pine as a precursor to mountain pine beetle attack. Z Angew Entomol 96:265–270

Webber JF, Gibbs JN 1989 Insect dissemination of fungal pathogens of trees. In: Wilding N, Collins NM, Hammond PM, Webber JF (eds) Insect-fungus interactions. Proc 14th Symp R Entomol Soc London in collaboration with Br Mycol Soc. Academic Press, New York London, 161–194

Weber NA 1979 Fungus culturing by ants. In: Batra LR (ed) Insect-fungus symbiosis: nutrition, mutualism, and commensalism. Halsted, Chichester, 77–116

Wheeler Q, Blackwell M 1984 Fungus-insect relationships: perspectives in ecology and evolution. Colombia Univ Press, New York, 514 pp

Whitney JH 1982 Relationships between bark beetles and symbiotic organisms In: Mitton JB, Sturgeon KB (eds) Bark beetles in North American forests: a system for the study of evolutionary biology. Univ Texas Press, Austin, 183–211

Wilding N, Collins NM, Hammond PM, Webber JF 1989 Insect-fungus interactions. Proc 14th Symp R Entomol Soc London in collaboration with Br Mycol Soc. Academic Press, New York London

Wingfield MJ 1983 Association of Verticicladiella procera and Leptographium terebrantis with insects in the Lake States. Can J For Res 13:1288–1245

Witcosky JJ 1985 The root-insect-black stain root disease association in Douglas-fir: vector relationships and implications for forest management. PhD Thesis, Oregon State Univ, Corvallis

Witcosky JJ, Schowalter TD, Hansen EM 1986a Hylastes nigrinus, (Coleoptera:Scolytidae), Pissodes fasciatus, and Steremnius carinatus (Coleoptera: Curculionidae) as vectors of black stain root disease of Douglas-fir. Environ Entomol 15:1090–1095

Witcosky JJ, Schowalter TD, Hansen EM 1986b The influence of precommercial thinning on the colonization of Douglas-fir by three species of root-colonizing insects. Can J For Res 16:745–749

Wong BL, Berryman AA 1977 Host resistance to the fir engraver beetle. 3. Lesion development and containment of infection by resistant Abies grandis inoculated with Trichosporium symbioticum. Can J Bot 55:2358–2365

Wood TG, Thomas RJ 1989 The mutualistic association between Macrotermitinae and Termitomyces. In: Wilding N, Collins NM, Hammond PM, Webber JF (eds) Insect-fungus interactions. Proc 14th Symp R Entomol Soc London in collaboration with Br Mycol Soc. Academic Press, New York London, 69–92

Wright LE, Berryman AA, Gurusiddaiah S 1979 Host resistance to the fir engraver beetle, Scolytus ventralis (Coleoptera: Scolytidae). 4. Effect of defoliation on wound monoterpenes and inner bark carbohydrate concentrations. Can Entomol 111:1255–1261

Wright LE, Berryman AA, Wickman BE 1984 Abundance of the fire engraver, Scolytus ventralis, and the Douglas-fir beetle, Dendroctonus pseudotsugae, following tree defoliation by the Douglas-fir tussock moth, Orgyia pseudotsugata. Can Entomol 116:293–305

Chapter 17

Water Relations in Trees and Responses to Infection

K. Suzuki

17.1 Introduction

Water is essential for plant life as a constituent of physiologically active tissues, a reagent in photosynthesis and hydrolytic processes. Various aspects of water relations of plants have been reviewed by Hsiao (1973), Duniway (1976), Ayres (1978), Kozlowski (1978), Talboys (1978), and others.

This chapter will provide a brief explanation of water status in trees and discuss the changes in plant water relations in response to environmental stresses and disease infection.

17.2 Water Balance in Trees

Most higher plants absorb water as a liquid from the soil, and move large amounts of water through the xylem to the atmosphere. The amount of water transported is determined by the water balance in trees, which is determined by the difference between water absorption and water loss (transpiration). Water balance in trees has been expressed in terms of moisture content, relative water content, and water potential (Kozlowski 1979).

Moisture content is the water content of tissues on a dry or fresh weight basis. The measurement is easily determined and widely used, but dry or fresh weight changes in tissues do not always indicate proportionate changes in physiological activity.

Relative water content (RWC), also called relative turgidity, is a comparison of the moisture content of tissue at a particular time with full turgidity as shown in the following equation:

$$\text{RWC} = (\text{fresh weight} - \text{oven dry weight}) / (\text{fully turgid weight} - \text{oven dry weight}) \times 100.$$

Water deficit (WD), often termed water saturation deficit (WSD) or saturation deficit (SD), is calculated in a similar manner:

$$\text{WD} = (\text{fully turgid weight} - \text{fresh weight}) / (\text{fully turgid weight} - \text{oven dry weight}) \times 100.$$

Direct determinations of the content of water in tissues have the merit of simplicity. Unfortunately, a given water deficit measured by percentage of water content does not represent the same water potential in leaves of different ages, species, or habitats. The measurement of water deficits in tissues is discussed in detail by Barrs (1968).

17.3 Water Potential

Probably the most satisfactory single measurement of plant water status is the measurement of water potential. Water potential is the description of water status in energy terminology. The components of water potential (ψ) of the cell are as follows:

$$\psi = \psi_s + \psi_p + \psi_m.$$

1. Osmotic potential or solute potential (ψ_s). A chemical free energy of a solution modified by the presence of solutes. In pure water, it is zero.
2. Pressure potential or turgor potential (ψ_p). A positive force exerted inward on the cell contents by the rigid cell wall. Pressure potential is usually calculated with the equation, rather than measured directly.
3. Matric potential (ψ_m). A force of absorption between the matrix surface and the water molecules. Usually matric potential is very small relative to ψ_s and ψ_p, therefore, it can be ignored. After disregarding ψ_m, the above equation becomes

$$\psi = \psi_s + \psi_p.$$

Transpiration of water from leaf cells produces a water potential gradient through the leaf tissue in response to which water moves from the vein. This results in a reduction of water potential in the xylem, in response to which water moves across the root cortex from the soil-root interface. Therefore, movement of water from the soil into the plant and, ultimately, the atmosphere is along gradients of decreasing water potential. This is usually described by the analogy of resistances in an electrical circuit, as shown in Fig. 17.1.

17.3.1 Apoplast-Symplast Concept

Münch (1930) introduced the concept of the pathway of water and solute movement through the plant. The cell walls, the intercellular spaces, and the lumina of dead cells should be considered as a single system and called the apoplast. This is the outside of the cell membranes, in a sense, the "dead" part of the plant. The ascent of sap in a plant would take place in the apoplast, particulary in the xylem portion. The rest of the plant, i.e., the "living" part of the plant, is called the symplast. The symplast is a system composed of all living cells connected with plasmodesmata in the plant. Each of these routes offers resistance to water flow in varying degrees (Salisbury and Ross 1978).

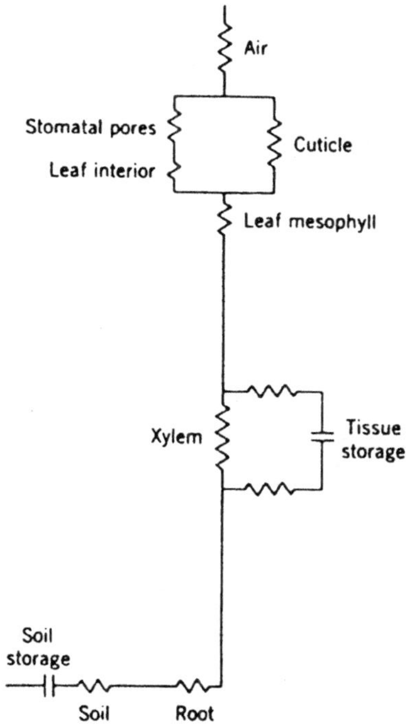

Fig. 17.1. Simplified diagram showing the flow of water through the soil-plant-atmosphere system analogous to the flow of electrical current through a series of resistances and capacitances (Kramer 1969)

17.3.2 Measurement of Components of Water Potential

Measurement of the water potential of leaves can be done by several methods (Barrs 1968); the pressure-bomb (chamber) technique, as developed by Scholander et al. (1965) has been widely used in studying the water relations of woody plants. This instrument can also be used to estimate water potential components, such as water potential (ψ_w), osmotic potential (ψ_s), and pressure potential (ψ_p), by means of the pressure-volume technique (Fig. 17.2).

Water potential versus reciprocal free water content (1/FWC) is used for graphic presentation of pressure-volume curves ($\psi_w = \psi_s + \psi_p$), as shown in Fig. 17.3 (Hinckley et al. 1980). As 1/FWC increases, ψ_w decreases curvilinearly to the point of incipient plasmolysis (ψ_w^{tlp}: turgor loss point), as a result of decreases in ψ_p. Beyond ψ_w^{tlp}, ψ_p drops to 0 and the relationship between ψ_w and 1/FWC becomes linear, dependent only on changes in ψ_s ($\psi_p = 0$, so $\psi_w = \psi_s$). This is the "osmotic line". Extrapolating the osmotic line back to 1/FWC = 1 gives the osmotic potential of the cells at full saturation (ψ_s^{sat}).

Pressure-volume analysis has also been used to estimate the bulk elastic modulus (ε), i.e., elastic changes in the cell walls after deformation. ε is

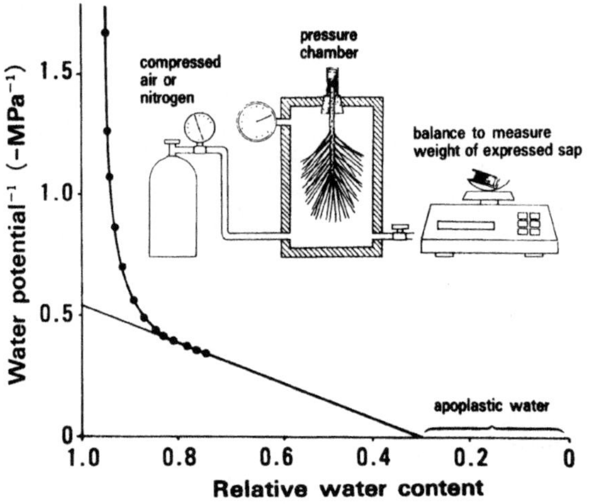

Fig. 17.2. Estimation of several parameters concerning the water relations of a pine seedling by the pressure-volume technique. Briefly, a branch is removed from a seedling and allowed to hydrate. The hydrated branch is then placed in the pressure chamber with the cut end protruding. Pressure is applied, and sap begins to exude from the cut end. The exuded sap is absorbed by dry tissue paper in a small glass tube and then weighed. Figure also shows a pressure-volume curve, which is at first a curved line, but with increasing pressure it then becomes linear. The straight line is called the osmotic line

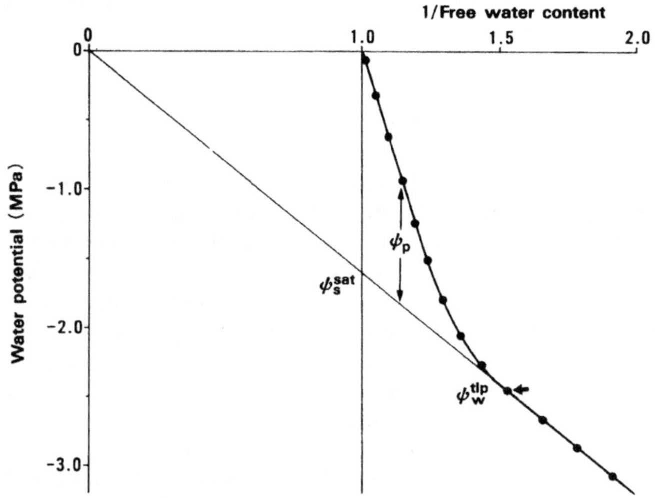

Fig. 17.3. A theoretical pressure-volume curve, where $\psi_w = \psi_s + \psi_p$. ψ_s^{sat} is the osmotic potential at full saturation and ψ_w^{tlp} is the water potential at the turgor loss point

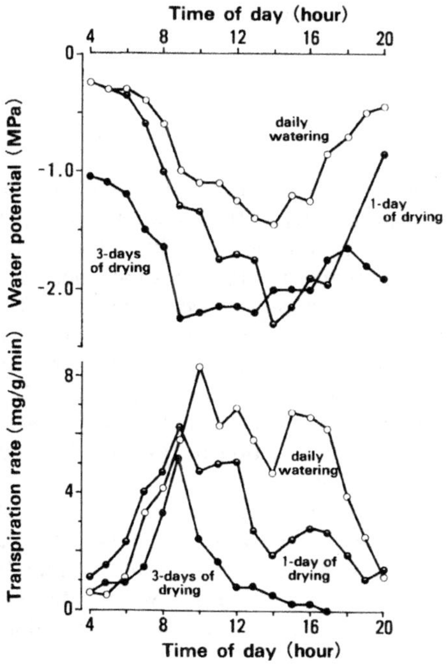

Fig. 17.4. Changes in transpiration and leaf water potential with time of day for 3-year-old potted seedlings of Japanese black pine, *Pinus thunbergii*, under different water regimes in a growth chamber. The quantitative measurement of transpiration is the loss of weight a few minutes after cutting. The rate of transpiration expressed as mg H_2O/g fresh weight per min is the average of several fascicles collected from the same branch at the same time

estimated from the relationship between the changes of ψ_p and relative cell volume (RCV) (Cheung et al. 1976).

$$\varepsilon = d\psi_p/dRCV, \; RCV = (V_o - V_e)/V_p$$

where V_o = symplasmic water volume at full turgidity, V_e = volume of water expressed from the sample, and V_p = symplasmic water volume at turgor loss.

The ε values obtained vary roughly linearly with the volume-averaged turgor (VAT) pressure. Though comparisons of ε's should be made at comparable VAT pressure, ε obtained near or at maximum VAT pressure (ε_{max}) may be a more reliable estimate.

17.4 Changes in Water Status as a Response to Environmental Stresses

Many environmental stresses affect forest, orchard, or urban trees under field conditions. The responses of trees vary with the severity of their environment.

Typical changes in plant-water relations in trees due to water deficiency and seasonal changes will be discussed.

17.4.1 Water Stress

Seedlings under water-stressed conditions show a progressive decrease in transpiration. Figure 17.4 shows the daily periodicity of transpiration and water potential for pine seedlings under different water regimes (Suzuki and Kiyohara 1978). In seedlings under well-watered conditions, transpiration increased rapidly after sunrise. The highest rate was generally reached within 2–3 h, falling later and reaching minimum values about 2:00 p.m. and increasing again in late afternoon. The transpiration again declined with sunset. Changes in water potential were roughly accompanied by changes in transpiration rate. The moderately water-stressed seedlings exhibited a pattern similar to the well-watered seedlings with time of day, but diurnal variation in transpiration was less, and water potential reached a lower value at midday.

The severely water-stressed seedlings exhibited remarkable decreases in transpiration. Soon after sunrise, the maximum rate of transpiration was attained and the rate of transpiration fell rapidly, ceasing completely during the middle of the day. The water potential shows the residual deficiency of water (Henckel 1964), i.e., the water potential before sunrise decreases to a low level and does not recover throughout the day.

The pressure-volume curves for the water-stressed pine seedlings after watering ceased show a characteristic tendency in response to severity of water stress, as shown in Fig. 17.5. The pressure-volume curve of the seedlings of Japanese black pine and lodgepole pine moved downward in proportion to time after watering ceased. At the same time, water potential at the turgor loss point (ψ_w^{tlp}) made a slow descent. When light chlorosis appeared in the leaves, a lowering of ψ_w^{tlp} disappeared. From this point, the pressure-volume curves moved upward, and then the ψ_w^{tlp} returned. At a later stage of severe water stress, the pressure-volume curve became quite different. The seedlings lost their tolerance for water stress entirely (Suzuki et al. 1987, Fukuda and Suzuki 1988c).

The results suggest that there is active osmotic adjustment in the leaves of both pine seedlings as a resistance mechanism to water deficiency. Such osmotic adjustment in response to water stress is considered to be related to the age of the leaves (O'Neill 1983). However, once chlorosis appeared in leaves, a lowering of the turgor loss point, i.e., active osmotic adjustment, stops. The bulk elastic moduli (ε) did not show any appreciable changes before chlorosis appeared in the leaves. Therefore, the pine seedlings did not show any loss of elasticity of cell walls or senescence of leaves.

These results suggest that osmotic adjustment in woody plants occurs in response to severe water stress. The extent of regulation seems to vary with species and environmental conditions (Hinckley et al. 1983, Teskey and Hinckley 1986).

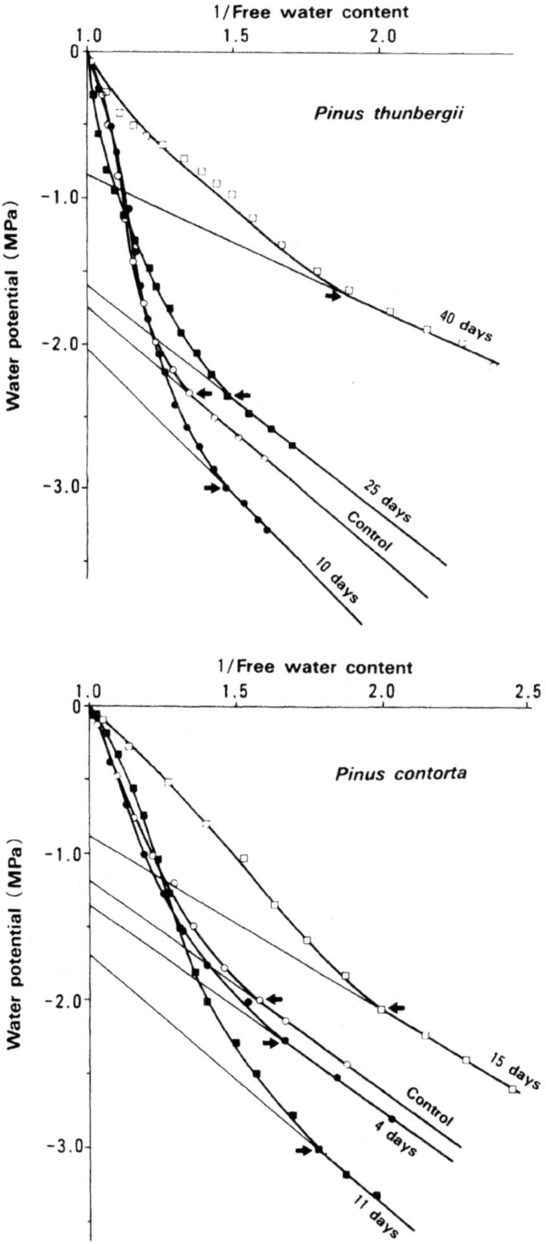

Fig. 17.5. Changes in pressure-volume curves for water-stressed 4-year-old seedlings of Japanese black pine, *Pinus thunbergii*, and lodgepole pine, *Pinus contorta*, after watering ceased. An *arrow* shows the turgor loss point (ψ_w^{tlp}) of the seedling

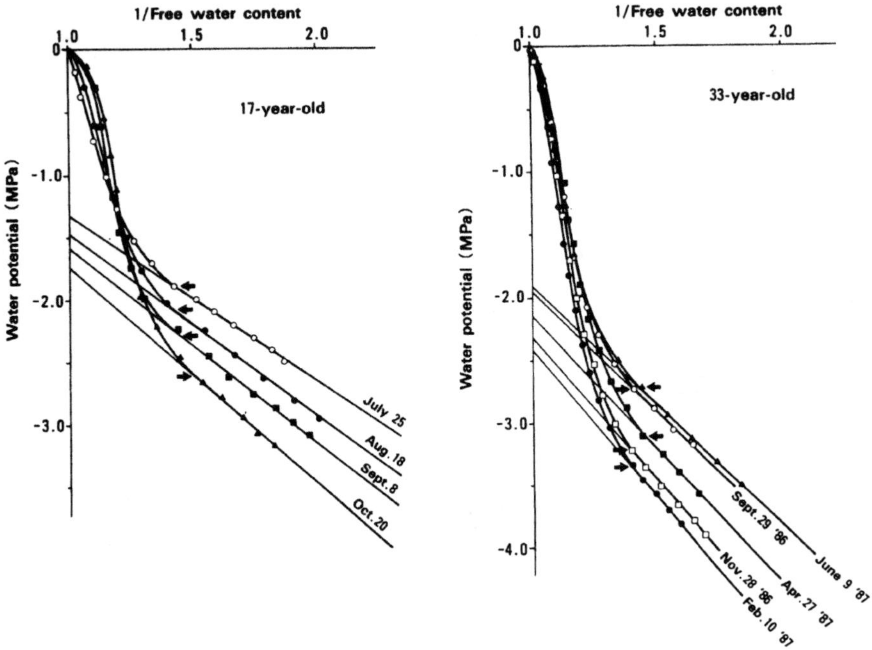

Fig. 17.6. Seasonal changes in pressure-volume curves for 17-year-old (*left*) and 33-year-old (*right*) Japanese red pine, *Pinus densiflora*. *Arrows* show turgor loss points (ψ_w^{tlp}) of the pine tree

17.4.2 Seasonal Changes in Water Relations

Seasonal changes in water relations are considered to play an important role physiologically for woody plants. Figure 17.6 shows changes in pressure-volume curves for Japanese red pine at different ages and sites. As to osmotic potentials, the osmotic line moved downward gradually as the season advanced. Therefore, osmotic potentials at full turgidity (ψ_s^{sat}) decreased. The degree of change seems to vary depending on tree age and site conditions (Fukuda and Suzuki 1988 a,b). Similar seasonal trends in osmotic potentials, with a minimum value during winter, were noted among forest trees, including deciduous and evergreen trees (Tyree et al. 1978, Roberts et al. 1980, Parker et al. 1982, Maruyama and Morikawa 1984, Teskey et al. 1984, Doi et al. 1986).

The decrease in ψ_w^{tlp} as the season advances is considered to be an osmotic adjustment in plant cells associated with increasing cold hardiness under field conditions (Doi et al. 1986, Sobrado 1986, Maruyama et al. 1988). Such a decrease in osmotic potential indicates an accumulation of cellular solutes during autumn and winter. The greater values of elasticity that contribute mechanically to freezing resistance are characteristics of the leaf cells in winter. Usually ε in evergreen species is higher than in deciduous species (Sobrado 1986).

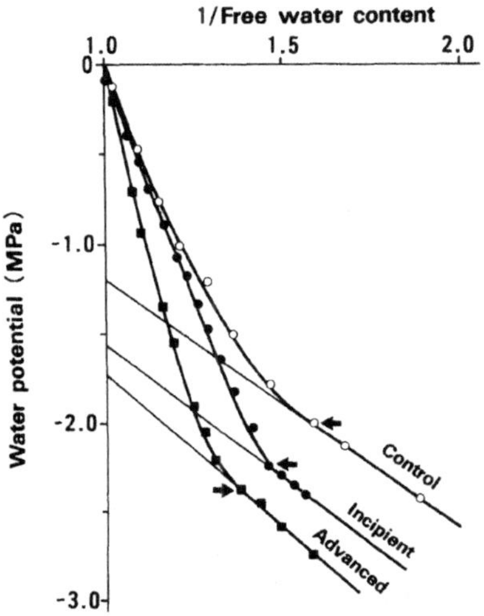

Fig. 17.7. Changes in pressure-volume curves for a 4-year-old lodgepole pine seedling, *Pinus contorta*, following infection with *Armillaria mellea*. *Incipient*: at incipient chlorosis; *Advanced*: at an advanced stage of disease development. Arrows show turgor loss points (ψ_w^{tlp}) of the seedling

17.5 Changes in Water Status as a Response to Disease Infection

Water status in trees has a profound effect on disease susceptibility and a close relationship with its development. It is usually difficult to monitor disease development accompanied by changes in water status. For example, root rot diseases and wilt diseases in forest trees have visible symptoms only in the advanced stage of disease development. To monitor plant water stress, measurement of water potential by a pressure chamber is currently widely used. Water potential itself does not indicate the level of water stress that a plant can withstand before damage occurs. However, by means of pressure-volume analysis, we can estimate several parameters concerning the water relations of the plant, as mentioned previously.

17.5.1 *Armillaria mellea*-Infected Lodgepole Pines

Armillaria mellea, sensu lato, is one of most important diseases of many woody plant species. The symptoms of the disease are resinosis, gradual decline with yellowish discoloration of foliage, or sudden death of the tree.

To monitor water stress of lodgepole pine after *Armillaria* infection, 4-year-old potted seedlings of *Pinus contorta* were used in a greenhouse. Inoculations

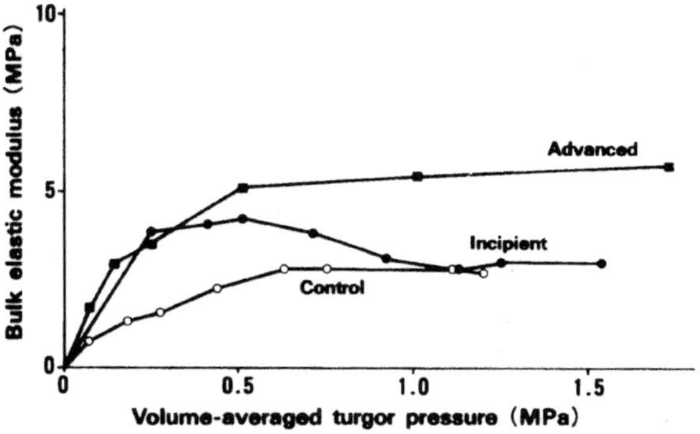

Fig. 17.8. The relation of the bulk elastic modulus (ε) with the volume-averaged turgor pressure for the lodgepole pine seedling shown in Fig. 17.7

were made by placing a branch segment infested with *A. mellea* beside a seedling in soil (Suzuki et al. 1987).

The first noticeable symptom appeared as an incipient chlorosis of the leaves about 40 days after inoculation. Once the symptom appeared, it developed very fast, and most of the leaves turned yellow to brown 1 week later. Figure 17.7 shows the pressure-volume curves for infected seedlings. Compared with the water-stressed seedling, a pressure-volume curve for a seedling showing an incipient chlorosis is similar to that of the seedling 11 days after watering ceased (Fig. 17.5). However, ψ_w^{tlp} of the infected seedling decreased slightly from −2 MPa to −2.2 MPa. At an advanced stage of disease development, the pressure-volume curve moved further downward, but ψ_w^{tlp} decreased only slightly, from −2.2 MPa to −2.4 MPa. This phenomenon suggests that there is some osmotic adjustment in the seedlings infected with *A. mellea*, and that the infected seedlings may be killed in a different manner than the water-stressed seedlings.

Changes in ε with the development of the disease varied with the volume-averaged turgor pressure (Fig. 17.8). ε at incipient chlorosis fluctuated between curves of the control and the advanced stage of disease development. ε_{max} increased to 4.5 MPa, 1.6 times that of the control. Subsequently, ε_{max} in the advanced stage of disease development increased to 5.8 MPa, two times that of the control. This phenomenon indicates that the cell walls of the infected seedling lose reversible elastic properties following infection with *A. mellea*.

17.5.2 Pine Wilt Caused by the Pine Wood Nematode

The pine wilt caused by the pine wood nematode, *Bursaphelenchus xylophilus*, is a serious disease of hard pines in Japan, and the pathogen is widely distributed throughout North America. Although this is not a fungal disease, pine wilt disease has been studied in relation to changes in water status, and these investigations should be of interest to those studying fungal interactions of trees. Symptom expression of pine wilt disease starts with a slight reduction in

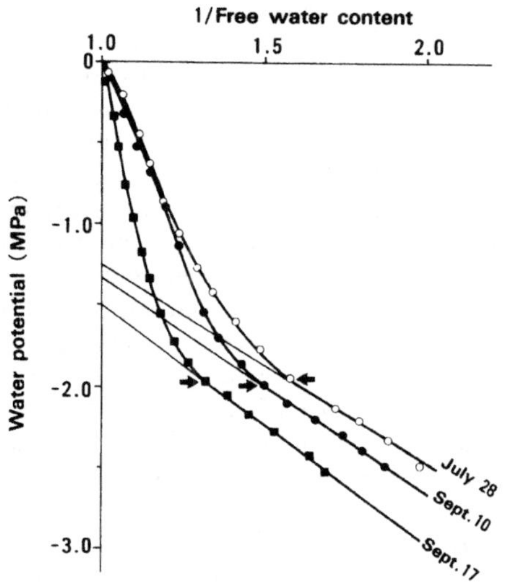

Fig. 17.9. Changes in pressure-volume curves for a 17-year-old Japanese red pine, *Pinus densiflora*, following inoculation with *Bursaphelenchus xylophilus*. An inoculation was made on July 28. Arrows show turgor loss points (ψ_w^{tlp}) of the pine tree

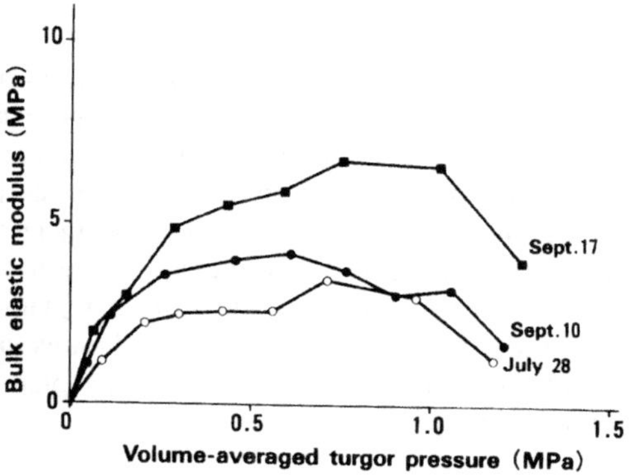

Fig. 17.10. The relation of the bulk elastic modulus (ε) with the volume-averaged turgor pressure for the Japanese red pine shown in Fig. 17.9

oleoresin exudation, followed by a decrease in leaf transpiration. Concurrently, 2- to 3-year-old needles exhibit chlorosis. Disease development is presumed to be proportional to the period of decrease in transpiration (Suzuki and Kiyohara 1978, Suzuki 1984). Pine trees invariably wilt 2–3 weeks after reduction of transpiration. Therefore, it is considered that the water status of pine trees plays an important role in disease development.

When inoculations were made on 17-year-old Japanese red pines (*Pinus densiflora*), typical symptoms were observed in the course of disease development, such as decreases in oleoresin exudation and transpiration rate. Discoloration of leaves was observed in late August, about 1 month after inoculation. Pressure-volume curves for a noninoculated control pine tree showed a decrease of ψ_w^{tlp} from −2 to −2.5 MPa in late summer as an osmotic adjustment in response to the climatic conditions (Fig. 17.6). However, the pressure-volume curves for the inoculated pine tree showed little change until September (Fig. 17.9). ψ_w^{tlp} had been around −2 MPa and did not move downward, thus indicating that the inoculated pine tree has no osmotic adjustment in leaves (Fukuda and Suzuki 1988a). In the middle of September, the slopes of the curvilinear portion of the pressure-volume curve became steeper, indicating that elastic changes in cell walls appear at a later stage of disease development.

Changes in ε of the inoculated pine tree are shown in Fig. 17.10. ε_{max} increased for the first time to 7.3 MPa at a later stage of disease development, a twofold increase from that of the early stage.

The significance of this is not yet known, but the results suggest that physiological changes in wilt symptoms of pine wilt disease is quite different from those in wilting induced by water deficiency. The water relations of pine wilt seem to be a unique phenomenon, whereby osmotic adjustment cannot be maintained during disease development.

17.6 Concluding Remarks

Changes in water relations occur in woody plants under field conditions and are influenced by various environmental stresses and diseases. However, there is much yet unknown regarding the water relation parameters in woody plants. A comparison of the pattern of changes in one water relation parameter as a function of changes in another parameter in plants subjected to artificially induced and pathogen-induced water stress may provide useful information as to the mechanisms of water relation changes in healthy and diseased plants (Misaghi 1982). The significance of the changes in plant-water relations in connection with the physiological and biochemical processes in woody plants will be an interesting subject for future research.

References

Ayres PG 1978 Water relations of diseases plants. In: Kozlowski TT (ed) Water deficits and plant growth V. Academic Press, New York London, 1–60

Barrs HD 1968 Determination of water deficits in plant tissues. In: Kozlowski TT (ed) Water deficit and plant growth I. Academic Press, New York London, 235–368

Cheung YNS, Tyree MT, Dainty J 1976 Some possible sources of error in determining bulk elastic moduli and other parameters from pressure-volume curves of shoots and leaves. Can J Bot 54:758–765

Doi K, Morikawa Y, Hinckley TM 1986 Seasonal trends of several water relation parameters in Cryptomeria japonica seedlings. Can J For Res 16:74–77
Duniway JM 1976 Water status and imbalance. In: Heitefuss R, Williams PH (eds) Encyclopedia of plant physiology, vol 4. Physiological plant pathology. Springer, Berlin Heidelberg New York, 430–449
Fukuda K, Suzuki K 1988a Changes of water relation parameters in pine-wood nematode-infested Japanese red pine. J Jpn For Soc 70:390–394
Fukuda K, Suzuki K 1988b Electrical resistance of wood tissue and water relation parameters of leaves in "over-year wilt" of pines induced by pine wilt disease. Ann Phytopathol Soc Jpn 54:625–628
Fukuda K, Suzuki K 1988c Water relation parameters of pine seedlings in pine wilt disease and water deficiency. Bull Tokyo Univ For 80:25–35
Henckel PA 1964 Physiology of plants under drought. Annu Rev Plant Physiol 15:363–386
Hinckley TM, Duhme F, Hinckley AR, Richter H 1980 Water relations of drought hardy shrubs: osmotic potential and stomatal reactivity. Plant Cell Environ 3:131–140
Hinckley TM, Duhme F, Hinckley AR, Richter H 1983 Drought relations of shrub species: assessment of the mechanisms of drought resistance. Oecologia 59:344–350
Hsiao TC 1973 Plant responses to water stress. Annu Rev Plant Physiol 24:519–570
Kozlowski TT 1978 How healthy plants grow. In: Horsfall JG, Cowling EB (eds) Plant disease. An advanced treatise III. Academic Press, New York London, 19–51
Kozlowski TT 1979 Tree growth and environmental stresses. Univ Washington Press, Seattle, 192 pp
Kramer PJ 1969 Plant and soil water relationships. A modern synthesis. McGraw-Hill, New York, 482 pp
Maruyama Y, Morikawa Y 1984 Seasonal changes of several water relations parameters in Quercus crispula, Betula ermani, and Abies homolepis. J Jpn For Soc 66:499–505
Maruyama Y, Morikawa Y, Hosogai H, Kanekawa T, Horiuchi T 1988 Tissue water relations of leaves of Cryptomeria japonica clones in early winter in relation to freezing resistance. J Jpn For Soc 70:518–524
Misaghi IJ 1982 Physiology and biochemistry of plant-pathogen interactions. Plenum, New York, 287 pp
Münch E 1930 Die Stoffbewegungen in der Pflanze. Fischer, Jena
O'Neill SD 1983 Role of osmotic potential gradients during water stress and leaf senescence in Fragaria virginiana. Plant Physiol 72:931–937
Parker WC, Pallardy SG, Hinckley TM, Teskey RO 1982 Seasonal changes in tissue water relations of three woody species of the Quercus-Carya forest type. Ecology 63:1259–1267
Roberts SW, Strain BR, Knoerr KR 1980 Seasonal patterns of leaf water relations in four co-occurring forest tree species: parameters from pressure-volume curves. Oecologia 46:330–337
Salisbury FB, Ross CW 1978 Plant physiology, 2nd edn. Wadsworth, Belmont, 422 pp
Scholander PF, Hammel HT, Bradstreet ED, Hemmingsen EA 1965 Sap pressure in vascular plants. Science 148:339–346
Sobrado MA 1986 Aspects of tissue water relations and seasonal changes of leaf water potential components of evergreen and deciduous species coexisting in tropical dry forests. Oecologia 68:413–416
Suzuki K 1984 General effect on water stress on the development of pine wilting disease caused by Bursaphelenchus xylophilus. Bull For Prod Res Inst 325:97–126
Suzuki K, Hiratsuka Y, Ayer WA 1987 Pathophysiological investigations on water potential components in lodgepole pine seedlings. Ann Phytopathol Soc Jpn 53:39–44
Suzuki K, Kiyohara T 1978 Influence of water stress on development of pine wilting disease caused by Bursaphelenchus lignicolus. Eur J For Pathol 8:97–107
Talboys PW 1978 Dysfunction of the water system. In: Horsfall JG, Cowling EB (eds) Plant disease. An advanced treatise III. Academic Press, New York London, 141–162
Teskey RO, Grier CC, Hinckley TM 1984 Changes in photosynthesis and water relations with age and season in Abies amabilis. Can J For Res 14:77–84

Teskey RO, Hinckley TM 1986 Moisture: effects of water stress on trees. In: Hennessey TC, Dougherty PM, Kossuth SV, Johnson JD (eds) Stress physiology and forest productivity. Nijhoff, Dordrecht, 9–33

Tyree MT, Cheung YNS, MacGregor ME, Talbot AJB 1978 The characteristics of seasonal and ontogenetic changes in the tissue-water relations of Acer, Populus, Tsuga, and Picea. Can J Bot 56:635–647

Chapter 18

Applications of Tissue Culture for Studying Tree Defense Mechanisms

M.E. OSTRY and D.D. SKILLING

18.1 Introduction

Using tissue culture to study plant pathogens is not a new practice. Soon after the demonstration of continuous culture of isolated plant tissues, White and Braun (1941) showed that secondary tumors on sunflower were free from *Agrobacterium*. It was Morel (1948) who first demonstrated the advantage of using tissue culture in the study of obligate fungi. The ability to grow obligate parasites in host tissue cultures was the major factor that stimulated many later studies of these host-parasite systems under aseptic conditions. The oldest practical application of tissue culture in disease control is freeing plants from viruses using meristem culture. Tissue culture of hybrid poplars was used to produce virus-free hybrid poplars (Berbee et al. 1976).

Several excellent reviews have been written on the various applications of aseptic plant culture for studying host-parasite interactions and selection for disease resistance (Earle 1982, Helgeson 1983, Miller and Maxwell 1983, Hammerschlag 1984, Daub 1986). Forest pathologists have also applied tissue culture techniques in studying a wide range of tree diseases (Amerson and Mott 1982, Diner and Karnosky 1987). The inability to regenerate complete plants from various types of tissue culture has hindered progress in using aseptic techniques with many tree species. There has been only one report of a successful regeneration system for a conifer from callus culture (Gladfelter and Phillips 1987). The progress made in vegetative regeneration in vitro of forest trees has recently been reviewed (Haissig 1989).

Most recently, interest in the use of tissue culture of tree species has centered on its application together with the rapidly developing methods of plant biotechnology. Many of the new techniques in molecular biology can now be applied to the study of tree diseases. For instance, gene expression in plant defense responses induced by pathogens (Collinge and Slusarenko 1987) is beginning to be studied. Genes that respond to wounding were investigated in poplars, and systemically accumulated mRNAs were demonstrated to be active (Parsons et al. 1989). Identification and ability to clone such wound-inducible genes, together with the recent development of a transformation system for poplars (Fillatti et al. 1987), offer promising possibilities for further research into host-parasite interrelationships in a tree species.

In this chapter, the literature on establishing dual cultures of tree tissues and fungal pathogens will be reviewed. Because of the lack of reports on the specific use of cell and tissue culture to study host defense mechanisms, the discussion will also include reports on the application of in vitro screening methods for detecting disease resistance. Many of these studies have provided rudimentary information on host resistance.

We have arranged the discussion according to the type of tissue culture system used and by the host-parasite system studied. Because of the lack of regeneration systems for many tree species in the past, most investigators have used callus cultures. However, significant progress has been made recently in using in vitro systems to study tree diseases, and progress will continue to accelerate as tissue culture systems are developed for more tree species.

18.2 Advantages and Limitations of Tissue Culture

At the outset, it is important to discuss the positive as well as the negative aspects of using tissue culture to study tree diseases and host defense mechanisms. Although progress has been made in plant tissue culture, much remains to be developed before tissue culture systems become routine (Krikorian 1988). This is especially true for most tree species. However, because of the size and long generation times of trees, using in vitro techniques to study disease resistance in trees can have greater advantages than using such techniques with other crop plants.

Precise control of the physical and chemical environmental conditions afforded by tissue culture is perhaps its greatest advantage to the study of tree diseases. The ability to uniformly challenge one or a few host cell types with a stressing agent without the presence of contaminating unwanted organisms results in a highly simplified experimental system in which specific events may be isolated for detailed study.

The most important potential disadvantage of using tissue culture is that expression of disease resistance in vitro may differ from responses ex vitro depending upon such factors as the developmental stage of the plant or the absence of static defense barriers such as intact cuticles. Cultured cells and tissues can be physiologically and, in some cases, genetically different from whole plants, so biochemical events may be quite different. Also, the phytohormones used in the tissue culture medium may influence responses to pathogens. In addition, for most forest tree diseases the genetics of the host and pathogen are poorly understood, complicating the application of in vitro studies of disease resistance.

18.3 Host-Parasite Tissue Culture Systems

18.3.1 Cell Suspension Culture

Few studies of tree diseases have been undertaken using cell suspensions of the host. This lack of research can be attributed to the limited knowledge of specific biochemical events in most tree pathosystems and lack of proven toxins, phytoalexins, or elicitors, substances that have been studied in other hosts using cell suspensions. The absence of reproducible regeneration systems for many tree species also reduces the utility of cell suspensions for some studies. Cell suspensions of sycamore, apple, and slash pine have been used to study the biological role of cell walls and lignin in the early stages of disease.

18.3.1.1 Sycamore – Hypersensitive Response

Using cell suspensions of sycamore (*Platanus*) derived from cambial tissues, Yamazaki et al. (1983) provided evidence that pectic-degrading enzymes could release pectic oligosaccharides from the cell walls that would inhibit protein synthesis, and cause cell death. This evidence indicated that the hypersensitive response defense mechanism may be induced by host cell wall fragments.

18.3.1.2 Apple – *Monilinia*

The early events leading to cell death and cell wall disruption after exposure to pectin lyase were studied by Hislop et al. (1979) and Keon (1985) using apple (*Malus*) cells derived from meristem callus cultures. Pectin lyase was thought to damage cells by the release of cellular hydrolases from the cultured apple cells. It was shown that during the early stages of injury, the cell wall integrity was maintained, and the degradation of the cell wall accompanied rather than preceded injury. The authors concluded that modifications of the cell wall and injury to the protoplast were linked and that injury involved a biochemical interaction between the cell wall and the protoplast.

18.3.1.3 Slash Pine – *Cronartium quercuum f.* sp. *fusiforme*

Suspension cultures of slash pine (*Pinus elliottii*), obtained with stem segments from 2-month-old seedlings, were used to study the responses of cells to the fusiform rust fungus *Cronartium quercuum* f. sp. *fusiforme* (Lesney 1989). Co-cultivation of the cell suspensions with the rust fungus resulted in a rapid clouding of the medium, inhibition of cell growth, and the presence of lignin-like compounds in the culture medium. Treatment of cell suspensions with chitin and chitosan also elicited the production of lignin-like compounds. It was suggested that this system would be useful in studying other resistance mechanism pathways in southern pines.

18.3.2 Callus Culture

Numerous investigators have used callus cultures of many tree species to study the infection process and to screen genotypes for resistance to pathogens. Callus culture systems have been used most frequently in the study of rusts caused by species of *Cronartium*. As pointed out earlier, a major prerequisite of such a system is that expression of resistance or susceptibility of undifferentiated cells must be similar to the whole plant reaction.

18.3.2.1 Juniper – *Gymnosporangium*

Axenic culture of a rust fungus was first accomplished with *Gymnosporangium*. Cultures were obtained by placing portions of telial galls onto a nutrient agar (Hotson and Cutter 1951). The first report of a tree callus culture infected by a plant rust was published in 1953 (Hotson 1953) in which infected callus from galls of cedar-apple rust on *Juniperus* was grown in vitro. In a later study, several *Gynosporangium* species were grown in dual culture with various rosaceous hosts (Cutter 1959).

18.3.2.2 White and Sugar Pine – *Cronartium ribicola*

Harvey (1967) grew western white pine (*Pinus monticola*) cambial tissues infected by white pine blister rust, caused by *Cronartium ribicola*, in vitro and observed aerial mycelium of the fungus. Pycniospores and aecia-like sori were observed on many cultures. With the ability to produce rust mycelium under controlled conditions, this technique was proposed as a model to study this host-parasite system (Harvey and Grasham 1969a) and was used to establish axenic cultures of *C. ribicola* (Harvey and Grasham 1974).

Intercellular rust mycelia were common, and monokaryotic and dikaryotic hyphae were observed with the nuclear condition being uniform within individual calluses (Harvey and Woo 1969). Haustoria were strongly attracted to host nuclei. A detailed histological study revealed a greater number of haustorial penetrations of single host cells in infected cultured host tissue than in natural tissues (Harvey and Woo 1971). These authors suggested that the results may be due to an increase in the production of metabolites utilized by the fungus or to a decrease in the defensive reactions of the host caused by the tissue culture environment.

Tissue culture was used by Koenigs (1968) in an attempt to develop a method for determining if *Cronartium ribicola* was alive in cankers on western white pine treated with antibiotics. Blister rust-infected nursery seedlings were surface sterilized and placed in a medium to induce callus, which was then examined for fungal colonization. The fungus was found to be alive in most of the cultures, suggesting that this type of propagation of the fungus would be useful for a variety of studies.

Healthy callus cultures of western white pine derived from needles grew slower but were colonized more rapidly by *C. ribicola* when inoculated with mycelium than callus cultures derived from healthy stem cortex and inoculated (Harvey and Grasham 1969b). The authors speculated that these cultures differed either because needles are the natural infection court of the mononucleate stage of the fungus and therefore more susceptible, or because the slower growth of the callus tissue allowed the more rapid rust proliferation.

Successful infection of callus cultures of *Pseudotsuga menziesii*, which is not normally a host of *C. ribicola*, demonstrated that factors governing host specificity may not be nutritionally related (Harvey and Grasham 1971).

Infection of callus cultures of western white pine by basidiospores of *C. ribicola* was reported by Harvey and Grasham (1970). The authors felt that although single basidiospores failed to produce infections, the mononucleate phase of the rust fungus (basidiospores and mycelium) is the most likely phase of the fungus to infect tissue cultures and could be used to develop clonal lines of the fungus.

Axenic cultures of *C. ribicola* and callus cultures of *Pinus monticola* were used by Robb et al. (1973) to study the ultrastructure of the fungus mycelium when grown in association with host tissue. They observed that hyphal walls were thinner at the host-parasite interface than in the culture remote from the host callus. In addition, the authors suggested that a capsular sheath formed at the host-parasite interface may contain substances toxic to host tissues or may protect the fungus from phenolic compounds in the pine tissues.

Callus cultures of *P. monticola* inoculated with axenic callus cultures of *C. ribicola* and *Ribes* leaves inoculated with aeciospores were used to study changes in ribonuclease activity (Harvey et al. 1974). Results revealed quantitative and qualitative changes in the enzyme unique to each host-rust combination, suggesting that the host plant plays a major role in determining the properties of the new RNase molecule formed.

Tissue culture was used to study the ultrastructure of pine cells inoculated with *C. ribicola* to eliminate the resin that is normally produced in wounded intact plants and which otherwise makes these types of studies difficult (Robb et al. 1975a,b). Cells under the influence of the fungal colony were observed to undergo changes suggestive of accelerated senescence. These degenerative changes in advance of the fungus were thought to have possibly been caused by a diffusible compound produced by the fungus. Changes in RNase activity were shown to be caused by one or more diffusable metabolite by placing a membrane between pine callus cultures and the rust fungus (Harvey 1979). Tannin deposits were common in and around infected pine cells, which the authors suggested may be a generalized defense mechanism. The capsular sheath around hyphae appeared to protect the fungus against these phenolic compounds released by the host.

The tolerance of pine cells and rust mycelium to pectic compounds and the role of these compounds in the host-parasite relationship in vitro were studied by Martin (1980). Pine cells were susceptible at lower concentrations of pectic compounds than the rust fungus and were more sensitive to pectin compounds of less structural complexity.

Cellular resistance of *Pinus lambertiana* to *C. ribicola* was demonstrated using pine callus cultures obtained from minced, germinated embryos from rust-resistant and -susceptible lines and inoculated with axenic rust cultures (Diner and Mott 1984). The hypersensitive resistance response was detected in the resistant lines by the depressed fungal growth and callus browning and necrosis.

18.3.2.3 Elm – *Ophiostoma*

Callus cultures of elm (*Ulmus*) established from clones resistant and susceptible to *Ophiostoma ulmi*, the cause of Dutch elm disease, were grown on media containing filtrates of shake cultures of *O. ulmi* (Hindal and McNabb 1973). Growth of callus cultures from susceptible clones was reduced more than the growth of callus from resistant elm clones.

In an attempt to develop a screening technique for elm cultivated in vitro, Roy and Neumann (1989) found that culture filtrates of *O. ulmi* incorporated into the media either inhibited cell division or stimulated callus production, depending upon the origin of the filtrate and the concentration used. They concluded that the tissue culture media can partly mask the toxic effect of the fungal culture filtrate.

Callus cultures initiated from cotyledons and hypocotyls of resistant and susceptible elm were used to study the ability of aggressive and nonaggressive isolates of *O. ulmi* in eliciting phytoalexin accumulation (Yang et al. 1989). Culture filtrates and cell fractions were found to elicit phytoalexin accumulation in callus. The nonaggressive isolate caused a more rapid accumulation of phytoalexin than the aggressive isolate in the first 12 h. However, both the aggressive and nonaggressive isolates elicited a greater accumulation of phytoalexin in the callus of a susceptible elm.

Culture filtrates of *O. ulmi* were used to screen callus cultures of elm. Resistant and susceptible elms were distinguished by the differential growth of the callus cultures when exposed to the culture filtrate (Pijut et al. 1990a). Calli of susceptible elm exposed to a culture filtrate of *O. ulmi* displayed mitochondrial and cytoplasmic disruption, while cells of resistant elm accumulated phenolic-like deposits that the authors suggested may be associated with a defensive mechanism against the toxic metabolites of the fungus (Pijut et al. 1990b).

18.3.2.4 *Citrus* – *Phytophthora*

Culture filtrates of *Phytophthora citrophthora*, the causal agent of trunk gummosis and root rot of *Citrus* spp., were used to determine if an in vitro selection technique could be used to identify resistant plants (Vardi et al. 1986). When grown on a medium amended with a culture filtrate of *P. citrophthora*, the sensitivity of calli obtained from *Citrus* lines of known resistance was in reverse order, and it was concluded that this selection technique could not be used. An auxin-like substance was detected in the culture filtrate.

18.3.2.5 Aspen – *Hypoxylon*

Clones of *Populus tremuloides* were screened for resistance to Hypoxylon canker, caused by *Hypoxylon mammatum*, by growing callus cultures on media amended with various concentrations of culture filtrates of the fungus (Baker

1986, Valentine et al. 1988). Callus inhibition varied among clones, and the authors concluded that the host response to the toxic metabolites in this in vitro bioassay was similar to responses observed in field inoculations using fungal mycelium. The callus bioassay measurements have been used for predicting field resistance of aspen clones, strengthening the conclusion that the toxic metabolites are important in the host-pathogen interaction.

18.3.2.6 *Eucalyptus – Phytophthora*

Using callus cultures of jarrah trees (*Eucalyptus marginata*) McComb et al. (1987) investigated the possibility of developing an in vitro bioassay for resistance against root rot caused by *Phytophthora cinnamomi*. Callus cultures were initiated from leaves or roots of in vitro shoot cultures obtained from mature trees expressing a range of responses to the pathogen in the field or from seedlings that had been inoculated with the fungus in greenhouse studies. Calli were inoculated with hyphae of the fungus, and the extent of hyphal growth, amount of callose formed, and extent of browning of calli were used to assess the expression of resistance of the various lines. Resistance and susceptibility of the jarrah lines were expressed in callus tissue, and the authors suggested that this technique, after further evaluation, may be useful for selection of resistant trees.

18.3.2.7 Chestnut – *Cryphonectria*

Callus cultures of chestnut (*Castanea*) have been used to study resistance to chestnut blight caused by *Cryphonectria parasitica*. Tannin content was one factor thought to influence colonization of callus (Grente and Sauret 1961). Using callus cultures obtained from American, hybrid, Japanese, and Chinese chestnut that represented a range of blight resistance, Hebard and Kaufman (1978) demonstrated a strong relationship between blight resistance and the hydrolyzable tannin in chestnut tissues. They concluded that the high tannin content of Chinese chestnut was responsible for its resistance to colonization by *Cryphonectria*. Colonization of the callus cultures, however, could not be used to distinguish resistant chestnut owing to differences in callus morphology and growth rate among the chestnut species tested, which were thought to influence the growth habit of the fungus and tannin content of the tissues, respectively.

18.3.2.8 Apple – *Venturia inaequalis*

Venturia inaequalis isolates did not colonize actively growing callus cultures of susceptible or resistant apple varieties but did overgrow senescing cultures, indicating a possible substance produced by the host tissue that was inhibitory to the fungus (Saad 1965).

Varieties of apple trees (*Malus*) susceptible to apple scab caused by *Venturia inaequalis* and apple varieties with the gene Vf for resistance were used to obtain callus cultures that were then inoculated with either conidia or mycelium of the fungus (Beech et al. 1985). The fungus grew on all apple varieties tested with no macroscopic changes in the tissues visible, thus the Vf gene for resistance was not expressed in the callus cultures. Ultrastructural interactions were also similar for susceptible and resistant varieties (Beech and Gessler 1986). One factor relating to the host-parasite interaction noted by the authors included evidence that perhaps cellulases and not mechanical pressure were involved in penetration of the host cell wall by the fungus. They also noted the presence of appositions that appeared to be incomplete papillae between the plasmalemma and cell wall.

18.3.2.9 Slash, Loblolly Pine – *Cronartium quercuum* f. sp. *fusiforme*

In vitro culture of *Cronartium quercuum* f. sp. *fusiforme* was obtained from rust-infected gall tissue of slash pine (*Pinus elliottii*), but the fungus failed to invade newly developed callus (Walkinshaw et al. 1965). Similarly, callus cultures of loblolly pine (*P. taeda*) inoculated with basidiospores of *C. quercuum* f. sp. *fusiforme* did not support intercellular or surface colonization (Jacobi et al. 1982). In vitro growth of the fungus was inhibited on media when grown in close association with loblolly pine callus (Jacobi 1982a). Using callus washes, Jacobi (1982b) demonstrated the presence of a toxic substance in resistant and susceptible loblolly pine and in susceptible slash pine callus that inhibited growth of the fungus when incorporated into the medium.

18.3.2.10 Larch – *Gremmeniella abietina*

In vitro responses of calli obtained from *Larix* species and balsam fir (*Abies balsamea*) inoculated with *Gremmeniella abietina*, the causal agent of Scleroderris canker, correlated with known ex vitro responses, suggesting that this rapid in vitro test may be useful for resistance screening (Abdul Rahman et al. 1987).

18.3.2.11 Miscellaneous Host–Parasite Systems

The level of inhibition of callus cultures obtained from several conifer and hardwood species against various fungal pathogens was assessed in vitro by Hrib and Rypacek (1978). It was found that, depending upon their degree of aggressiveness, the growth of the fungi was either inhibited or stimulated by spruce callus. Spruce (*Picea*), beech (*Fagus*), and birch (*Betula*) callus either stimulated or inhibited various fungi, depending upon whether the fungus normally attacked the heartwood or sapwood of hosts (Hrib and Rypacek 1981).

Phaeolus schweinitzii was inhibited to a greater degree by pine and Douglas-fir (*Pseudotsuga menziesii*) callus than by *Larix* callus, correlating with their known field resistance to the fungus (Hrib and Rypacek 1983). The inhibition was greater if the callus cultures were initiated 2 weeks before the fungus cultures were initiated, indicating that secondary metabolites may be involved in host resistance.

Callus from black spruce (*Picea mariana*) and balsam fir (*Abies balsamea*) was inoculated with two endophytic fungi to study the resulting tissue reaction (Johnson and Whitney 1988). Cells exposed to the fungi exhibited a thickening of cell walls, indicating lignification of affected cells.

Callus cultures of Sitka spruce (*Picea sitchensis*) initiated from the cambium of roots were used to study the host responses to the presence of 16 species of decay and mycorrhizal fungi (Woodward and Pearce 1988). Most of the fungi tested were not inhibited by the callus. The extent of inhibition was not related to the known pathogenicity of the fungi but could be influenced by the levels of growth regulators in the tissue culture medium. Cell wall thickening, lignification, and suberization were observed in cell layers adjacent to the fungal inoculum. The authors cautioned that although this system may be useful in the study of decay fungi and host responses, care must be taken not to misinterpret results due to the influence of growth regulators in the medium on host responses.

Aeciospores of *Peridermium pini*, the cause of resin top disease, were used successfully to inoculate callus cultures of several pine species (Pei 1989). Intercellular mycelia and haustoria were observed in infected tissues.

Axenic culture of *Peridermium harknessii*, the cause of western gall rust, was first accomplished using galls from lodgepole pine (*Pinus contorta* var. *latifolia*) (Allen and Blenis 1988). Callus tissue grew around the explant and intercellular mycelium was observed, but haustoria were rare. Although successful colony production was low, the authors stated that further modifications of the media and growth conditions were being made and that the potential availability of clonal lines of inoculum would be useful for resistance testing.

Two colony types of *P. harknessii* that varied in their morphological features were isolated from galls on ponderosa pine (*P. ponderosa*) (Lundquist et al. 1990a). These axenic cultures were used to study the stages in nuclear development to obtain a better understanding of the life cycle of the fungus. In an attempt to develop an early resistance screening technique, aeciospores of *P. harknessii* were applied to the surface of callus cultures initiated from buds of ponderosa pine (Lundquist et al. 1990b). Although the spores germinated, no infection of host tissue was observed.

Callus cultures obtained from embryos of loblolly (*Pinus taeda*), shortleaf (*P. echinata*), and Virginia pine (*P. virginiana*) were screened for resistance to the littleleaf disease pathogen, *Phytophthora cinnamomi* (Jang and Tainter 1989a). The in vitro resistance reaction was evaluated by the growth of the fungus on the calli, amount of intracellular hyphae and cytological changes of host cells, and the surface reaction of calli. Resistance ex vitro was correlated to several in vitro responses, such as less fungal growth, fewer intracellular hyphae, and necrosis and presence of phenolics in callus cells.

18.3.3 Embryo Culture

The in vitro culture of gametophytic tissue is potentially valuable because germplasm can be screened at an early stage and because the production of haploids affords the possibility of studying the expression of recessive traits. Of the limited amount of work reported using embryo culture, screening for resistance and study of the early host-parasite reactions have been the topics of investigation.

18.3.3.1 Loblolly Pine – *Cronartium quercuum* f. sp. *fusiforme*

Sterile embryos of loblolly pine (*P. taeda*) with various levels of resistance were inoculated with basidiospores of *Cronartium quercuum* f. sp. *fusiforme* to study host responses and to determine if those responses correlated with field resistance (Gray and Amerson 1983). Rapid necrosis characterized the incompatible host response and correlated with field disease resistance of those lines. The inoculum level used and time of assessment after inoculation influenced the extent of necrosis. An ultrastructure study of loblolly hypocotyls inoculated with *C. quercuum* f. sp. *fusiforme* characterized the infection and early colonization of host tissues (Gray et al. 1983). It was concluded that infection of loblolly pine embryo hypocotyls in vitro was the same as field infection of seedlings and mature trees (Gray and Amerson 1985). Necrosis showed a high family correlation with field resistance in a study of the possible use of an in vitro screening technique. It was suggested that tissue culture techniques may be suitable for rapidly assessing resistance in a tree improvement program (Frampton et al. 1983).

18.3.3.2 White and Sugar Pine – *Cronartium ribicola*

The hypersensitive resistance reaction to infection by *Cronartium ribicola* was observed within 2 weeks after inoculation of sugar pine (*Pinus lambertiana*) embryos with axenic cultures of the fungus (Diner and Mott 1982a). Embryos of susceptible genotypes, however, exhibited progressive disease development. Unlike needles protected by a thick cuticle, hypocotyls were penetrated directly through the thin cuticle (Diner and Mott 1982b).

18.3.3.3 Loblolly, Shortleaf, and Virginia Pine – *Phytophthora*

Results of in vitro inoculation of loblolly (*Pinus taeda*), shortleaf (*P. echinata*) and Virginia pine (*P. virginiana*) plantlets obtained from embryonic cotyledons with *Phytophthora cinnamomi* correlated with results of ex vitro inoculations of seedlings (Jang and Tainter 1989b). The authors suggested that the tissue culture system may be useful in a tree breeding program for littleleaf disease resistance.

18.3.4 Shoot Culture

Shoot cultures of several tree species have been used to study host responses to pathogens and to screen genotypes for resistance to several fungal diseases.

18.3.4.1 White Pine – *Cronartium ribicola*

Shoot cultures of western white pine (*Pinus monticola*) were inoculated in vitro with vegetative hyphae of *Cronartium ribicola* (Diner and Mott 1985). Cloned shoots were obtained from adventitious buds on excised cotyledons, and the fungus was applied to primary needles and wounds on stem apices of 12-week-old shoot cultures. Penetration of needles depended upon age of the inoculum, and stem infection depended upon the site of the wound. Characteristic intracellular haustoria were formed. The authors suggested that because both the host and pathogen could be cloned, the genetic and biochemical regulation of infection and resistance could be studied using this technique.

18.3.4.2 Loblolly Pine – *Cronartium quercuum* f. sp. *fusiforme*

Shoot cultures of loblolly pine (*Pinus taeda*) were successfully inoculated with basidiospores of *Cronartium quercuum* f. sp. *fusiforme* (Jacobi et al. 1982). Infection was usually by direct penetration of hypocotyls of 7- to 12-day-old shoot cultures. According to the authors, the pathogen and host interacted as they do under natural conditions.

Shoot cultures growing in the presence of loblolly pine callus washes or callus cultures were more resistant to colonization by *C. quercuum* f. sp. *fusiforme* than controls (Jacobi 1982b). The inhibitory factor was water-soluble and became systemic in the treated seedlings.

18.3.4.3 Larch – *Gremmeniella* and *Mycosphaerella*

Adventitious shoot cultures, 7- to 11-week-old, of several species of *Larix* differing in susceptibility were inoculated in vitro with vegetative hyphae of *Gremmeniella abietina* to determine if differential resistance responses could be obtained (Abdul Rahman et al. 1987). The fungus sporulated on the infected cultures within 1 month, and differential responses were evident within and between the clones.

Larch shoot cultures were also used to determine if an in vitro screening system could be used to determine the susceptibility of larch species and larch from different seed sources with in species to the needlecast fungus *Mycosphaerella laricina* (Ostry et al. 1990). Inoculation of 7-week-old cultures resulted in disease reactions after 6 weeks that were similar to results obtained in previous field trials.

18.3.4.4 Aspen – *Hypoxylon*

Shoot cultures of aspen (*Populus tremuloides*) obtained from a single cotyledon and growing the rest of the seedling allowed study of the in vitro and ex vitro reactions of the tested genotypes to the toxin produced by *Hypoxylon mammatum* (Wann 1985). The use of this system of selection and regeneration of resistant plants was proposed for inclusion in a tree improvement program (Einsphar and Wann 1985).

Axillary buds from aspen were used to obtain shoot cultures that were subjected to various levels of water stress to study the effects on infection by ascospores of *H. mammatum* (Belanger et al. 1989b). It was found that infection by ascospores was stimulated when plantlets were under moisture stress and may be related to levels of proline or other amino acids (Belanger et al. 1990). Clonal differences in response were observed; however, when the tissue culture bioassay was compared to other measures of disease, it was found that the bioassay did not correlate with disease incidence in the field (Belanger et al. 1989a).

Shoot cultures of several aspen crosses from parents with different responses to field inoculum were used to determine the pattern of inheritance of the toxin response (Kruger and Manion 1990a). Progeny response could not be predicted and was inconsistent. Clonal responses of shoot cultures to ascospores and to toxic metabolites were detected, but the sensitivity to one bioassay was not correlated with sensitivity to the other (Kruger and Manion 1990b).

18.3.4.5 Apple – *Gymnosporangium*

Inoculation of axillary and adventitious shoot cultures of *Malus* sp. with the cedar-apple rust fungus *Gymnosporangium juniperi-virginianae* resulted in responses similar to known ex vitro responses of a resistant and susceptible cultivar (Joung et al. 1987). Pycnial lesions developed on the susceptible cultivar within 10 days of inoculation, but no symptoms developed on the resistant cultivar.

18.3.4.6 Papaya – *Phytophthora*

Partial resistance in papaya (*Carica papaya*) to *Phytophthora palmivora*, which causes fruit rot, trunk canker, and root rot, was detected in shoot cultures inoculated with sporangial suspensions through stem wounds (Sharma and Skidmore 1988). In vitro responses were similar to whole plant responses in the field. Significant differences in several variables were detected among host-pathogen combinations, and host specificity was observed among the fungal isolates.

18.3.4.7 Peach – *Leucostoma*

Excised shoot tips of peach (*Prunus persica*) were used to detect a phytotoxin of *Leucostoma* by exposing shoot cultures to different fractions of cell-free filtrates of the fungus (Svircev et al. 1989). Peach canker-like symptoms developed on shoots treated with a specific fraction of the filtrate.

18.3.4.8 Red Pine – *Fusarium*

In vitro culture of red pine (*Pinus resinosa*) was used to study the infection process of *Fusarium oxysporum*, the cause of a damping-off and root rot disease (Farquhar and Peterson 1989). Possible host defense barriers such as wall appositions and osmiophyllic materials were observed infrequently, and the authors suggested that the culture conditions may have been responsible for the lack of such features.

Pinus resinosa seedlings grown in vitro and inoculated with the ectomycorrhizal fungus *Paxillus involutus* were more resistant than control seedlings to infection by *Fusarium oxysporum* (Duchesne et al. 1988). It was found that the observed disease suppression was associated with an increase in fungistatic chemicals in the rhizosphere of the inoculated seedlings. This sterile test tube culture system was later used to characterize the early effects of *P. involutus* on *F. oxysporum* (Farquhar and Peterson 1990). A rapid effect on sporulation and hyphal growth of *F. oxysporum* was observed. The sterile system provided results similar to a soil-culture system also used by the investigators.

18.3.5 Regenerated Plants

There have been two reports in which trees regenerated from tissue culture have expressed different resistance reactions to pathogens than the nontissue cultured plants from which they were obtained.

18.3.5.1 Hybrid Poplar – *Septoria musiva*

Tissue culture of hybrid *Populus* clones susceptible to leaf spot caused by *Septoria musiva* was used to obtain plants with increased resistance to the disease (Ostry and Skilling 1988). Adventitious shoots of several clones of hybrid poplars were induced from callus cultures, rooted in the greenhouse, and tested for resistance to Septoria leaf spot using an in vitro leaf disk bioassay. This was the first demonstration of somatic variation in disease resistance in a tree species, but the cause for the variation obtained was not determined.

Somaclonal variation can potentially produce isogenic stocks of parental lines that can be valuable in studying resistance among regenerated plants (Evans and Sharp 1986) However, variation from tissue culture can also be a major problem (De Klerk 1990).

18.3.5.2 Cottonwood – *Melampsora*

Somaclonal variation for resistance to two races of leaf rust caused by *Melampsora medusae* was obtained in eastern cottonwood (*Populus deltoides*) regenerated from leaf callus (Prakash and Thielges 1989). Plants were recovered with more or less resistance than the original clone and this variation was race-specific, but complete resistance was not observed. The authors suggested that tissue culture-generated variation may be useful to increase partial resistance in cottonwood to leaf rust and to produce cogenic lines that can be used to study mechanisms of the host-pathogen interaction.

18.4 Future Applications

Manipulation of trees and fungi in vitro, coupled with the emerging tools of molecular genetics, provides many future opportunities to study host defense mechanisms and may eventually lead to selection for resistance at the cellular level. More effort is needed in identifying marker genes in trees, and breeding experiments will be required to fully understand the genetics of host resistance, pathogen virulence, and their interactions (Valentine 1989).

Genetic engineering research is providing novel methods for the introduction of foreign genes. Recombinant DNA technology permits new gene combinations to be constructed that will potentially allow the study of gene regulation in disease resistance and will possibly provide resistance in breeding lines much more quickly than using traditional breeding procedures. Recently, for example, *Agrobacterium* was used in an attempt to transform hybrid *Populus* with the proteinase inhibitor gene pin-2 (P.I. II) to increase pest resistance using this wound-inducible plant defense gene system (Klopfenstein et al. 1989, Heuchelin et al. 1990). It was shown that a major pathogen of poplars, *Septoria musiva*, produces proteinase and that proteinase activity was inhibited by purified P.I. II under assay conditions (Sillick et al. 1989).

Examples of applications of biotechnology, including cell and tissue culture, to increase disease resistance in trees were recently reviewed (Hubbes 1989). The increasing ability to clone tree genotypes and the new biotechnologies that allow study of both the host and pathogen will provide greater opportunities to use in vitro approaches to the study of host-parasite interactions and to develop techniques for in vitro screening and selection for disease resistance.

18.5 Summary

Based on the literature reviewed, there have been many successful studies using dual cultures of trees and fungal pathogens. However, it is also clear that many of the reports have provided only preliminary results that have not yet been either duplicated in other laboratories or, as many of the authors suggest, been applied to the study of specific host-parasite interactions or incorporated into tree improvement programs.

Although there are many advantages of using a tissue culture system to study resistance, there are also many limitations such as the possible lack of recognition features or stimuli that may limit the use of in vitro systems to study specific events and host responses.

There have been encouraging reports of a number of host-pathogen systems where results of in vitro resistance screening and host responses have been similiar to ex vitro reactions. These screening techniques, however, need additional field verification before they can be practically applied to tree improvement efforts.

Also encouraging are recent reports of the detection and recovery of putative somaclonal variants among plants regenerated from tissue cultures that have increased disease resistance, but much remains to be learned about the nature of this variation, its control, and its inheritance.

Tissue culture techniques are now being developed for many commercially important tree species. In vitro approaches to the study of host-pathogen interrelationships will continue to be applied with greater success than in the past using the knowledge gained from the investigations reviewed in this chapter.

References

Abdul Rahman NN, Diner AM, Skilling DD, Karnosky DF 1987 In vitro responses of conifer adventitious shoots and calli inoculated with Gremmeniella abietina. For Sci 33:1047–1053

Allen EA, Blenis PV 1988 Axenic culture of Endocronartium harknessii. Mycologia 80:120–123

Amerson HV, Mott RL 1982 Phytopathology and tissue culture alliances. In: Bonga JM, Durzan DJ (eds) Tissue culture in forestry. Nijhoff, Dordrecht, 208–230

Baker SS 1986 Testing Populus tremuloides tissue culture response to Hypoxylon mammatum metabolites as an in vitro bioassay for Hypoxylon canker resistance. MS Thesis, SUNY Coll Environ Sci For, Syracuse NY, 135 pp

Beech I, Defago G, Gessler C 1985 Growth of Venturia inaequalis on apple tissue cultures. Bot Helv 95:121–124

Beech I, Gessler C 1986 Interaction between Venturia inaequalis and apple callus tissue cultures: an electron microscopic study. J Phytopathol 116:315–322

Belanger RR, Falk SP, Manion PD, Griffin DH 1989a Tissue culture and leaf spot bioassays as variables in regression models explaining Hypoxylon mammatum incidence on Populus tremuloides clones in the field. Phytopathology 79:318–321

Belanger RR, Manion PD, Griffin DH 1989b Hypoxylon mammatum ascospore infection of Populus tremuloides clones: effects of moisture stress in tissue culture. Phytopathology 79:315–317

Belanger RR, Manion PD, Griffin DH 1990 Amino acid content of water-stressed plantlets of Populus tremuloides clones in relation to clonal susceptibility to Hypoxylon mammatum in vitro. Can J Bot 68:26–29

Berbee JG, Omuemu JO, Martin RR, Castello JD 1976 Detection and elimination of viruses in poplars. In: Intensive plantation culture. US For Serv Gen Tech Rep NC-21, 85–91

Collinge DB, Slusarenko AJ 1987 Plant gene expression in response to pathogens. Plant Mol Biol 9:389–410

Cutter VM 1959 Studies on the isolation and growth of plant rusts in host tissue cultures and upon synthetic media. I. Gymnosporangium. Mycologia 51:248–295

Daub ME 1986 Tissue culture and the selection of resistance to pathogens. Annu Rev Phytopathol 24:159–186

De Klerk GJ 1990 How to measure somaclonal variation. Acta Bot Neerl 39:129–144

Diner AM, Karnosky DF 1987 Tissue culture application to forest pathology and pest control. In: Bonga JM, Durzan DJ (eds) Cell and tissue culture in forestry, vol 2. Specific principles and methods: growth and developments. Nijhoff, Dordrecht, 351–373

Diner AM, Mott RL 1982a A rapid axenic assay for hypersensitive resistance of Pinus lambertiana to Cronartium ribicola. Phytopathology 72:864–865

Diner AM, Mott RL 1982b Direct inoculation of five-needled pines with Cronartium ribicola in axenic culture. Phytopathology 72:1181–1184

Diner AM, Mott RL 1984 Cultured cells of white pine show genetic resistance to axenic blister rust hyphae. Science 224:407–408

Diner AM, Mott RL 1985 In vitro inoculation of western white pine tissue culture propagules with vegetative hyphae of Cronartium ribicola. Phytopathology 75:1130–1131

Duchesne LC, Peterson RL, Ellis BE 1988 Interaction between the ectomycorrhizal fungus Paxillus involutus and Pinus resinosa induces resistance to Fusarium oxysporum. Can J Bot 66:558–562

Earle ED 1982 Application of plant tissue culture in the study of host-pathogen interactions. In: Tomes DT, Ellis DE, Harney PM, Kasha KJ, Peterson RL (eds) Application of plant cell and tissue culture to agriculture and industry. Univ Guelph, Ont, 45–62

Einspahr DW, Wann SR 1985 Use of tissue culture techniques in a hardwood tree improvement program. In: Schmidtling RC, Griggs M (eds) Proc 18th S For Tree Improv Conf, May 21–23, 1985, Long Beach, MS, 33–41

Evans DA, Sharp WR 1986 Applications of somaclonal variation. Bio/Technology 4:528–532

Farquhar ML, Peterson RL 1989 Pathogenesis in fusarium root rot of primary roots of Pinus resinosa grown in test tubes. Can J Plant Pathol 11:221–228

Farquhar ML, Peterson RL 1990 Early effects of the ectomycorrhizal fungus Paxillus involutus on the root rot organism Fusarium associated with Pinus resinosa. Can J Bot 68:1589–1596

Fillatti JJ, Sellmer J, McCown B, Haissig B, Comai L 1987 Agrobacterium mediated transformation and regeneration of Populus. Mol Gen Genet 206:192–199

Frampton LJ Jr, Amerson HV, Weir RJ 1983 Potential of in vitro screening of loblolly pine for fusiform rust resistance. In: Proc 17th S For Tree Improv Conf, June 6–9, 1983, Athens GA, 325–331

Gladfelter HJ, Phillips GC 1987 De novo shoot organogenesis of Pinus eldarica Medw in vitro I. Reproducible regeneration from long-term callus cultures. Plant Cell Rep 6:163–166

Gray DJ, Amerson HV 1983 In vitro resistance of embryos of Pinus taeda to Cronartium quercuum f. sp. fusiforme: ultrastructure and histology. Phytopathology 73:1492–1499

Gray DJ, Amerson HV 1985 In vitro colonization and resistance of loblolly pine embryos infected with the fusiform rust fungus. In: Henke RR, Hughes KW, Constantin MJ, Hollaender A (eds) Tissue culture in forestry and agriculture. Plenum, New York, 257–270

Gray DJ, Amerson HV, Van Dyke CG 1983 Ultrastructure of the infection and early colonization of Pinus taeda by Cronartium quercuum formae speciales fusiforme. Mycologia 75:117–130

Grente J, Sauret S 1961 Epreuve de la résistance à l'encre et à l'endothia sur des cultures de tissus de clones de châtaignier. Ann Epiphyties 12:61–63

Haissig BE 1989 Status of forest tree vegetative regeneration for biotechnology. Am Biotech Lab 7:48–51

Hammerschlag FA 1984 In vitro approaches to disease resistance. In: Collins GB, Petolino JG (eds) Applications of genetic engineering to crop improvement. Nijhoff, Dordrecht, 453–490

Harvey AE 1967 Axenic culture of western white pine cambial explants infected with blister rust. Phytopathology 57:1005

Harvey AE 1979 Pine tissues affected by blister rust despite dialysis barrier between the organisms in culture. US Dep Agr For Serv Res Note INT-265, 7 pp

Harvey AE, Chakravorty AK, Shaw M, Scrubb LA 1974 Changes in ribonuclease activity in Ribes leaves and pine tissue culture infected with blister rust, Cronartium ribicola. Physiol Plant Pathol 4:359–371

Harvey AE, Grasham JL 1969a Growth of the rust fungus Cronartium ribicola in tissue cultures of Pinus monticola. Can J Bot 47:663–666

Harvey AE, Grasham, JL 1969b The relative susceptibility of needle- and stem-derived white pine tissue cultures to artificial inoculation with mycelium of Cronartium ribicola. Can J Bot 47:1789–1790

Harvey AE, Grasham JL 1970 Inoculation of western white pine tissue cultures with basidiospores of Cronartium ribicola. Can J Bot 48:1309–1311

Harvey AE, Grasham JL 1971 Inoculation of a nonhost tissue culture with Cronartium ribicola. Can J Bot 49:881–882

Harvey AE, Grasham JL 1974 Axenic culture of the mononucleate stage of Cronartium ribicola. Phytopathology 64:1028–1035

Harvey AE, Woo JY 1969 Some cytological characteristics of Cronartium ribicola in tissue cultures of Pinus monticola. (Abstr) Phytopathology 59:12

Harvey AE, Woo JY 1971 Histopathology and cytology of Cronartium ribicola in tissue cultures of Pinus monticola. Phytopathology 61:773–779

Hebard FV, Kaufman PB 1978 Chestnut callus-cultures: tannin content and colonization by Endothia parasitica. In: Proc Am Chestnut Symp, Jan 4–5, 1978, Morgantown WV, 63–70

Helgeson JP 1983 Host-pathogen interactions in vitro. In: Helgeson JP, Deverall BJ (eds) Use of tissue culture and protoplasts in plant pathology. Academic Press, New York London, 9–38

Heuchelin SA, McNabb HS Jr, Klopfenstein NB, Thornburg RW 1990 Agrobacterium mediated transformation of Populus X Euramericana "OGY" using the proteinase inhibitor II gene (Pin 2) to increase pest resistance (Abstr). Phytopathology 80:976

Hindal DF, McNabb HS Jr 1973 Interactions of tissue cultures of Ulmus and an isolate of Ceratocystis ulmi. In: 2nd Int Congr Plant Pathol, Sept 5–12, Minneapolis, Abstr 0105

Hislop EC, Keon JPR, Fielding AH 1979 Effects of pectin lyase from Monillinia fructigena on viability, ultrastructure and localization of acid phosphatase of cultured apple cells. Physiol Plant Pathol 14:371–381

Hotson HH 1953 The growth of rust in tissue culture. Phytopathology 43:360–363

Hotson HH, Cutter VM 1951 The isolation and culture of Gymnosporangium juniperi-virginianae Schw. upon artificial media. Proc NAS 37:400–403

Hrib J, Rypacek V 1978 The growth response of wood-destroying fungi to the presence of spruce callus. Ceska Mykol 32:55–60

Hrib J, Rypacek V 1981 A simple callus test to determine the aggressiveness of wood-destroying fungi. Eur J For Pathol 11:270–274

Hrib J, Rypacek V 1983 In vitro testing for the resistance of conifers to the fungus Phaeolus schweinitzii (Fr) Pat on callus cultures. Eur J For Pathol 13:86–91

Hubbes M 1989 Methods of biotechnology for the development of disease-resistant plant material for energy plantations. In: Proc Int Energy Agency Task II Joint Worksh Willow breeding and biotechnology development, Long Ashton Res Stn, Bristol, UK, Sept 28–Oct 6, 1988, 93–104

Jacobi WR 1982a Growth medium for dual cultures of loblolly pine callus and Cronartium fusiforme. Phytopathology 72:136–138

Jacobi WR 1982b Inhibition of Cronartium fusiforme by loblolly pine callus. Phytopathology 72:143–146

Jacobi WR, Amerson HV, Mott RL 1982 Microscopy of cultured loblolly pine seedlings and callus inoculated with Cronartium fusiforme. Phytopathology 72:138–143

Jang JC, Tainter FH 1989a A tissue culture system for studying disease resistance to Phytophthora cinnamomi (Abstr). Phytopathology 79:1221

Jang JC, Tainter FH 1989b Development of in vitro and in vivo screening techniques to detect littleleaf disease resistance (Abstr). Phytopathology 79:1221

Johnson JA, Whitney N 1988 A preliminary study of conifer tissue reaction to the presence of endophytic fungi using issue culture, light and electron microscopy techniques (Abstr). Can J Plant Pathol 10:366

Joung H, Korban SS, Skirvin RM 1987 Screening shoot cultures of Malus for cedar-apple rust infection by in vitro inoculation. Plant Dis 71:1119–1122

Keon JPR 1985 Cytological damage and cell wall modification in cultured apple cells following exposure to pectin lyase from Monilinia fructigena. Physiol Plant Pathol 26:11–29

Klopfenstein NB, Heuchelin SA, McNabb HS Jr, Thornburg RW, Hall RB, Hart ER 1989 Transformation of hybrid Populus with the proteinase inhibitor II gene (Abstr). Phytopathology 79:1004

Koenigs JW 1968 Culturing the white pine blister rust fungus in callus of western white pine. Phytopathology 58:46–48

Krikorian AD 1988 Plant tissue culture: perceptions and realities. Proc Indian Acad Sci (Plant Sci) 98:425–464

Kruger BM, Manion PD 1990a Inheritance of response of Populus tremuloides tissue culture plantlets to bioassays of toxic metabolites of Hypoxylon mammatum (Abstr). Phytopathology 80:120

Kruger BM, Manion PD 1990b Relationship between in vitro ascospore and toxic metabolite bioassays of Populus tremuloides tissue culture plantlets (Abstr). Phytopathology 80:977

Lesney MS 1989 Growth responses and lignin production in cell suspensions of Pinus elliottii "elicited" by chitin, chitosan or mycelium of Cronartium quercuum f. sp. fusiforme. Plant Cell Tissue Organ Cult 19:23–31

Lundquist JE, Walla JA, Tuskan GA 1990a Description of two colony types of Peridermium harknessii. In: Proc 3rd Int IUFRO Rusts of pines Working Party Conf, Sept 18–22, 1989, Banff, Alberta (in press)

Lundquist JE, Walla JA, Tuskan GA 1990b Progress in developing early screening techniques for resistance of ponderosa pine to western gall rust. In: Proc 3rd Int IUFRO Rusts of pines Working Party Conf, Sept 18–22, 1989, Banff, Alberta (in press)

Martin NE 1980 Interactions of pectins and tissues of Cronartium ribicola and Pinus monticola in culture. US Dep Agr For Serv Res Pap INT-250, 5 pp

McComb JA, Hinch JM, Clarke AE 1987 Expression of field resistance in callus tissue inoculated with Phytophthora cinnamomi. Phytopathology 77:346–351

Miller SA, Maxwell DP 1983 Evaluation of disease resistance. In: Evans DA, Sharp WR, Ammirato PV, Yamada Y (eds) Handbook of plant cell culture, vol 1. Techniques for propagation and breeding. MacMillan, New York, 853–879

Morel G 1948 Recherches sur la culture associée de parasites obligatoires et de tissus végétaux. Ann Epiphyties 14:1–112

Ostry ME, Pijut PM, Skilling DD 1990 Screening larch in vitro for resistance to Mycosphaerella laricina (Abstr). Phytopathology 80:1065

Ostry ME, Skilling DD 1988 Somatic variation in resistance of Populus to Septoria musiva. Plant Dis 72:724–727

Parsons TJ, Bradshaw HD Jr, Gordon MP 1989 Systemic accumulation of specific mRNAs in response to wounding in poplar trees. Proc Natl Acad Sci USA 86:7895–7899

Pei MH 1989 Inoculation of callus tissues of Scots, Corsican pine and three varieties of mountain pine with Peridermium pini. Mycol Res 93:503–507

Pijut PM, Domir SC, Lineberger RD, Schreiber LR 1990a Use of culture filtrates of Ceratocystis ulmi as a bioassay to screen for disease tolerant Ulmus americana. Plant Sci 70:191–196

Pijut PM, Lineberger RD, Domir SC, Ichida JM, Krause CR 1990b Ultrastructure of cells of Ulmus americana cultured in vitro and exposed to the culture filtrate of Ceratocystis ulmi. Phytopathology 80:764–767

Prakash CS, Thielges BA 1989 Somaclonal variation in eastern cottonwood for race-specific partial resistance to leaf rust disease. Phytopathology 79:805–808

Robb J, Harvey AE, Shaw M 1973 Ultrastructure of hyphal walls and septa of Cronartium ribicola on tissue cultures of Pinus monticola. Can J Bot 51:2301–2305

Robb J, Harvey AE, Shaw M 1975a Ultrastructure of tissue cultures of Pinus monticola infected by Cronartium ribicola. I. Prepenetration host changes. Physiol Plant Pathol 5:1–8

Robb J, Harvey AE, Shaw M 1975b Ultrastructure of tissue cultures of Pinus monticola infected by Cronartium ribicola. II. Penetration and post-penetration. Physiol Plant Pathol 5:9–18

Roy G, Neumann P 1989 Effect of an Ophiostoma ulmi culture filtrate on Ulmus americana callus formation (Abstr). Can J Plant Pathol 11:198

Saad AT 1965 The culture of apple callus tissue and its use in studies on pathogenicity of Venturia inaequalis. PhD Thesis, Univ Wisconsin, Madison, 111 pp
Sharma NK, Skidmore DI 1988 In vitro expression of partial resistance to Phytophthora palmivora by shoot cultures of papaya. Plant Cell Tissue Organ Cult 14:187–196
Sillick JM, McNabb HS Jr, Thornburg RW 1989 Extracellular proteinases of Septoria musiva (Abstr). Phytopathology 79:1005
Svircev AM, Biggs AR, Miles N, Chong C 1989 Isolation, purification and characterization of a phytotoxin from liquid cultures of Leucostoma persoonii and L. cincta (Abstr). Phytopathology 79:1154
Valentine F 1989 Biotechnology and breeding disease resistant poplar clones. In: Proc Int Energy Agency Task II Joint Worksh Willow breeding and biotechnology development, Long Ashton Res Stn, Bristol, UK, Sept 28–Oct 6, 1988, 112–127
Valentine F, Baker S, Belanger R, Manion P, Griffin D 1988 Screening for resistance to Hypoxylon mammatum in Populus tremuloides callus and micropropagated plantlets (Abstr). In: Ahuja R (ed) Somatic cell genetics of woody plants. Kluwer, Dordrecht, 181
Vardi A, Epstein E, Breiman A 1986 Is the Phytophthora citrophthora culture filtrate a reliable tool for the in vitro selection of resistant Citrus variants? Theor Appl Genet 72:569–574
Walkinshaw CH, Jewell FF, Walker NM 1965 Callus culture of fusiform rust-infected slash pine. Plant Dis Rep 49:616–618
Wann SR 1985 In vitro isolation and propagation of mammatoxin-resistant aspen. PhD Thesis, Inst Pap Chem, Appleton, WI, 125 pp
White PR, Braun AC 1941 Crown gall production by bacteria-free tumor tissue. Science 94:239–241
Woodward S, Pearce RB 1988 Responses of Sitka spruce callus to challenge with wood decay fungi. Eur J For Pathol 18:217–229
Yamazaki N, Fry SC, Darvill AG, Albersheim P 1983 Host-pathogen interactions: XXIV. Fragments isolated from suspension-cultured sycamore cell walls inhibit the ability of the cells to incorporate [^{14}C]leucine into proteins. Plant Physiol 72:864–869
Yang D, Jeng RS, Hubbes M 1989 Mansonone accumulation in elm callus induced by elicitors of Ophiostoma ulmi, and general properties of elicitors. Can J Bot 67:3490–3497

Chapter 19

Immunocytochemistry of Fungal Infection Processes in Trees

R.A. Blanchette and A.R. Abad

19.1 Introduction

Immunocytochemical techniques to detect specific cell wall components, proteins, and other macromolecules have improved our understanding of cell architecture and provided information on precise locations of enzymes or other antigens in cells of animals, plants, and microorganisms. During the past few years, techniques and procedures for locating and identifying antigens in tissues have been perfected, and improvements in biological electron microscopy have made immunocytochemistry an accurate and routine method of cytological study. The number of investigations using immunocytochemistry to locate and identify various antigens in plant tissues has increased dramatically in recent years (Knox 1982, Herman 1988). However, only a few studies have applied these techniques to detect fungal infections and metabolites produced in host-pathogen interactions (Atkey and Wood 1987, Blanchette 1987, Ouellette and Benhamou 1987). The use of immunocytochemistry to study defense mechanisms of trees to pathogen attack has great potential to provide important new information about host-parasite interactions. The results that have been published to date demonstrate the success of immunocytochemical investigations to visualize subcellular processes. These studies are an indication of the important contributions that immunocytochemistry can make to our knowledge of how plants defend themselves against pathogens. It will undoubtedly be an active area for future research.

To detect a specific antigen within tissues, antibodies are needed that have antigenic properties to the enzyme or other macromolecule of interest. The antibodies are labeled with a fluorescent compound (e.g. fluorescein isothiocyanate) for use in light microscopy or with an electron dense marker (e.g. colloidal gold particles) for electron microscopy (DeMey 1983a, Sternberger 1986). Primary antisera may be labeled and directly used to identify antigens, or a more sensitive method (and one that is routinely used) is to use unlabeled primary antibodies followed by a secondary labeled antibody. The second antibody is raised to the IgG of the animal donating the first antibody. Another indirect labeling method employs the use of Protein A (Roth et al. 1978). Protein A, derived from the cell wall of *Staphylococcus aureus*, binds to the F_c portion of immunoglobulin molecules (VanNoorden and Polak 1983). Protein A is labeled with an appropriate marker and used to localize the primary antibody (Bendayan 1984). More recently, Protein G, an immunoglobulin IgG-binding molecule, has been used to bind gold particles to primary antibodies (Bendayan 1987). The tagged protein is linked to the antigen within the tissue and locations of the complex are observed with electron microscopy.

The localization of macromolecules by immunocytochemistry is dependent upon antisera that are specific to a particular antigen. Conventional antisera are obtained via the production of polyclonal antibodies in chickens, rabbits, sheep, or goats. These antibodies are heterogeneous and usually have specificity to a wide range of determinants. Monoclonal antibodies, or the antibodies raised by the hybridoma method, are produced by spleen cells of a mouse that have been fused with myeloma cells. Monoclonal antibodies are homogeneous, and usually have specific affinity for a single antiginic determinant (epitope). Methods used for antibody production have been reviewed previously, and procedures are routine (Bellanti 1985, DeMay 1983b, Kennett 1980). Preparations of various antibodies, as well as colloidal gold, Protein A, and other conjugates, are available commercially.

19.2 General Immunocytochemical Procedures

The most appropriate procedure for immunocytochemical experiments will vary with each specific objective. The method used will be dependent upon the type of tissue and antigen of interest. There are many different procedures available for sample preparation and selection of the technique that will provide the best results requires knowledge of antibody-antigen interactions, cellular ultrastructure, and mechanisms of observation (Knox 1982, Raikhel et al. 1984, Herman 1988).

Fixation is a very important procedure that preserves cellular constituents and protects subcellular detail. The type of fixation, however, can drastically affect the antigenic sites (Knox 1982, Bendayan 1984, Bendayan et al. 1987). Osmium tetroxide has been a routinely used fixative in sample preparation for electron microscopy because of enhanced preservation of membrane structure in biological samples. However, osmium tetroxide may destroy antigenic sites and is often not compatible with immunocytochemical labeling (Bendayan and Zollinger 1983). Other fixatives such as formaldehyde and glutaraldehyde are often used, or in some cases no fixation is used and tissues are frozen (Tokuyasu 1980, Raikhel et al. 1984, Herman 1988).

Resins for embedding plant tissues include the hydrophobic epoxy resins, such as Spurr's (Spurr 1969) and Epon (Luft 1961), and the hydrophilic acrylic resins, such as LR white (Newman et al. 1983) and Lowicryl (Bendayan 1984). The hydrophilic resin Lowicryl is photopolymerized at low temperatures and excellent antigen retention has been reported (Bendayan 1984, Bendayan et al. 1987). LR white resin is cured by heat and infiltrates samples well but results in reduced amounts of labeling in comparison to Lowicryl (Herman 1988). Although the methacrylate or acrylic resins are required for preservation of some antigens (Bendayan and Shore 1982, Bendayan et al. 1983), they unfortunately have several drawbacks when used for woody tissues. Infiltration of the resin can be inadequate, the embedded sample is often brittle, and microtomy is difficult. Poor resolution is also associated with the nonepoxy resins. Recently, Quetol 651 resin has been used successfully to embed woody tissues for ultra-

Table 19.1. Lowicryl KM[a] embedding protocol for TEM[b]

Procedure	Time
Fixation at 4°C:	
Sample is placed in one of the following:	2–4 h
2% Glutaraldehyde in PBS	
1% Glutaraldehyde +4% paraformaldehyde in PBS	
8% Paraformaldehyde in PBS	
Rinse twice in PBS	10 min
Dehydration:	
30% Methanol at 4 °C	5 min
50% Methanol at 4 °C	5 min
70% Methanol at 4 °C	5 min
90% Methanol at 4 °C	30 min
Infiltration at −20 °C:	
90% Methanol: Lowicryl (1:1 ratio)	60 min
90% Methanol: Lowicryl (1:2 ratio)	60 min
Lowicryl	60 min
Lowicryl	8–10 h
Polymerization:	
Quick change of Lowicryl	
Embed in small gelatin capsule at −20 °C with ultraviolet light (30 cm distance)	5–7 days

[a] Lowicryl formula = 0.15 g initiator, 4.0 g cross-linker, 2.60 g monomer. Chemische Wnete Lowi BmgH, Federal Republic of Germany.
[b] Bendayan (1984); Bendayan et al. (1987); Carlemalm et al. (1982); Kellenberger et al. (1980); Roth et al. (1981).

structural study (Abad et al. 1988), and a modified Quetol resin has proven to be a good medium-temperature epoxy resin for immunocytochemistry (Abad unpublished). Protocols for preparing samples for transmission electron microscopy and immunocytochemistry used are provided in Tables 19.1–19.3.

Sections obtained from embedded tissues can be used for immunocytochemical labeling and observed with light or electron microscopy. The general procedures used for labeling are presented in numerous publications that provide information on postembedding methods (Knox 1982, VanNoorden and Polak 1983, Bendayan 1984, Powell 1986, Smit and Todd 1986, Bendayan et al. 1987, Blanchette et al. 1989).

19.3 Detection of Fungi in Woody Tissues

Serological methods have been used by many investigators to detect pathogens in host tissues (Gugerli 1983). Although these methods do not involve direct microscopic observation of the antigen in host tissues, antibodies and serological methodology are needed for immunocytochemical studies. Cytochemical investigations can be easily applied to visualize the location and distribution of

Table 19.2. L.R. white[a] embedding protocol for TEM[b]

Procedure	Time
Fixation:	
Sample is placed in one of the following:	2–4 h
2% Glutaraldehyde in PBS	
1% Glutaraldehyde +4% paraformaldlehyde in PBS	
8% Paraformaldehyde in PBS	
Rinse twice in PBS	
Dehydration:	
50% Ethanol	15 min
70% Ethanol	15 min
100% Ethanol	15 min
100% Ethanol	15 min
Infiltration:	
Ethanol in L.R. White (1:1 ratio)	60 min
L.R. White	60 min
L.R. White	60 min
L.R. White	60 min
Polymerization:	
Sample placed in small gelatin capsule	24–48 h at 50 °C

[a] L.R. White hard mixture; London Resin Co., Basing Stoke, UK.
[b] Bendayan et al. (1987); Newman and Hobot (1987); Newman et al. (1983).

Table 19.3. Quetol[a] water-embedding protocol for TEM[b]

Procedure	Time
Fixation:	
1.5% Glutaraldehyde in cacodylate buffer at pH 7.2, 0.1 M	90 min
Rinse twice in buffer	10 min
Osmium tetroxide (1%) in buffer	60 min
Rinse in water 2x	10 min
Dehydration:	
50% Acetone	10 min
75% Acetone	10 min
95% Acetone	20 min
95% Acetone	30 min
Infiltration:	
95% Acetone/Quetol (1:1)	60 min
1% Water in 100% Quetol	60 min
1% Water in 100% Quetol	90 min
Polymerization:	
Quick change of resin	–
48 °C	12–36 h

[a] Quetol 651, Nissin EM Co. Ltd., 8–19 Tomisha-cho, Shinjyuku, Tokyo, Japan.
[b] Modified technique of Abad et al. (1988).

pathogens within tissues once antibodies have been raised and basic serological procedures completed.

Development of an enzyme-linked immunosorbent assay (ELISA) has facilitated the rapid detection of bacteria (Cother and Vruggink 1980), fungi (Casper and Mendgen 1979, Aguelon and Dunez 1984, Palfreyman et al. 1987, Goodell et al. 1988a,b), and viruses (Bar-Joseph and Garnsey 1981) within plants and insect vectors (Jellison 1986). In addition to detecting specific organisms, the technique can also provide a quantitative determination of microbial colonization. Applications of ELISA to detect forest pathogens have been realized in recent years. In conifers, rapid diagnosis of Scleroderris canker (caused by *Ascocalyx abietina*) has been possible using twigs displaying symptoms of the disease (Liese and Gotlieb 1982). The efficiency of the technique was reported to be 85% compared to a 10% efficiency for traditional isolation procedures where segments of infected twigs are cultured in growth media.

The rapid and accurate detection of a seedborne pathogen, *Sirococcus strobilinus*, in spruce seeds has also been accomplished with ELISA (Mitchell and Sutherland 1986). The technique was reported to detect as little as 25 ng fungal protein in extracts from 100 seeds. This sensitive assay was far superior to traditional methods of isolating fungi from seeds due to the large number of samples that can be processed within a short time. Another advantage was the increased specificity of the assay. Monoclonal antibodies to *S. strobilinus* are very specific and do not interact with nonrelated commensal fungi found on spruce seed (Mitchell 1986). Using conventional techniques to test spruce seeds, these other commercial fungi often mask the presence of *S. strobilinus*, reducing the accuracy obtained when isolation procedures are used.

Another important application of ELISA has been to detect wood-colonizing stain and decay fungi. The detection of incipient stages of decay in living trees and forest products has been extremely difficult to accomplish with accuracy (Blanchette 1987). Detecting early stages of decay by brown rot fungi in wood products is extremely important, since only a slight weight loss, 1 to 2%, results in a concomitant loss of over 50% in some strength properties (Wilcox 1978). Techniques based on color indicators (Cowling and Sachs 1960, Eslyn 1979), sonic and electrical resistance devices (Inwards and Graham 1980, Shortle 1982), and fluorescent stains (Fukazawa et al. 1976, Krahmer et al. 1982) have not been adequate for rapidly detecting the earliest stages of decay. Identification of microscopic changes in wood cells during incipient stages of brown rot decay, with conventional light microscopy, has also been unsuccessful. Serology and immunocytochemical applications, however, are ideally suited to help solve this problem.

Polyclonal antisera produced to cell walls and extracellular filtrates of *Postia placenta*, a brown rot fungus, provided results in detecting and quantifying the fungus within wood (Goodell et al. 1988, Jellison and Goodell 1988a, 1989). No cross-reaction was observed with nondecay fungi that also were present in the substrate. The current work to produce monoclonal antibodies to various wood-destroying fungi should provide additional advantages to identify specific fungal species in wood, quantify the extent of colonization, and determine the amount of cell wall damage (Jellison and Goodell 1986).

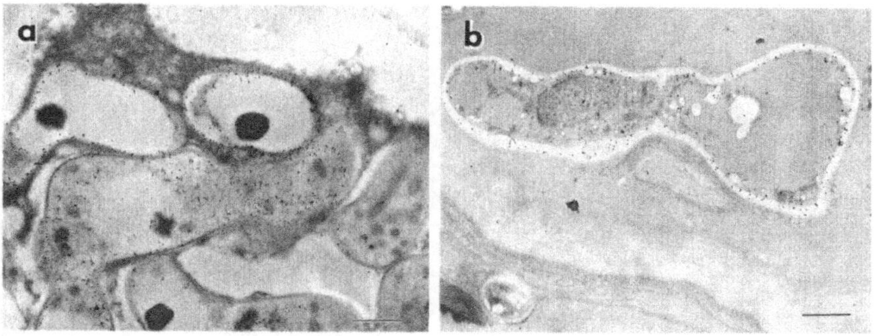

Fig. 19.1a,b. Immunocytochemical detection of mycorrhizae in roots of *Rhododendron*. **a, b** Sections of root treated with antisera to *Clavaria* (**a**) or *Pezizella* (**b**) and gold-labeled showing positive reaction in hyphae. *Bar* = 1 μm (Mueller et al. 1986)

ELISA has also been used to study the colonization of stain fungi within wood (Breuil et al. 1988a,b). A species of *Ophiostoma* was detected at concentrations as low as 0.1 g/ml from 1-mg samples of wood. The test could differentiate this sapwood-staining fungus from other genera of wood-colonizing fungi but could not discriminate different *Ophiostoma* species. In addition to diagnostic applications, the fungal colonization of blue stain within wood could be quantified (Breuil et al. 1988a). Procedures such as this one have great value to monitor growth of specific fungal species within wood, such as biological control species that are being used to prevent colonization by or to eradicate stain or decay fungi in wood (Bruce and King 1983, Cease et al. 1989).

Immunocytochemical detection of fungi has been used in a variety of ways to observe fungi in roots (Schmidt et al. 1974, Mueller et al. 1986), stems (Zollfrank et al. 1987, Goodell et al. 1988b) and foliage (Suske and Acker 1988). Antisera prepared to various fungi (i.e., *Clavaria* and *Pezizella*) growing in the vicinity of *Rhododendron* plants were used to determine the possible mycorrhizal association of these fungi to *Rhododendron* (Fig. 19.1). Indirect colloidal gold labeled antisera to the two fungi were able to detect and differentiate the fungi in and around root tissues. This study showed that more than one fungus can be associated with roots, and demonstrated clearly the effectiveness of using immunocytochemical techniques to identify mycorrhizal fungi that are associated with trees and shrubs. These results are an improvement over early attempts to identify mycorrhizae associated with roots using direct labeling methods (Schmidt et al. 1974). Cross-reactions between *Pisolithus tinctorius* and nonmycorrhizal fungi were evident in this investigation. The difficulties encountered working with mycorrhizae and problems identifying mycorrhizal fungi associated with roots for use in immunocytological work (Chap. 15) could be overcome with serological and immunocytochemical studies. The production of specific antibodies to mycorrhizal fungi, such as those obtained by Mueller et al. (1986), will contribute significantly to advances in studying fungal symbiotic associations of roots.

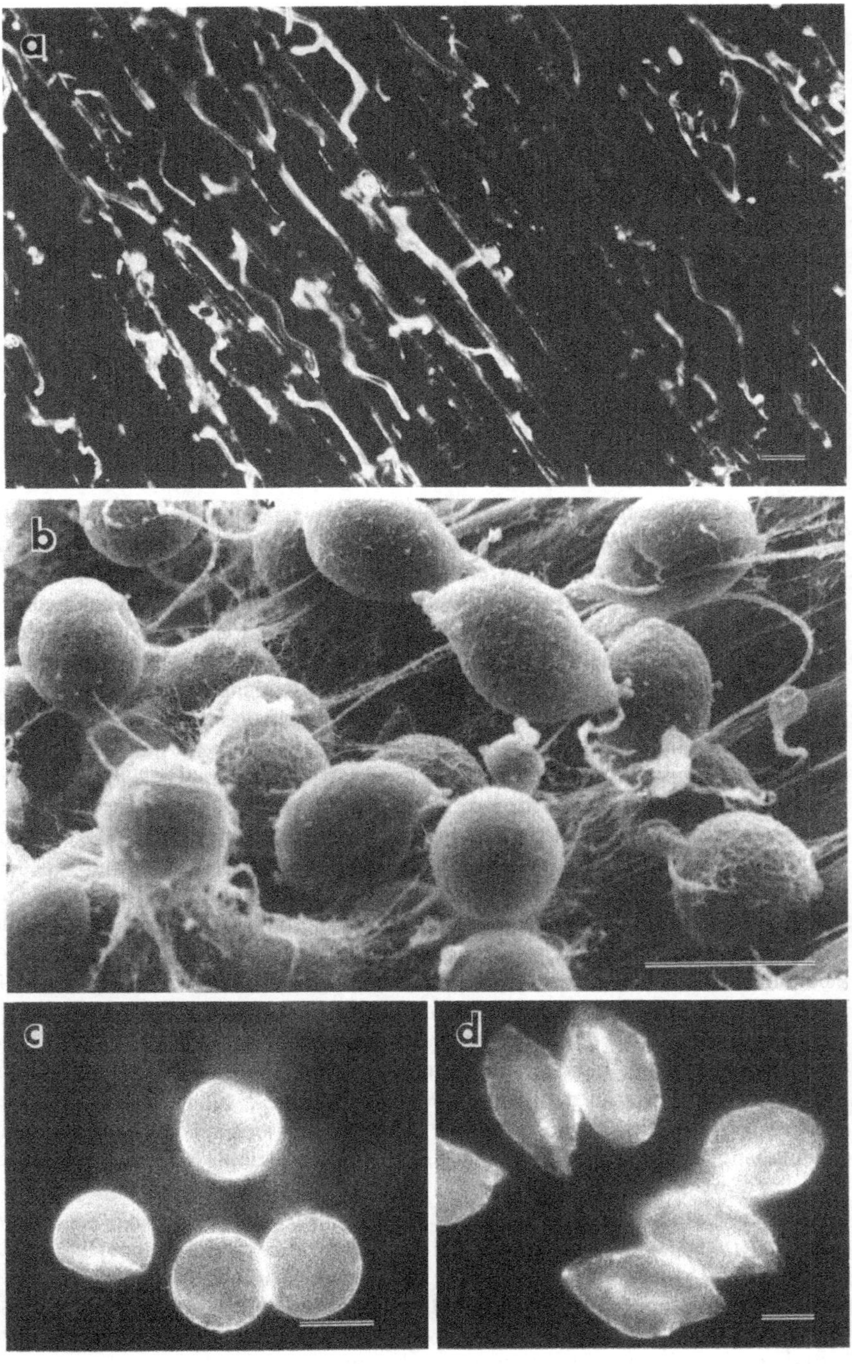
a
b
c
d

In foliage, *Lophodermium piceae* has been detected in asymptomatic spruce needles using polyclonal antiserum to *L. piceae* and immunogold labeling techniques (Suske and Acker 1988, 1989). Endophytic hyphae of many other fungi are often found within needles simultaneously infecting the tissue, and it was difficult to distinguish *Lophodermium* from these other fungi. The specificity of the antisera was improved by these investigators by cross-absorption of the antisera with mycelial extracts of different fungal species that are commonly found in spruce needles.

Immunofluorescent methods have been used to identify hyphae or propagules of various pathogens to obtain information on dispersal, survival, and early stages of infection (Malajczuk et al. 1975, MacDonald and Duniway 1979). Specific antibodies to zoospores and cysts of *Phytophthora cinnamomi* were used to selectively observe the spatial distribution of the propagules and role of zoospores in the development of the disease (Fig. 19.2b–d). One of the obstacles to progress in controlling *P. cinnamomi* and understanding isolate specificity and pathogenicity is a reliable method to monitor the effectiveness of control measures and to identify different species and strains of *Phytophthora* (Hardham et al. 1985, 1986). Monoclonal antibodies from different cell lines of *P. cinnamomi* have been shown to be isolate-, species-, and genus-specific to certain compounds (Hardham et al. 1986). The availability of this type of antisera will have great potential for use in identification of *Phytophthora* species (Gallegly 1983) and also in the separation of possible races of *Phytophthora* with differing avirulence genes (Wycoff et al. 1987). The immunofluorescence microscopy reported by Hardham et al. (1986) demonstrates the effectiveness of immunocytochemical probes to characterize and map certain antigens on surfaces of hyphae and propagules (Fig. 19.2b–d).

Detection of wood-destroying fungi in wood has also been facilitated by immunofluorescence microscopy. Antisera to *Postia placenta* (a brown rot fungus) had an affinity for both mono- and dikaryotic isolates (Fig. 19.2a). Hyphae tagged with fluorescein isothiocyanate-conjugated antiserum were easily observed within xylary cells with little interfering autofluorescence from the surrounding cells (Jellison and Goodell 1988). In another study (Zollfrank et al. 1987), two major root rot pathogens of trees, *Armillaria species* and *Heterobasidion annosum*, were differentiated in Norway spruce with indirect

Fig. 19.2a–d. Applications of fluorescent antibody microscopy. **a** Hyphae of a brown rot fungus colonizing wood after labeling with FITC-labeled antisera to a cell wall fraction of *P. placenta*. Hyphae, tagged with fluorescence, are easily identified within the lumina of the cells. **b** Zoospores and cysts of the fungus *Phytophthora cinnamomi*. Zoospores are ovoid and cysts are spherical. **c** Selective fluorescent staining of cysts using monoclonal antibodies that are specific antigens to cyst surfaces. Cysts fluoresce but zoospores do not. **d** Fluorescent staining of zoospores with monoclonals specific to specific antigens of zoospore surfaces. This antisera labels zoospores but not cysts. **a**, **c**, **d** Fluorescence microscopy; **b** SEM. *Bar* = 10 μm (**a** Courtesy of K. Swengle', B. Goddell, and J. Jellison, University of Maine, unpublished; **b–d** from Hardham et al. 1986) immunofluorescence labeling. Antisera to each fungus were able to detect the respective root rot fungus in host tissue without cross-reaction. Antibodies to *Armillaria* were not specific enough to separate different species and reacted positively with *A. borealis*, *A. bulbosa*, and *A. ostoyae*. The authors postulated that additional investigations to distinguish species of *Armillaria* by cross-absorption techniques or monoclonal antibodies to more specific antigens will provide the needed antisera to differentiate species with accuracy.

19.4 Host-Parasite Interactions

19.4.1 Localization of Enzymes

Subcellular localization of enzymes has been advanced by the use of immunoenzymatic techniques. It is possible to visualize with high resolution fungal enzymes within and on surfaces of fungal cells and extracellularly in cells of the host. One area that has received considerable attention involves the ultrastructural localization of enzymes produced by wood-destroying fungi. In recent years, great efforts have been made to ascertain the major extracellular enzymes produced during lignocellulose degradation. Interest in enzymes from fungi in the Basidiomycotina has apparently increased due to the potential biotechnological applications of lignin-degrading enzymes for industrial purposes (Kirk et al. 1983, Eriksson et al. 1990). The discovery of several enzymes involved in lignin degradation and advances in protein purification have generated new enzymes for immunological investigation (Eriksson et al. 1990). Investigations using antibodies to specific isozymes of lignin peroxidase have shown the extracellular distribution of the enzyme in xylary cells colonized by various white rot fungi and intracellularly within hyphae (Strebotnik et al. 1988, Blanchette et al. 1989, Daniel et al. 1989) (Fig. 19.3). The penetration of enzyme into the woody cell wall layers was different among the various white rot fungi tested (Blanchette et al. 1989, Daniel et al. 1989). Fungi that caused a simultaneous decay of all cell wall components exhibited labeled enzyme around the circumference of the eroded cell wall, whereas fungi that were more selective for lignin removal showed a diffuse penetration of enzyme into the inner layers of the wall. Other degradative enzymes involved in wood decomposition have also been investigated in other immunogold studies involving white rot fungi. Laccase was found to penetrate beech wood cells during decay by *Trametes versicolor*, suggesting enzymatic diffusion into the wood (Evans and Fraser 1987). Anti-xylanase immunogold labeling has demonstrated the presence of xylanase within birch wood decayed by a variety of white rot fungi and a brown rot fungus (Blanchette et al. 1989). Intense labeling was evident on hyphae and slime layers that encase the hyphae as well as in wood cell walls decayed by white and brown rot fungi. The greatest concentration of enzyme was found during early stages of degradation when xylan was apparently being attacked. In advanced stages of decayed wood, the xylan content had been depleted and little xylanase labeling was evident.

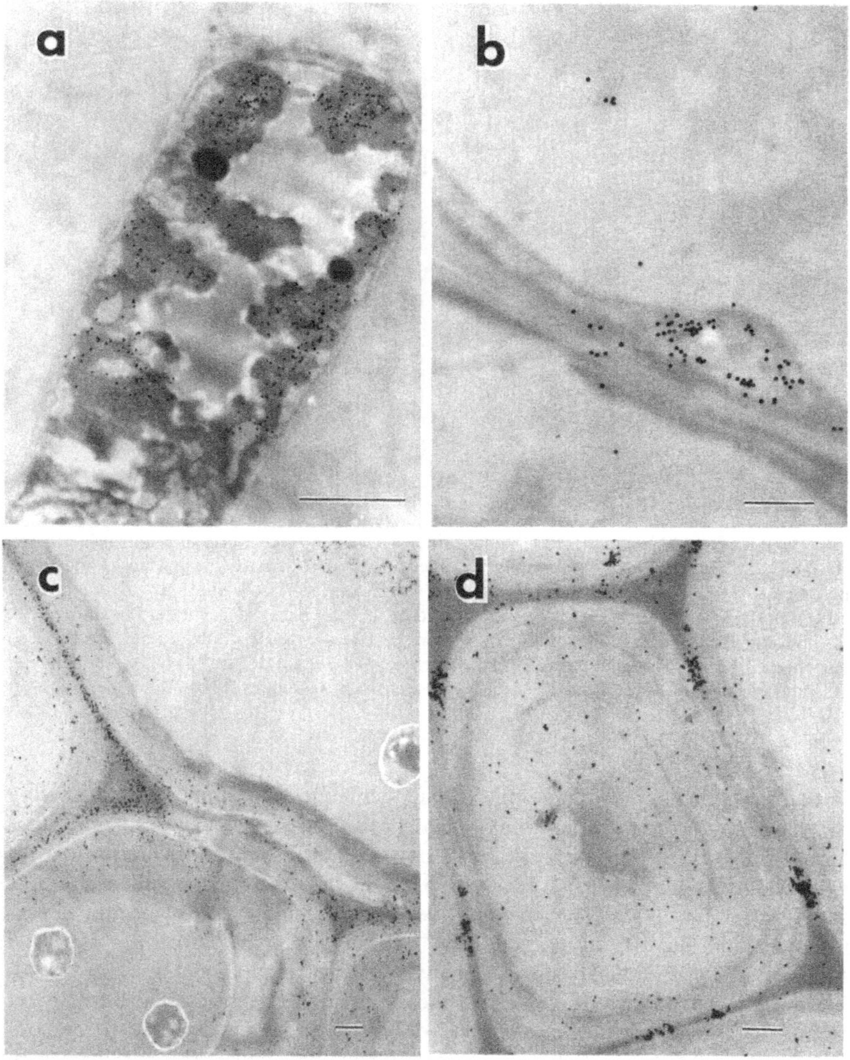

Fig. 19.3a–d. Immunogold labeling of hyphae and decayed wood with antisera to lignin peroxidase. **a** Gold particles are concentrated within hyphae of *Phanerochaete chrysosporium* showing sites of enzyme. **b** Hypha of *Trametes versicolor* in decayed wood of *Betula* with moderate gold labeling around the hypha and at sites of eroded cell walls. **c**, **d** Cells with various stages of delignification caused by *Phellinus pini* showing sites of labeled enzyme within the woody cell wall. Gold particles are concentrated within cell walls at sites where the middle lamellae were being degraded. Transverse sections, TEM. *Bar* = 1 μm (Blanchette et al. 1989, 1990)

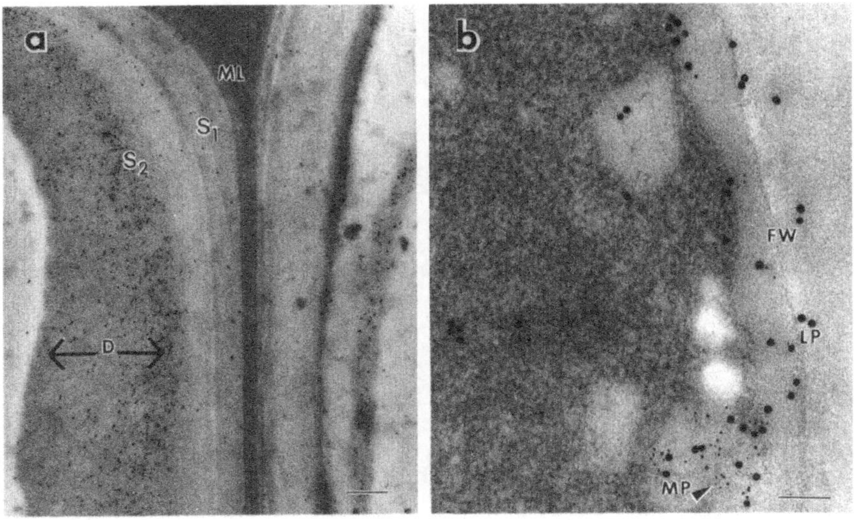

Fig. 19.4a,b. Double immunogold labeling of peroxidases in birch wood and fungal hyphae. **a** Extracellular localization of manganese-dependent peroxidase (*small* gold particles) and lignin peroxidase (*large* gold particles) in birch wood samples degraded by *Phanerochaete chrysosporium*. Both enzymes were able to diffuse into decayed secondary walls (*D*). **b** Cytoplasmic localization of manganese-dependant peroxidase (*MP*) and lignin peroxidase (*LP*) enzymes in hyphae of *Phanerochaete chrysosporium*. *FW* Fungal cell wall; *ML* middle lamella; S_1 and S_2 secondary wall layers; TEM. *Bar* = 0.5 μm in **a**; 0.1 μm in **b** (Daniel et al. 1990)

Selecting the appropriate preparation techniques for immunogold labeling that are suitable for the antibody-antigen reaction of interest is an important factor. Some of the problems that can be encountered are illustrated in one of the first investigations using immunocytochemistry to detect lignin peroxidase activity (Garcia et al. 1987). These investigations failed to show lignin peroxidase activity outside of hyphal cells. Although lignin peroxidase is easily obtained in extracellular filtrates of many white rot fungi (Eriksson et al. 1990), no immunogold labeling occurred in these previous investigations. Subsequent studies using modified techniques and different antisera were able to demonstrate extracellular lignin peroxidase around hyphae and within decayed substrates (Strebotnik et al. 1988, Blanchette et al. 1989, Daniel et al. 1989). It has now been possible to localize the enzyme and monitor the progressive stages of enzyme movement within cell walls of wood. Recent investigation has shown the distribution of two different enzymes, lignin peroxidase and manganese-dependent peroxidase, in the same micrographs by using two different antisera and differently sized gold particles (Daniel et al. 1990). The double immunogold labeling showed the spatial relationship of the two enzymes within the hyphae and wood using 5- and 15-nm gold particles to show manganese-dependent peroxidase and lignin peroxidase, respectively (Fig. 19.4).

Another application of immunocytochemistry to visualize the accessibility of various woody tissues to enzymes has involved the impregnation of substrates with enzyme followed by immunogold labeling to ascertain the extent of enzyme diffusion into cell walls of wood, pulp fibers, or other plant materials (Daniel et al. 1989, Strebotnik and Messner 1990). These techniques provide information regarding degradative changes that occur in cell walls, resulting in increased porosity. The size of the enzyme that diffuses into the wall reflects the pore size due to fungal attack. Proteins, such as lignin peroxidase, myoglobin, and ovalbumin with molecular weights ranging from 46000 to 16500 daltons, have been used to show the changes that occur in cell walls and the size of pores that could be used by degradative enzymes to reach the innermost regions of cell wall layers. Sound wood provides little accessibility to enzymes by inflitration, but wood decayed by fungi exhibits different degrees of cell wall alteration with increased pore size and various extents of protein infiltration (Daniel et al. 1989, Strebotnik and Messner 1990).

There are numerous examples of immunocytological localization of enzymes in nonwoody plant tissues that could have application in studies involving defense mechanisms of trees and shrubs. A suberization-associated anionic peroxidase that functions in wound healing of potato tubers has been detected using immunofluorescence (Espelie et al. 1986). This enzyme was found on the inside of the periderm in wounded potato tubers and is apparently involved in the deposition of aromatic compounds associated with suberin formation. Similar enzymes that affect suberization in wounds of woody plants (see Chap. 2) need to be used to advance our knowledge of wound compartment formation.

Pectinesterases, enzymes that cause a demethylation of pectin and are a prerequisite for action by polygalacturonases and pectatelyases produced by *Phytophthora infestans*, have been used to produce antibodies and locate the enzymes in hyphae of *Phytophthora* (Forster and Mendgen 1987). This study showed the presence of these enzymes in vesicles and dictyosomes of hyphal tips. In regions of older hyphae, the enzyme was found to accumulate at the plasma membrane, demonstrating that subapical regions of *Phytophthora* hyphae were not as permeable to enzymes as the hyphal tips. Since pectin is a major component of primary cell walls, immunocytochemical studies of pectinolytic enzymes by plant pathogenic fungi should be a valuable aid to investigations of infection processes. Pectin enzymes may also induce host resistance mechanisms, and could be used to help elucidate changes that take place during pathogenesis (West et al. 1985).

An enzyme, 5-endo pyruvylshikimic acid 3-phosphate synthase (EPSP) has been detected in cells adapted to growing in the presence of the herbicide glyphosate (Smart and Amrhein 1987). The herbicide is an inhibitor of the shikimate pathway and cultured cells grown in the herbicide overproduce EPSP. Immunogold labeling showed that this enzyme was located intracellularly in the plastids. Enzymes that affect pathways, such as the shikimate pathway, lead to the biosynthesis of aromatic compounds. These compounds are of great interest in the study of host-parasite interactions. In addition, interest in genetically engineered trees that are resistant to glyphosate make enzymes, such as EPSP,

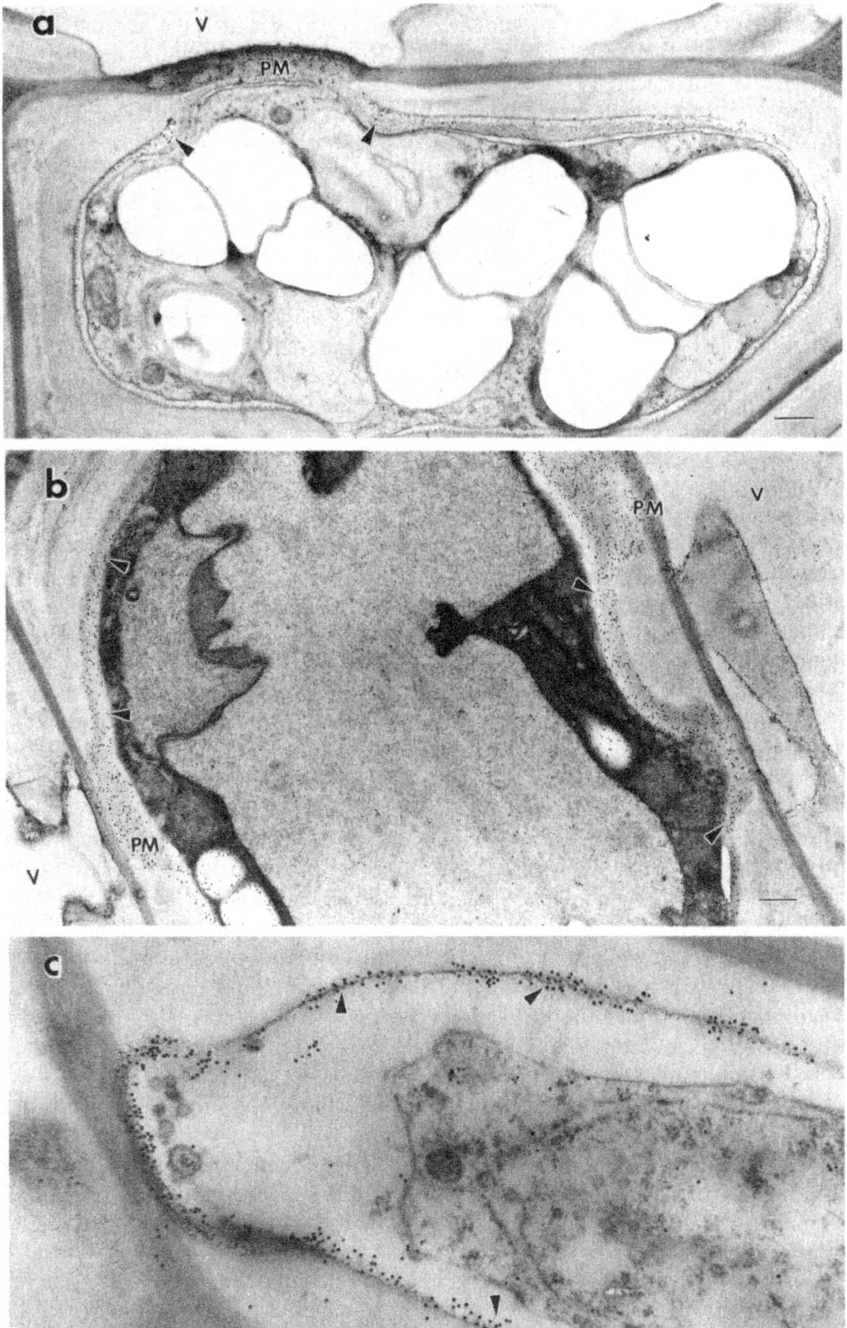
a
V
PM
b
PM
V
PM
V
c

of particular importance in understanding the mode of action of herbicides, and in the successful expression of resistance to this herbicide in woody plants.

19.4.2 Localization of Toxins

Toxins have been implicated in symptom development in host plants of many fungal pathogens. One disease that has received considerable attention has been Dutch elm disease, caused by *Ophiostoma ulmi*. An overview of defense mechanisms in elm to *O. ulmi* is presented in Chapter 13.

Toxic compounds produced by *O. ulmi* are small polypeptides bound to polysaccharides of various molecular weights (Salemink et al. 1965, Strobel et al. 1978, Takai 1978). Polyclonal and monoclonal antibodies to these phytotoxic glycopeptides and immunocytochemical techniques have been used to locate the toxin subcellularly in the pathogen and host (Benhamou et al. 1985a,b, Ouellette and Benhamou 1987, Svircev et al. 1988). Infected elm seedlings showed the toxin on pit membranes and the innermost wall of parenchymal cells (Benhamou et al. 1985a,b) (Fig. 19.5). The presence of the toxin in cells not colonized by hyphae provided evidence for a diffusion of the toxin into new cells. As disease symptoms progressed, the intensity of gold labeling increased (Fig. 19.5). In fungal cells, the presence of labeling varied from cell to cell depending on age, but cell walls of older cells had the greatest concentrations. In another study (Svircev et al. 1988), the toxin cerato-ulmin was detected on the surface of vegetative hyphae, synnemata, perithecia, and ostiola hairs of aggressive *O. ulmi* isolates. Differences in labeling intensities were found among aggressive and nonaggressive isolates. Since accumulation of toxin on cell surfaces was of prime interest in this investigation, scanning electron microscopy was used to detect protein A-gold complexes to polyclonal antiserum of cerato-ulmin. The spatial relationship of toxin to various hyphal structures and their distribution in hosts have been demonstrated in previous investigations (Benhamou et al. 1985a,b, Ouellette and Benhamou 1987, Svircev 1988), thus serving as models for further work on application of immunocytochemical techniques to localize toxins in woody tissues attacked by other pathogens.

←

Fig. 19.5a–c. Elm seedlings 1 (**a**), 3 (**b**), and 4 days (**c**) after inoculation with *Ophiostoma ulmi* and labeled with antisera against a phytotoxic glycopeptide obtained from *O. ulmi*. **a** Parenchyma cell adjoining a vessel (*V*) with gold particles over distended pit membrane (*PM*) and in adjacent areas of the parenchyma cell inner layer (*arrowheads*). **b** Parenchyma cell adjacent to two vessels (*V*) showing labeling evenly distributed over pit membranes (*PM*) and over inner layer of the parenchyma cell (*arrowheads*). **c** Parenchyma cell is extensively disorganized and organelles are no longer recognizable. The inner wall of the parenchyma cell is distended and is densely labeled (*arrowheads*). Transverse sections, TEM. *Bar* = 3 μm (Benhamou et al. 1985a)

19.4.3 Localization of Growth Regulators and Plant Defense Elicitors

Plant growth regulators are important in many different developmental processes and host-parasite interactions (Sequeira 1973). Although these compounds have relatively low molecular weights, methods have been developed to conjugate cytokinins, abscisic acid, and other low-molecular weight compounds to bovine serum albumin to raise polyclonal or monoclonal antibodies (Mertens et al. 1983, Szelke 1983, Zavala and Brandon 1983). Localization of plant hormones in the past has been difficult due to their extremely low concentrations in plant tissues. Immunocytochemistry has provided a mechanism to detect these compounds within tissues even when the concentration is exceedingly low. Immunogold labeling of corn root tips has shown the distribution of cytokinins (Zavala and Brandon 1983). This phytohormone was found in cytoplasmic regions of the cells but rarely in nuclei, vacuoles, or cell walls. The most intense labeling occurred in cells of the root cap and meristermatic regions surrounding the quiescent center (area with cells displaying few mitotic divisions).

The use of hybridoma clones with a high degree of specificity for growth regulators (e.g., abscisic acid and cytokinins), or elicitors of plant defense mechanisms (e.g., ethylene, benzoic acid) should provide powerful tools for use in studies of pathogen interactions with trees. The relationship of plant stress, moisture loss, and the ability of abscisic acid to trigger stomatal closure strongly suggests that further cytochemical study will have great merit (Mertens et al. 1983). The role of ethylene during pathogen invasion and colonization is also of significance in plant-microorganism interactions. Ethylene is produced as an early response by plant tissues to attack from pathogens and appears to induce a variety of defense mechanisms (Boller 1982, Toppan et al. 1982, Sequeira 1983, Anderson et al. 1985). Recently, ethylene was found to be induced by fungal endoxylanase (Fuchs et al. 1989). Immunocytochemical techniques have been used to localize endo-1,4-xylanase activity in wood (Blanchette et al. 1989), and methodology could easily be adapted to study this enzyme as an elicitor of ethylene production in host tissues.

Proteins that elicit a defense response in plants to infection by fungi can also be identified within plant cells using immunocytochemistry. An extracellular protease produced by *Nectria galligena* has been implicated in the accumulation of benzoic acid in apple (Rey and Noble 1984). The benzoic acid imparts antifungal activity. Anti-protease labeling with immunofluorescence and immunogold techniques showed that the enzyme may be secreted into an extracellular matrix in apple tissue. The presence of the enzyme in advance of the fungus demonstrates that protease of *N. galligena* can diffuse through intercellular spaces, cross cell walls, and contact physiologically active plasmalemma of host cells. The protein apparently elicits benzoic acid production in advance of the fungus. The antifungal properties of benzoic acid produce an active host defense mechanism.

Various plant glycoproteins, such as hydroxyproline-rich glycoprotein, are produced in response to infection by many fungal pathogens (Esquerre-Tugaye et al. 1979, Clark et al. 1981, Mazau and Esquerre-Tugaye 1986). The accumulation of glycoprotein has been considered a general defense mechanism, and enhancement or inhibition of hydroxyproline-rich glycoproteins has been associated with increased or decreased host resistance (Toppan et al. 1982, Benhamou et al. 1990). Using immunocytochemistry it was possible to determine the location of accumulated hydroxyproline-rich glycoproteins in tomato roots attacked by *Fusarium oxysporum* f. sp. *radicis-lycopersici* (Benhamou et al. 1990). Accumulations of glycoprotein were initiated after pathogen attack and occurred concomitantly with tissue damage. The occurrence in wall appositions, formed in response to infection, suggests that they are involved in defense against the invading pathogen. The immunological study by Benhamou et al. (1990) has shown the spatial distribution of hydroxyproline-rich glycoproteins in plant cells during different stages of infection and the association of these proteins to structural defense barriers. The possible role of glycoproteins as matrix substances for subsequent deposition of lignin and phenolic compounds would suggest that ultrastructural localization of glycoproteins has potential application to better understand papillae formation and other host defense mechanisms (Whitmore 1978, Ride 1983). The applications of the techniques previously shown in other host-pathogen systems, such as those of Benhamou et al. (1990), should contribute to our knowledge of structural, biochemical, and molecular processes involved in tree defense mechanisms against fungi.

19.5 Conclusions

Immunocytochemical studies have evolved over recent years to a specialized science that has immense potential to help answer many basic questions about how plants defend themselves against pathogens. There are many instances where purified compounds and enzymes produced by fungi are available, and polyclonal and monoclonal antibodies have been produced for use in biochemical and molecular studies. However, they have not been utilized as extensively as warrented for immunocytochemistry. New cooperative studies among plant biologists, electron microscopists, biochemists, and immunologists will allow the wider application of available antisera for use in diverse interdisciplinary investigations. The foundations have been set for using immunocytochemistry to study tree diseases and it will undoubtedly be an exciting, productive area of study in the future.

References

Abad AR, Cease KR, Blanchette RA 1988 A rapid technique using epoxy resin Quetol 651 to prepare plant tissues for ultrastructural study. Can J Bot 66:667–682

Aguelon M, Dunez J 1984 Immunoenzymatic techniques for detection of Phoma exigua in infected potato tissues. Appl Biol 105:463–469

Anderson JD, Chalutz E, Mattoo AK 1985 Induction of ethylene biosynthesis by cell wall digesting enzymes. In: Key JL, Kosuge T (eds) Cellular and molecular biology of plant stress. Liss, New York, 263–273

Atkey PT, Wood DA 1987 Immunogold labelling of fungal enzymes. In: Grange JM, Fox A, Morgan NL (eds) Immunological techniques in microbiology. Blackwell, Oxford, 111–121
Bar-Joseph M, Garnsey SM 1981 Enzyme-linked immunosorbent assay (ELISA): principals and applications for diagnosis of plant viruses. In: Maramorosch K, Harris KF (eds) Plant diseases and vectors. Academic Press, New York London, 35–58
Bellanti JA 1985 Immunology, basic processes. Saunders, Philadelphia, 214 pp
Bendayan M 1984 Protein A-gold electron microscopic immunocytochemistry: methods, applications, and limitations. J Electron Microsc Tech 1:243–270
Bendayan M 1987 Introduction of the protein G-gold complex for high-resolution immunocytochemistry. J Electron Microsc Tech 6:7–13
Bendayan M, Nanci A, Kan FWK 1987 Effect of tissue processing on colloidal gold cytochemistry. J Histochem Cytochem 35:983–996
Bendayan M, Reddy MK, Hashimoto T, Reddy JK 1983 Immunocytochemical localization of fatty acid metabolizing heat-stable and heat-labile enoyl-CoA hydratases in liver and renal cortex. J Histochem Cytochem 31:509–516
Bendayan M, Shore GC 1982 Immunocytochemical localization of mitochondrial proteins in the rat hepatocyte. J Histochem Cytochem 30:139–147
Bendayan M, Zollinger M 1983 Ultrastructural localization of antigenic sites on osmium fixed tissues applying the protein A-gold technique. J Histochem Cytochem 31:101–109
Benhamou N, LaFontaine JG, Joly JR, Ouellette GB 1985a Ultrastructural localization in host tissues of a toxic glycopeptide produced by Ophiostoma ulmi, using monoclonal antibodies. Can J Bot 63:1185–1195
Benhamou N, Mazau D, Esquerre-Tugaye MT 1990 Immunocytochemical localization of hydroxyproline-rich glygoproteins in tomato root cells infected by Fusarium oxysporum f. sp. radicis-lycopersici: study of a compatible interaction. Phytopathology 80:163–173
Benhamou N, Ouellette GB, LaFontaine JG, Joly JR 1985b Use of monoclonal antibodies to detect a phytotoxic glycopeptide produced by Ophistoma ulmi, the Dutch elm disease pathogen. Can J Bot 63:1177–1184
Blanchette RA 1987 New directions in forest products pathology. Can J Plant Pathol 9:361–369
Blanchette RA, Abad AR, Cease KR, Farrell RL, Lovrien RE, Leathers TD 1990 Enzyme immunocytochemistry and ultrastructural localization of cell wall components by enzyme-gold complexes. In: Kirk TK, Chang H-M (eds) Applications of biotechnology in pulp and paper manufacture. Butterworths, Stoneham, MA, 67–79
Blanchette RA, Abad AR, Farrell RL, Leathers TD 1989 Detection of lignin peroxidase and xylanase by immunocytochemical labeling in wood decayed by Basidiomycetes. Appl Environ Microbiol 55:1457–1465
Boller T 1982 Ethylene-induced biochemical defenses against pathogens. In: Waring PF (ed) Plant growth substances. Academic Press, New York London, 303–312
Breuil C, Seifert KA, Yamada J, Rossignol L, Saddler JN 1988a Quantitative estimation of fungal colonization of wood using an enzyme-linked immunosorbent assay. Can J For Res 18:374–377
Breuil C, Yamada J, Seifert KA, Saddler JN 1988b An enzyme linked immunosorbent assay (ELISA) for detecting staining fungi in unseasoned wood. J Inst Wood Sci 11:132–134
Bruce A, King B 1983 Biological control of wood decay by Lentinus lepideus (Fr.) produced by Scytalidium and Trichoderma residues. Mat Org 18:171–181
Carlemalm E, Garavito RM, Villiger W 1982 Resin development for electron microscopy and an analysis of embedding at low temperature. J Microsc 126:123–143
Casper R, Mendgen K 1979 Quantitative serological estimation of a hyperparasite: detection of Verticillium lecanii in yellow rust infected wheat leaves by ELISA. Phytopathology 94:89–91
Cease KR, Blanchette RA, Highley TL 1989 Interactions between Scytalidium species and brown or white-rot basidiomycetes in birch wood. Wood Sci Technol 23:151–161
Clark JA, Lisker N, Lamport DTA, Ellingboe AE 1981 Hydroxyproline enhancement as a primary event in the successful development of Erysiphe graminis in wheat. Plant Physiol 67:188–189
Cother EJ, Vruggink H 1980 Detection of viable and nonviable cells of Erwinia carotovora var. atroseptica in inoculated tubers of var. bintje with enzyme-linked immunosorbent assay (ELISA). Potato Res 23:133–135

Cowling EB, Sachs IB 1960 Detection of brown rot with Osmium tetroxide stain. For Prod J 10:594–596

Daniel G, Nilsson T, Petterson B 1989 Intra- and extracellular localization of lignin peroxidase during the degradation of solid wood and wood fragments by Phanerochaete chrysosporium by using transmission electron microscopy and immuno-gold labeling. Appl Environ Microbiol 55:871–881

Daniel G, Petterson B, Nilsson T, Volc J 1990 Use of immunogold cytochemistry to detect Mn(II)-dependent and lignin peroxidases in wood degraded by the white rot fungi. Phanerochaete crysosproium and Lentinula edodes. Can J Bot 68:920–933

DeMey J 1983a Colloidal gold probes in immunocytochemistry. In: Polak JM, VanNoorden S (eds) Immunocytochemistry, practical applications in plant pathology and biology. Wright, London, 82–112

DeMay J 1983b Raising and testing antibodies for immunocytochemistry. In: Polak JM, VanNoorden S (eds) Immunocytochemistry, practical applications in pathology and biology. Wright, London, 43–52

Eriksson K-E, Blanchette RA, Ander P 1990 Microbial and enzymatic degradation of wood and wood components. Springer, Berlin Heidelberg New York, 416 pp

Eslyn WE 1979 Utility pole decay. Par 3: Detection in pine by color indicators. Wood Sci Technol 13:117–126

Espelie KE, Franceschi VR, Kolattukudy PE 1986 Immunocytochemical localization and time course of appearance of an anionic peroxidase associated with suberization in wound-healing potato tuber tissue. Plant Physiol 81:487–492

Esquerre-Tugaye MT, Lafitte C, Mazau D, Toppan A, Touze A 1979 Cell surfaces in plant-microorganism interactions. II. Evidence for the accumulation of hydroxyproline-rich glycoproteins in the cell wall of diseased plants as a defense mechanism. Plant Physiol 64:320–326

Evans CS, Fraser MA 1987 Ultrastructural localization of ligninolytic enzymes of Coriolus versicolor. In: Odier E (ed) Proc Lignin enzymic and microbial degradation Symp, Paris, INRA Publ, Versailles, Fr, 218–222

Forster H, Mendgen K 1987 Immunocytochemical localization of pectinesterases in hyphae of Phytophthora infestans. Can J Bot 65:2607–2613

Fuchs Y, Saxena A, Gamble HR, Anderson D 1989 Ethylene biosynthesis-inducing protein from cellulysin is an endoxylanase. Plant Physiol 89:138–143

Fukazawa K, Imagawa H, Doi S 1976 Histochemical observations of decayed cell wall using ultraviolet and fluorescence microscopy. Res Bull Coll Agr Hokkaido Univ 33:101–114

Gallegly ME 1983 New criteria for classifying Phytophthora and critique of existing approaches. In: Bartnicki-Garcia S, Tsao PH (eds) Phytophthora: its biology, taxonomy, ecology and pathology. Am Phytopathol Soc, St Paul, 149–165

Garcia S, Latge JP, Prevost MC, Leisola M 1987 Wood degradation by white-rot fungi: cytochemical studies using lignin peroxidase-immunoglobulin-gold complexes. Appl Environ Microbiol 53:2384–2387

Goodell B, Daniel G, Jellison J, Nilsson T 1988a Immunolocalization of extracellular metabolites from Poria placenta. Int Res Group Wood Preserv Doc IRG/WP/1361

Goodell B, Jellison J, Hosli JP 1988b Serological detection of wood decay fungi. For Prod J 38:59–62

Gugerli P 1983 Use of enzyme immunoassay in phytopathology. In: Avrameasmeds S (ed) Immunoenzymatic techniques. Elsevier, Amsterdam, 369–384

Hardham AR, Suzaki E, Perkin JL 1985 The detection of monoclonal antibodies specific for surface components on zoospores and cysts of Phytophthora cinnamomi. Exp Mycol 9:264–268

Hardham AR, Suzaki E, Perkin JL 1986 Monoclonal antibodies to isolate-, species-, and genus-specific components on the surface of zoospores and cysts of the fungus Phytophthora cinnamomi. Can J Bot 64:311–321

Herman EM 1988 Immunocytochemical localization of macromolecules with the electron microscope. Annu Rev Plant Mol Biol 39:139–155

Inwards RD, Graham RD 1980 Comparing methods for inspecting Douglas-fir poles in service. Am Wood Preserv Assoc 76:283–287

Jellison J 1986 Immunoenzymatic detection of plant virus on aphid stylets. Phytopathol Z 117:181–184
Jellison J, Goodell B 1986 Production of monoclonal antibodies to fungal metabolites. Int Res Group on Wood Preserv Doc IRG/WP/1306
Jellison J, Goodell B 1988 Immunological detection of decay in wood. Wood Sci Technol 22:293–297
Jellison J, Goodell B 1989 Inhibitory effects of undecayed wood and the detection of Postia placenta using the enzyme-linked immunosorbent assay. Wood Sci Technol 23:13–20
Kellenberger E, Durrenberger M, Villiger W, Carlemalm E, Wurtz M 1987 The efficiency of immunolabel on Lowicryl sections compared to theoretical predictions. J Histochem Cytochem 35:959–969
Kennett RH 1980 Monoclonal antibodies. Hybridomas: a new dimension in biological analyses. Plenum, New York London, 423 pp
Kirk TK, Jeffries TW, Leatham GF 1983 Biotechnology: applications and implications for the pulp and paper industry. TAPPI J 66(5):45–51
Knox RB 1982 Methods for locating and identifying antigens in plant tissues. Tech Immunocytochem 1:205–238
Krahmer RL, DeGroot RC, Lowell EC 1982 Detecting incipient brown rot with fluorescence microscopy. Wood Sci 15:78–80
Liese AL, Gotlieb AR 1982 Utilization of the ELISA technique for the diaganosis of two fungal diseases: Scleroderris canker of conifers and downy mildew of sunflower. Phytopathology 72:263
Luft JH 1961 Improvements in epoxy resin embedding methods. J Biophys Biochem Cytol 9:409–414
MacDonald JD, Duniway JM 1979 Use of fluorescent antibodies to study survival of Phytophthora megasperma and P. cinnamomi zoospores in soil. Phytopathology 69:436–441
Malajczuk N, McComb AJ, Parker CA 1975 An immunofluroescence technique for detecting Phytophthora cinnamomi Rands. Aust J Bot 23:289–309
Mazau D, Esquerre-Tugaye MT 1986 Hyroxyproline-rich glycoprotein accumulation in the cell walls of plants infected by various pathogens. Physiol Mol Plant Pathol 29:147–157
Mertens R, Deus-Neuman B, Weller EW 1983 Monoclonal antibodies for the detection and quantitation of the endogenous plant growth regulator, abscisic acid. FEBS Lett 160:269–272
Mitchell LA 1986 Derivation of Sirococcus strobilinus specific monoclonal antibodies. Can J For Res 16:939–944
Mitchell LA, Sutherland JR 1986 Detection of seed-borne Sirococcus strobilinus with monoclonal antibodies in an enzyme-linked immunosorbent assay. Can J For Res 16:945–948
Mueller WC, Tessier BJ, Englander L 1986 Immunocytochemical detection of fungi in the roots of Rhododendron. Can J Bot 64:718–723
Newman GR, Hobot JA 1987 Modern acrylics for post-embedding immunostaining techniques. J Histochem Cytochem 35:971–981
Newman GR, Jasani B, Williams ED 1983 A simple post-embedding system for rapid demonstration of tissue antigens under the electron microscope. Histochem J 15:543–555
Ouellette GB, Benhamou N 1987 Use of monoclonal antibodies to detect molecules of fungal plant pathogens. Can J Plant Pathol 9:167–176
Palfreyman JW, Bruce A, Button D, Glancy H, Vigrow A, King B 1987 Immunological methods for the detection and characterization of wood decay basidiomycetes. In: Houghton DR, Smith RN, Eggins HOW (eds) Biodeterioration 7. Elsevier, Amsterdam London, 709–713
Powell MJ 1986 Cytochemical techniques for the subcellular localization of enzymes in microorganisms. In: Aldrich HC, Todd WJ (eds) Ultrastructure techniques for microorganisms. Plenum, New York, 421–445
Raikhel NV, Mishkind M, Palevitz BA 1984 Immunocytochemistry in plants with colloidal gold conjugates. Protoplasma 121:25–33

Rey MEC, Noble JP 1984 Subcellular localization by immunocytochemistry of the extracellular protease produced by Nectria galligena Bres. in infected apple tissue. Physiol Plant Pathol 25:323–336

Ride JP 1983 Cell walls and other structural barriers in defense. In: Callow JA (eds) Biochemical plant pathology. Wiley & Sons, New York, 214–236

Roth J, Bendayan M, Carlemalm E, Villiger W, Garavito M 1981 Enhancement of structural preservation and immunocytochemical staining in low temperature embedded pancreatic tissue. J Histochem Cytochem 29:663–671

Roth J, Bendayan M, Orci L 1978 Ultrastructural localization of intracellular antigen by the use of the protein A-gold complex. J Histochem Cytochem 26:1074–1081

Salemink CA, Rebel H, Kerling LCP, Tchernoff V 1965 Phytotoxin isolated from liquid culture of Ceratocystis ulmi. Science 149:202–203

Sequeira L 1973 Hormone metabolism in diseased plants. Annu Rev Plant Physiol 24:253–380

Schmidt EL, Biesbrock JA, Bohlool BB, Marx DH 1974 Study of mycorrhizae by means of fluorescent antibody. Can J Microbiol 20:137–139

Shortle WC 1982 Decaying Douglas-fir wood? ionization associated with resistance to a pulsed electric current. Wood Sci 15:29–32

Smart CC, Amrhein N 1987 Ultrastructural localization by protein A-gold immunocytochemistry of 5-enolpyruvylskikimate acid 3-phosphate synthase in a plant cell culture which overproduces the enzyme. Planta 170:1–6

Smit J, Todd WJ 1986 Collodial gold labels for immunocytochemical analysis of microbes. In: Aldrich HC, Todd WJ (eds) Ultrastructure techniques for microorganisms. Plenum, New York, 469–516

Spurr AR 1969 A low-viscosity epoxy resin embedding medium for electron microscopy. J Ultrastruct Res 26:31–43

Sternberger LA 1986 Immunoctyochemistry. Wiley & Sons, New York, 524 pp

Strebotnik E, Messner 1990 Accessibility of sulfite pulp to lignin peroxidase and marker proteins with respect to enzymatic bleaching. J Biotechnol 13:199–210

Strebotnik E, Messner K, Foisner R, Petterson B 1988 Ultrastructural localization of ligninase of Phanerochaete chrysosporium by immunogold labeling. Curr Microbiol 16:221–227

Strobel G, VanAlfen N, Hapner KD, McNeil M 1978 Some phytotoxic glycopeptide from Ceratocystis ulmi, the Dutch elm disease pathogen. Biochem Biophys Acta 538:60–75

Suske J, Acker G 1989 Endophytic needle fungi: Culture, ultrastructure and immunocytochemical studies. In: Schulze E-D, Lange OL, Oren R (eds) Forest decline and air pollution. Springer, Berlin Heidelberg New York, 121–136

Suske J, Acker G 1989 Identification of endophytic hyphae of Lophodermium piceae in tissues of green, symptomless Norway spruce needles by immunoelectron microscopy. Can J Bot 67:1768–1774

Svircev AM, Jeng RS, Hubber M 1988 Detection of cerato-ulmin on aggressive isolates of Ophiostoma ulmi by immunocytochemistry and scanning electron microscopy. Phytopathology 78:322–327

Szelke M 1983 Raising antibodies to small peptides. In: Polok JM, VanNoorden S (eds) Immunocytochemistry, practical applications in pathology and biology. Wright, London, 53–68

Takai S 1978 Cerato-ulmin, a wilting toxin of Ceratocystis ulmi. Cultural factors affecting cerato-ulmin production by the fungus. Phytopathol Z 91:157–158

Tokuyasu KT 1980 Immunochemistry on ultra-thin frozen sections. Histochem J 12:381–403

Toppan A, Roby D, Esquerre-Tugaye MT 1982 Cell surfaces in plant microorganism interactions. III. In vivo effect of ethylene on hydroxyproline-rich glycoprotein accumulation in the cell wall of diseased plants. Plant Physiol 70:82–86

VanNoorden S, Polak JM 1983 Immunocytochemistry today, techniques and practice. In: Polak JM, VanNoorden S (eds) Immunocytochemistry, practical applications in pathology and biology. Wright, London, 11–42

West CA, Moestra P, Jin DF, Lois AF, Wickham KA 1985 The role of pectic fragments of the plant cell wall in the response to biological stress. In: Key JL, Kosuge T (eds) Cellular and molecular biology of plant stress. Liss, New York, 335–350

Whitmore FW 1978 Lignin-protein complex in cell walls of Pinus elliotii: amino acid constitutents. Phytochemistry 21:315–318

Wilcox WW 1978 Review of the literature on the effects of early stages of decay on wood strength. Wood Fiber 9:252–257

Wycoff KL, Jellison J, Ayers AR 1987 Monoclonal antibodies to glycoprotein antigens of a fungal plant pathogen, Phytophthora megasperma f. sp. glycinea. Plant Physiol 85:508–515

Zavala ME, Brandon DL 1983 Localization of a phytohormone using immunocytochemistry. J Cell Biol 97:1235–1239

Zollfrank U, Sautter C, Hock B 1987 Fluorescence immunohistochemical detection of Armillaria and Heterobasidion in Norway spruce. Eur J For Pathol 17:230–237

Subject Index